# Advances in Solar Energy

**An Annual Review of Research and Development**

## Volume 6

A Continuation Order Plan is available for this series. A continuation order will bring delivery of each new volume immediately upon publication. Volumes are billed only upon actual shipment. For further information please contact the publisher.

# Advances in Solar Energy

## An Annual Review of Research and Development

## Volume 6

Edited by

**Karl W. Böer**
*University of Delaware*
*Newark, Delaware*

**AMERICAN SOLAR ENERGY SOCIETY, INC.**
Boulder, Colorado • Newark, Delaware
and
**PLENUM PRESS**
New York • London

The Library of Congress has cataloged this title as follows:

Advances in solar energy.—Vol. 1 (1982)–     —New York: American Solar Energy
  Society, c1983–
    v. ill.; 27 cm.
  Annual.
  ISSN 0731-8618 = Advances in solar energy.


  1. Solar energy—Periodicals.   I. American Solar Energy Society.
TJ809.S38                        621.47′06—dc19                         85-646250
                                                                  AARC 2 MARC-S
Library of Congress                       [8603]

ISBN 0-306-43727-9

# Foreword

In Volume 6 of the *Advances in Solar Energy* we have specifically targeted for a review the rich experience of the Power Utilities. Their hands-on experience in a large variety of means to employ solar energy conversion and to evaluate the technical and economical feasibilities is of great importance to their future use. In designing the lay-out for this volume, we wanted to collect all relevant information, including success *and* failures and wanted to emphasize the lessons learned from each type of experiment.

The publication of such a review now has the advantage of a settled experience in the first phase of solar involvement of the utility industry with a large amount of data analyzed. We are confident that this information will be of great value to direct the future development of the solar energy mix within this industry.

We have added to this set of reviews three articles which deal with the most promising high-technology part of solar energy conversion using exclusively solid state devices: solar cells. The development over the last two decades from barely 10% to now in excess of 30% conversion efficiency is breathtaking. In addition, the feasibility of economic midrange efficient thin-film technology holds the promise of opening large scale markets in the near future. This field will enter head-on competition for large power generation with more conventional technology.

These articles are written by the key experts in the field and provide a wealth on data, critical assessments of the present state-of-the-art and contain a long list of literature for further in-depth studies.

I greatly appreciate the assistance of the Editors and referees of the articles and the willingness of the authors to follow the many constructive suggestions.

My special thanks go to Ms. Linda Abrantes for her dedicated work in type-setting the manuscript and to Mr. John Holowka for assisting Linda in producing the camera-ready copy in the University of Delaware's Publication Office, and to the University of Delaware for their support. My personal appreciation also goes to the Chairman and the Board of Directors of ASES for their encouragement in our continued effort to document the advances in solar energy.

Karl W. Böer

# Contents

Chapter 1  **The U.S. Electric Utility Industry's Activities in Solar and Wind Energy: Survey and Perspective**                                    1

Edgar A. DeMeo and Peter Steitz

Chapter 2   The Status of Solar Thermal Electric Technology   219

Richard J. Holl and Edgar A. DeMeo

## Chapter 3   High Efficiency III-V Solar Cells       394

### John C.C. Fan, Mark B. Spitzer and Ronald P. Gale

xix

# About the Authors

Dr. Edgar A. DeMeo is the Solar Power Program Manager at the Electric Power Research Institute (EPRI), Palo Alto, California. Prior to joining EPRI in 1976, he was on the Engineering faculty at Brown University. From 1967 to 1969, he was an Officer Instructor in the Science Department at the U.S. Naval Academy. While at EPRI, Dr. DeMeo has placed primary emphasis on the assessment and development of solar and wind power for use by the electric utility industry.

Dr. DeMeo received the B.E.E. degree from Rensselaer Polytechnic Institute in 1963 and the Sc.M. and Ph.D. degrees in Engineering in 1965 and 1968, respectively, from Brown University.

John C. C. Fan is Chairman and Chief Executive Officer of Kopin Corporation. Prior to founding Kopin, headed the MIT Lincoln Laboratory's Electronic Materials Group. Holds more than two dozen patents and author of over 170 papers in the fields of semiconductors, photovoltaic cells, solid-state electronics and optical devices. He edited three books.

Dr. Fan received his BSEE from the University of California, Berkeley and his M.S. and Ph.D. in applied physics from Harvard University.

Ronald P. Gale is a Vice President of Kopin Corporation, managing the photovoltaics product group. He has been with Kopin since its founding in 1985, developing high-performance solar cells. Previously, he spent seven years as a Member of the Technical Staff at the M.I.T. Lincoln Laboratory in Lexington, MA, working on the growth of III-V compounds by chemical vapor deposition.

Dr. Gale received the Ph.D. degree in materials science and engineering from the Massachusetts Institute of Technology in 1978. He received the B.S. and M.Engr. degrees in applied and engineering physics from Cornell University in 1972 and 1973.

Dr. Richard Holl has been a consultant on solar energy to the Electric Power Research Institute since 1986. He was branch Manager and then Chief Program Engineer of Advanced Solar Energy for McDonnell Douglas from 1973 to 1986; concentrating on development of utility-scale solar thermal electric power systems. Dr. Holl directed research on space nuclear power for McDonnell Douglas from 1962 until 1973 and commercial nuclear power for Allis Chalmers from 1956 to 1962.

Dr. Holl received B.S. and M.S. degrees in physics from the University of Wisconsin and a Ph.D. in engineering from the University of Michigan.

**Richard L. Mitchell** is a project manager for the Polycrystalline Thin Film Program and the manager of the New Ideas for Photovoltaics Program at the Solar Energy Research Institute (SERI). Both programs involve support for educational and industrial research efforts in the development of copper indium diselenide and cadmium telluride photovoltaics. Mitchell has been involved in the field of photovoltaics for over seven years, has written over fifteen technical papers in the field. He graduated from Colorado School of Mines with a M.S. in Physics in 1974.

**Ronald A. Sinton** was born in Denver, Colorado, in 1959. He received the B.S. degree in engineering physics from the University of Colorado, Boulder, in 1981, and the M.S. and Ph.D. degrees in applied physics from Stanford University in 1984 and 1987, respectively. His dissertation research involved the optimization of silicon solar cells for use in highly-concentrated sunlight. Since 1987, he has continued to research the device physics and fabrication of high-efficiency single-crystal silicon solar cells as a Research Associate at Stanford University.

**Mark B. Spitzer** is the Manager of III-V Processing. He is the author of 58 publications on Photovoltaics and related topics. He received the Ph.D. in Physics from Brown University and B.A. with distinction in Physics from Boston University. He is a Member of the American Physical Society and the IEEE.

**Peter Steitz**, P.E., is a consultant and also serves as Assistant General Manager and Director of Power Supply and Operations at Wisconsin Public Power, Inc. SYSTEM. Prior to assuming his present position in June of 1985, he was employed for 12 years with Burns & McDonnell Engineering Company of Kansas City, Missouri, where he served a broad range of electric utility industry clients in a variety of assignments. He has authored and presented numerous reports, articles and papers on matters of bulk electric power supply and alternative electric power technologies. Previously Mr. Steitz served four years as an officer in the U.S. Navy.

Mr. Steitz received B.S. degrees in Electric Engineering and Naval Science from the University of Wisconsin in 1968 and an MBA from the University of Missouri in 1977. He is a member of the National Society of Professional Engineers and the Institute of Electrical and Electronic Engineers.

**Richard Swanson** was born in Davenport, Iowa. He received his BSEE and MSEE from Ohio State University and the Ph.D. in Electrical Engineering from Stanford University in 1974. In 1976, he joined the faculty at Stanford. Since then he has been actively involved in photovoltaics research. His research has produced a new type of silicon concentrator cell, the point-contact cell, which holds the record conversion efficiency of 28 percent. Professor Swanson has published over 100 technical papers and is a member of the Electronic Materials Committee of the AIME and the Power Generation Committee of the IEEE Power Engineering Society.

**Kenneth Zweibel** is the manager of the Polycrystalline Thin Film (PTF) Program at the Solar Energy Research Institute (SERI). The PTF is involved with assisting U.S. industry to develop copper indium diselenide and cadmium telluride as successful PV options for large-scale applications. Zweibel has been involved with PV for over ten years. He has made numerous public presentations on the topic, has been quoted extensively, and is also the author of two books: "Basic Photovoltaic Principles and Methods" and "Harnessing Solar Power: The PV Challenge" (1990; also to be published by Plenum). He has written over thirty technical and popular articles on PV and has been interviewed for others that have appeared in Scientific American, New York Times, Los Angeles Times, Washington Post, Wall Street Journal, Science, Discover, Audubon, National Public Radio, and more. He resides with his wife and two small children in Denver, Colorado. He is a graduate in Physics of the University of Chicago (1970).

# CHAPTER 1

# The U.S. Electric Utility Industry's Activities in Solar and Wind Energy
## Survey and Perspective

Edgar A. DeMeo and Peter Steitz

## 1.1 Abstract

U.S. electric utilities have played a key role in the massive effort undertaken world-wide in the early 1970's to evolve technically-feasible and cost-effective renewable energy systems. Hundreds of utilities from all parts of the U.S. have conducted or participated in literally thousands of different activities and projects including resource assessments, technical and economic analyses, basic research and development, component and systems experiments and field tests, customer assistance and information programs, and various business ventures. In part as a result of the aggregate effort by utilities, several solar and wind power technologies are approaching economic competitiveness in bulk power use.

The U.S. utilities became involved in solar and wind for various reasons, some as a result of their own initiative and curiosity and others in response to consumer and political pressures. Many pursued these technologies even in situations where neither technical or economic feasibility prospects were encouraging. They did so independently, in cooperation with other utilities and with the Electric Power Research Institute (EPRI), and in cooperation with other sectors including the solar and wind power industries, government, universities, and their own customers. A major part of the aggregate utility experience with the solar and wind technologies has been gained through the EPRI Solar Power Program. This experience is documented in approximately 170 EPRI technical reports and numerous papers and articles (See bibliography of EPRI Solar Power Program Publications at the end of this chapter).

By the 1980's, changes in external factors began to discourage utility involvement in solar and wind energy. This appears to be a primary reason why utility activity in some of the solar and wind technologies trended downward in the early 1980's. Another key reason is that in the aftermath of the 1973 oil embargo, a great deal of activity was initiated by the U.S. utilities in response to the "crisis" atmosphere of the time. Much of this activity was not thoroughly planned and lacked a clear purpose and logic. Although many utilities had high hopes for these early efforts, progress was often slow and disappointments were encountered. By the early

1980's, perspective had begun to emerge; and while the better-conceived activities and programs endured, others fell by the wayside or were channeled toward more meaningful directions. As one might expect, the best planned activities were usually the most successful and are most likely to have lasting value.

Solar thermal systems (heating and cooling of buildings, electric power, industrial process heat) were among the initial renewable technology activities of many utilities. These technologies seemed relatively simple and straight-forward, and activity was encouraged by federal programs providing cost-sharing opportunities and tax incentives. However, by the early 1980's, after understanding began to emerge that solar heating and cooling's main impacts were to increase awareness of energy conservation and energy-efficient design practices and after the technical and economic risks associated with most solar-thermal electric systems had become apparent, utilities shifted their attention more to wind power and photovoltaics (PV).

Utility interest in wind power also grew rapidly in that early period, coupled primarily with the large, megawatt-scale wind turbine programs of the federal government. As problems with the large turbines became apparent and the wind power focus shifted to the smaller machines in the independent-power-producer arena, utility interest waned. Within the past several years, however, as awareness of successes with mid-size turbines in the post-tax-credit era has increased, utility interest in wind power is again on the upswing. Meanwhile, utility interest in PV has grown steadily since the early 1980s, to the point where, for several years now, PV has received the greatest level of attention among all the emerging renewable energy technologies. As a result of continuing technical advances in both wind and PV, increased utility confidence is emerging in the suitability of their use in grid-connected applications; particularly in locations where it can be shown that the solar and wind resources are such that a significant capacity value, in addition to an energy value, can be derived from a plant's output. Also, as a result of the striking progress achieved by the Luz organization, renewed interest has recently been shown by Southwest utilities in solar-thermal electric power technology in the form of the solar-fossil-hybrid trough concept being employed successfully by that organization at its SEGS plants in Southern California.

As information and experience have been gained by utilities, they have become more selective and sophisticated in their solar and wind activities. During the latter part of the 1980's, a few utilities began a transition from involvement in solar and wind power on an assessment- and experimentation-level basis toward involvement with the more promising of these technologies through serious business ventures. Most telling, however, utilities in areas with good solar and wind resources are beginning to view solar and wind technologies as viable power generation options, some to the point where they are factoring these technologies into their generation planning. This extensive and continuing involvement of U.S. utilities in solar and wind energy is all the more remarkable upon realization that most of the incentives afforded to developers and independent power producers, such as renewable energy tax credits and favorable energy sales conditions, have not been available to the utilities.

This chapter provides a comprehensive overview of the nature and extent of the electric utility industry's activities in solar and wind energy technologies. The implications in terms of prospects for the *energy-significant* application of these technologies by electric utilities and their customers are discussed. (By *energy-significant*, we mean a total contribution to the nation's energy consumption of at least one percent.)

## 1.2 Background

### 1.2.1 Introduction

The energy crisis of the early 1970's spawned extensive efforts, both in the United States and throughout the world, aimed at research, development, testing and demonstration of renewable energy technologies. Public and private investment worldwide in the development and deployment of emerging renewable power technologies has exceeded the equivalent of U.S. \$5 billion. As a result of this aggregate effort, prospects for these technologies have been clarified substantially, and only a small number continue to be pursued with vigor. The principal emerging renewables are solar-thermal electric power, wind power, and photovoltaic power. These technologies have posted impressive gains over the past ten years through persistent engineering development efforts that, particularly for wind and solar-thermal trough systems, have had the benefit of financial incentives, such as federal and state tax credits. A great deal of the progress achieved has been carried out in the private sector with substantial, critically-important entrepreneurial drive aimed at marketplace success.

Among the key participants in these development efforts have been the U.S. electric utilities. Collectively, they have undertaken thousands of renewable energy projects and activities, some independently and others in conjunction with entities such as the Electric Power Research Institute (EPRI), the U.S. Department of Energy (DOE), universities, manufacturers, independent power producers, and the utilities' own customers. As a result of these efforts, electric utilities have acquired a substantial base of experience and knowledge concerning these technologies. And, based upon technology advances achieved in recent years, there is growing optimism within the electric utility industry that solar and wind energy systems will be well on their way toward becoming *energy-significant* by the end of this decade.

The requirement for large land areas is often raised as a constraint on the ability of solar and wind power technologies to play an *energy-significant* role in the nation's energy future. It is true that in heavily populated urban locations, available area, including rooftops, is insufficient to allow but a minor contribution to local energy needs from solar or wind energy systems. But in many U.S. locations with good solar and wind resources, land availability is not a major issue. Mature solar power technology is expected to generate approximately 40 megawatts (MW) per $km^2$ (100 MW per square mile), an amount comparable to coal power plants when mining is included. Mature wind power plants are expected to generate approximately 8 to 35 MW per $km^2$ (20 to 90 MW per square mile). Since vast areas of land (thousands of square miles) are available in much of the Southwestern U.S. and in other reasonably sunny parts of the country for solar power production, and in the Great Plains and elsewhere for wind power generation, it is clear these power technologies have the potential to generate far more than just a few percent of the nation's energy requirements. It has been estimated that 22,000 square miles of land, or about 20% of the area of Nevada, is all the land that would be needed using mature solar technology to generate an amount of electricity equivalent to the total current annual production from all sources in the United States (DeMeo, 1989).

### 1.2.2  Purpose and Scope

This chapter seeks to provide the reader with a fundamental understanding of the scope and depth of the U.S. electric utility industry's activities in solar and wind energy, the utilities' motivations for undertaking these efforts, the status of these efforts, and prospects for the future. In the context of this chapter, solar and wind technologies are taken to be those emerging renewable power technologies that have received substantial development attention and funding since the early 1970's, but have not yet become firmly established in the commercial marketplace. Not included are those renewable technologies that are well-established, such as hydroelectric power and direct combustion of wastes from agriculture, lumbering, and municipal refuse. The technologies covered in this chapter include solar heating and cooling (SHAC), solar-thermal power generation, wind power, photovoltaics, solar-thermal process heat and ocean-thermal energy conversion (OTEC). The chapter draws upon interviews with individuals representing some 36 utilities and utility-related organizations, reports and other materials provided by these individuals, and reports and publications prepared by EPRI, DOE, the Solar Energy Research Institute (SERI), and other organizations.

Although we have tried to make this chapter comprehensive in its overview, the reader should recognize that the sheer size of the electric utility industry and the breadth and depth of the past and current utility involvement in solar and wind energy technology is such that only a portion of the total number of utility projects and activities is described or even mentioned. However, for the most part these are either the key projects and activities or ones representative of similar efforts by a number of utilities. Thus, although we do not mention nearly all utilities and activities, we believe that the utilities, activities and experiences described will provide the reader with a good understanding of the electric utility industry involvement and experience with solar and wind technologies.

Also, in reading this chapter, we suggest the reader attempt to maintain a sense of the history that led to the 1973 oil embargo and that has brought us to where we are today. An important realization is that the 20th century has been and continues to be an era of great and dynamic change. Indeed, it is becoming evident that human societies have reached the point where their actions can, and are, profoundly influencing the global environment. In addition, it is important to realize that the major forces which have acted to shape this history and bring about the many changes, some of which have tended to advance and others to retard the development of solar and wind energy systems, cannot be fully controlled by any one individual, company, industry or even nation.

### 1.2.3  Factors Influencing Utility Applications of Solar and Wind Energy

The U.S. electric utility industry, although extensive in size, is not monolithic. It is, instead, comprised of several thousand individual utilities, each with its own unique situation in terms of power and energy resources and customer needs. Ultimately, it is people or groups of people within these utilities who make the decisions concerning the mix of energy sources the utilities will integrate into their power systems or propose to customers for installation on the cusomer-side of the meter as conservation or load-management (demand-side) options. These decisions are being made in increasingly complex and ever-changing economic, technical, and competitive environments and usually are subject to regulatory scrutiny and approvals. Thus, the electric utility industry experience with solar and wind energy systems should not be viewed in isolation, but rather within the framework of the environment and structure of the electric utility industry and the relevant internal and external factors and pressures influencing individual utilities within that industry.

### 1.2.3A Utility Characteristics

The 1989 Edition of the Electrical World Directory of Electric Utilities (McGraw-Hill, 1989) lists 3,221 utilities including 208 investor-owned companies, 1,810 municipal systems, 934 rural electric cooperatives, 77 public power districts, 9 irrigation districts, 37 federal systems, 71 state and 5 county systems, and 70 utilities falling into other categories. These utilities range in size from systems as large as Commonwealth Edison Company of Chicago, Illinois, which reported a 1988 summer peak of 15,683 megawatts (MW) and 24,487 MW of generating capacity, to systems as small as the Isabel, Kansas Electric Department, which has no generating capacity and reported a summer peak of about 350 kW. Investor-owned utilities account for about 79 percent (DOE, 1989) of the electricity generated in the United States with the balance produced by publicly-owned entities. Most investor-owned utilities are relatively large (over 1,000 MW) and vertically integrated to include the generation, transmission, and retail distribution functions within the same company. While some publicly-owned systems, especially the larger municipals, are also vertically integrated, most are not. The majority of municipal and cooperative utilities are distributors only. They own little or no generation and purchase most, if not all, of their power requirements at wholesale from either a privately-owned company, a generation and transmission cooperative, a municipal wholesale power supply agency, or one of the federally-owned systems, such as the Tennessee Valley Authority (TVA) or Bonneville Power Administration (BPA).

In addition to the variation in size, ownership, and function, utilities also vary in terms of such factors as customer mix, load characteristics and load growth rate, service area size, existing generation mix, the availability and cost of new generation capacity, the availability and cost of fuel, solar and wind resource availability, the cost and availability of financing, and organizational structure and staffing. Two key points need to be made about these factors. One is that a utility's ability and proclivity to pursue solar and wind technologies is influenced by these and other internal and external factors. The other is that the factors are continually in transition. For example, oil and coal prices rose dramatically after the 1973 Oil Embargo and continued to climb until the early 1980's. Since that time, oil and coal prices have declined substantially in terms of constant dollars. Similarly, utility demand growth dropped off precipitously during the early 1980's, coinciding with an economic recession. Many utilities were caught in the midst of large capacity construction programs and left with very substantial and costly capacity surpluses. For a number of utilities, the 1980's were a period of financial hardship. Only within the last few years has load growth shown significant resurgence, providing evidence of a growing economy. Meanwhile, the financial condition of the industry, as a whole, has improved. Generation capacity surpluses are disappearing, and power shortages are already developing in some parts of the country.

### 1.2.3B Emergence of Least-Cost Planning

While load growth is once again on the increase, albeit at a slower rate than historically, planning within the electric utility industry has shifted from the traditional focus on capacity expansion plans to the development of "least-cost" plans which incorporate both demand-side (conservation and load management) and supply-side (capacity expansion) alternatives. Under least-cost planning (also referred to as integrated planning), a utility examines a broad range of supply and demand-side alternatives and selects an "optimal" integrated plan designed to meet the electricity needs of its customers on a least-cost basis. In many cases, least-cost planning is being mandated by state regulatory agencies. Proponents of least-cost planning believe that it will encourage substitution of conservation, load management, and renewable energy options for new fossil-fuel generation capacity.

An emerging concept in least-cost planning is the incorporation of societal concerns that are difficult to quantify in traditional utility economic analyses, (such as environmental impacts, resource depletion, and employment) into the process of evaluating resource options. In a 1989 order adopting uniform guidelines for least-cost planning, the Wisconsin Public Service Commission directed electic utilities to recognize external social, political and environmental costs, which were referred to as "NEEDS," or costs "not easily expressed in dollars." The Wisconsin Commission also directed that utilities give a 15% credit to "noncombustion" options in their economic analyses in order to promote the use of energy options that do not discharge pollutants into the air. Incorporating such considerations into the planning process could significantly improve the relative ranking of solar and wind power technologies in the outcomes of utility assessments of resource alternatives. The utility regulatory agencies in Massachusetts, Oregon, Virginia and other states have also begun to require utilities to factor such external costs into their planning. These initiatives appear to represent only the beginning of such efforts by state regulatory agencies and much debate on the issue remains (Burkhart, 1989).

### 1.2.3C Competition Among Resource Alternatives

Although formally integrated into the planning processes of many utilities under least-cost planning, conservation and load management programs are not new to utilities. In fact, the pursuit of such programs was an initial reaction of many utilities to the energy crises of the 1970's. What has changed is that the technologies, the industrial base and the products available to support demand-side programs have evolved, since the early 1970's, to the point where utilities can indeed impact load growth rates and load shapes through the implementation of such programs. Also, a number of new or improved generation options are becoming available for utilities to factor into their planning. These include, besides solar and wind options, such alternatives as clean coal combustion technologies, more dependable and efficient combustion turbines, combined-cycle generating plants, cogeneration plants, compressed-air energy storage plants, battery storage plants, and solid-waste burning plants. And, spurred on by the provisions of the Public Utility Regulatory

Policies Act of 1978 (PURPA), private developers of cogeneration and other power production facilities (also referred to as independent power producers or IPP's) have emerged as significant elements of the power supply picture within the electric utility industry. Some utilities have already been offered more capacity by IPPs than they can use. Competitive bidding as a means for utilities to obtain new capacity and energy from plants built by IPPs is another evolving concept.

As a further complication, the Federal Energy Regulatory Commission (FERC) recently has been considering various options for restructuring of the electric utility industry to promote competition including possible deregulation of the generation function and modification of the rules under which utilities provide access to their transmission systems. Natural gas, once out of the picture in new plants for all but peaking use due to the constraints imposed by the Fuel Use Act of 1978 (Fuel Use Act), is again a significant factor in utility planning as a result of increased gas availability, lower gas prices and the repeal of the Fuel Use Act.

As the electricity supply picture tightens going into the 1990's, many utilities are looking to relatively low-cost, gas-fired combustion turbines (about $300 to $400/kW capital cost in 1989 dollars) for peaking service (use during high demand periods) or efficient gas-fired combined-cycle plants (about $600 to $700/kW capital cost in 1989 dollars) for intermediate duty service (use during moderate to high demand periods) for their next increments of generating capacity. Not only do such plants have relatively low capital costs, but they are relatively easy to site, have short construction lead times, and emit relatively few pollutants. Also, many U.S. utilities have an excess of baseload capacity resulting from construction programs completed in the 1980's. For these utilities, peaking or intermediate duty capacity is needed to balance their generation mixes.

### 1.2.3D  Innovation in the Electric Utility Industry

The role of the U.S. utilities in solar and wind energy development also needs to be understood in the context of innovation within the electric utility industry. The industry's history of innovation dates back to September 4, 1882, when Thomas Edison threw a switch energizing the Pearl Street Generating Station in New York City, the world's first electric power plant. Since that time, the primary emphasis of utilities has been on meeting the growing demand for electricity of customers, demand which often grew so rapidly that it strained the ability of utilities to install sufficient capacity to keep the lights on. Innovations were made to satisfy customer demands, improve reliability and lower costs. These innovations resulted in the evolution of our modern power systems.

However, despite the many technological advances achieved by the electric utility industry, electric utilities cannot be characterized as research and development organizations. Rather, they are primarily planning, construction management, operating and customer service organizations. Most utilities employ architect-engineering firms, private contractors, and equipment vendors to assist with the design and construction of new generating and transmission facilities. As a result, the innovations in the electric utility industry have come not only from utilities, but also from engineering firms, construction services firms and equipment manufacturers. The bulk of the R&D capability has resided with the equipment manufacturers who developed new products in response to their utility customers' demands and their own perceptions of market needs.

The formation of EPRI in 1972, as an outgrowth of the Edison Electric Institute's (EEI's) Research Committee, provided electric utilities with an improved mechanism for cooperation in the pursuit of research and development. Prior to the formation of EPRI, research was conducted mainly by companies individually, by two or more companies working together, or through an industry association committee such as the EEI's Research Committee. Often it was conducted in cooperation with a vendor or, as in the case of the development of nuclear power, in cooperation with the federal government. Thus, R&D in the electric utility industry has historically been fragmented and evolutionary in nature. Many players were involved in the development of modern power systems and in the progress achieved over the 100-year history of the industry.

It also is germane that the electric utility industry has historically tended to be conservative in adopting innovations. This conservatism stems to a large extent from the desire of utilities to remain competitive and from their commitment to providing reliable, low-cost service to their customers. These constraints should not be underestimated in terms of their importance in the selection of generation options for utilities. Utilities typically require power generation and transmission facilities to have at least a thirty-year life, to provide reliable service, to be practical and cost-effective to operate and maintain, and to enable a utility to provide affordable, low-cost service to the consumer. Also, such facilities must be capable of being integrated appropriately with the utilities' power systems in terms of operability, power quality, and safety. This means that the equipment must be rugged and relatively easy to maintain, yet sophisticated enough to function well within a utility system. Meeting these multiple objectives poses a formidable challenge for any new generation technology which must compete for a market niche with mature technologies that have evolved over decades.

### 1.2.3E  Utility Perspective on Solar Technologies

Finally, given the many generation and demand-side technology options that exist, utilities must make choices concerning which technologies they will incorporate into their resource plans. In making these choices, utilities have no intrinsic bias against solar, wind or any other renewable technologies. Some utilities considered and investigated solar and wind energy for their power generation possibilities decades before the recent energy crises. Arizona Public Service Company (APS), for example, hosted an international solar energy exhibition in Phoenix in 1954. This meeting provided the genesis for the establishment of the International Solar Energy Society (ISES). And, Detroit Edison, Public Service Electric and Gas of New Jersey, and several other utilities pursued wind energy as a generation option during the 1930s. However, the technological hurdles and high costs associated with the extraction of energy from the sun and wind were viewed as substantially exceeding the potential benefits of pursuing these technologies. Thus, no major efforts to evolve commercially viable systems were undertaken at that time by utilities.

The 1973 Oil Embargo and its aftermath provided new incentives for utilities to pursue solar and wind energy. The accompanying gasoline lines and high energy prices vaulted "energy" to prominence as a major economic and political issue for the American public. At about the same time, public concerns about the environmental effects of using coal and nuclear energy to produce electricity were growing. Meanwhile, electricity consumption continued to increase at a rate of about 7 percent per year, placing ever greater demands on utilities for new generating capacity. Utilities responded to the growing demand with the construction of new generating capacity, primarily coal and nuclear plants. The construction of this new capacity, often with expensive new pollution controls and safety features, during a period of high inflation, high interest rates, and frequent construction delays, when combined with shortage-inspired fuel cost increases, resulted in a reversal of the electric utility industry's historic trend toward lower costs per kilowatt-hour of electricity produced. As retail electric rates rose and environmental concerns mounted, utilities came under a barrage of public criticism. Among other things, they were criticized for not pursuing more vigorously various forms of renewable energy, especially solar and wind energy and relying so heavily on fossil fuels and nuclear energy. Some advocates for solar and wind wanted utilities to develop and deploy solar and wind technologies on a massive scale through crash programs.

In response to such pressures, utilities began to reexamine solar and wind. Although most did not consider solar and wind power to be *the* solution to the nation's energy problems, at least not in the short term, many viewed solar and wind as resources with the potential for helping to supplement other energy technologies on an energy-significant basis, but mainly in the longer term. They recognized that pursuing the evolution of solar and wind technologies suitable for utility system and customer applications would involve a costly and protracted process with uncertain outcomes. Nonetheless, these utilities began to actively pursue the development of such energy systems in a variety of ways.

### 1.2.4 Overview of Utility Activities in Solar Energy

A broad perspective on the nature and extent of the electric utility activities in solar energy can be gained by reviewing statistics compiled in annual surveys conducted by EPRI during the period 1976 to 1983 (EPRI AP-3665-SR).* Even though EPRI has not updated the survey since 1983, the period covered by the surveys serves to illustrate the overall trends.

Table 1.1 and Table 1.2 summarize the electric utility solar energy activities for the period 1976 to 1983 for the categories of solar heating and cooling, wind power, solar thermal electric power, photovoltaics, biofuels, solar-thermal process heat, and ocean-thermal energy conversion. These are the major categories of solar energy research and development that have been pursued within the U.S. Although included in Table 1.1 and Table 1.2 and several of the figures which follow, biofuels are not covered by this article, as previously mentioned.

For each category of activity, Table 1.1 shows the number of utilities participating, and Table 1.2 the number of activities and the percent of the total activities. It should be noted that in Table 1.1 the total number of utilities participating for

---

* Also, EPRI ER-321-SR, ER-649-SR, ER-966-SR, ER-1299-SR, AP-1713-SR, AP-2516-SR, and AP-2850-SR. Note: References for citations of EPRI reports on solar and wind power are provided in the bibliography of EPRI reports following the list of references at the end of this chapter.

**Table 1.1:** Number of Utilities Participating in Solar Energy Activities (Source: EPRI AP-3665-SR)

| Category | 1976 | 1977 | 1978 | 1979 | 1980 | 1981 | 1982 | 1983 |
|---|---|---|---|---|---|---|---|---|
| | | | | Year of Survey | | | | |
| Solar Heating and Cooling | 98 | 124 | 142 | 154 | 158 | 165 | 156 | 139 |
| Wind Power | 20 | 25 | 40 | 53 | 91 | 110 | 102 | 104 |
| Solar Thermal Electric Power | 8 | 18 | 26 | 34 | 54 | 53 | 47 | 26 |
| Photovoltaics | 5 | 10 | 23 | 24 | 32 | 40 | 41 | 41 |
| Biofuels | - | - | 7 | 19 | 28 | 30 | 27 | 32 |
| Solar Thermal Process Heat | - | - | - | - | 13 | 14 | 13 | 9 |
| Ocean-Thermal Energy Conversion | - | - | - | - | 4 | 4 | 5 | 4 |
| Other | 10 | 18 | 15 | 15 | - | - | - | - |
| Total Utilities[1] | 116 | 150 | 165 | 180 | 236 | 236 | 235 | 184 |

[1] Total utilities is less than the sum of utilities by category since individual utilities may have had projects in more than one category.

**Table 1.2:** Trends in Electric Utility Solar Energy Actitivies (Source: EPRI-3665-SR)

| Category | 1976 | 1977 | 1978 | 1979 | 1980 | 1981 | 1982 | 1983 |
|---|---|---|---|---|---|---|---|---|
| | | | | Year of Survey | | | | |
| | Number of Activities | | | | | | | |
| Solar Heating and Cooling | 238 | 360 | 442 | 511 | 516 | 545 | 538 | 484 |
| Wind Power | 29 | 34 | 56 | 84 | 152 | 191 | 182 | 195 |
| Solar Thermal Electric Power | 11 | 28 | 42 | 52 | 57 | 64 | 61 | 56 |
| Photovoltaics | 5 | 12 | 30 | 32 | 48 | 68 | 74 | 74 |
| Biofuels | - | 3 | 10 | 23 | 34 | 42 | 39 | 47 |
| Solar Thermal Process Heat | 6 | 11 | 13 | 15 | 18 | 20 | 18 | 9 |
| Ocean-Thermal Energy Conversion | - | - | 1 | 4 | 8 | 7 | 10 | 5 |
| Other | - | - | - | 14 | 6 | 6 | 8 | - |
| Total Activities | 289 | 448 | 594 | 735 | 839 | 943 | 930 | 870 |
| Category | Percent of Total Activities | | | | | | | |
| Solar Heating and Cooling | 82 | 80 | 75 | 70 | 61 | 58 | 58 | 56 |
| Wind Power | 10 | 8 | 9 | 11 | 18 | 20 | 20 | 22 |
| Solar Thermal Electric Power | 4 | 6 | 7 | 7 | 7 | 7 | 7 | 6 |
| Photovoltaics | 2 | 3 | 5 | 4 | 6 | 7 | 8 | 9 |
| Biofuels | - | 1 | 2 | 3 | 4 | 4 | 4 | 5 |
| Solar Thermal Process Heat | 2 | 2 | 2 | 2 | 2 | 2 | 2 | 1 |
| Ocean-Thermal Energy Conversion | - | - | - | 1 | 1 | 1 | 1 | 1 |
| Other | - | - | - | 2 | 1 | 1 | 1 | - |

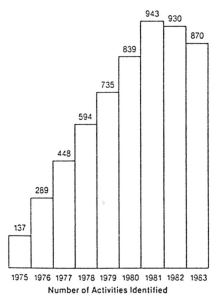
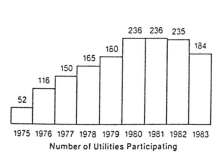

**Figure 1.1:** Electric utility activity in solar and wind energy 1975-1983 (Source: EPRI AP-3665-SR).

all activities is less than the sum of the utilities participating by category because many utilities reported activities in more than one category. Figure 1.1 provides a graphical presentation of the overall trends in terms of the total number of activities and the number of utilities participating. Figure 1.2 and Figure 1.3 provide similar breakdowns by category of activity.

Table 1.1 and Figure 1.1 indicate that the total number of utilities participating in solar energy and the total number of activities reported grew rapidly after 1975 and peaked in 1981 at 236 and 943, respectively. By 1983, the number of utilities participating had declined to 184 and the number of activities to 870. The decline in number of participants and activities in the early 1980s generally coincides with changes in several key external factors which occurred at the same time and influenced utility involvement in solar energy activities. These include a decline in utility load growth which left many regions of the country with substantial capacity surpluses, the stabilization and eventual drop in oil prices, and the reduction in federal funding of solar energy activities.

Also, as can be seen in Table 1.1 and Table 1.2 and Figure 1.2 and Figure 1.3, the trends varied by category of solar energy activity. It is clear that SHAC dominates the overall statistics, accounting for over 70% of the total activities in the 1970s but declining to 56% by 1983. The peak level of reported activity in SHAC occurred in 1981 with 165 utilities participating in 545 projects. Thus, the decline in the overall level of activity in the early 1980s can be largely explained by the trends in SHAC. Significant declines in terms of the number of utility participants and number of projects were also recorded for solar-thermal electric power, process heat, and ocean-thermal energy conversion. Meanwhile, the categories of wind power, photovoltaics

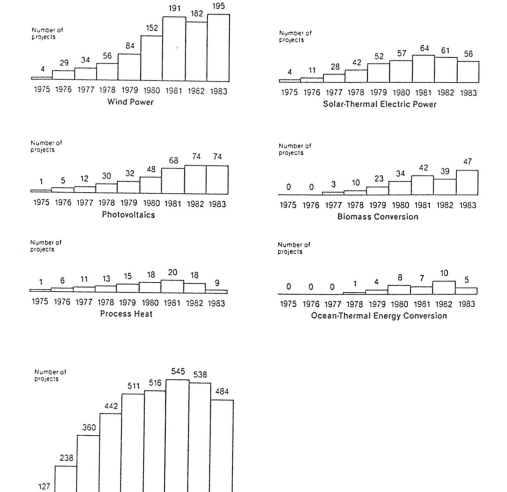

**Figure 1.2:** Electric utility solar projects by category 1975-1983 (Source: EPRI AP-3665-SR).

and biofuels appeared to plateau rather than decline. Since that time, the level of utility activity in photovoltaics has continued to grow while the level of activity in wind, although involving fewer utilities, has remained about the same. The level of activity in the other technology categories has declined substantially.

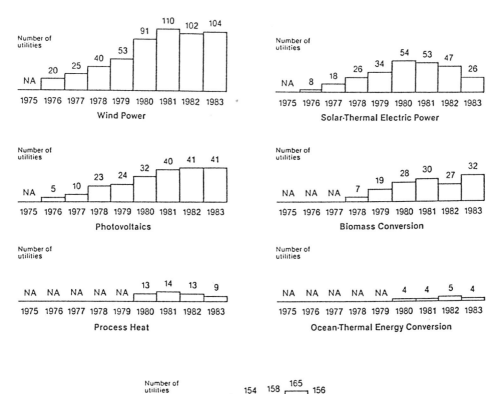

**Figure 1.3:** Number of utilities participating in solar projects 1975-1983 (Source: EPRI AP-3665-SR).

Table 1.3 provides a breakdown of the utility activities in 1983 for each category by the type of activity, and Table 1.4 a similar breakdown by source of sponsorship and use sector. As can be seen in these tables, the largest share of the activities (34%) involved field tests and demonstrations, and most of the activities (69%) were sponsored by the utilities themselves (in house). Figure 1.4 shows that activities were pursued by utilities in all regions of the country represented by the major electric utility reliability councils. Table 1.5 shows that the activities involved all categories of utility ownership.

The trends in utility industry involvement in the solar and wind technologies over the past fifteen years are represented quite well by the Figure 1.2 and Figure 1.3 data. The following sections describe in detail (a) representative project and program activities; (b) the rationale for these activities; (c) the reasons for increases and decreases in activity level in the respective technology areas; and (d) key findings,

Table 1.3: Breakdown by Type of Activity for Electric Utility Solar Projects in 1983 EPRI Survey (Source: EPRI AP-3665-SR)

| Type of Activity | Photo-voltaics | Solar-Thermal Electricity Production | Biofuels | Ocean-Thermal Energy Conversion | Wind Power | Solar Heating and Cooling | Process Heat | Total | % of Total |
|---|---|---|---|---|---|---|---|---|---|
| Resource and Siting Assessments | 1 | 1 | 2 | 0 | 71 | 27 | 0 | 102 | 12 |
| Integration and Assessment Studies | 13 | 12 | 17 | 2 | 23 | 34 | 4 | 105 | 12 |
| System Design and Evaluation | 9 | 26 | 4 | 2 | 6 | 10 | 0 | 57 | 7 |
| Component Development and Testing | 5 | 8 | 7 | 1 | 2 | 10 | 1 | 34 | 4 |
| Field Test and Demonstration | 42 | 9 | 16 | 0 | 42 | 184 | 4 | 297 | 34 |
| Monitoring | 4 | 0 | 0 | 0 | 48 | 186 | 0 | 238 | 27 |
| Incentive, Sales, Leasing and Educational Programs | 0 | 0 | 1 | 0 | 3 | 33 | 0 | 37 | 4 |
| Total | 74 | 56 | 47 | 5 | 195 | 484 | 9 | 870 | |
| Percent of Total | 9 | 6 | 5 | 1 | 22 | 56 | 1 | 100 | |

lessons learned and recommendations resulting from these activities. The technology areas are covered generally in chronological order relative to the progression of technology emphases in time, except that technologies that have received only a minor level of attention are discussed last.

# 1.3 Solar Heating and Cooling

## 1.3.1 Introduction

Solar heating and cooling involves the use of solar energy for the production of hot water and/or the heating or cooling of buildings. As noted above, SHAC has been the most ubiquitous and, for the majority of utilities, also the initial form of involvement in solar energy. The utility SHAC projects have included both active and passive systems and systems designed for both commercial and residential applications. By the early 1980s utility project activity in SHAC had begun to decline, suggesting that utilities were attaining the level of SHAC system understanding needed for rational planning and for effective customer information programs. Both active and, especially, passive SHAC systems are viewed by the utility industry as mainly demand-side measures. Recognizing the relationship of solar heating and cooling to conservation, EPRI in 1980 transferred SHAC from its Solar Power Program to its demand-side management program.

## 1.3.2 Experiences of Individual Utilities

To gain further insights into the extent and nature of the utility industry experience with solar heating and cooling, it is illustrative to examine the activities and experiences of a few specific utilities.

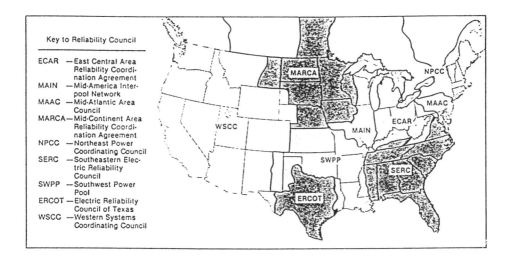

Breakdown by Reliability Council and Technology

| Reliability Council | Photovoltaics | Solar-Thermal Electricity Production | Biofuels | Ocean-Thermal Energy Conversion | Wind Power | SHACOB | Process Heat | Total | Percent of Total |
|---|---|---|---|---|---|---|---|---|---|
| ECAR | 3 | 0 | 0 | 0 | 11 | 57 | 0 | 71 | 8 |
| MAIN | 1 | 0 | 0 | 0 | 13 | 33 | 0 | 47 | 5 |
| MAAC | 3 | 0 | 2 | 0 | 7 | 27 | 2 | 41 | 5 |
| MARCA | 3 | 1 | 10 | 0 | 17 | 36 | 2 | 69 | 8 |
| NPCC | 4 | 0 | 3 | 1 | 43 | 59 | 0 | 110 | 13 |
| SERC | 14 | 2 | 10 | 3 | 9 | 60 | 1 | 99 | 11 |
| SWPP | 4 | 4 | 4 | 0 | 10 | 28 | 0 | 50 | 6 |
| ERCOT | 0 | 3 | 0 | 0 | 7 | 16 | 0 | 26 | 3 |
| WSCC | 42 | 46 | 18 | 1 | 78 | 168 | 4 | 357 | 41 |
| Total | 74 | 56 | 47 | 5 | 195 | 484 | 9 | 870 | |

**Figure 1.4:** Breakdown of electric utility 1983 solar activity by geographic area (Source: EPRI AP-3665-SR).

## 1.3.2A Long Island Lighting Company

Long Island Lighting Company (LILCO) began its activities in solar and wind energy in the early 1970s. Long Island's insolation resource, according to LILCO, is about 5% below the U.S. average but at the top for New York State. By 1976, LILCO already had six activities under way in SHAC. The 1983 EPRI survey (EPRI

**Table 1.4:** Breakdown by Sponsorship and Use Sector For Electric Utility Solar Activities in 1983 EPRI Survey (Source: EPRI AP-3665-SR)

| | Photo-voltaics | Solar-Thermal Electricity Production | Biofuels | Ocean-Thermal Energy Conversion | Wind Power | Solar Heating and Cooling | Process Heat | Total | % of Total |
|---|---|---|---|---|---|---|---|---|---|
| **Project Cosponsorship** * | | | | | | | | | |
| In-House | 64 | 33 | 45 | 5 | 179 | 468 | 8 | 802 | 69 |
| Private | 17 | 15 | 7 | 1 | 13 | 23 | 1 | 77 | 6 |
| Federal Government | 16 | 35 | 3 | 2 | 30 | 28 | 0 | 114 | 10 |
| University | 5 | 4 | 5 | 0 | 15 | 28 | 5 | 62 | 5 |
| Utility Group | 5 | 21 | 7 | 0 | 15 | 14 | 1 | 63 | 5 |
| EPRI | 0 | 2 | 1 | 0 | 5 | 8 | 0 | 16 | 2 |
| State/Local Government | 6 | 2 | 2 | 0 | 10 | 13 | 0 | 33 | 3 |
| **Use Sector** | | | | | | | | | |
| Residential | 22 | 0 | 2 | 0 | 57 | 299 | 0 | 380 | 44 |
| Commercial/Industrial | 21 | 8 | 5 | 0 | 18 | 120 | 1 | 173 | 20 |
| Agricultural | 0 | 1 | 6 | 0 | 8 | 3 | 7 | 25 | 3 |
| Utility | 21 | 42 | 30 | 4 | 44 | 11 | 1 | 153 | 17 |
| Multiple | 10 | 5 | 4 | 1 | 68 | 51 | 0 | 139 | 16 |
| Total | 74 | 56 | 47 | 5 | 195 | 484 | 9 | 870 | 100 |

\* Some projects have several cosponsors

**Table 1.5:** Breakdown of Electric Utility 1983 Solar Activity by Utility Ownership (Source: EPRI AP-3665-SR)

| Ownership | Number of Utilities | Percent of Total | Number of Projects | Percent of Total |
|---|---|---|---|---|
| Investor-Owned | 104 | 57 | 644 | 74 |
| Municipal | 22 | 12 | 71 | 8 |
| Federal/State | 17 | 9 | 82 | 10 |
| Cooperative | 41 | 22 | 73 | 8 |
| Total | 184 | 100 | 870 | 100 |

AP-3665-SR) reported seven SHAC activities for LILCO including studies, residential and commercial demonstrations, and residential and commercial monitoring.

One of the studies involved an integration and assessment effort that LILCO helped to initiate with local government and industry in Suffolk and Nassau counties. The effort included evaluation of SHAC systems, examination of their effectiveness and utility compatibility, dissemination of information about SHAC systems to homeowners, and promotion of energy conservation methods.

LILCO's earlier SHAC residential demonstrations investigated the viability of residential solar space heating by retrofitting homes with solar collectors or solar assisted heat pumps and various heat storage systems. LILCO also hosted the construction of five SHAC demonstration homes as part of an EPRI project (Arthur D. Little, Inc., 1977). The project combined solar energy with air-to-air heat pumps, thermal storage, and load management. Performance data from the experimental homes was used to validate a computer program to determine the preferred SHAC systems for any area of the United States. Data collection was completed in March 1981, at which time the thermal storage and load management systems were deactivated.

One of LILCO's residential monitoring projects involved data collection and analysis for hot water systems installed in 600 residences. The solar systems included two 1.95 m$^2$ (21-ft$^2$) collectors and a 454 liter (120 gallon) storage tank. Additional water heating was provided by off-peak electricity. All systems were monitored for hot water use and energy efficiency for two years and some for five years. The data showed a 50% cost savings over conventional fuels. The yearly solar contribution averaged 44%.

LILCO reports that it is no longer active in the area of SHAC. According to LILCO, SHAC systems appeared attractive in the late 1970's because of rising oil prices and tax credits and because about 80% of the 700,000 homes in Long Island were heated by oil. SHAC systems promised to be cost-effective had oil prices continued to rise. However, by 1988 the price of heating oil had dropped to $0.25 per liter ($0.95 per gallon) compared to $0.34 per liter ($1.30 per gallon) in 1980. In real terms, the current price is even lower and, in fact, lower than the price of heating oil in 1973. Thus, although LILCO's SHAC program was technically successful, the costs of installation and operation of SHAC systems were found to substantially exceed the economic benefits.

### 1.3.2B Northern States Power Company

Northern States Power Company (NSP) initially embarked on activities in solar and wind energy to obtain data needed to evaluate these technologies and to properly factor them into NSP's resource planning. The SHAC projects of NSP included a component development and testing program, residential demonstrations, and monitoring of a commercial installation. In one residential demonstration, NSP constructed a 97.5 m$^2$ (1050-ft$^2$) split-level home in the Minneapolis suburb of Brooklyn Park. The design used 4.72 cm (12-inch) insulated walls. A large south-facing, triple-glazed window and a heat-recovery ventilation system were other features of the house. The cost to heat this home through a typical Minnesota winter was estimated at $50 to $100. The house was sold in June 1983, and energy use was monitored for three years.

In a December, 1987 report (NSP, 1987), NSP reported on the results of a solar demonstration program mandated by the Minnesota Public Utilities Commission in October, 1984 as part of an agreement involving the licensing of Unit III of NSP's Sherco Generating Station. Under this program, NSP was to finance the installation of 100 solar domestic hot water (DHW) systems and monitor their performance. NSP also contributed $63,000 to the Minnesota Solar Industries Guild for the establishment of a solar installation certification program. The main objective of the effort was to determine the performance, reliability, and cost effectiveness of solar DHW systems to purchasers and NSP ratepayers.

In May 1985, during the middle of the program, the Minnesota Legislature retroactively repealed the state's 20% solar tax credits effective January 1, 1985. Also, at the end of 1985 the federal solar tax credits for SHAC systems expired and were not renewed by Congress. Both of these actions greatly hampered NSP's efforts to finance and monitor the 100 solar systems to be installed under this program. Only 24 systems were eventually installed.

According to the report (NSP, 1987), NSP found that the average DHW system saved its purchaser $134 annually at 1987 electricity prices. The average payback of the systems was projected to be 35 years without the 40% Federal tax credits, and 21 years with the tax credits. The present values of the DHW systems to NSP's ratepayers were estimated to range from –$210 to –$570. NSP concluded that solar water heating systems are not cost effective in its service area for either the participating customers or NSP's ratepayers, especially without tax credits.

### 1.3.2C  Pacific Gas & Electric Company

The Pacific Gas & Electric Company's (PG&E's) interest in solar and wind energy was a direct consequence of the shock of the 1973 Oil Embargo. The Company's initial response was to focus on conservation including formulating standards, aiding with building weatherization and insulation, and supporting the installation of solar heating systems especially for hot water and swimming pools. The 1976 EPRI survey (ER-321-SR) identified 11 SHAC projects for PG&E and the 1983 survey (AP-3665-SR) 12 projects. The projects listed in the 1983 survey include a resource and siting assessment, operation of a test facility, three commercial demonstration projects and two commercial monitoring activities, five residential monitoring activities, and an incentive program.

PG&E's resource and siting assessment project involved installation of long-term monitoring stations for the accumulation of detailed solar insolation data from various locations within PG&E's service territory. This effort continues today and has given PG&E a solid understanding of the solar energy resource available for prospective solar power plants in its service area. The operation of the test facility involved testing active and passive SHAC equipment, systems, and concepts. Information obtained was used to evaluate manufacturers' claims, to determine regional applicability, and to assess operational compatibility of various components.

One of PG&E's commercial demonstration projects involved the use of solar collectors to operate a 2,700 kg (3 ton) absorption air conditioner. The system utilized 46 m$^2$ (500-ft$^2$) of evacuated-tube collectors and produced 7 °C (45 °F) chilled water for cooling. In another commercial demonstration, PG&E cosponsored the installation of 60 roof-mounted solar collectors to preheat approximately 1.1 million liters (300,000 gallons) of water at a San Francisco aquarium. In one of the two commercial monitoring projects, the solar water-heating systems used by two dairies to supplement their electric water-heating systems were monitored by PG&E. In the other PG&E monitored the performance of a solar space-heating system for a 163 m$^2$ (1,750-ft$^2$) office building at Stanislaus State University.

In one of the residential monitoring projects, 400 single-family systems and 75 multi-family systems were monitored. In a second, PG&E monitored 1,200 solar water-heating systems, and in a third, 20 solar water-heating systems. PG&E also collected load data from the homes of customers participating in one of the residential water heating programs, one of PG&E's residential time-of-use programs, and from respondents to a survey. Data collected were to be used to develop possible rate incentives and other programs to encourage use of solar energy.

In addition, as an incentive to build passive solar homes, PG&E offered builders up to $1,000 per home. Awards were made on the basis of the energy savings of each design. Every home had to qualify as a premium energy conservation home; that is, it had to use 50–75% less energy than a conventional home for space and domestic hot water needs. The pilot program began in July 1979 and was completed at the end of 1981.

The aggregate experience from these projects has given PG&E a sufficient understanding of these systems to respond to its customers' inquiries. Very little research and development activity in the area of SHAC is being conducted by PG&E at this time. The activity that remains primarily involves public information-related effort. Since the technology involved with SHAC is commercially available, PG&E feels it should be up to consumers, builders, and the SHAC industry to pursue the incorporation of the design and installation of such systems in new residential and commercial structures.

### 1.3.2D  Tennessee Valley Authority

The Tennessee Valley Authority (TVA) also conducted a very extensive SHAC program in the 1970s and early 1980s. The TVA efforts in SHAC were conducted as part of a conservation program begun in 1974. In the 1976 EPRI survey, TVA reported five SHAC activities and by the time of the 1983 Survey, a total of 24 activities. The latter include two integration and assessment studies, two system design evaluations, one component development and testing project, one field test and demonstration, eight residential demonstration activities, five commercial demonstrations, two incentive programs, and two educational programs. The last four of these programs and the status of TVA's SHAC activities are summarized below:

### 1.3.2D.1  Incentive Program - Various Locations, TN

TVA offered financing for solar water heaters in a two-stage program. The first stage included the design, installation, and low-interest financing of 1,000 solar water heaters to supplement existing electric water heaters. The second stage provided for the financing of at least 11,000 solar water-heating systems. The home of each interested consumer was surveyed by TVA to determine the feasibility of installing a solar water heater. Consumers were then offered a choice of approved solar systems and a selection of qualified installation components. After installation, the system was inspected by TVA before payment was to be made to the installer. By March 1983, 1,899 installations were completed. TVA offered consumers 20-year monthly loan payments through their electric bills for both stages of the program.

### 1.3.2D.2  Incentive Program - Chattanooga, TN

Commercial and industrial customers were provided with technical and design assistance to improve energy efficiency in buildings, both new construction and retrofit. Services offered included computer modeling, thermal performance calculations, recommendations for energy-saving strategies, and cost-effectiveness studies. The program provided private-sector design professionals and their clients with direct access to proven passive solar concepts and technologies. In addition, the program promoted these concepts through education and training activities. By March 1983, 300 applications for assistance had been received and processed, and 170 of the applicants received significant assistance from TVA. A partial follow-up survey of the technical and design assistance projects identified 52 projects completed as new construction or retrofits. It also confirmed that one-third of the designers were applying the newly-learned passive solar concepts to other new buildings they were designing.

### 1.3.2D.3  Status of TVA SHAC Efforts

As an agency of the federal government, the TVA has been at the forefront of national efforts related to solar energy. The level of activity at TVA reached its peak in the late 1970's, after President Carter in his 1978 State of the Union Address challenged TVA to become the nation's solar showcase. The TVA program plateaued after the 1980 Presidential election and then suffered a rollback. Because of substantial capacity surpluses, TVA recently concluded that it does not need renewables until the next century. Accordingly, energy conservation programs, including SHAC, have been scaled back.

### 1.3.2E  Public Service Electric & Gas

Public Service Electric and Gas (PSE&G) has also had extensive experience with SHAC including solar water heaters, pool heaters, and space heaters. PSE&G installed, monitored and operated commercially available hot water, space and pool heating systems; and analyzed their performance with the help of consultants. In conjunction with the New Jersey Institute of Technology, PSE&G also investigated more cost-effective alternatives to the conventional glass and metal panel construction of solar collectors. After concluding these demonstrations, the company actively engaged in the sale and installation of solar water heaters. The utility offered a full five-year warranty including parts and labor for the systems. PSE&G also trained contractors in the proper installation of the systems. However, only five systems were ever sold.

PSE&G also published information brochures, and PSE&G personnel gave hundreds of talks at meetings throughout its service area. Slide shows for PSE&G personnel to use were prepared through PSE&G's Speaker's Bureau. This effort continued from 1975 to 1983. The effort continues through PSE&G's Energy Conservation Center, a customer service which offers rebate programs and other incentives. Despite these efforts, only a few customers have received rebates for solar water heaters. PSE&G indicates that the interest in such systems has nearly vanished.

PSE&G estimates that there may be as many as 5,000 solar water heaters in the state, but probably many of them are no longer functioning or work very poorly. Most manufacturers have left the field; at one time New Jersey was the third largest manufacturer of solar hot water collectors in the U.S. Many firms were involved. When oil prices subsided and the tax credits were eliminated, these firms left the business. Many installations failed after costing thousands of dollars. Meanwhile, the support infrastructure had disappeared making it difficult to obtain service for such installations. In some cases, PSE&G endeavored to help customers locate contractors who could assist them with their solar hot water systems, but the abrupt disappearance of the industry made this difficult. The market was apparently artificial, induced by tax credits and often served by unscrupulous individuals and firms.

### 1.3.3  Additional Input From Other Utilities

#### 1.3.3A  Carolina Power & Light Company
The activities of Carolina Power & Light Company (CP&L) in the area of SHAC included working with an EPRI computer model designed to simulate the effects of passive solar features on a home. CP&L also worked with the North Carolina Alternative Energy Corporation in instrumenting and monitoring 24 solar water heaters. It was determined that the system achieved a significant energy displacement, but that fuel cost savings produced only a 10-year payback. CP&L concluded that most customers would want a shorter, 5-year or less, payback in order to justify the expense of a solar water heating system.

#### 1.3.3B  Florida Power Corporation
According to Florida Power Corporation (FPC), solar hot water systems were prevalent in Florida in the 1920s and 1930s. They disappeared when natural gas became available for hot water heating. In the mid-1970s, many manufacturers, distributors and installers of solar hot water heating systems appeared in Florida. For awhile, business boomed, aided by tax credits. The incentives of high gas and oil prices and tax credits disappeared during the 1980s, as did interest in the installation of such systems.

#### 1.3.3C  Wisconsin Power & Light Company
Solar activities at Wisconsin Power & Light Company (WP&L) began around 1978 with a solar water-heater program. After a pilot program involving selected customers, WP&L began to market solar water heaters to customers. Field personnel were trained to sell them, and contractors were engaged to accomplish the installations. The program lasted through 1981, and involved installation of about 100 systems. The marketing disappeared in the early 1980 because the units were too expensive compared with gas or even electric water heating. WP&L still maintains a supply of parts and feels obligated to provide help if a customer has problems.

WP&L also became involved with demonstrations of passive solar homes at several locations. WP&L created brochures and other materials to answer questions for customers. Customer interest in passive solar homes continues. In 1983–1984, WP&L introduced its "Good Sense Home" concept, which emphasizes a total building energy efficiency concept. The focus is on good building practices.

#### 1.3.3D  Georgia Power Company
In 1981, Georgia Power Company (GPC) began construction of a demonstration home designated as FUTURE I. It was intended to showcase possibilities for future homes in terms of energy conservation and use of solar energy. FUTURE I had super insulation, passive solar heating, and a 4 kW PV array. The passive heating features produced an annual average of 55% of the required heat for the building. In practice, this was excessive. Although the heat was advantageous during winter,

the summertime heat collection was more than desirable. The PV system produced 4,800 kWh/yr, which was 22% of the annual requirements for the home.

FUTURE I was on display for 18 months during which time about 125,000 visitors passed through it. Afterward, the home was occupied for two years by a family while data collection continued. The experiment was discontinued in 1985. GPC then removed the PV array and instrumentation, and sold the house.

The general conclusion was that the passive heating system should have been designed to provide about 33% of the heating requirements. The PV system worked well throughout the 3.5 years of testing and provided power of suitable quality. There were no inverter problems during the test period.

GPC also constructed 3 other homes designated FUTURE II, which incorporated various types of super insulation. One had a solar heater for domestic hot water, but there were no other solar features on any of the homes.

### 1.3.4  Lessons Learned and Conclusions

Many utilities had involvements and experiences with SHAC similar to those described above. The aggregate experience with thousands of SHAC installations has given utilities a sufficient understanding of these systems to provide useful information and assistance to their customers on an ongoing basis. From their experiences with activities in SHAC, utilities have learned that significant amounts of energy can be saved with well-designed solar hot water systems and through the application of building design and construction practices which seek to optimize energy efficiency. The technologies are known, and the required hardware and systems are generally commercially available. However, although passive SHAC can be cost-effective in many locations, the costs of installing and maintaining solar water heating and active SHAC systems currently exceed the energy cost savings from such systems by a substantial margin in most locations.

As can be surmised from the number of activities, the large number of installations included in many individual activities, the high degree of customer interaction and the associated labor intensiveness, and the very large number of utilities that have been involved, the aggregate U.S. utility industry investment in the development of a solid perspective in SHAC technology and applications has been very substantial. Measured in terms of manpower, funding, management attention, internal and external expectations, and organizational creativity and entrepreneurship, this investment may still exceed the aggregate utility investment in the electric power related solar technologies. Many would argue that the results of this extensive activity are not commensurate with the level of effort expended. Indeed it is likely that some utility industry executives still harbor a measure of skepticism relative to renewable energy activity in general as a result of their experiences with SHAC programs, especially since many of these programs were fueled by public pressures with associated high expectations that, to a large degree, were not fulfilled. However, these programs have resulted in three primary benefits whose value should not be underestimated:

- SHAC activities contributed substantially to our understanding, as a nation, of the importance and benefits of attention to energy efficiency, conservation measures, and sound engineering and architectural principles in the design and construction of residential and commercial buildings.

- A solid understanding emerged of the technologies and application principles associated with SHAC. This is available for use now in situations where SHAC is warranted, and in the future should conventional energy costs trend substantially upward.

- In part through their early extensive activities in SHAC, utilities have developed a deeper perspective on the importance of thorough planning in renewable energy activities, including well-conceived objectives and sound logic aimed at meaningful results, even when public pressures and the political climate demand involvement without delay.

## 1.4 Solar-Thermal Electric Power

### 1.4.1 Introduction

On May 3, 1978, Sun Day, President Jimmy Carter visited Albuquerque, New Mexico where he met Public Service Company of New Mexico (PNM) President Jerry Geist who presented him with a proposal (PNM, 1978) for the repowering of PNM's Reeves Station Unit No. 2, a 50 $MW_e$ gas- and oil-fired generating unit, using solar energy. The proposal included a development plan that estimated a market potential among major Southwestern utilities of 58 repowered units. Although there was considerable enthusiasm at PNM and several other utilities in the Southwest about the solar-fossil repowering concept, none of the proposed repowering projects was ever built. However, a number of other solar-thermal power projects were constructed, and utility participation in these ranged from ownership and operation to purchasing project output.

Utilities took an early interest in the application of solar-thermal energy for power production because the technology in many respects uses components and systems similar to those used in conventional power plants. Solar-thermal electric power technology involves the concentration of solar flux, using tracking mirrors or lenses, onto a receiver where the solar energy is absorbed as heat that is subsequently converted into electricity using a heat engine and a generator. Many variations in the manner in which the solar energy is concentrated and converted to electric power in such systems are possible. The two primary forms of solar-thermal technology are central receivers and distributed receivers. Central-receiver systems use fields of two-axis-tracking mirrors called heliostats to focus sunlight onto a tower-mounted receiver. Distributed receivers employ point-focusing parabolic dishes or line-focusing parabolic troughs to concentrate sunlight. In addition to central receivers and distributed receivers, the category of solar-thermal electric power is defined here to include salt-gradient solar ponds.

Solar-thermal power plants also may be stand-alone or hybrid (solar-fossil fuel) systems. Most of the solar-thermal field experiments and demonstrations in which utilities have been involved are solar-only. However, by far the most successful solar-thermal installations in the world to date have been the privately-funded SEGS plants of Luz International, which are commercial plants using a solar-fossil-hybrid parabolic trough concept.

This section focuses on the major solar-thermal activities that have involved utilities in the funding and/or conduct of design, development and testing efforts, and in particular on the results obtained, lessons learned and perspectives developed from those activities. More thorough discussion of these activities, along with coverage of other solar thermal field test activities worldwide, can be found in the article entitled "The Status of Solar Thermal Electric Technology" in Chapter 2 of this Volume.

### 1.4.2  Overview of Utility Activity

Based on information compiled in the EPRI surveys, the number of utilities involved in solar-thermal electric power peaked in 1980 at 54 utilities while the number of activities peaked in 1981 at 64, as previously shown in Table 1.1 and Table 1.2 and Figure 1.2 and Figure 1.3. By 1983, the number of participants reported had dropped to 26 and the number of projects to 56. According to Figure 1.4, most of the 56 reported activities in solar-thermal power in 1983 were located in the Southwest. Table 1.4 shows that most solar-thermal power projects were cooperative efforts involving one or more utilities, the federal government, and/or private funding sources.

### 1.4.3  Central Receivers

The heliostats in central-receiver systems reflect solar energy to heat a working fluid circulating through a tower-mounted receiver. The principal advantage of central-receiver systems is their ability to deliver energy at very high temperatures. The heat-transfer fluid in the receiver could be steam, air, molten-nitrate salt, or sodium at temperatures ranging from 540 °C (1,000 °F) to over 1,000 °C (2,000 °F). Steam would be a secondary working fluid used to drive a turbine to produce electric power if salt or sodium is used as the primary working fluid in the receiver. Figure 1.5 provides a conceptual drawing for a central-receiver power plant employing water-steam as the working fluid.

Central-receiver power systems have been evaluated by utilities both as stand-alone generating plants and as hybrid systems integrated with fossil-fuel generating plants. Commercial central-receiver systems would most likely produce power in the range of 50 MW or higher.

The most well-known central-receiver installation is the Solar One Plant operated by Southern California Edison (SCE) from 1982 to 1988. Other key utility involvements in central-receiver research and development activities include the utility/Sandia National Laboratories, Albuquerque (Sandia) molten-salt receiver experiments, the SCE Solar 100 proposal, the Arizona Public Service (APS) and Pacific Gas & Electric (PG&E) proposals to develop a 30 to 100 MW central-receiver power plant, and the central-receiver repowering studies performed by a number of Southwest utilities in the late 1970s and early 1980s. In fact, the central-receiver repowering studies had the most utility participants of any of the solar thermal activities and were also among the earliest activities undertaken by utilities in the solar thermal area. A list of key solar-thermal, central-receiver projects in which utilities have been actively involved is presented in Table 1.6.

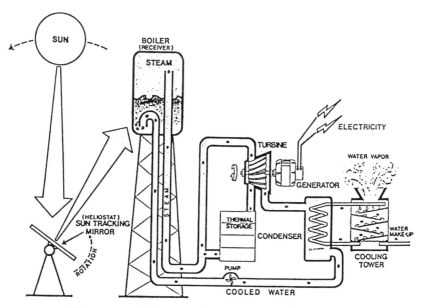

**Figure 1.5:** Solar central receiver water-steam system.

### 1.4.3A  Central-Receiver Repowering Studies

The following is a discussion of PNM's central-receiver repowering proposal and similar repowering proposals which were developed by several other Southwest utilities in the late 1970s and early 1980s.

### 1.4.3A.1  Public Service Company of New Mexico Proposal

In 1974, PNM conceived the concept in which an existing power plant would be modified to allow replacement of all or a portion of the fossil fuel (oil or natural gas) with solar energy. To accomplish this, PNM proposed locating a solar-thermal, central-receiver system adjacent to an existing fossil-fuel power plant. During periods of adequate insolation, the heliostats would focus sunlight on the tower-mounted receiver which would then provide steam to be piped to the power plant for operation of the steam turbines. During periods of inadequate insolation, existing fossil-fired boilers would be utilized to generate steam. The technical name given this concept was "solar-hybrid repowering." Initial studies performed by the PNM System Planning Department indicated that a high percentage of the fossil fuel normally burned by such a repowered unit would be saved.

Encouraged by these studies and a market survey which indicated a potential for widespread application of the concept throughout the Southwest, PNM prepared an unsolicited proposal (PNM, 1976), for a technical and economic assessment of the solar hybrid repowering concept. This proposal was submitted to the Energy Research and Development Administration (ERDA, the forerunner of DOE) in December of 1976. It proposed a two-phased project consisting of a one-year assessment study followed by implementation of the solar-hybrid repowering concept. Phase I, the assessment study, included the preparation of a conceptual design and the performance of detailed economic analyses. The potential benefits to PNM were estimated as

**Table 1.6:** Key U.S. Solar Thermal Central Receiver Power Projects and Participating Utilities

---

Central Receiver Experiments/Demonstrations

- 10 MWe  -  Solar One - Water/Steam, Barstow, CA, Southern California Edison and Los Angeles Department of Water and Power

- 750 kWe  -  Molten Salt Electric Experiment, CRTF, Sandia National Laboratories, Albuquerque, NM, Arizona Public Service, Pacific Gas and Electric, Public Service Company of New Mexico, Southern California Edison, and Electric Power Research Institute

- 5 MWth  -  Molten Salt Subsystem/Component Tests, CRTF, Sandia National Laboratories, Albuquerque, NM, Southern California Edison, and Arizona Public Service

Central Receiver Studies/Activities

- Various  -  Repowering Studies and Designs
  - - Arizona Public Service, El Paso Electric, Public Service Company of New Mexico, Public Service Company of Oklahoma, Sierra Pacific Power Company, Southwestern Public Service, Texas Electric Service Company, and West Texas Utilities

- 100 MWe  -  Solar 100 - Molten Salt, Commercial Design, Southern California Edison

- 30 MWe  -  Carrisa Plains, Liquid Sodium, Pre-Commercial Design, Pacific Gas and Electric

- 100 MWe  -  Molten Salt, Pre-Commercial Design, Pacific Gas & Electric and Arizona Public Service

---

well as those to other utilities in the Southwestern U.S. The proposal also included an extensive survey of the potential market for solar-fossil hybrid systems in the Southwest. The proposal requested $831,000 to be expended over twelve-months. On September 30, 1977, DOE and PNM entered into a letter contract based on the proposal. The Phase I assessment study was completed in September 1978 (PNM, 1978a).

Phase II was to include the construction and operation of a 25 MW$_e$ pre-commercial system at PNM's Reeves Station Unit No. 2 in order to achieve a 50% repowering of that unit. PNM proposed to provide the land for the solar field and to fund certain required modifications to the unit and pay for all operation and maintenance costs. Figure 1.6 is a schematic of PNM's concept. The conceptual design adapted the DOE-developed design for the Solar One plant at Barstow, California. It was planned to utilize the Solar One technology pending the development of more advanced and cost-effective hardware in order to minimize program costs and technical risks. Two major subcontractors were selected to assist PNM in the execution of the proposed program. Westinghouse Electric Corporation was to provide system engineering and project integration, and Bechtel Corporation was to provide architect-engineer and construction services.

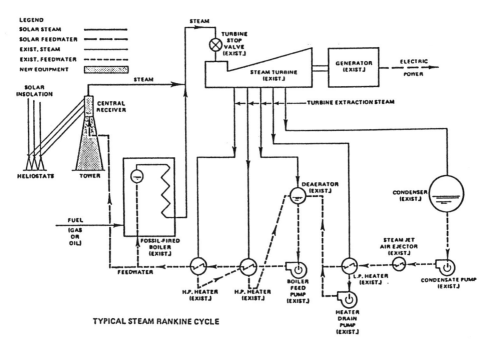

**Figure 1.6:** Solar repowering concept proposed in 1978 by Public Service Company of New Mexico.

PNM viewed the solar-hybrid repowering concept as helping to conserve dwindling oil and natural gas reserves, reduce the nation's foreign oil dependence, achieve the earliest commercialization of solar components, and facilitate earliest direct utility participation in commercializing solar energy. Other perceived advantages of the plan were that it would extend the life and use of existing generation equipment, utilize present generation sites wherever adjacent land is available, reduce fossil fuel emissions at existing sites, diversify the utility's generation mix, establish solar-fossil hybrid operating principles, and minimize backup requirements for early solar-thermal electric plants.

Realizing that its proposal would require the mass production of heliostats, PNM contacted eight potential heliostat manufacturers and received commitments for development of the manufacturing capability needed to meet the repowering goal. PNM also developed a utility program plan for commercialization of the repowering concept and identified potential financial incentives from federal, state, and local governments. The total package was intended to allow the DOE to proceed immediately with a plan of commercializatoin at multiple sites.

A total of 55 utilities indicated their support of the plan. Houston Lighting and Power Company, Public Service Company of Oklahoma, Arizona Public Service Company, Nevada Power Company, Tucson Gas & Electric, and others wrote supportive letters to PNM indicating their strong interest in the solar-hybrid repowering concept for possible application to their own units (PNM, 1978).

### 1.4.3A.2  Other Central-Receiver Repowering Proposals

Following the completion of PNM's Phase I effort, DOE issued a solicitation under its Utility Repowering/Industrial Retrofit Program. Subsequently, a number of other southwestern utilities submitted proposals for central-receiver repowering of existing fossil-fuel stations. Each proposed effort was to begin with a DOE-funded conceptual design study, including the preparation of cost estimates and economic value analyses. DOE funded studies were performed on the following utility systems:

- Arizona Public Service Company
- El Paso Electric Company
- Public Service Company of Oklahoma
- Sierra Pacific Power Company
- Southwestern Public Service Company
- Texas Electric Service Company
- West Texas Utilities Company

The studies were performed by different contractors, each using different economic evaluation techniques. The economic results of the seven studies (Table 1.7) indicated a significant cost-benefit range for the proposed projects (Zaininger, 1980).

Based on the proposals submitted to DOE, a preliminary design contract was awarded to El Paso Electric Company (El Paso) in October 1982 for solar repowering of one-half of Newman Station Unit 1, an 82 MW oil/gas-fired unit. El Paso planned to have the repowered unit in operation by 1986 if further funding had been approved. However, the federal government did not approve the necessary funding, apparently due mainly to curtailments in the federal budget for solar energy research. Also, the emergence of generating capacity surpluses in the Southwest and a return of more stable energy prices rendered the concept economically unattractive, at least for the near term, and diminished any urgency to conduct such an experiment, especially in view of other solar thermal experiments, such as Solar One, already in progress.

The recent APS/PG&E/DOE studies, discussed later in the section, considered the possibility of repowering the APS Saguaro station. Also, APS has completed a DOE-funded preliminary design for a 60 $MW_e$ molten salt module at its Saguaro power plant. However, we are aware of no significant ongoing activity in the area of solar-fossil, central-receiver repowering.

### 1.4.3B  Solar One

Solar One, shown in Figure 1.7, is a 10 $MW_e$ solar thermal, central-receiver pilot power plant located near Barstow, California. It was operated by Southern California Edison from April 1982 through September 1988. Solar One was a cooperative effort between the U.S. Department of Energy (DOE) and The Associates, which included SCE, the Los Angeles Department of Water and Power (LADWP), and the California Energy Commission. U.S. Department of Energy participants during the test and evaluation phase included: Sandia National Laboratories (Livermore, CA), McDonnell Douglas, Rocketdyne, Martin Marietta, and Stearns Roger. The Southern California Edison Company had responsibility for the plant's operation and maintenance.

**Table 1.7:** Summary of Seven Economic Analyses of Solar-Hybrid Repowering

|  | EPE | PSO | APS*** | SPS | SPP | WTU | TESCO* |
|---|---|---|---|---|---|---|---|
| Utility Cost/Value | 1.5 - 3.6 | 2.6 - 11.2 | 3.2 - 5.0 | 1.6-4.4 | 1.0 | 0.9 - 3.9 | 8.7 |
| Capacity Credit | Yes | Yes | No | Yes | Yes | No | No |
| Fuel Displacement Cost |  |  |  |  |  |  |  |
| Oil ($/MBtu) | 12 - 4** |  | 4.36 - 4** |  | 6.7 - 12.3** | 2.2 - ?* |  |
| Gas ($/MBtu) | 4.5 - 2.5* | 2.8 - 2.5* |  | 1.95 - 2.82* | 5.5 - 6.8** |  | 2.5* |
| Solar Performance |  |  |  |  |  |  |  |
| Life (Years) | 30 | 15 | 10 - 15 | 30 | 30 | 30 | 7 |
| Capacity Factor (%) | 22 | 15 | 26 | 22 | 59 | 22 | 27 |
| Solar Cost ($/kW) | 2917* | 2167** | 2470** | 2034* | 2061** | 2330* | 3520** |

\*   $1980
\*\*  $1985
\*\*\* Utility System Representation

Source: Zaininger, 1980.

## 1.4.3B.1 Plant Description

Solar One converts solar energy into electricity using water/steam as the working fluid for both energy collection and power conversion. The pilot plant design element drawing, Figure 1.8, depicts the Solar One operating concept. Sunlight is reflected by the 1,818 heliostats located in the collector field onto a tower-mounted receiver (boiler) which absorbs the solar energy and converts the water to superheated steam at 1,450 psi and 510 °C (950 °F). The steam is directed to a conventional turbine generator which produces electrical energy or to the thermal storage system where the steam is used to heat a combination of heat transfer oil, rocks and sand contained in the thermal storage tank. The heat transfer oil can then, at a later time, be used to extract the energy from the thermal storage tank to produce moderate pressure and temperature steam at 450 psi and 304 °C (580 °F) which can be directed to the turbine generator for the production of electrical energy.

The main plant process systems were operable in eight operating modes which emulated any of the operating modes being given consideration for future central-receiver plants. The plant demonstrated successful operation in each of the eight

**Figure 1.7:** Solar One.

operating modes including automatic computer-controlled mode transitions (Lopez, 1988).

### 1.4.3B.2  Project Funding and Costs

Breakdowns of the sources of funds for the design and construction of Solar One and the plant construction cost are provided in Table 1.8 and Table 1.9. The design and construction cost of Solar One was $141.4 million or about $14,100 per kW. The total project cost through completion of construction was $143.6 million with $19.6 million of that provided by SCE, $4.0 million by LADWP, and the balance by DOE. While the cost of Solar One on a $/kW basis was substantially higher than the current cost of conventional power plants (the cost of a new coal-fired power plant today is about $2,000 per kW), it must be remembered that Solar One was a small, one-of-a-kind pilot plant and, thus, its costs were much higher than might be expected for a commercial plant. A better comparison is provided by the projected cost of Solar 100, which is discussed later in this section. SCE projected Solar 100, a proposed 100 MW plant, to cost $400 million or $4,000 per kW.

### 1.4.3B.3  Milestones and Status

Major milestone dates for development of the Solar One project include:

- Project authorized: 1975
- Detailed design started: 1978

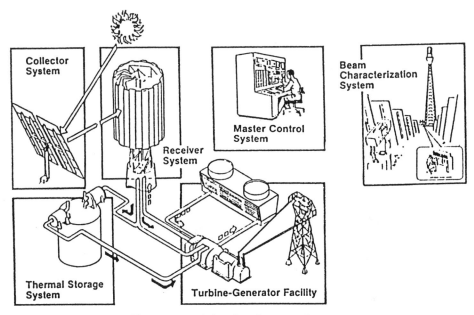

**Figure 1.8:** Solar One Systems elements.

- Start of construction: September 1979
- Initial turbine roll: April 1982
- Test and evaluation phase: August 1982 - July 1984
- Power production phase: August 1984 to September 1988
- Current Status: Inactive (shutdown)

Solar One completed a two-year test and evaluation phase jointly managed by SCE and Sandia for DOE, on July 31, 1984. Solar One then began a three-year power production phase which ended on July 31, 1987. From August, 1987 through September, 1988 the plant was operated on a five-day per week basis under a joint agreement between the project sponsors. The plant was managed throughout the power production phase by SCE. The energy produced by the plant was used by SCE and LADWP. Performance data acquired during the test and evaluation phase proved the solar central-receiver concept to be technically feasible. Data acquired during the power production phase is being used to estimate the technical feasibility and cost effectiveness of future larger solar central-receiver power plants.

**1.4.3B.4  Achievement of Design Goals**

The following were the principal plant design parameters for Solar One:

- Maximum net capability: 10.8 $MW_e$
- Maximum net daily energy production: 112 $MWh_e$ (summer solstice), 48 $MWh_e$ (winter solstice)
- Maximum electrical energy generation from thermal storage: 28 $MWh_e$ at a rate of up to 7 $MW_e$
- Average annual net energy production: 24,000 $MWh_e$

**Table 1.8:** Solar One Construction Funding Sources ($1,000)

| | U.S. DOE | Project Participants | | Total |
| --- | --- | --- | --- | --- |
| | | Southern California Edison | L.A. Dept. of Water & Power | |
| Design and Construction of Solar Facilities: heliostats, tower, thermal storage, master control system | $120,000 | | | $120,000 |
| Design and Construction of Non-Solar Facilities: electrical generation equipment | | $ 17,840 | $3,660 | $ 21,500 |
| Participant Services: Land | | $ 250 | | $ 250 |
| Plant integration and startup: technical support and plant operating procedures | | $ 520 | $ 130 | $ 650 |
| Project Reviews and Information Dissemination | | $ 964 | $ 241 | $ 1,205 |
| Total Design and Construction Cost | $120,000 | $ 17,840 | $3,660 | $141,500 |
| Total Participant Services | | $ 1,734 | $ 371 | $ 2,105 |
| Total Costs | $120,000 | $ 19,574 | $4,031 | $143,605 |

Source: Lopez, 1988.

System design requirements which were met or exceeded include the delivery of 10.8 MW$_e$ net from receiver steam, the delivery of 7 MW$_e$ net from thermal storage, and the delivery of 28 MWh$_e$ from thermal storage. Two design requirements which were not met are the delivery of 10 MW$_e$ net for 7.8 hours on the most favorable day of the year (summer solstice) and for 4 hours on the least favorable day of the year. According to SCE, a major reason for the plant's failure to achieve these design output goals is the fact that these goals assumed the plant would be built with 2,000 heliostats. In fact, only 1,818 heliostats were installed at the plant to keep the cost of construction within the $120,000,000 budget. The combined effects of low direct insolation, heliostat outage, mirror soiling, receiver efficiency and turbine efficiency are other key factors which prevented the plant from meeting these two design

**Table 1.9:** Solar One Construction Cost (Millions)

| Item | Cost | Percent |
|------|------|---------|
| Solar Facility Design Cost | 31.2 | 22 |
| Collector Field Fabrication and Construction | 40.0 | 28 |
| Receiver Fabrication and Construction | 23.4 | 17 |
| Thermal Storage Fabrication and Construction | 12.0 | 8 |
| Plant Control System | 3.0 | 2 |
| Beam Characterization System | 1.0 | 1 |
| Miscellaneous Support Systems | 9.4 | 7 |
| Total Solar Facility Design / Fabrication / Construction Cost | 120.0 | 85 |
| Turbine / Generator Design / Construction Cost | 21.5 | 15 |
| Total Plant Cost | $141.4 | 100 |

Source: Lopez, 1988.

goals. The plant's design rating was premised on 100% equipment availability and optimized thermal performance under conditions which were not attainable. This is evidenced by the results shown in Table 1.10.

#### 1.4.3B.5 Energy Production

During the testing phase of Solar One (April 1982 through July 1984), cumulative gross energy generation approximately equaled cumulative parasitic power requirements (station usage). Starting August 1, 1984, plant operations emphasized power production. During the first three years of the power production phase, the plant generated 42,780 $MWh_e$ (gross) and delivered 27,656 $MWh_e$ to the SCE system (Table 1.11). The plant capacity factor during this period averaged 10.5% versus the 27.4% anticipated based on design-phase projections. During the last 12 months of the plant's operation (October 1987 to September 1988), with the plant being operated only on week days, the net power output of the plant was 6,725 $MWh_e$, equivalent to a capacity factor of 7.7%.

#### 1.4.3B.6 Operating Availability

One reason the plant was not able to produce its design electrical output was the unrealized design conditions. Another is that the solar energy availability was

**Table 1.10:** Design Versus Actual For Key Solar One Performance Parameters (Tests Conducted June 21, 1984)

| | Performance Paramaters | |
| --- | --- | --- |
| | Design | Actual |
| Receiver Absorptivity (%) | 95 | 90 |
| Receiver Steam Outlet Pressure* | 102 / 1,450 | 91 / 1,300 |
| Receiver Steam Outlet Temperature ( °C / °F ) | 510 / 950 | 413 / 775 |
| Heliostats in Service (%) | 100 | 96 |
| Heliostat Cleanliness (%) | 100 | 85 |
| Heliostat Mirror Corrosion (%) | 0 | 0.029 |
| Heliostat Reflectivity Surface (%) | 100 | 97.9 |
| Heliostat Beam Spillage (%) | 0 | unknown |

Source: Lopez, 1988.

* kg per cm² / PSI

less than expected, and weather outages precluded plant operation on 40% of the days. In addition, plant power output projections did not provide for the $1\frac{1}{2}$ hours of time required for plant startup and shutdown, which effectively reduced the plant availability about 12% on winter days and 8% on summer days. Plant power output projections also failed to recognize that on days when the usable solar insolation was of less than 2 hours duration, the plant could not be operated cost effectively, i.e., auxiliary power consumption would be greater than power output for these short periods of available solar energy. Scheduled plant outages to evaluate component integrity coupled with forced outages due to equipment failures further reduced plant availability another 10%.

Thus, the operating characteristics of the plant, combined with plant equipment outages and weather-related outages, contributed to the significantly lower-than-expected power generation. Plant availability is shown in Figure 1.9 for the period August 1984 through July 1986. The plant was "on line" for only 40% of the total daylight hours during these two years (EPRI GS-6573).

**Table 1.11:** Solar One Power Production During Initial Three Year Production Phase

|           | 1985      | 1986      | 1987      |
|           | MWH (Net) | MWH (Net) | MWH (Net) |
|-----------|-----------|-----------|-----------|
| August    | 883       | 1,776     | 1,160     |
| September | 896       | 1,187     | 1,127     |
| October   | 591       | 799       | 982       |
| November  | 323       | 205       | 696       |
| December  | -41       | 186       | 162       |
| January   | 421       | 577       | -66       |
| February  | -21       | 345       | 3362      |
| March     | -44       | 723       | 612       |
| April     | 1,014     | 1,057     | 852       |
| May       | 1,230     | 1,427     | 1,034     |
| June      | 1,407     | 875       | 1,443     |
| July      | 835       | 1,188     | 1,535     |
| Totals    | 7,302     | 10,454    | 9,900     |

Source: Lopez, 1988.

### 1.4.3B.7 Parasitic Power Consumption

In designing Solar One, it was recognized that the auxiliary systems could be designed to minimize parasitic power consumption. However, it was assumed by the plant's designers that Solar One, as a pilot plant, would be evaluated primarily on the basis of the performance of production-related systems. Accordingly, the auxiliary system designs were not designed to minimize power consumption during inactive periods. In effect Solar One was designed for full load continual operation. The design was not optimized to reduce parasitic power consumption during off-peak periods.

Although SCE's operating personnel made considerable progress during Solar One's test and operation phases in reducing parasitic power requirements and

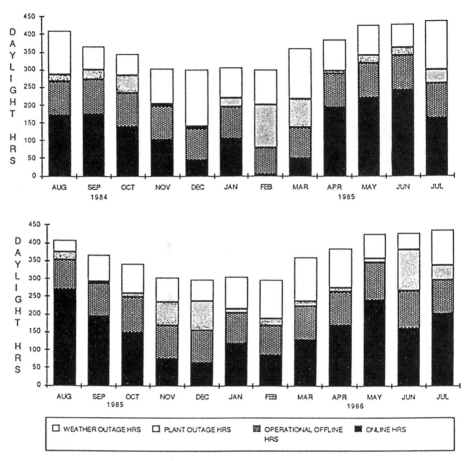

**Figure 1.9:** Solar One availability August 1984–July 1986 (Source: EPRI GS-6573.)

improving net energy generation, the parasitic power consumption remained high in proportion to the plant's gross energy production. A key lesson learned was that future central-receiver plants must be designed with consideration of power consumption during inactive periods. This is equally true of any cycling thermal plant.

### 1.4.3B.8 Startup and Shutdown

Starting Solar One requires breaking condenser vacuum and purging the system with nitrogen gas. The operators learned how to break vacuum at night and purge the system with nitrogen before sealing the condenser. Maintaining the turbine seals required operation of several auxiliaries–condensate pump, cleaning plant steam exhauster, bearing cooling water heat exchanger, etc. The amount of parasitic power increases greatly if seals are maintained. This can be alleviated in future plants by installing jockey water pumps. For Solar One, this was not cost-effective except for the service water pump.

**Table 1.12:** Solar One Annual Operating and Maintenance Budget (1986 Dollars)

|  | Cost | Percent |
|---|---|---|
| Labor | $1,368,600 | 46 |
| Material | 289,800 | 10 |
| Contract | 272,800 | 9 |
| Miscellaneous Expenses | 45,000 | 2 |
| Administrative and General | 996,922 | 33 |
| Total O & M Budget | $2,973,122 | 100 |

Source: Lopez, 1988

Also, the thermal storage system was used for several years to generate auxiliary steam. However, this system was no longer used following a fire on August 30, 1986. Thereafter, auxiliary steam was generated by the electric boiler. Initially, the electric boiler was somewhat unreliable, but the operation was modified, so that it was operated only in the morning for startup, and sometimes in the evening shutdown. Previously, the electric boiler operated continually and experienced failures of heater elements, electrical contacts, etc. It was not only unreliable, but consumed about 1 MW of power, doubling the parasitic-power consumption.

### 1.4.3B.9 Operation and Maintenance

Operation and Maintenance (O&M) expenses were another area of concern at Solar One. The value of the plant's net energy production was insufficient to cover SCE's O&M expenses which are summarized for one year in Table 1.12. In the first three years of the power production phase, SCE and LADWP contributed $500,000 of the annual operating expense and DOE the balance (about $2,500,000). In the last year of the plant's operation, the plant operated only five days per week with a reduced staff. The differential between the energy produced and cost of operation was shared by SCE, DOE and EPRI (about $400,000).

To minimize O&M expenses, SCE reduced the Solar One staff and arranged to share staff with the neighboring Cool Water Generation Station (Table 1.13). Initially, there were 39 operating personnel on site. For seven-day operation, staffing was eventually pared to 18 operators. By the last year of operation the staff was down to 14 people for 5 days per week operation of the plant. The Solar One experience highlights the importance of minimizing staff requirements and overall

**Table 1.13:** Solar One Plant Staffing

|  | 1985 | 1987 |
|---|---|---|
| Supervisor of Operations & Maintenance | 1 | 0 |
| Engineer | 1 | 1/2 |
| Shift Supervisors | 4 | 1 |
| Maintenance Foreman | 1 | 1 |
| Maintenance Planner | 1 | 0 |
| Stenographer | 1 | 0 |
| Accounting Clerks | 2 | 1/2 |
| Control Operators | 5 | 6 |
| Assistant Control Operators | 4 | 4 |
| Plant Equipment Operators | 2 | 0 |
| Instrument Technicians | 4 | 2 |
| Electricians | 2 | 1 |
| Boiler & Condenser Mechanics | 2 | 1 |
| Chemical Technician | 1 | 1/2 |
| Security Officer | 1 | 0 |
| Total | 32 | 17 1/2 |

Cool Water O&M Support - Part Time

Station Management
Test Technician "A"
Welder
Machinist

Source: Southern California Edison

O&M expenses while maximizing energy production in a commercial central-receiver system. SCE has estimated a staffing requirement of 45 people for a 100 MW plant.

### 1.4.3B.10  Receiver System

The receiver, consisting of six preheat and 18 once-through subcritical boilers, demonstrated good overall performance. The steam quality produced by the receiver is particularly noteworthy. During simulated cloud passages, as an example, all heliostats that were reflecting energy were removed from the receiver and then restored to the receiver after 3 minutes. Steam temperature deviation during the simulated transition resulted in only an 8 °C (15 °F) deviation from its designed 510 °C (950 °F) setpoint (Lopez, 1988).

The only significant problems experienced with the receiver included tube failures and degradation of the black paint that was applied to the receiver tubes to improve energy absorptivity. The tube failures are attributable to metal fatigue caused by daily startups, and shutdowns and startups caused by cloud passages. During its first five years of operation, Solar One recorded over 1,500 startups. Tube failures declined following efforts to relieve tube stress concentration including repair of receiver panel expansion guides and installation of receiver panel edge tube radial shields, as well as the selective removal of flow distribution orifices to moderate panel metal temperature gradients. The paint degradation was expected to occur, and its absorptivity degraded from its design of 96% to 87%. The receiver panels were repainted to restore their absorptivity in December, 1985 (Lopez, 1988).

### 1.4.3B.11 Collector System

The collector system performed better than anticipated. On occasion all 1,818 heliostats were available for service in spite of the low maintenance priority given the heliostats (plant components whose failure would cause a plant shutdown were given a higher maintenance priority). The high level of heliostat availability is attributed to the initiative of on site personnel in diagnosing and modifying components to improve reliability (Lopez, 1988).

The operation of Solar One also demonstrated the importance of washing heliostats on a routine basis to maintain their reflectivity. As an example, degradation of heliostat reflectivity from 91% to 89% is equivalent to the loss of 40 heliostats from service. The washing system used at Solar One was modified many times. Beginning initially with a brush and water spray system, it evolved into an all-spray system. SCE found that rapid and frequent washing is more important than achieving absolute cleanliness by means of a periodic washing and scrubbing. In other words, it is better to achieve a 12% increase in reflectivity using a wash only procedure rather than a 13% increase with a more time consuming wash and scrub. Ideally, the field should be washed at least once a month. Because the plant was manpower limited, the actual washing was performed less frequently. During rainy periods, SCE found there was no need for washing. The project developed an automated heliostat wash truck with which a single operator could wash all 1,818 heliostats in a 10-day wash period (Lopez, 1988).

SCE found no pitting of the heliostat surfaces from sand in the desert environment. The wind driven sand was found to stay close to the ground.

### 1.4.3B.12 Thermal Storage System

The thermal storage system met all of its design objectives. Because direct use of receiver steam in the turbine generator provides a turbine-cycle efficiency of 35%, the plant did not normally generate electrical power with the thermal storage system which only provided a 25% turbine-cycle efficiency. More commonly, the thermal storage system was used for the generation of auxiliary steam for plant start-up, component preheating to minimize thermal-cycling and to provide steam blanketing of heat exchange equipment during inactive periods. Use of the thermal storage system for these services proved to be both reliable and cost effective (Lopez, 1988).

### 1.4.3B.13  Control System

Solar One was a showcase for modern digital control system technology. It was unique in the electric utility industry at the time in that the plant was automatically controlled by a master control system. The master control system includes a total of five computers which supervise the operation of the plant's 1,940 microprocessors. There are very few analog controls, dedicated switches, control knobs, and meters in the control room. Information on plant operation was provided to the operators on color-graphic video displays and the operator interaction with the system was through keyboards, light pens, function keys and function switches.

According to SCE, its operators were very pleased with operation of the Solar One power plant. The capabilities of these operators were typified by one observer's experience during a visit to the plant's control room. While in the control room, an alarm sounded. A large cloud, with a very sharp edge, drifted over the Solar One plant site. The power output of the plant dropped from about 7 MW to 0.7 MW very rapidly, and the turbine tripped off line. Although the turbine tripped, the operators kept all the auxiliary systems operating. The cloud moved away quickly, and within a few minutes the operators had the plant back on line. Thus, within an interval of a few minutes, the plant went through a complete cycle – on line, tripped, and restart. The observer was amazed at how smoothly this entire operation had been handled by the plant's operators.

Solar One demonstrated that modern computer control technology can be successfully utilized in the electric utility industry. The control system, coupled with the design of other plant systems, provided a plant power turndown ratio of 20 to 1, which is far superior to the turndown ratio of a conventional steam plant (more typically about 5 to 1). In addition, the control system facilitated continuity of the plant operation during cloud transients. The Solar One control system allowed the operators to devote more of their time to maintaining plant availability and efficiency; energy output was thereby increased (Lopez, 1988).

### 1.4.3B.14  Summary

In evaluating the results of Solar One, it is important to recognize that the plant was not designed to be a commercial power plant, but rather a small-scale, experimental pilot plant intended to provide technical information applicable to larger, commercial-scale facilities. As a pilot project, Solar One has been a technical success in that it operated with reasonable reliability and has provided performance and operating information critical to the feasibility assessment and design of any future commercial solar plants of this type. The 10 MW$_e$ size of the plant allows a meaningful interpretation of performance data without being masked by the losses and parasitics as in other, smaller experiments. The six years of operation by SCE provide a sound basis for selecting an operations and maintenance approach and predicting many of the costs.

The future of Solar One, now that the energy production phase is completed, has not been determined. Various ideas have been proposed including conversion of Solar One to an advanced salt receiver system. The concept is that existing Solar One facilities, such as the heliostat field, represent a considerable investment which could be used to avoid additional expenditures at a new site.

### 1.4.3C Molten-Salt Electric Experiment (MSEE)

#### 1.4.3C.1 Project Description

The MSEE was a 750 $kW_e$ system-level test operated by utility personnel from May 1984 until July 1985 at the Sandia Central Receiver Test Facility (CRTF) in Albuquerque, New Mexico. Approximately half of the MSEE funding requirement and other support was supplied by a consortium of utilities (Arizona Public Service, Pacific Gas & Electric, Public Service Company of New Mexico and Southern California Edison), EPRI and industries. DOE, through Sandia, supplied the balance as well as on-site labor and existing CRTF facility hardware. The objective of the MSEE was to provide a system-level test of a molten salt receiver as well as pumps, piping, valves and balance-of-plant equipment.

Due to budgetary constraints, many existing CRTF facilities were used in the construction of the MSEE. This resulted in a system which was sized or scaled proportionally for only certain aspects of a "commercial" system with the balance of the system remaining unscaled or "distorted." Closely simulated parameters included molten salt temperatures and properties (including salt flow, leakage and freezing), steam-side temperatures and pressures, heat transfer coefficients and the analytical aspects of systems-level operation and control.

#### 1.4.3C.2 Project Results

The principal objectives of the experiment were met. The MSEE contributed substantially to knowledge and understanding of the potential and problems of a molten-nitrate salt, central-receiver solar power plant. The experiment demonstrated the technical feasibility of the molten-salt, central-receiver concept. No technical barriers that would preclude further development of the technology were identified. Molten salt was shown to be an effective, low-cost heat transfer fluid for energy collection and storage. The thermal-storage configuration effectively decouples solar energy collection from power production. This allows the collection function to follow solar availability and power production to follow user demand. The experiment's distributed digital controls operated successfully. Many operating sequences were automated and further automation is possible, offering the potential to minimize operator requirements for future plants. The experiment results also indicate that reasonably rapid startup can be accomplished without excessive loss of collectible solar energy (EPRI GS-6577, Vol. 1).

On the other hand, the high melting point of the salt created substantial problems in maintaining all receiver loop piping and equipment at the temperature required for startup. The major problems were associated with trace heating, insulation, wind protection, the extensive instrumentation required for temperature monitoring, high parasitic power, and general system complexity. Net positive power production was not demonstrated, but major losses have been identified. Equipment reliability under cyclic operation was poor. These problems must be considered for future development. It was concluded that a large-scale experiment, preceded by needed component development, will be required to confirm the technical performance for a commercial plant (EPRI GS-6573).

### 1.4.3D  Molten Salt Subsystem/Component Test Experiment (MSS/CTE)

The Molten Salt Subsystem Component Test Experiment (MSS/CTE) was a hardware development and test program conducted at the CRTF. The objective of the program was to resolve technical uncertainties related to components and subsystems of molten salt central receiver technology. The goals were to test an advanced molten salt receiver and to demonstrate that commercially available pumps and valves are suitable for molten salt applications.

The MSS/CTE had three major parts: (1) a model receiver test, (2) a valve seal bench test and (3) a pump and valve test in molten salt test loops. Receiver testing was conducted during 1987.

The project was cost-shared by DOE and a consortium of utilities and industry. Contributing utility sponsors were Arizona Public Service Company and Southern California Edison Company. A description of the receiver test follows.

### 1.4.3D.1  Project Description

The MSS/CTE receiver test program was directed primarily toward the testing of a scaled-down commercial receiver and was not designed as a full-system experiment because there was no electric power production capability. The receiver was retrofit to the MSEE system. It was a 5 MWth molten-salt, C-shaped cavity designed to incorporate features of a 60 $MW_e$ commercial plant design. Most of the subsystems were the same as for the MSEE except that the turbine, which had failed during the MSEE project, was bypassed to heat rejection equipment to condense the steam produced by the steam generator. Two different panel designs, one by Babcock & Wilcox and one by Foster Wheeler Development Corporation, were incorporated into the receiver. The panels differed mainly in their supports and header arrangements.

### 1.4.3D.2  Project Results

Most of the test program was devoted to thermal loss testing and receiver control testing for use in subsequent simulation of receiver operation. Consequently, only limited data were available for solar noon and partial day performance, and no data were available for longer-term performance. The test results indicate that the receiver performance was within the expected range. The experiment also provided test data for correlation with thermal models, and demonstrated several methods for receiver thermal conditioning and for early morning start-up (EPRI GS-6573).

The MSS/CTE program also demonstrated improved trace heater reliability over earlier test programs. However, reliability was still a problem, and further quality control improvements, both in the heaters and their installation, are required. It was learned that system performance analyses require more of an overall systems approach rather than a subsystem-by-subsystem approach in order to improve the accuracy of performance calculation. Such analyses must also consider reduced-temperature energy collection during the early morning, late afternoon and during cloud transients. For example, rapid early morning start-up can be achieved; however, a warm-up heliostat aiming strategy is needed and this requires considerable development. To simplify operations and allow integrated sequencing and interlocking, a single master computer to control the subsystem computers (e.g. process control, heliostat control, and trace heater control) should be provided. Also, future designs require better coordination between designers and controls engineers to

**Figure 1.10:** Solar 100 conceptual drawing. Source: Southern California Edison.

resolve issues relating to strategies for controlling receiver stresses. These issues have major implications for receiver operations, procedures, controls, annual performance and lifetime (EPRI GS-6573).

### 1.4.3E  Large Scale Conceptual Design Studies/Proposals

In addition to participating in Solar One and the molten salt electric experiments at the CRTF, Southern California Edison also advanced a proposal for the development of a large-scale central receiver power plant. Similar proposals were developed by Pacific Gas & Electric and Arizona Public Service.

### 1.4.3E.1  SCE Solar 100 Proposal

Solar 100 was to have been a molten-salt, central-cavity-receiver generating station consisting of two heliostat-field, central-receiver modules using a conventional steam reheat cycle to produce nominally 100 $MW_e$ (net) of generating capacity at a 60 percent annual capacity factor. Southern California Edison, McDonnell Douglas Corporation, and Bechtel Power Corporation released a final report addressing Solar 100 on August 3, 1982 (SCE, 1982). The heliostat fields were to be installed in a staggered manner so that the project would be more financeable. The first heliostat field module was to have been installed by July 1986 and the second module by July 1987. The concept is illustrated in Figure 1.10 and Figure 1.11.

The Solar 100 project was initially delayed due primarily to financing, technical risk and power marketing considerations. By 1985, SCE had decided not to pursue further the Solar 100 project and, instead, merged its effort on the project into a cooperative study effort involving APS, PG&E and DOE. Subsequently, SCE

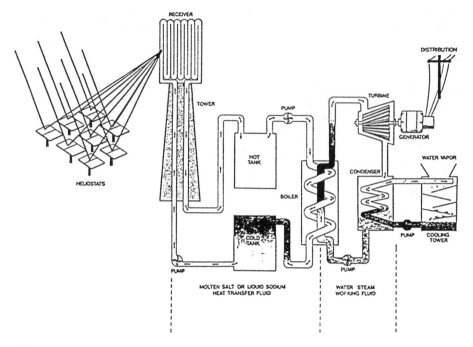

**Figure 1.11:** Solar 100 central-receiver system. Source: Southern California Edison.

determined that it did not want to be involved in developing solar energy projects on its own, preferring instead to purchase power and energy from solar plants developed by others. As a result, in October 1986 SCE withdrew from the joint effort with APS, PG&E and DOE. Key factors in SCE's decision were an excess of generation capacity and lower than projected load growth and fuel prices.

### 1.4.3E.2 APS/PG&E Conceptual Designs

PG&E also developed an independent proposal for a central receiver project in the early 1980's. This project, referred to as the Carrisa Plains Project, was initially in competition with SCE's Solar 100 Proposal. It called for the construction at Carrisa Plains, located in PG&E's service territory, of a 30 MW$_e$ cavity-type sodium receiver, solar-thermal power plant. The improved heat transfer characteristics of a sodium receiver would have allowed the receiver to be smaller than either a salt or water/steam receiver. Initially, sodium was also contemplated for use in the proposed plant's storage system, but subsequently, after concerns about mixing of salt and sodium had been addressed, the concept was modified to provide for a salt storage system. The Carrisa Plains Project was also delayed due to technical risks and financing uncertainties.

In the Spring of 1986 (prior to SCE's withdrawal) APS, PG&E, and SCE, along with their solar industry partners, submitted unsolicited proposals to DOE for cost-shared studies to: (1) identify the solar central receiver design most feasible for commercial-scale operation in the 1990s; and (2) develop an R&D plan, including system experiment(s), for the resolution of technical issues in a cost-effective manner.

After SCE withdrew, APS and PG&E continued to work with DOE to develop a consensus approach for commercializing central-receiver technology.

The studies were performed under cooperative agreements between DOE and the utilities, and work was started in the fall of 1986. A Utility Coordination Board (UCB) was established to coordinate the study efforts and to reach a consensus on the steps that would need to be taken to commercialize solar central receivers. To aid the UCB in making its technical recommendations, a Technical Committee was also formed. This committee consisted of representatives from DOE, EPRI, Sandia National Laboratories, the central-receiver technical community, and the utility industry. The intent of the UCB was to reach a consensus and make a unanimous recommendation to DOE and the utility industry regarding the results of this joint study (U.S. DOE, 1987).

The studies were performed in two phases. In Phase I, the utilities performed parallel studies to develop a conceptual design for a commercial-scale plant and to define a preliminary technology development approach. A portion of the system-concept-comparison effort was funded by EPRI in cooperation with the utility study teams. At the conclusion of Phase I a technology concept representing the preferred approach to commercialization in the 1990s was recommended by the utilities and a group of expert advisors. Phase II provided designs for alternative system experiments including construction cost estimates, risk assessments, and performance and energy cost estimates.

Agreement on a preferred commercial design was reached in August, 1987. This design uses molten salt as the heat transfer medium and is for a plant in the 100 to 200 $MW_e$ size range (APS, September 1988 and November 1988; PG&E, August 1988). The final study recommendation was to build a demonstration plant in order to move the central receiver closer to commercialization. Potential sites identified for such a demonstration plant are at Carrisa Plains in PG&E's service territory, APS's Saguaro Power Plant located about 30 miles north of Tucson and Solar One (PG&E, October 1988). No action on this recommendation has been taken by DOE or the utilities.

### 1.4.3E.3 The PHOEBUS Project

PG&E is also a participant (with Bechtel Corporation) in a European consortium named PHOEBUS. This consortium was formed in 1986 by a number of companies and organizations that have been actively engaged in research, development and construction of solar central-receiver plants for electricity production. The consortium's objective is to finance, design, construct and commission a 30 $MW_e$ central-receiver power plant. This size is considered necessary by the participants to confirm the operation and performance of the plant and its potential for reaching commercial production costs. Initial activities in this project included a review of the accomplishments and problems encountered in previous full system experiments. This information was used to make a preliminary system concept evaluation.

### 1.4.3F Additional Utility Perspectives on Central Receivers

The perceptions of utilities concerning the outlook for central-receiver technology range from qualified optimism toward pessimism.

### 1.4.3F.1 Arizona Public Service

As a utility that has been involved in many solar central-receiver projects, Arizona Public Service is among the more optimistic concerning the long term prospects for central-receiver technology. APS's experience includes both projects involving solar central-receiver repowering as well as projects involving solar-only central-receiver concepts. The repowering efforts involved the Saguaro Power Plant which has two 100 MW oil-fired steam units. Based upon its experience, APS is pessimistic about the near term economic feasibility prospects of central-receiver repowering at Saguaro, even though some interest persists among the utilities. On the other hand, APS remains convinced that the solar-only central-receiver concept is an attractive option in the long term and hopes to consider this central-receiver concept as an option in the future.

According to APS, a major advantage of the central receiver is that the concept is amenable to storage. One outcome of APS's joint studies with PG&E has been an agreement on the heat transfer fluid. PG&E championed liquid sodium for a number of years, but the two utilities have agreed that molten salt is the preferred working fluid. There was concern at APS about the costs and risks of handling sodium. A central-receiver experiment in Spain that operated with sodium (the 500 kW$_e$ IEA-CRS project) experienced a bad fire in 1986, terminating the experiment. Some of APS's personnel had personal experience with molten salt as a coolant for a chemical reactor. This left APS without the concerns that were felt relative to the use of sodium.

However, the selection of molten salt over sodium was made mainly on the basis of economics, according to APS. Sodium has a much higher heat transfer coefficient and heat capacity than molten salt. This strongly influences the design of the receiver, which is a major cost element of the plant. Sodium allows a much smaller receiver. Also, salt has a penalty for pumping costs, since the working fluid must be elevated several hundred feet.

However, these cost penalties for molten salt appeared to be more than offset by the anticipated problems and attendant costs of operating a sodium-cooled system. One concept proposed was to only use sodium as receiver coolant and molten salt throughout the rest of the system. However, the sodium-salt heat exchanger provides a very formidable and costly engineering challenge.

While sodium is not ruled out for the long term, the practical immediate issue is the need to continue the research. Unfortunately, this technology does not have the possibility for small experiments. Central receivers must rely on economies of scale, just as large power plants do. That makes it very difficult to proceed with technology development. Demonstration of a large-size plant (e.g., 100 MW$_e$) suitable for utility application would require hundreds of millions of dollars. Funding at that level is not available at present. Therefore, the strategy is to identify market conditions that will justify additional hardware testing. There is little interest in constructing another plant the size of Solar One using a different technology. The aim is to demonstrate a plant at a size large enough for utility use. At present, the minimum size for a commercial plant is nominally 100 MW. Although the first commercial plant might be 100 MW, the 5th or 6th plant probably would be 200 MW in size, which might become a standard modular size for such units.

Thus, with a 10 MW unit at Barstow, and 100 MW representing a significant advance of the technology, the problem is how to go from one size to the next. The required technology development program was the subject of Phase II of the APS/PG&E study for DOE.

One option is to retrofit Solar One for molten salt. While this might be a minimum cost approach, it might not be the most cost-effective in terms of the results obtained. For the proposed molten-salt system at the Saguaro Plant, APS is considering 30 MW increments. This could lead to a 100 MW plant ultimately. Some kind of staging process would keep the teams together and active.

APS views such projects as insurance in case of another energy crisis. Unless the central-receiver technology base is maintained, it will be necessary to reestablish it in the event of need. APS has put most of its emphasis on the central receiver as opposed to other solar **thermal** technologies. APS also follows other technologies. For example, at the STAR facility (described later), APS would be interested in testing a dish thermal system.

### 1.4.3F.2 San Diego Gas and Electric

SDG&E has been involved in the central receiver concept through its participation in EPRI's solar power advisory committees and has also performed internal evaluations of solar thermal power. SDG&E's view is that central receiver technology is not very promising for widespread use in the United States. Only in a few sunny parts of the country, such as Arizona, New Mexico and Southern California, is the technology likely to be viable. The technology has inherent problems. A major impediment to central receivers is the frequent startup and shutdown and associated thermal cycling stresses. The entire system is vulnerable to outage because of the need to store and transfer high temperature fluids. Leaks tend to occur and materials may freeze in a molten salt system unless temperatures are kept high (200 °C) overnight. Protection from freezing involves the installation of heat trace on piping, another cause of parasitic losses. The entire area of parasitic losses poses significant problems for central receiver systems.

Among the important lesson learned by SDG&E is that jumping into a large project without smaller demonstrations should be avoided. There is a need for hands on experience, and it is impossible to learn all that is needed just from engineering descriptions. Also, everything possible should be done to keep costs down, even for a demonstration project. One of the issues involved in Solar One is the $14,000/kW cost. Even though Solar One is described as a *demonstration* project, utility planners still quote that cost. Gas peaking turbines can be installed for about $400/kW. The question is: how can costs be reduced sufficiently to make the central-receiver technology competitive?

### 1.4.3F.3 Public Service Company of New Mexico

PNM had the greatest of hopes that its proposed solar-fossil hybrid repowering concept would be funded for design and construction by DOE, but when the proposal reached the federal government, matters slowed. In PNM's view, many in the government had their own solutions and programs to solve the energy crisis. The proposal received a great deal of scrutiny; some apparently perceived the plan as a threat to the Barstow pilot plant. There was disagreement within the industry as to which was preferable. Many preferred a stand-alone concept - a completely new facility. Suddenly, here was an aggressive utility in New Mexico coming up with a proposal for a 25 MW repowering project and with eight other utilities prepared to make similar proposals.

PNM's plan was that as each plant was constructed, it would receive less subsidy than the last, so that there would be enough heliostats and other components constructed that eventually there would be a transition to commercial feasibility status at about the sixth unit. In PNM's view, its proposal not only challenged the concept of Solar One, but promised to take the technology to the point where (at least on paper) it appeared to be commercially feasible.

Although PNM's proposed 25 MW repowering project was 2.5 times larger than Solar One, it was projected to cost less. PNM's proposal was projected to cost less because it would have used an existing power generation plant. Also, it would not require hot oil storage, dual-admission steam turbines, and other unique features at Solar One.

DOE, however, elected to delay funding of PNM's project. There was a furor over the possibility of granting PNM's request for funding on the basis of an unsolicited proposal when other companies might be interested in doing this also. DOE ultimately elected to fund a number of studies, discussed previously. By the time DOE requested and evaluated proposals, three years had elapsed.

The delay took the process into the early 1980's. After the change in administration in Washington, federal funding for new, large-scale alternative energy projects was eliminated. The new administration's policy was to focus on research rather than to assist in demonstration and commercialization efforts. A commitment had already had been made to construct Solar One. In general, the major effort during the early 1980's was to keep alive the demonstration projects to which the federal government had made commitments, but projects such as the utility repowering projects were suddenly no longer of interest.

### 1.4.3F.4 Texas Utilities Electric Company

Texas Utilities Electric Company (TUE) began its participation in solar power as result of a DOE solicitation in March 1978 for proposals for solar repowering. After reviewing the request and considering several options, TUE decided to submit a proposal. The TUE plants considered for repowering were the Morgan Creek and the Permian Basin steam-electric stations. Both then were gas-fired. Texas Electric Service Company (now a division of TUE) was the utility that took the action, which occurred in March 1979.

The DOE solicitation preferred that the utility be the prime contractor, and that state or local government be involved to assure that public concerns would be properly addressed. A site-specific conceptual design was to be developed, including an economic evaluation. Also, a market survey for potential retrofit applications of the technology proposed for repowering was to be conducted. Other aspects of the solicitation were that the potential for utility cost-sharing with DOE for the repowering construction be specified; all data be made available to the public; and that any contract awarded would be on a cost plus fixed-fee basis.

TUE elected to team with Rockwell International to study the repowering of the Permian Basin steam-electric station, near Monahans (west Texas). Rockwell developed the proposal and submitted it as the prime contractor. The proposal was for a study of the feasibility of repowering a 115 MW generating unit. The proposal was accepted, and the project began in September 1979. The contribution by Texas Electric Service was a review of the Rockwell design work. Also, TUE was a member of the utility advisory committee on repowering in conjunction with Sandia National Laboratories. The final report on the Permian Basin station repowering study was completed on July 15, 1980. The report included the economic findings and other key results and a schedule for development including final design, construction, and operation.

Historically, the Barstow project went forward, but none of the repowering projects was ever implemented. Although TUE became a member of the utility advisory committee for the Barstow project, TUE elected not to proceed with the repowering effort after considering the technical and economic feasibility, and the availability of funding. The project costs were high for the time, and even today, with an estimated capital cost of $111 million ($2,200/kW). This compares with today's costs of $1,500 to $2,000/kW for new baseload plants, and as little as $300 to $400/kW for combustion turbines. As part of its resource planning, TUE annually considers renewable power generation, including solar thermal. The cost of solar thermal units, either stand-alone or fossil-hybrid, is believed to currently be substantially higher than that of other available resources. Photovoltaics is the renewable technology followed most closely by TUE.

### 1.4.4 Distributed Receiver Technology

Utilities have also been active in the area of distributed receiver technology. As indicated previously, distributed receiver systems consist of one or more point-focus or line-focus collectors, each of which utilizes a concentrator to focus the sun's energy and a receiver to absorb the energy and convert it to heat. Point-focus modules using two-axis tracking concentrate sunlight onto a receiver located at the focal point of a parabolic reflector (dish). The solar energy heats fluid circulating through the receiver and this hot fluid is pumped to a central collection point for conversion to electricity or for use in process heat applications. Alternatively, electric power can be generated by a small engine mounted at the focal point of the dish. Each dish is a unit that can function independently or as part of a group of dishes. A single parablic dish, 15 meters in diameter, can achieve temperatures in excess of $1,100\ ^\circ$C $(2,000\ ^\circ$F$)$ and efficiently produce up to 30 kW of electricity.

Trough systems concentrate sunlight onto a receiver tube positioned along the focal line of the reflecting parabolic trough collector. A heat-transfer fluid flowing through the tube is heated to temperatures in the range of 100 °C to 390 °C (212 °F to 735 °F). The trough usually rotates about one axis to follow the sun. Thermal storage systems can be used to accumulate thermal energy for use during cloudy weather or at night. Currently, storage system capacities can range from buffer storage for short time intervals, such as during cloud passage, to as long as six continuous hours (U.S. DOE, 1987).

Among the key projects involving distributed receiver technology in which utilities have been involved, either directly or indirectly as power purchasers, are Georgia Power Company's Solar Total Energy Project (STEP or Shenandoah Project); the LaJet Solarplant 1 dish-electric installation located in Warner Springs, California;

**Table 1.14:** Key U.S. Distributed Receiver Power Projects and Participating Utilities

---

- 400 kWe   -- Shenandoah - Synthetic Oil/Steam, Cogeneration, SKI Dishes, Georgia Power Company

- 4.9 MWe   -- Solarplant 1 - Warner Springs CA, Water/Steam, Central Turbine Generators, LaJet Dishes - San Diego Gas & Electric

- 24 kWe   -- Vanguard - Advanco Dish w/United Stirling engine, Palm Desert, CA, Southern California Edison

- 25 kWe   -- MDC/USAB Dish Stirling - Southern California Edison, Georgia Power Company

- 245 MWe* -- SEGS I-VIII - Solar-Fossil Hybrid Steam. Constructed and operated by Luz International. Output sold to Southern California Edison Company.

     * Planned output is 650 MWe by end of 1993 for SEGS I-XIII. Construction of SEGS IX (80 MWe) started in February 1990 and is planned for completion by the fall of 1990. San Diego Gas & Electric has executed an agreement with Luz International to begin purchasing the output of a future SEGS unit beginning in 1994.

---

the Vanguard I dish-Stirling engine module experiment conducted at Rancho Mirage, California; and the McDonnell-Douglas Corporation/United Stirling AB effort to develop dish-electric systems for commercial application, and the SEGS plants of Luz International, which are parabolic trough systems. These projects are described below. Also discussed is the Crosbyton, Texas fixed-bowl system. A list of the key U.S. distributed receiver projects is presented in Table 1.14.

### 1.4.4A Georgia Power Company's Shenandoah Project

The Solar Total Energy Project (STEP) located at the Georgia Power Company Solar test site at Shenandoah, Georgia is an industrial solar total energy full-system experiment employing dish concentrators. Sponsored by DOE and Georgia Power Company, STEP is configured as a hybrid system with a gas-fired heater. The electrical rating is 400 kW$_e$. The project was inititated in 1977 and startup occurred in 1982. The overall DOE objectives for STEP were to obtain engineering and development experience on this concept as preparation for subsequent commercial-size applications, assess the interaction of solar energy technology with the application environment, narrow the prediction uncertainty of the cost and performance of solar total energy systems, expand solar engineering capability and experience with large-scale hardware systems, and disseminate information and results. An aerial view of the project is shown in Figure 1.12. This project and its results are summarized in a report prepared by Georgia Power Company (1988) for DOE.

**Figure 1.12:** Georgia Power Company's solar total energy project.

### 1.4.4A.1 Plant Description

STEP uses the solar energy collected to partially meet the electrical, air conditioning, and process steam needs in an industrial application. The energy collection processes uses a silicone heat transfer fluid which is circulated through the receiver tubes of parabolic dish collectors. The solar energy collected is normally supplemented with energy provided by a natural gas-fired heater and then delivered to steam generator heat exchangers to produce superheated steam to drive a conventional turbine/generator set. Steam is extracted from the turbine to provide process steam for knitwear pressing, and the low-pressure exhaust steam is used in an absorption chiller to produce chilled water for air conditioning.

Startup of the system normally begins with the fossil-fuel heater. This heater is required in order to meet the early morning energy needs of the plant prior to the availability of solar-derived energy. The fossil-fuel heater provides all of the initial thermal energy input to the steam generator. The collector field warmup is accomplished by circulating the heat transfer fluid through the collector field until it reaches operating temperature. Once the operating temperature is reached, the hot heat transfer fluid is combined with the output from the fossil-fueled heater for delivery to the steam generator. At this point the system is operating in the hybrid mode. As the solar input to the system increases, the fossil-fuel heater contribution may be reduced to maintain a load-following steam production rate. The plant was originally provided with a thermal storage tank which gave additional operating flexibility; however, this tank was removed from service when it became necessary to increase the fluid pressure above the tank pressure limit.

The plant can also operate in the solar-only mode; however, a size mismatch between the collector field and the thermal utilization subsystem results in continuous part-load operation and poor performance. Consequently, the solar-only operating mode is rarely used.

### 1.4.4A.2 Project Results

Extensive performance test data were collected during a 30-day commercial operations test and a 14-day continuous operations test conducted in the summer of 1985. The results of this testing indicated a mismatch between the collector field and the thermal utilization system. Throughout the testing, the operations were dominated by the fossil-fuel heater with the collector field supplying only 25% and 18% of the total thermal input to the system for the 30-day and 14-day test periods, respectively. For solar-only operation, the plant has only been able to generate 135 KW$_e$ gross and about 22 KW$_e$ net under ideal steady-state conditions compared to the plant rating of 400 KW$_e$ (Fair, 1985 and Hicks, 1985). According to Georgia Power, a major reason for this is that the number of solar collectors was limited to 114, versus the 192 originally planned, due to budget constraints.

Although it is a predominantly fossil-fueled system, efficiencies are poor compared to conventional generating plants. The poor efficiencies for fossil-fueled operation are primarily due to low heater efficiencies and the heater location in the heat transfer fluid loop rather than in the water/steam loop. To enhance operation and improve performance, a natural gas heater was added. The system, as installed and operated, is not capable of achieving any economic advantages over conventional energy systems (EPRI GS-6573).

### 1.4.4A.3 Major Accomplishments

Among the major accomplishments of the Shenandoah STEP project are that it advanced the technology of parabolic dish collectors, demonstrated the solar total energy concept, and provided engineering and operating data which will allow future system capabilities to be predicted with better accuracy. The project also identified system deficiencies on a small scale prior to the commitment of resources for a large commercial-size system, and showed the importance of matching solar collector capacity to system load (Georgia Power, 1988).

### 1.4.4A.4 Lessons Learned

Among the major lessons learned from the STEP project are that partial load efficiencies of fossil fuel supplemental heaters must be carefully evaluated in optimizing the system design. Low fossil fuel heater efficiencies at partial loads significantly reduced overall plant efficiency. Adequate margin must be provided between the nominal operating temperature and the upper limit of the heat transfer fluid. Lack of tight temperature control capability necessitated a significant reduction in the field outlet temperature from the design temperature. Also, the centralized control and instrumentation system originally installed for the plant is not recommended for future systems. A system using the distributed control concept (an "expert" control system) was installed in 1988. The· closed-loop collector tracking control system originally installed was found to be inadequate and required replacement with an open-loop tracking method (GPC, 1988).

### 1.4.4B LaJet Solarplant 1

The LaJet Solarplant 1 is a privately-funded, 4.9 MW$_e$ solar-thermal electric generating facility located near Warner Springs, California. The Solarplant 1 project (Figure 1.13) was developed by the LaJet Energy Company. The project sold electricity to San Diego Gas & Electric Company (SDG&E) for a period of time beginning on July 31, 1984. Full rated operation was expected in early 1985 but apparently never achieved. The plant's 700 parabolic dish concentrators are each made up of 24 1.5-meter diameter individually-focused, polymer-film, stretched-membrane modules. The sun's rays are focused on receivers mounted at each dish's focal point. The thermal energy collected is removed from the receivers using water. The water circulating through the receiver/heat exchanger leaves the concentrator as a 274 °C water/steam mixture, is recycled through superheating concentrators to 371 °C, and then is transported through insulated piping to two turbine generators. The main turbine has a capacity of 3.7 MW$_e$ and the second turbine, with a capacity of 1.2 MW$_e$, is used for start-up, shutdown, low insolation and peak generation periods.

The stretched-membrane concentrators used at Solarplant 1 are lightweight and inexpensive, but of questionable durability under high insolation and wind loading conditions. According to SDG&E, the expected life of the mylar membranes was three years, but SDG&E does not know how close the membranes came to achieving this life span before significant deterioration occurred. Also, the double-walled, cavity-type receivers reduce thermal flux by using water/steam heat exchangers embedded in molten nitrate salt. Although this receiver design provides a thermal buffer and energy storage capability, system thermal inertia is increased by overnight freezing of the molten salt.

Although sufficient data for an independent assessment are not available, project results do not appear encouraging. Significant equipment and operational problems have been encountered, particularly related to daily cycling and excessive startup time. It is reported that startup after an extended shutdown can take half a day (McGlaun, 1987).

During 1987, LaJet formed a joint venture with Cummins Engine to convert Solarplant 1 to a solar/diesel combined cycle. Two 1 MW$_e$ diesel generator sets with exhaust heat recovery were installed. The recovered heat was to be used to keep the steam headers and turbines warm. This may mitigate some of the operational problems that the system experienced (McGlaun, 1987).

SDG&E engineers see promise in the concept of lightweight, low-cost solar tracking assemblies such as used by LaJet, especially in desert environments. To prove successful, these reflectors must retain a high reflectivity throughout their design life under moderate winds (20–40 mph). They must also be able to survive high winds (100 mph), hail and sand storms. SDG&E engineers also see point-of-focus heat engines, such as a Stirling or Brayton cycle, as having more promise for use with two-axis tracking assemblies than the central steam turbine concept. Use of

**Figure 1.13:** LaJet Solarplant 1.

tracker-mounted heat engines avoids the problems associated with miles of exposed steam piping.

### 1.4.4C Vanguard I

Vanguard I was a 24 kW$_e$ prototype solar dish-Stirling engine module located at Rancho Mirage, California. Sponsors of the project included DOE, Advanco (under a cooperative agreement with DOE), Southern California Edison Company, and EPRI (extended test and evaluation). The test site was SCE property located adjacent to SCE's Santa Rosa substation. The module is shown is Figure 1.14. Operational tests were conducted from February 1984 through July 1985. The initial objectives of the Vanguard project were to:

- Identify an early market for dish systems and provide an implementation plan for marketing a dish-Stirling system.
- Determine all functional, performance, cost/schedule and programmatic requirements for the dish-Stirling module.

**Figure 1.14:** Vanguard Dish Stirling Module (SCE's Santa Rosa Substation in background).

- Design, fabricate and test a prototype (Vanguard I) for the selected market.
- Prepare a plan for commercial product development.

### 1.4.4C.1 Project Description

The Vanguard I module used a parabolic dish manufactured by Advanco to concentrate solar energy onto the receiver of a United Stirling, AB, Stirling engine mounted at the focus of the dish. The Stirling engine was coupled to an induction generator which supplied power to SCE's power system. The dish-Stirling system is unique among solar-thermal power generation systems in that it requires only a few minutes from the time the module starts to track the sun until useful power is produced. This is because the dish-Stirling system is close-coupled and has very little thermal inertia (no long runs of piping or thermal storage systems to heat up each day). Throughout the day, the dish tracks the sun automatically. The hydrogen gas pressure within the engine/receiver is varied to maintain a constant receiver temperature as the insolation varies (Holgersson, 1984).

Temporary shadowing of the dish by a cloud results in a power transient during which the input is suddenly terminated and then suddenly resumed. During cloud passage, the dish continues to track the sun. When insolation returns, the engine temperature and speed increase automatically and power production is resumed (Washom, 1984).

**Table 1.15:** Vanguard Measured Performance (18 month test program)

| | |
|---|---|
| Maximum Net Efficiency: | 29.4 % |
| Average 24 Hr Net Efficiency: (15 day power production test) | 22.7 % |
| Average On-Sun Net Efficiency: (18 mo., operating days) | 22.8 % |
| Average 24 Hr Net Efficiency: (18 mo., operating days) | 18.5 % |
| Average 24 Hr Net Efficiency: (18 mo., all days) | 9.7 % |
| Equipment Availability (18 mo): | 72.0 % |

Source: EPRI AP-4608

### 1.4.4C.2  Project Results

This was a highly successful project which produced a great deal of useful information. It established the potential viability of the dish-Stirling system concept for solar thermal electric power production. The performance of the Vanguard I module set numerous records for solar-to-electricity conversion efficiency. These are included in Table 1.15, which summarizes the module's performance throughout the 18-month test program (EPRI GS-6573).

This high performance is mainly due to the module's short startup time and rapid response to insolation transients. High conversion efficiency, good part load efficiency and low parasitics also contribute to the favorable results. The performance is all the more impressive in that it was achieved by an initial prototype unit during its first 18 months in the field.

On the other hand, engine maintenance requirements and costs are a major uncertainty of this system concept. A great deal more data from a larger number of engines operating over a longer period of time will be required to make a maintenance assessment with any confidence (EPRI GS-6573).

A commercial 30 MW$_e$ installation was being planned as the next step after the Vanguard I. However, during the course of this project, United Stirling, AB, manufacturer of the power conversion system, entered into an exclusive joint venture agreement with the McDonnell Douglas Corporation for dish-Stirling system development and commercialization. As a result, no follow-on activities to Vanguard I by the Advanco organization were possible.

### 1.4.4D  MDC/USAB Dish-Stirling System Tests

Southern California Edison and Georgia Power Company (GPC) tested the McDonnell Douglas Corporation (MDC)/United Stirling, AB (USAB) dish-Stirling system, a 25 KW$_e$ module employing the USAB 4-95 Stirling engine from 1985 until 1988. A picture of an MDC/USAB dish-Stirling module is shown in Figure 1.15. The MDC/USAB project started in November 1983 and terminated in June 1986 because of the delay perceived in the timing of a commercial market caused by the decline in world oil prices. Six complete modules were built, and hardware was produced for two additional modules, extra test engines and spares. Three modules were tested at the McDonnell Douglas solar test site in Huntington Beach, California starting in December 1984. Single modules were installed in 1985 at SCE's Solar One site and at Georgia Power Company's Shenandoah solar test site. These companies continued operation of the modules through 1988 (EPRI GS-6573). A third module was installed at Nevada Power Company, but as explained below, never tested. The system design and rights and much of the hardware were bought by SCE following termination of the MDC/USAB project.

### 1.4.4D.1 Description

The MDC/USAB dish-Stirling module used a parabolic dish concentrator to power a Stirling engine mounted at the focus of the dish. The concentrator was designed for low-cost under mass production. The engine was a Mark II USAB 4-95 unit (The Mark I version was used in the Vanguard I project). Operation of the MDC/USAB dish Stirling module is virtually the same as for the Vanguard I system except that no starter motor is used. Power from the grid turns the generator as an induction motor. As the Stirling engine warms up, it powers the motor/generator; when 1800 RPM is passed, it delivers positive power to the grid (Hallett et al., 1985).

### 1.4.4D.2 Test Results

Excellent performance was achieved for this module in tests at SCE and GPC. The SCE dish-Stirling module was operated on the same site with Solar One in Daggett, California. A comparison of the Solar One and dish-Stirling performances, normalized to the collector area, is given in Table 1.16 for the first six months of 1986. As can be seen, the net efficiency (sunlight to electricity) of the dish-Stirling system (14.8%) was much higher than that of Solar One (5.7%), even though both systems have comparable theoretical thermal-to-electric conversion efficiencies, and Solar One was in its fourth year of operation while these were the first six months in the field for the dish-Stirling module. The difference in the performance can be attributed to differences in thermal inertia, length of time required for startup, response to insolation transients, operation on intermittent cloudy days, and parasitic energy consumption.

### 1.4.4D.3 Summary of SCE Experience

According to SCE, the dish Stirling's performance indicates it to be the optimum solar-to-electric conversion technology. The unit's demonstrated high efficiency, environmentally benign and modular characteristics make it an early candidate for commercialization. However, due to the existing excess generation capacity in the southwestern United States, and less-than-expected system load growth rates, SCE does not deem it appropriate to continue development of the dish-Stirling or any other generation resource for the present.

**Figure 1.15:** McDonnell Douglas/USAB Stirling dish (located at Daggett, California adjacent to Solar One).

**Table 1.16:** Performance Comparison Solar One Central Receiver vs. MDC/USAB Dish Stirling Solar One Test Site (Jan. - June 1986)

|  | Insolation ( kWh r/m$^2$ ) | Net Electricity ( kWh/m$^2$ ) | |
|---|---|---|---|
|  |  | Solar One | Dish Stirling |
| January | 160.1 | 8.0 | 29.1 |
| February | 153.0 | 6.3 | 7.1 |
| March | 184.3 | 10.0 | 32.6 |
| April | 220.7 | 14.7 | 35.0 |
| May | 265.4 | 19.8 | 40.1 |
| June | 267.2 | 12.2 | 41.0 |
| Total ( 6 months ) | 1250.7 | 71.0 | 184.9 |
| Net Efficiency |  | 5.7 % | 14.8 % |

Source: EPRI GS-6573.

### 1.4.4D.4 Summary of Georgia Power Experience

The GPC experience with the MDC/USAB dish-Stirling was also favorable. However, GPC believes that in the future consideration should be given to adapting the system as a hybrid generator with natural gas firing to supplement the solar energy. This would ensure the availability of power in the late afternoons when GPC's peaks occur and would also facilitate dispatch by allowing GPC's system operators to call upon the capacity whenever it is needed. Such an arrangement is viewed as potentially improving the attractiveness of the dish-Stirling relative to PV for utility applications. Hybrid operation would allow dish-Stirling systems to displace combustion turbine capacity for peak power generation. GPC is pursuing the development of hybrid Stirling engines with other firms. The company is also interested in pursuing the free-piston engine concept for this application.

### 1.4.4D.5 Nevada Power Experience

Nevada Power also received one of the MDC/USAB dish-Stirling units tested by Southern California Edison and Georgia Power Company. The cost to Nevada Power for participation in the project was to be $150,000 plus site preparation costs (approximately $70,000). The site was completed and the dish installed and in the check-out phase when the MDC/USAB effort was terminated. McDonnell Douglas offered to leave the dish with Nevada Power, but the company declined because it lacked people with the training needed to maintain the system and expected the operation and maintenance requirements for the system to be substantial.

McDonnell Douglas refunded that portion of the $150,000 paid by Nevada Power up to the time the project was canceled, but Nevada Power was forced to absorb the site preparation costs. Nevada Power indicated that while the project was still in progress, the technicians attending the unit were frequently away at other sites (SCE and Georgia Power) leaving the Nevada Power unit without maintenance support, once for a period of two months. As a result, the project was never completed. The experience left Nevada Power's management very skeptical about such projects.

### 1.4.4E  Solar Electric Generating System (SEGS)

The SEGS plants are a group of commercial solar-fossil hybrid electric plants employing trough collectors. Up to 25 percent of the thermal input of the plants comes from natural gas. The first two plants (SEGS I and II) are located at Daggett, California adjacent to Solar One. The next five plants (SEGS III through VII) are located at Kramer Junction, California and an eighth unit, SEGS VIII, at Harper Dry Lake, California. The electrical output of all plants is sold to the Southern California Edison Company. An aerial view of the SEGS III through VII is shown in Figure 1.16. The following discussion is based primarily on information provided by Luz (Kearney, 1989).

### 1.4.4E.1  Plant Descriptions

The initial 14 $MW_e$ unit (SEGS I) was placed in service in 1984 and occupies 67 acres of land. SEGS II, with a capacity of 30 $MW_e$, was completed in 1985 at the same location using 150 acres.

SEGS III through SEGS VII are rated 30 $MW_e$ each and occupy approximately 1,000 acres at their Kramer Junction site, located about 40 miles west of Daggett.

Construction was completed in late 1989 on an 80 $MW_e$ unit (SEGS VIII) at Harper Dry Lake, a 2,250 acre site which is located about 30 miles northwest of Daggett. Five additional 80 $MW_e$ units are planned for installation by the end of 1993. SEGS VIII is already operating, and construction was started in February 1990 on SEGS IX. To minimize capital requirements, these units are being constructed on expedited schedules. According to Luz, SEGS VIII took only 9 months to construct, and the construction of SEGS IX is expected to require only 7.5 months. SEGS I–VIII account for more than 90 percent of the electricity currently produced directly from solar energy in the United States. When all planned units are completed, the combined capacity of the 13 SEGS units will be over 650 $MW_e$, enough electricity to meet the residential needs of approximately one million people.

### 1.4.4E.2  Purpose and Participants

The SEGS plants are a commercial venture whose major objective is to produce electricity for sale at a profit under long term contracts. The plants were designed, financed, constructed and are now being operated by Luz International. The SEGS plants are privately-owned under both limited partnerships and co-tenants as the basic ownership structures. The investors in the SEGS plants including banking, savings and insurance firms, utilities and large private investors. The plants benefit

**Figure 1.16:** Aerial view of SEGS plants, Kramer Junction, California.

from economic subsidies including favorable power purchase agreements with SCE under PURPA regulations, federal and state tax incentives, and low interest loans from major vendors. All power produced from SEGS I through VIII is being sold to SCE. However, San Diego Gas & Electric recently signed a 30-year contract with Luz International for the purchase of the output, beginning in 1994, of another 80 MW plant to be built at Harper Dry Lake (Electric Utility Week, 1989).

A list of the first eight SEGS plants and their basic characteristics is given in Table 1.17.

### 1.4.4E.3 System Design

The plants consist of the following primary subsystems:

- Line-focus parabolic trough solar field
- Heat transfer fluid (HTF) system
- Reheat steam turbine/generator
- Waste heat discharge
- Water supply and treatment
- Natural gas-fired steam boiler or HTF heater

SEGS I through VII are designed to produce full electrical power with steam generated using either solar energy or natural gas. In SEGS VIII, the boiler is replaced by a gas-fired HTF heater which can independently heat the full flow rate of the HTF to design operating temperature. The plants are operated to provide

**Table 1.17:** SEGS Plants I–VIII

| SEGS Plant Number | Rating (MWe, Net) | Date In Service | SCE Power Purchase Agreement |
|---|---|---|---|
| I | 13.8 | 12 / 84 | Negotiated |
| II | 30.0 | 12 / 85 | " |
| III | 30.0 | 12 / 86 | Standard Offer #4* |
| IV | 30.0 | 12 / 86 | " |
| V | 30.0 | 9 / 87 | " |
| VI | 30.0 | 9 / 88 | " |
| VII | 30.0 | 12 / 88 | " |
| VIII | 80.0 | 12 / 89 | Standard Offer #2** |

\* Standard Offer #4 provides fixed energy payment rates for the first ten years; capacity payment rates are fixed for the duration of the project. The on-peak period energy payment rates are particularly high, notably during the key summer months from June through September. (Source: EPRI GS-6573)

\*\* Energy payment rates under Standard Offer #2 are revised quarterly to reflect current conditions. Capacity payment rates are fixed as in Standard Offer #4.

maximum power during the period when SCE's purchased energy rates are the highest. The highest rates occur from 12 noon to 6 p.m. during weekdays from June through September, a period which coincides well with the ability to operate at full power on solar input only. Natural gas is used to supplement solar energy in periods of low isolation, such as cloudy days, or when operation in the evening hours is desired. Approximately 70 percent of the total annual electrical output is supplied by solar energy alone.

The basic component of the solar field is the Solar Collector Assembly (SCA). Each SCA has its own parabolic trough solar collector, positioning system, and local control system. The parabolic trough solar collector is a mirrored-glass reflector which focuses direct radiation on an efficient evacuated receiver, or heat collection element (HCE). SEGS VI, for example, contains 800 SCAs, each having a mirror aperture of 235 m$^2$ (2,530-ft$^2$). The HTF in the SCAs is heated from 304 ° to 390 °C (579 ° to 734 °F).

Luz Industries Israel has developed and manufactured three generations of solar collector assemblies, the LS-1 (Luz System 1), LS-2 (Figure 1.17) and, the newest

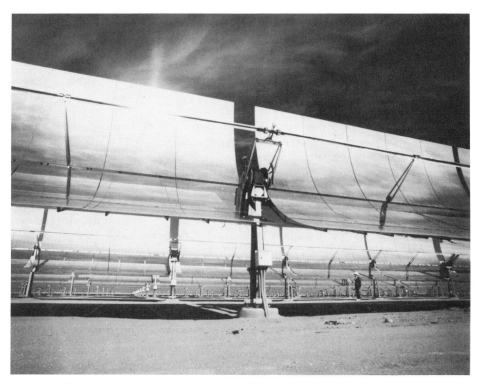

**Figure 1.17:** SEGS parabolic troughs, Kramer Junction, California.

model, the LS-3. The LS-2 collector has extensive field experience in the SEGS II through SEGS V plants; as of the end of 1987, SEGS had 1.8 million $m^2$ (19.4 million-$ft^2$) of collecting surface in operation.

The power block consists of a reheat steam-turbine cycle utilizing a conventional steam-driven turbine-generator and associated equipment. The condenser cooling water circulates to a mechanical draft cooling tower located near the turbine. Water is provided to the SEGS projects either from wells or, in the case of the Kramer Junction site, from the California aqueduct system. Plant demineralizers condition the water, and the blowdown from the cooling tower sump is sent to evaporation ponds. Total water usage is about 450 acre-feet per year for a 30 MW plant.

### 1.4.4E.4 Operating Results

Luz has made considerable progress improving the operating efficiency and reliability of the SEGS plants. By increasing the solar field outlet temperature to 391 °C (735 °F) and utilizing a reheat turbine cycle, the thermodynamic efficiency has been increased from about 30 percent in the early plants to 38 percent in SEGS VI and VII. In addition, the parabolic trough technology has been advanced with the introduction of the LS-3 collector system to further decrease costs and improve solar field performance.

Luz reports that SEGS III–VII are performing well and providing reliable service. The first two SEGS plants are performing at less than design levels and need additional improvements. Luz expects to see a continuous evolution of the SEGS technology leading to both increased operating efficiencies and reduced capital costs. As the plant size increases and as the technology evolves, Luz expects the plants to become even more competitive. According to Luz, it can now generate power for an average of about 12 cents per kWh for SEGS III through VIII and expects its new plants (SEGS VIII through X) to produce electricity for about 8–9 cents per kWh, assuming current costs for natural gas (Kearney, 1989).

The major technical issues associated with the 1988 performance of the SEGS Kramer Junction solar field are: effectiveness and frequency of mirror washing, Heat Collection Element tube breakage, and operational considerations. Efforts undertaken by Luz to deal with these issues appear to be effective (Kearney, 1989).

The success achieved by the SEGS plants is evidenced by the planned expansion of SEGS to over 650 MW. Interest in the concept employed in the SEGS plant is growing, and Luz has been in communication with several southwest utilities about the possible application of its solar-fossil hybrid approach in other locations.

### 1.4.4E.5  Summary

Solar thermal parabolic trough systems are now available commercially in a natural-gas-hybrid configuration. Despite long-term potential judged to be lower than that of the higher-flux-concentration dish and central-receiver systems, the trough system has come a long way over the past several years on the path toward commercial success. Key reasons for this progress offer important lessons for all involved in the development of new energy technologies. First, the firm that developed the technology, Luz, shouldered most of the financial risk as a business venture. Of course there have been subsidies in the form of tax credits and favorable energy sales contracts. But the firm's success hinged on the ability to provide hardware that performed in accordance with specifications and the expectations of private sector investors. Hence a powerful entrepreneurial motivation has been operative. In contrast, much of the development work funded through directed programs of the government and other agencies, such as EPRI, has been conducted on a cost-plus-fixed-fee basis, with very little cost and risk sharing by the industrial contractors performing the work. These organizations were rewarded for work done on a "best-efforts" basis, rather than for success in achieving solid commercial production objectives.

Second, the trough effort had access to an extensive technological data base, much of it funded by the federal government (see the discussion of industrial process heat efforts in a later section of this article); and Luz made full use of this information. Third, the technology has been amenable to affordable development steps that have required about one year each. This is to be contrasted with, for example, the central-receiver technology, where each development step at the system level is far more costly and requires several years for completion. And fourth, all of the development steps were conducted within the same organization by essentially the same group, allowing lessons learned from each step to be incorporated immediately into subsequent steps.

### 1.4.4F  Crosbyton Fixed-Bowl System

Another solar-thermal experiment, somewhat related to the dish concept, was a unique system demonstrated at Crosbyton, Texas. The Crosbyton Power & Light

System participated with Texas Tech University, E-Systems, and Foster Wheeler Energy Corporation in a DOE-sponsored solar-thermal demonstration unit using the fixed-mirror distributed-focus concept. Construction of the system was completed in January 1980, and initial operation of the system occurred on January 23, 1980 and continued until May 1983 when the system was dismantled. The experiment used a fixed-bowl to focus sunlight on a receiver which moved with the insolation focal point of the bowl as the sun moved across the sky. The bowl was 20 m (65-ft) in diameter and steam was produced in the temperature range of 425 to 540 °C (800 to 1000 °F) at pressures of 56 to 70 kg/cm$^2$ (800 to 1000 psi). Plans called for a next phase in which 61 m (200-ft) diameter dishes would be employed in a 5 MW power system. Although the experimental results apparently verified performance predictions, the operators had many problems maintaining the mirrors in the reflective surface. These had to be frequently adjusted to close tolerances. The concept appears to have significant disadvantages compared to dish systems, and the authors are aware of no follow-up effort to further pursue this concept.

### 1.4.5 Solar Ponds

Another thermal generation option which has been considered as a possible power generation source by utilities is the solar pond. There are several types of solar ponds, but the type which appears to have the most potential for energy production, particularly on a large scale, is the salt gradient solar pond.

### 1.4.5A Structure of a Salt Gradient Pond

A salt gradient solar pond is a shallow body of water with three distinct zones, as shown in Figure 1.18. The top layer or the upper convecting zone is typically 0.2 to 0.5 m (0.7 to 1.6-ft) deep. It consists of uniform, relatively low salinity water (0–6% by weight) with a temperature that is in equilibrium with the ambient air and the heat flux from below. Below this is the non-convecting gradient zone in which both temperature and salinity increase with depth. This zone is about 0.75 m to 1.5 m (2.5 to 5-ft) thick and helps to insulate the underlying heat storage or bottom convecting zone. This last zone is usually between 1.0 and 3.0 m (3.3-ft to 10-ft) thick, and consists of brine with high salinity (15%–25%) often approaching saturation levels. Short-wave solar radiation penetrates through the upper and middle zones into the bottom convecting zone and raises its temperature. The high density of the bottom zone suppresses mixing with the upper layers, thus enabling temperatures to reach high values (as much as 40 °C to 70 °C or 104 to 158 °F above ambient). Energy can then be extracted from this zone. The study of solar ponds was begun in Israel in 1958. Several solar ponds of various sizes have been constructed in the U.S. (EPRI AP-3842).

### 1.4.5B Utility Activity in Solar Ponds

### 1.4.5B.1 Tennessee Valley Authority

In the early 1980s, the Tennessee Valley Authority constructed and instrumented a 4,000 m$^2$ (one-acre) solar pond to demonstrate the technical and economic feasibility of using salt-gradient ponds for industrial process heat. A secondary objective was to determine whether pond temperatures and heat extraction rates would be

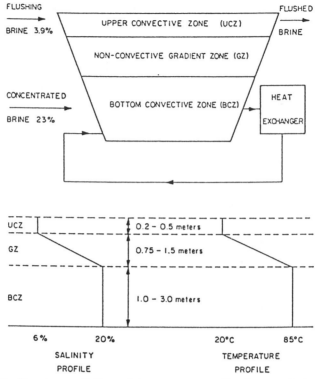

**Figure 1.18:** Structure, salinity and temperature profiles for a typical solar pond.

suitable for electric power generation. Testing and evaluation began in 1982 and continued for three years.

While pond temperatures in excess of 93 °C (200 °F) were achieved, and heat was successfully extracted at the rate of 2,600,000 BTU/h (limited by pump and heat exchanger capacity) without adversely impacting the gradient, TVA concluded that the salt-gradient solar pond was not an economically viable source of process heat or electricity for the Southeast. Also, the results showed that where rainfall exceeds evaporation, handling of excess brine presents a key technical problem unless the excess can be discharged to the environment.

A leak at the pond demonstrated the need for a leak detection system and/or a secondary liner if releases to the environment are to be avoided. The added cost, however, makes the concept even more economically infeasible. A positive aspect of the leak at the TVA pond was that it was demonstrated that a leak could be detected and successfully repaired at reasonable cost.

It was also found that a large amount of dust, pollen and algae can collect in solar ponds, reducing the ability of the pond to collect and store energy. A simple process of inplace clarification of the pond water was developed, which allowed a 50% increase in pond water clarity. This significantly increased the energy collection of the pond.

### 1.4.5B.2 Los Angeles Department of Water and Power

In March, 1981, the Los Angeles Department of Water and Power announced plans (The Solar Thermal Report, 1981) to construct the nation's first solar pond power plant at Owens Lake to generate 210 kW of electricity by mid-1983 at Owen Lake. The Owens Lake site, located 200 miles north of Los Angeles, was selected because it provides a unique combination of fresh and brine water, salt deposits, and abundant insolation. The pond was to have been 61,000 m$^2$ (15-acres) in size and 3.5 m$^2$ (11.5-ft) deep. Solar energy stored in the heavier, highly saline water at the pond bottom was expected to reach 85 to 100 °C (185 to 212 °F) before being pumped to a heat exchanger where it would evaporate a working fluid (Freon). The pressurized vapor would then have been used to power a turbine generator.

The project, as conceived, was mainly intended as a research and development project to provide data for U.S. utilities interested in solar pond technology. The plant also was expected to serve as a training facility for solar pond operators and allow development of solar pond operating and maintenance techniques. The plant, which was estimated to cost $2.75 million, was never constructed. LADWP had requested funding from the California Energy Commission (CEC) for the project. The project was in competition for funding with other projects. LADWP decided against proceeding, after the project was not selected for funding by the CEC.

### 1.4.5B.3 Southern California Edison

The Southern California Edison Company also investigated the possibility of constructing a solar pond. SCE worked with ORMAT Turbines, an Israeli firm which developed much of the technology of solar ponds in Israel. SCE and ORMAT Turbines announced the installation of four 12 MW solar pond units at Danby Lake in Southern California beginning in 1985. The proposed development would have been a private venture with SCE buying the output of the solar pond plant. However, the project did not go forward, apparently due in large part to declining energy prices and the repeal of solar energy tax credits.

### 1.4.5B.4 El Paso Electric

Probably the most successful salt-gradient solar pond in the U.S. has been the El Paso Solar Pond Project. This project was started in 1983 and was a cooperative effort involving several organizations including the U.S. Bureau of Reclamation, the Texas Energy and Natural Resources Advisory Council, and the El Paso Electric Company. The project was designed to take advantage of a favorable application and site at a food processing plant in El Paso. A 0.8 acre water storage pond (previously used for fire protection) was converted to a solar pond by late 1984. In the summer of 1985, the pond was operated at 72 °C (162 °F) and became the first solar pond in the world to supply process heat. During the summer of 1986, the solar pond was operated at 85 °C (185 °F) and hot water was delivered as boiler feedwater to the food plant.

A 100 kW organic Rankine cycle engine-generator was installed in the summer of 1986. By September 1986, with the pond at 85 °C (185 °F), the project generated the first electricity produced by a solar pond in the United States. The solar pond has continued to supply electricity during summer months and thermal energy during winter months since that time. In the summer of 1987, a 24-stage, falling film, low-temperature desalination unit was installed at the site, and the El Paso

Solar Pond became the first in the U.S. to ·produce drinking water from brackish water. The production unit is powered by thermal and electric energy from the solar pond and produces 19,000 liters (5,000 gallons) per day of drinking water at 50 ppm dissolved solids from brackish water at 12,000 ppm (Reid and Swift, 1989).

### 1.4.5C  Feasibility of Solar Ponds for Power Generation

A study completed in 1985 for EPRI by the Massachusetts Institute of Technology (MIT) (EPRI AP-3842) concluded that solar ponds are not generally an economic option for utility-owned power generation in the U.S. In performing the study, the investigators gathered information on solar ponds worldwide. They selected a 50 MW system at the Salton Sea in California as a base case for estimating energy production costs. After assessing published information on solar pond theory and operation, the analysts consolidated information on the thermal energy-conversion equipment, construction, and maintenance needs of the base case system, its available resources and potential environmental effects. They then estimated levelized busbar energy costs for the base case by combining predicted capital costs, operation and maintenance costs, performance figures, and financial information. Other estimates accounted for possible variations in the base case, such as the use of single-membrane liners, the purchase of salt, and reduction of heat exchanger size.

The study confirmed that the basic physics and engineering of solar ponds are established and that no technological breakthrough is necessary to operate salt-gradient systems. Pond construction can use conventional methods to clear the site and form dikes. Power plant designs are conventional also, except that they might substitute refrigerant vapors for steam. Units of 2.5 to 5 MW at the Dead Sea (and the El Paso Project results) demonstrate the technical feasibility of producing electric power with this technology.

The uncertainties of a new technology and site-specific variables make it difficult to predict the costs of solar pond energy production accurately. At favorable sites - those having salt, high insolation, and no need for lining - costs were estimated to range from 19 to 24 cents/kWh (1983 dollars). When solar ponds require polymer liners or the purchase and transport of salt, however, costs become prohibitive. Land requirements are also high. A baseload unit would require one acre for every 8 to 16 kW produced, and a peaking unit, one acre for every 30 to 60 kW. Estimates suggest overall system efficiencies of approximately 1%.

These results indicate that utility-owned and -operated ponds are not likely to be economic in the United States in the near term, except in very special situations.

### 1.4.5D  EPRI Solar Thermal Program

EPRI has been involved with solar thermal technology since the mid-1970s. EPRI's involvement has included numerous studies, assessments, technical evaluations, experiments, test programs, workshops and meetings. During the late 1970s and early 1980s, EPRI supported extensive development and testing efforts for central receivers employing high-temperature air or gas as the working fluid in a fossil-hybrid Brayton-Cycle power-generation system. Despite early success that included the inaugural receiver test at the DOE's Central Receiver Test Facility (CRTF), these efforts were terminated in 1982 before definitive results could be obtained. The primary reason was EPRI's realization that, with increasing program costs and substantial reduction in related and complementary DOE and utility programs, it could not sustain the needed effort.

EPRI has published 18 technical reports on its solar thermal activities, which are listed in the bibliography at the end of this chapter. Topics covered in the reports range from the "Physiological Effects of Redirected Solar Radiation" (ER-651) to the "Performance of the Vanguard Solar Dish-Stirling Engine Module" (AP-4608). The two most recent reports address the "Molten Salt Solar Electric Experiments" (GS-6577) and the "Status of Solar Thermal Technology" (GS-6573). The level of activity in solar thermal in EPRI's Solar Power Program has declined since the early 1980s as the emphasis within the Solar Power Program has shifted more and more toward photovoltaics. However, EPRI plans to remain active in solar thermal, monitoring progress and pursuing appropriate activities as opportunities arise.

### 1.4.6 Summary and Conclusions About Solar Thermal Power

The electric utilities have gained an extensive amount of knowledge about solar-thermal electric power technology through their participation in various activities and projects. Their most significant findings have been that the solar-thermal electric power technologies are technically feasible and are capable of producing *energy-significant* quantities of electric power for utility systems located in the Southwest. However, both capital and operation and maintenance costs of solar-thermal power plants are high, and, for most concepts, reducing these costs to levels required for commercially feasible plants would involve large and costly development programs. The single, striking exception is the solar-fossil hybrid concept being successfully employed on a commercial basis by Luz International at the SEGS plants. Luz appears to have achieved commercial success using available energy price incentives with an internally-funded technology development effort. As discussed above, highly-effective entrepeneurial motivation and the trough technology's amenability to development have combined to bring about this impressive success.

In general, the high capital costs of solar-thermal plants are associated with the substantial amount of hardware, including structures and mechanical and electrical systems, required to efficiently accomplish the sunlight-to-electricity conversion. The costs of producing the components of solar-thermal power plants are fairly well understood, and the potential for cost reduction can be estimated with a reasonable level of confidence because solar thermal plants are comprised of mechanical systems using variations of conventional hardware.

Operation and maintenance costs of solar-thermal plants per kWh of energy production tend to be high due to:

- The quantity of solar energy collection equipment and number of systems that need to be maintained.
- The relative complexity of the equipment and systems compared to other solar technologies.
- Fatigue and deterioration of materials and components due to stress from thermal cycling.

For the above reasons, the prospects for major breakthroughs that alone would very substantially reduce solar-thermal capital and operation and maintenance costs appear remote. Cost reductions will come mainly through scaling of plants to optimal size and automating systems wherever cost effective to reduce operating manpower requirements.

In addition to the high capital and operating and maintenance cost, utilities have found that the central receiver system experiments have not met energy production projections and that the cost of central receiver system experiments is very high. Although a commercial plant's capital costs per kW could be expected to be much lower, the difference between the capital cost of Solar One and the capital cost of currently available generation is a large gap to overcome in the transition from pilot plant to commercial-scale facility.

Also, of the several central receiver system experiments conducted in the U.S. and abroad, Solar One was the only one to produce a positive net output of electrical energy; however, that output was substantially less than projected. While there are good explanations for the low energy production of the central-receiver system experiments, their below-expectations performances raise concerns about the commercial viability of more advanced systems using molten salt or sodium. At the same time, the inherent size and complexity of central-receiver systems makes any development and demonstration programs, which could help remove uncertainties, both lengthy and costly. By contrast, the SEGS program was able to proceed successfully through several generations of technology in a relatively short period of time, implementing design modifications in successive plants.

In addition to the above general findings and conclusions from the experimental central receiver activities, two more specific conclusions can be drawn that have important implications for all sun-tracking power plant technologies; and, in the case of the second, for *all* bulk power plant technologies. First, the experience with heliostat fields has been very favorable, suggesting that tracking systems for mirrors, dishes, or photovoltaic arrays are likely to have minimal operating and maintenance requirements. Second, experience with modern, sophisticated digital control systems has also been favorable. Utility power plant operations personnel who exercised these control systems were quick to learn their operation and appreciate their capabilities. They also indicated a desire to see more widespread use of digital control systems throughout the utility industry.

A recent EPRI study (EPRI GS-6573, which is summarized in the article entitled "The Status of Solar Thermal Technology" in Chapter 2 of this volume) evaluates the central-receiver, dish and trough systems for utility-scale electricity generation in solar-only applications. Of the three concepts, the highest demonstrated performance has been achieved by the dish systems, which to date have also received the least amount of development support. Even the first dish-engine modules deployed achieved outstanding performance and reasonably good availabilities. In contrast to the central-receiver, the small size of the dish system module allows system development and product improvement in smaller, less costly steps. On the other hand, maintenance costs are very high and overcoming these costs will probably require development of a low-maintenance advanced engine concept.

The central-receiver concept in utility-scale systems is also judged to have substantial potential for attractively low energy costs. However, as discussed above, major uncertainties in cost and performance remain that require favorable resolution before commercial introduction can commence.

Compared to the central-receiver and dish systems, the assessment of the potential for trough systems is relatively straightforward. The construction of solar-thermal plants using trough systems involves technology well within the state-of-the-art; no technology breakthroughs are required. Cost reductions can be achieved through economy-of-scale optimization and incremental component and system improvements.

The major weakness of the trough systems is that the energy cost potential in the solar-only mode is a factor of two higher than for either central-receiver or dish systems. This would suggest that in the solar-only mode, the trough technology is the least attractive for electric power generation of the three technologies. However, the achieved energy cost of the trough is within the potential energy cost range of either the central receiver or dish systems, and there is no guarantee that dish or central-receiver systems will actually achieve their potential energy costs.

Although the utilities have no direct experience with power generation using troughs, the successes being achieved at Luz indicate that in solar-fossil hybrid applications, the trough technology does indeed have good potential. It is apparent that the reliability and dipatchability afforded by the hybrid configuration is a key element of Luz's success. It is likely any future solar plant using trough technology would involve solar-hybrid operation.

The experiences at Luz, the repowering studies proposed by the utilities from the Southwest in the late 1970s and early 1980s, and Georgia Power Company's assessment of its dish-Stirling experience, suggest that the solar-fossil hybrid option also should be given serious consideration for any future central-receiver or dish system developments. Hybrid operation reduces the need for thermal cycling of equipment and improves plant availability. It also improves the capacity credit which utilities can assign to solar-thermal plants, avoiding duplicate conventional facilities to provide backup when the solar plant is not in operation. In addition, it spreads the costs of operation and maintenance and parasitic energy over more kilowatt-hours.

Overall, utilities view solar-thermal power technologies as being applicable primarily in the Southwest where insolation is relatively abundant. In the near term, the SEGS-type trough systems have the best prospects of any technology for bulk power generation from solar energy. The other solar-thermal technologies discussed, assuming favorable resolution of technical and economic uncertainties, may achieve *energy-significant* applications in utility systems after the year 2000.

## 1.5 Wind Power

Electric utilities have played a key role in helping to achieve the successes attained in the U.S. since 1973 in the development of wind power technology. The other major players have been the federal government and the wind power industry.

The electric utilities became active in wind power following the 1973 oil embargo. According to the EPRI surveys, 20 utilities reported 29 activities in the area of wind power by 1976, a number which grew to 104 utilities and 195 activities by 1981 (Table 1.1 and Table 1.2). These activities involved a mixture of project cosponsorships and use sectors (Table 1.4) and were conducted by utilities throughout the U.S., with the majority of the activity in the northeast and western states (Figure 1.4). The electric utilities' activities in wind power have included wind resource assessments, technical analyses and feasibility studies, testing of both large and small wind turbines, and the development and operation of their own experimental wind power stations. In addition they have cooperated with independent wind power producers through the development of wind power station interconnection guidelines and apparatus and the purchase of electricity from these stations, and they have acquired favorable wind sites and in some cases have leased such sites to wind power developers.

At the time of the 1973 oil embargo, no wind power technology suitable for electric utility applications existed. Since that time, wind power technology has advanced rapidly, especially within the past decade. As of 1989, there were approximately 17,000 commercial wind turbines installed in the U.S., mostly in California, with a total installed capacity rating of about 1,500 MW (Schaefer, 1989). Most of these machines have been installed in three California locations: the Altamont Pass, east of San Francisco; the Tehachapi Mountains, north of Los Angeles; and the San Gorgonio Pass, east of Los Angeles. As can be seen in Table 1.18, approximately 90% of the total wind capacity in the world is located in the U.S.

Many of the early wind turbines and projects did not lead to commercial success, particularly those involving large, megawatt-scale machines. However, much of the activity over the past decade has been successful, and the experience gained has demonstrated that well-maintained wind power stations using mid-sized machines (ratings of about 100 kW to 300 kW) can be economically attractive in areas with good wind resources and moderate to high conventional electric energy production costs. Although the technology is not yet mature, wind turbines have been developed to the point where utilities with access to suitable wind resources and sites can factor wind power into their long-range generation expansion plans.

Most of the megawatt-scale wind turbine projects had substantial utility industry and EPRI involvement. While a great deal of valuable experience was obtained from these activities, their lack of overall technical and economic success has left many utility managers with the general impression that wind power technology is of little value to utilities. Meanwhile the successes with smaller machines have received little attention from utility executives because they have occurred in the independent power sector outside the utility operations' arena, and because they have in many cases been over shadowed by press coverage of problems with poorly engineered machines and of tax credit abuses that, unfortunately, did occur. Further, because of the PURPA legislation, utilities have been forced to purchase electricity at elevated prices from wind power developers that in some cases has not been needed. Many felt wind power would quietly disappear after the available tax credits expired.

Nevertheless, many within the utility industry are realizing that (a) significant technical and economic successes have occurred in the wind industry; (b) even though federal tax credits for wind expired in 1985 and California tax credits expired in 1986 (while federal tax credits for solar electric technologies are still in effect), new wind turbines are still being installed; (c) attractive wind resources exist in their service territories; and (d) further advances in wind power technology are likely to occur. Consequently, a number of utilities are quietly reevaluating their positions with respect to wind power, and some have actually invested, through

**Table 1.18:** Estimated 1988 World-Wide Wind Generated Electricity

| | Energy Output, Million kWh | World Percent | Cumulative Percent | Total MW Installed, Dec 1988 |
|---|---|---|---|---|
| CALIFORNIA | | | | |
| Altamont (PG&E[1]) | 993 | 44.7 | 44.7 | 734 |
| Altamont (DWR[2]) | 4 | 0.2 | 45.9 | 6 |
| Tehachapi (SCE[3]) | 484 | 21.8 | 66.7 | 397 |
| San Gorgonio (SCE[3]) | 346 | 15.6 | 82.2 | 267 |
| HAWAII | 51 | 2.3 | 84.5 | 29 |
| OTHER U.S.A. | 5 | 0.2 | 84.7 | 7 |
| DENMARK | 290 | 13.0 | 97.8 | 190 |
| OTHER EUROPE | 20 | 0.9 | 98.7 | 12 |
| OTHER | 30 | 1.3 | 100.0 | 18 |
| WORLD TOTAL | 2,223 | | | 1,660 |

[1] Sold to Pacific Gas & Electric Company
[2] Sold to California Department of Water Resources
[3] Sold to Southern California Edison Company

Source: Smith et al, 1989

unregulated arms, in California wind power stations. If technical and economic progress continue, it is likely that utilities with good wind resources will increase their involvement with wind power significantly over the coming decade.

A 1986 workshop sponsored by EPRI in cooperation with the Solar Energy Research Institute (SERI) examined the prospects and requirements for the geographic expansion of wind power usage in the U.S. A key workshop result was the finding that there are a number of areas in the United States with substantial wind resources and that determining the potential for development of these resources for wind power will require local, detailed wind measurements. Another finding was that there are no technological barriers to more widespread usage of wind power.

The workshop also concluded that the greatest potential for advancement of wind power technology is in the application of power electronics to enable variable speed wind turbine operation and to integrate and improve wind power station control and protection. Although wind power stations were seen as posing no health hazard, it was the consensus of the workshop participants that public and utility acceptance of the technology will require strong educational activities in all aspects of wind power (EPRI AP-4794).

This section reviews the history and status of wind power development in the U.S., examines the activities and experiences of a number of specific utilities, describes the efforts of EPRI in the area of wind power, and summarizes the lessons learned and implications for the future.

## 1.5.1 History of Wind Power Development

### 1.5.1A Early Activity

Wind energy has been used to produce electricity since the late 1800s. By the 1920s and 1930s, wind machines were being widely used for electricity generation in rural areas of the United States. However, when the Rural Electrification Administration (REA) brought lower-cost electric utility service to farms and small villages in the 1930s and 1940s, the U.S. market for wind turbines all but disappeared. By 1973, there were only a few firms selling wind turbines in the U.S. (DOE, 1985). These were small machines not intended for utility bulk-power applications.

In the late 1920s and early 1930s, Detroit Edison, Public Service Electric and Gas of New Jersey (PSE&G) and several other utilities considered the development of a full-scale, wind-powered generating plant with the Madaras Rotor Power Corporation. The work was prompted by the successful crossing of the Atlantic in 1925 of an odd-looking ship employing two large magnus-effect rotors. Under the magnus-effect principle, as each rotor spun about its vertical axis, air flowing across the rotor from the side was spun around the front of the rotor, creating a suction and pulling the rotor and the ship forward (Figure 1.19). Using this concept, between 15 and 40 railroad cars were to have been propelled around a closed track (Figure 1.20). Each car was to have a generator in its base, and as the wind forced the cars around the track, power was to be generated at the rate of one megawatt per car (Pratt, 1976). PSE&G actually installed and tested a 90-foot Madaras rotor in 1932. The system had been in operation for about 18 months when it was destroyed in a hurricane (Roman, 1988). Due to technical problems which limited system efficiency, the idea for such a power plant was scrapped (Pratt, 1976).

In 1939, the S. Morgan Smith Company hired Palmer C. Putnam to head a team of scientists in developing a large-scale wind power generator of more conventional design. The result was a 125-megawatt, 60-hertz wind generator erected in 1941 on Grandpa's Knob, a 2,000-foot peak in Vermont. The unit had a 175-ft, two bladed rotor mounted on a 110-ft galvanized-iron tower (Figure 1.21). Between 1941 and 1945, the Smith-Putnam wind turbine operated successfully, feeding power into the power system of the Central Vermont Public Service Company. Then in March 1945, one of the 8-ton blades broke off, a result of stress cracking and improper welding. About $1.25 million had been invested in the project up to that time, and

**Figure 1.19:** Flettner Rotor Ship, employing Magnus-Effect rotors, crossed the Atlantic Ocean in 1925. Source: Putnam, 1948 (Photo provided by Detroit Edison).

the Smith Company was reluctant to spend more. With the loss of the blade, the Smith-Putnam experiment was terminated (Putnam, 1948).

The success initially achieved with the Smith-Putnam Experiment prompted the Federal Power Commission (FPC, the forerunner of the FERC) to engage Percy H. Thomas to develop a design for a large-scale wind turbine. Thomas responded by designing a 6,500 kW, twin-rotor turbine mounted on slender-latticed steel tower welded in triangular configuration. The rotors were each 200 feet in diameter and the turbines and tower exceeded the Washington monument in height. A subject of debate in the early years of the Eisenhower Administration, the conceptual turbine design failed to achieve federal budget approval for the design and construction of a prototype and was eventually shelved (Wahrenbrock, 1979). For the next two decades, no significant efforts were made to advance wind power technology.

### 1.5.1B  Developments Since 1973

The modern revival of wind power technology began in the early 1970s. The federal program was the dominant force until about 1981. The federal program's primary emphasis was on the evolution of multimegawatt wind turbines. These large machines were believed to offer scale economies and other advantages over smaller machines. The premier development effort in large wind turbines was carried out by the National Aeronautics and Space Administration (NASA) Lewis Research Center under sponsorship of the DOE. The DOE-NASA program, begun under

**Figure 1.20:** Conceptual drawing of Madaras Rotor Power Plant which was never constructed. Source: Madaras Rotor Power Company Report (picture provided by Detroit Edison).

other agencies in the early 1970s, involved development by several contractors of a series of progressively larger wind turbines. There were similar efforts in Europe.

The earliest efforts of the federal government were the NASA-designed MOD-0 and -OA wind turbine generators. These were relatively small machines, 100 and 200 kW$_e$, respectively, whose basic goal was engineering proof of concept. As R&D machines, they proved to be valuable tools and a number of significant lessons were learned (DiGiovacchino, 1982). MOD-OA machines, Figure 1.22, were installed and tested by DOE at Clayton, New Mexico in March 1978 – operated by the Clayton Municipal Electric System; Culebra Island, Puerto Rico in January 1979 – operated by the Puerto Rico Electric Power Authority; Block Island, Rhode Island in October 1979 – operated by the Block Island Power Company; and at Kahuku Point in Oahu, Hawaii in May 1980 – operated by the Hawaiian Electric Company.

The next step was the MOD-1, Figure 1.23, which was designed and manufactured by General Electric Company. The MOD-1 was installed in Boone, North Carolina in July 1979 and operated by DOE in conjunction with the Blue Ridge Electric Membership Corporation. With its 2,000 kW$_e$ rating, the MOD-1 demonstrated the impact of a large machine on the electrical grid and the local environment. The MOD-1 was also an important step in verifying the analytical tools for structural dynamics and load prediction. Significant contributions were made to the understanding of wind turbine noise, TV interference and grid interaction of multi-megawatt wind turbine generated power (DiGiovacchino, 1982).

**Figure 1.21:** Smith-Putnam Windgenerator, built in 1941, delivered 1.25 MW into the Central Vermont Public Service Power Systems prior to a blade failure in 1945. Source: Putnam, 1948 (Photo provided by Detroit Edison)

**Figure 1.22:** 200 kW MOD-OA Wind Turbine.

The next generation of large wind turbines, the MOD-2, designed by Boeing Corporation (Boeing), appeared in 1981. Generally, these machines were the product of aerospace technology. They combined a 300-ft diameter rotor connected to the drive train through a unique elastomeric teeter bearing with a light-weight,

Figure 1.23: 1 MW MOD-1 Wind Turbine.

epicyclic transmission. The machines were mounted on light-weight tubular towers. Three MOD-2s were placed in operation at Goodnoe Hills, near Goldendale, Washington, in 1981 for evaluation on the Bonneville Power Administration network; a fourth unit was installed at the Medicine Bow, Wyoming, site of a U.S. Bureau of Reclamation project to evaluate wind energy integration with hydroelectric power from the Colorado River Storage Project. Pacific Gas and Electric Company also tested a fifth MOD-2 turbine at a site in Solano County, California, northeast of San Francisco.

Design and construction of the MOD-2 turbines was one of the more notable success stories of the federal government's solar energy programs in the late 1970s. But problems that surfaced during testing and operation of the MOD-2s had important implications for the future of large wind turbines. In June 1981, after about seven months of operation, the first of the Goodnoe Hills MOD-2s suffered a damaging overspeed event resulting from a hydraulic valve failure (caused by contaminated hydraulic fluid) during a test of the emergency shutdown system. The unit's generator and other components were badly damaged, placing the turbine on the disabled list for nearly nine months and requiring modifications to the other MOD-2s at the site (Moore, 1984).

About 18 months later, the same unit shut itself down during another test; this time, technicians found a large crack running two-thirds of the way around the thick, low-speed steel shaft that supported the 100-ton rotor. All other MOD-2s were idled for the next 7 months while engineers studied the problem. The 15-ft (5 m) drive shafts on all three Goodnoe Hills MOD-2s were replaced with new, sturdier designs, keeping the units down for most of 1983. Major modifications, including the addition of a crack detection system, were required on the MOD-2s in Wyoming and in California (Moore, 1984).

Problems such as those encountered with the MOD-2s are not unexpected in any large technology R&D program, but the implications fueled concerns that too much complexity had been built into large wind turbines and that they were further from commercial availability than once believed by some proponents. However, the MOD-2s were never intended as mature commercial wind turbines. Rather, they were viewed within the federal R&D program as the steppingstone to a more advanced, third generation turbine that could generate energy at low cost. This third-generation turbine was called the MOD-5 (Moore, 1984).

Two design concepts of the MOD-5 were considered: a 7.3 MW, 400-ft rotor (122 m) MOD-5A to be designed by General Electric Company; and a 7.2 MW, 420-ft rotor (128 m) MOD-5B to be designed by Boeing. The MOD-5B was later downsized to 3.2 MW, and the rotor reduced to 320-ft (98 m) - slightly longer than a MOD-2 rotor. Under a plan adopted by DOE-NASA in 1980, each contractor was to have one MOD-5 turbine ready for testing in 1984. Both designs were to incorporate existing concepts employed in the MOD-2, as well as insights gained in operating the second-generation machines, such as the effects of structural loading on the fatigue life of major components. Shutdown of the MOD-2s during much of the MOD-5 design period, however, prevented some important feedback. The MOD-5s were also to feature significant design innovations, such as variable-speed generation, intended to lower the cost of energy from MOD-2 projections and reduce structural loading (Moore, 1984).

As often happens with technology R&D, however, politics and economics worked to alter the outlook for continued development of large turbines. First, a new administration with a markedly different philosophy regarding the role of the federal government in the energy marketplace took charge in Washington, D.C.; DOE's wind energy budget in 1983 dropped from a level of about $60 million the previous year to less than $20 million. As a result, the MOD-5 development effort was transformed from a government demonstration project to a cost-sharing arrangement with the turbine suppliers and potential host utilities. Second, world oil prices – a fundamental incentive to exploit wind – fell, eroding the perceived urgency for wind and other renewable energy technology deployment (Moore, 1984).

Finally, General Electric, citing these reasons as well forecasts of reduced utility load growth and the uncertain extension of federal wind energy tax credits, withdrew in late 1983 from a contract with Hawaiian Electric Company to demonstrate the MOD-5A on Oahu's Kahuku Point. General Electric said it had unsuccessfully sought to interest dozens of other utilities and small power producers in purchasing MOD-5s and therefore concluded there was no present or near-term market for the turbine. Other firms pulled back from the large wind turbine business as well (Moore, 1984). However, discussions continued between DOE-NASA and Boeing leading eventually to the installation of a MOD-5B turbine at Kahuku Point in Hawaii in August 1987. As discussed later in this section, this machine was acquired in 1988 and is now being operated by Hawaiian Electric Renewable Systems, a subsidiary of Hawaiian Electric Industries which is the parent of Hawaiian Electric Company.

In summary, the federal government's experience with the first large-scale prototypes revealed serious problems, particularly in the understanding of costs and of cyclic loads and their origins. Long downtimes of the large machines slowed maturation of the technology. Capital costs were projected to be high even for volume production – $1,800 to $2,000 per kW versus $800 to $1,200 per kW for smaller units (Lynette, 1990). Questions regarding practicality of use and size of market arose and, in turn, made it doubtful that the production volumes upon which early cost estimates were based would ever be reached. The significant risks that surfaced made it difficult to justify the additional long-term research commitments that would have been required to continue evolution of large machine technology. In general, the challenge would have been to achieve, simultaneously, gains in structural margins and reductions in weight, complexity and cost through new generations of prototypes (EPRI AP-1317, AP-1641, AP-1959, AP-2456, and AP-2796).

Meanwhile, as a result of incentives provided by the Public Utility Regulatory Policies Act of 1978 (PURPA) and various federal and state tax credits, a major commercialization of smaller wind turbines began in 1981. Wind power development companies sprang up virtually overnight to take advantage of the tax credits, the requirement under PURPA that utilities purchase energy at their avoided costs and the ability of the developers to purchase machines on credit (Lynette, 1988). The development companies installed large numbers of machines and sold the energy produced to electric utilities. The activity was concentrated in California where the business climate was most favorable, and it grew from 144 machines totaling 7 MW in 1981 to the cumulative total of approximately 15,600 machines installed by 1989 with a combined rating of about 1, 400 MW (Lynette, 1989).

Federally-sponsored design and development effort was reflected in many of the wind turbines commercialized by wind turbine manufacturers in the early 1980s. Private U.S. development efforts included Jay Carter, Fayette, and U.S. Windpower. The original U.S. Windpower machines were based on a design concept developed at the University of Massachusetts, under an ERDA/DOE subcontract. Of course, continual improvements give the present U.S. Windpower machines very little similarity to the original design. Federally sponsored development was also associated with designs commercialized by Enertech, Energy Sciences Inc. (ESI), Northern Power Systems, FloWind, Windtech, Arbutus, and Dynergy.

The developers of wind power plants came mainly from the financial community although some, such as U.S. Windpower and Jay Carter, were also wind turbine manufacturers. U.S. manufacturers dominated the wind turbine industry through 1984, providing most of the wind turbines purchased by the developers. Many of these manufacturers were poorly managed and under-financed. The wind turbine products they sold were generally not ready for mass deployment. Meanwhile, federal efforts on smaller machines were affected by budget reductions just as large turbine programs were, and little development of advanced mid-scale machines was undertaken in the 1980s. This was a major omission that is just now being addressed by private industry and the federal government. However, at the time the market for wind turbines was so strong that almost any turbine produced could be sold. The result was that many U.S. machines failed or achieved only marginal performance. By 1985, developers were turning to foreign wind turbines, with the exception of U.S. Windpower, which continued to rely on its own machines.

When many of the U.S.-built turbines began to fail, the under-capitalized U.S. manufacturers did not have the resources to effect repairs required under warranties given to developers, and a number of them went bankrupt. The termination of federal tax credits at the end of 1985 compounded the industry's problems. As a result, of approximately 15 original U.S. wind-turbine manufacturers, only four are still financially viable, and as of the end of 1989, U.S. Windpower is the only one still manufacturing significant quantities of U.S.-made machines (Lynette, 1988).

The approach taken by foreign manufacturers has been to produce a more heavy-duty and somewhat less complex machine. Machines produced outside the U.S. are generally three-bladed, upwind, rigid-rotor turbines that weigh approximately 2.5 times as much per kW as U.S. machines. Although the availability (fraction of total time a machine is operating or ready to operate) of foreign wind turbines was very high (0.95+) for the first few years of operation, a number of them have begun to exhibit fatigue-related problems. This has particularly been true for wind turbines sited in areas with high turbulence, such as the Tehachapi Mountains and San Gorgonio Pass (Lynette, 1988).

### 1.5.2 Wind Power Technology Status

The commercial activity of the last decade illuminated the various advantages of the smaller machines, including lower costs per kW, design simplicity, ease of installation and repair, and rapid reliability maturation. Installed turbine ratings crept upward from about 50 kW to about 100 kW on the average, some as high as 330 kW. While some benefits may be derived from going to even higher ratings, increasing machine size does not seem essential from an economic standpoint. Leading developers have gained an understanding of the benefits of centralized monitoring, control, maintenance management and data acquisition that can be obtained if their machines are deployed in more fully integrated wind power generating stations.

As a result, the availability of individual wind turbine units in leading commercial wind power stations rose rapidly and has been steady at about 0.95 over the last few years. This high availability is expected to continue as more operating time accrues, barring any major unforeseen component wear-out or fatigue-related structural problems. However, the success of wind-turbine technology does not hinge on a precise understanding of fatigue life. Ultimately, fatigue life determines the interval between component overhauls, which in turn affects operation and maintenance costs. Thus, fatigue life is not expected to be a showstopper for use of the technology in areas with sufficiently attractive wind regimes.

Operation and maintenance costs have continued to fall and are averaging about 1.2¢/kWh for large wind power stations. Installed costs for wind power stations have fallen dramatically to attractive levels and are now about $1,000 per kW, and up, including the balance of systems (wiring, transformers, switchgear, road, installation, etc.) (Lynette, 1989).

As might be expected, performance varies with the quality of the wind resource. EPRI has monitored turbine performance at Altamont, San Gorgonio and Tehachapi (GS-6256) and found that the turbines in the windiest area, San Gorgonio, consistently averaged 26–27% annual capacity factors, while those at Altamont averaged 19–22%. Variations in the capacity factor of turbines at a given site illustrate the importance of micrositing. For example, one group of 58 turbines at San Gorgonio averaged capacity factors of 35–39% over a four year period, whereas another group of 20, located only a few miles away, averaged 14–15% (Schaefer, 1989).

In general, wind power stations using the mid-size machines look attractive as a future utility generation option. In contrast to the experience with the smaller wind turbines, the experimental multimegawatt wind turbines exhibited little improvement in reliability and poor availabilities because of their size, complexity, and demanding logistic requirements. The prevailing viewpoint is that it is more cost-effective to develop wind turbine technology from the bottom up than from the top down; i.e., to begin with the small machines and then go on to larger ones instead of trying to streamline the multimegawatt machines.

### 1.5.3 Experiences of Individual Utilities

Many utilities have been involved in wind power activities and their experiences have varied greatly. Although it is not possible here to describe the experiences of all of these utilities, it is instructive to examine the experiences of a few specific utilities. This section describes some of the unique activities and experiences of Pacific Gas & Electric Company, Southern California Edison Company, Hawaiian Electric Company, Bonneville Power Administration, Northern States Power Company, the Santa Clara California Electric Department, Green Mountain Power Company, and several other utilities.

**Figure 1.24:** Altamont Pass wind turbines. Source: U.S. Windpower.

### 1.5.3A  Pacific Gas & Electric Company

In several important respects, PG&E's experience with wind power is unsurpassed by that of any utility in the world. One reason is that located within PG&E's service territory is Altamont Pass, site of the world's largest aggregation of wind turbines and wind power plants (Figure 1.24). At the end of 1988, about 7,800 wind turbines totaling 740 MW of wind-turbine capacity had been installed at Altamont (Smith et al., 1989). The 1988 electric energy production at Altamont Pass was 997,000 MWh, 45% of the world wind energy production (See Table 1.18). Most of the output at Altamont is being purchased by PG&E. There are a few wind turbines connected to the PG&E system that are not in Altamont Pass, but they produced less than 1% of PG&E's wind energy in 1988. However, this is changing, as U.S. Windpower is presently constructing a 600-turbine, 60 MW wind power plant in Solano County (Smith et al., 1989).

Another reason PG&E's experience is noteworthy is that from September 1982 until November 1988, PG&E owned, operated and maintained a 2.5 MW MOD-2 wind turbine in Solano County, California. In addition, PG&E maintains a wind program consisting of various assessment and development projects. Major projects have included the Altamont Pass Evaluation, Solano Wind Turbine Evaluation, Wind Resource Assessment Study, and Wind Farm Array Effects Study (Steeley et al., 1986).

### 1.5.3A.1  Altamont Wind Power Performance

PG&E evaluated the 1988 performance of the Altamont wind power plants. Figure 1.25 shows the quarterly Altamont wind power plant output from 1982

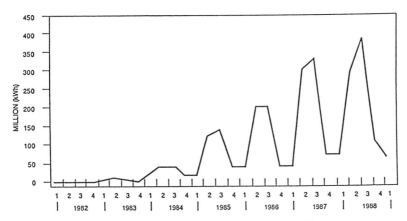

**Figure 1.25:** Quarterly Altamont Wind Power Plant output.

through the first quarter of 1989. Energy production has increased every year, and most of the output comes in the second and third quarters of each year. Part of the year-to-year increase in output is due to the increase in the installed capacity.

Figure 1.26 shows the number of new turbines installed in Altamont each year, and gives the cumulative number of turbines. The peak year for turbine installation was 1984, when the energy tax credits still existed and 2,200 new turbines were added. By end of December 1988 about 86 sq. kilometers (34 square miles) been developed with an average of one wind turbine for each 1.2 hectares (3 acres) of land. This amounts to 8.6 MW per sq. kilometer (22 MW per square mile) of developed area. Even though the energy tax credits are no longer available, construction of new wind turbines continues although at a much slower rate. It appears that some Altamont wind plant construction will continue, due almost entirely to U.S. Windpower. However, the physical size limitations of the windy area in Altamont Pass make it unlikely that the installed capacity will ever exceed 900 MW (Smith et al., 1989).

Figure 1.27 shows the quarterly capacity factors at Altamont. (The quarterly capacity factor is derived by dividing the average energy output each quarter by the average installed capacity for the quarter including the capacity of plants that were non-producing during the period.) Figure 1.27 shows a clear trend of improvement in performance over time, except from 1985 to 1986 when the average annual capacity factor remained the same at about 12%. PG&E believes, based on its wind monitoring data, that 1986 was a low wind year with about 10% less energy in the wind than the average for the previous four years. The Altamont annual capacity factor continued its trend of improvement by increasing to 15.1% in 1987 and 15.8% in 1988, even though winds appear to have been slightly less energetic in 1988 than in 1987. The improvement in capacity factor in 1988 was due to a wind plant coming back on line, and the replacement of several hundred non-operating 50 kW units with new 100 kW units by U.S. Windpower (Smith et al., 1989).

Although PG&E's monitoring of the Altamont Pass wind power plants since 1984 has found several positive trends, the Altamont wind power plants face challenges (Smith et al., 1989):

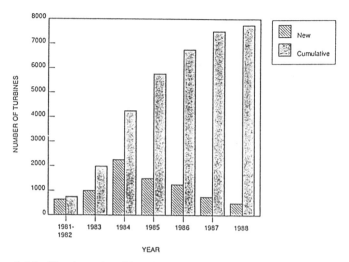

**Figure 1.26:** Number of turbines in Altamont. Source: Smith et al., 1989.

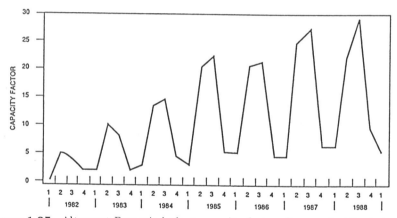

**Figure 1.27:** Altamont Pass wind plant capacity factor. Source: Smith et al., 1989.

- The European turbines, which were previously thought to be very reliable, have developed fatigue-related failure problems, especially with their fiberglass blades.
- Oil prices are low, and thus PURPA-mandated avoided costs (for new contracts) remain below those attainable even with the best available new wind plants.
- The fit of the wind to PG&E's needs, as measured by the Altamont Load Carrying Capability, was only half as good in 1988 as it was in 1987.
- The pace of development at Altamont has slowed substantially, and it appears that almost all of the very good wind areas in the Altamont area have been developed.

### 1.5.3A.2  Performance of PG&E Owned and Operated MOD-2

PG&E owned and operated the Solano MOD-2 turbine, shown in Figure 1.28. The turbine began operation in April 1982, and it produced more than 15,000 MWh

**Figure 1.28:** PG&E 2.5 MW MOD-2 installation.

of energy prior to its dismantling in November 1988. Of the five MOD-2 turbines, PG&E's produced the most energy (Smith et al., 1989).

The purpose of the Solano MOD-2 demonstration was to determine the technical and economic viability of large wind turbines in PG&E's service territory. PG&E had planned to purchase additional MOD-2 turbines and install them on PG&E's property in Solano County if the demonstration had proven the machine to be reliable and cost-effective. Unfortunately, the MOD-2 was not sufficiently dependable and trouble-free to meet PG&E's criteria. Major problems were encountered, such as cracks in the low-speed shaft discussed earlier in this section, and difficulties in

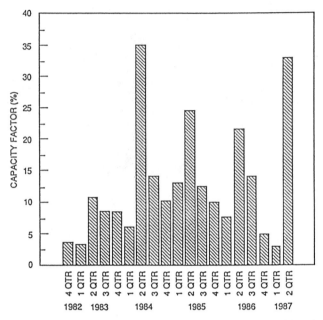

**Figure 1.29:** Quarterly capacity factors for the PG&E MOD-2 turbine Source: Ilyin et al., 1987.

lubricating the bearing supporting the adjustable-pitch blade tips. These problems reduced the turbine's reliability and availability and also increased operation and maintenance costs significantly. The cost of operation and maintenance averaged about 5 cents per kWh for the last three years of operation. Since the cost of operating the turbine exceeded the value of its electrical energy output, it was dismantled and removed from the site December, 1988 (Smith et al., 1989).

Had there been a number of co-located turbines rather than only one, operation and maintenance could certainly have been carried out more efficiently and at lower cost. However, the problems encountered with the single unit uncovered sufficient technical risks that investment in a number of additional units would have been imprudent.

While operating, the MOD-2 achieved annual capacity factors ranging from 7.7 percent in 1983 to 17.0 percent in 1987 (Weinberg et al., 1988). This is well below the design-level projections of 34%. The overall capacity factor was 13.3% between January 1983 and September 1986. PG&E estimated that if a 90% turbine availability had been maintained, the projected annual capacity factor of 34% would have been attained (Steeley et al., 1986). In the second quarter of 1987, the MOD-2 had a capacity factor of 33% as shown in Figure 1.29. (Overall, the Altamont Pass wind power plants had a capacity factor of 24% during the same period.) In July of 1987 the turbine produced over 1,000 MWh, representing a 56% capacity factor and a record for the unit (Ilyin et al., 1987).

According to PG&E, the Solano MOD-2 wind turbine made it possible for PG&E and Boeing engineers and technicians to modify, test and improve alternative large-scale turbine design features. The basic layout, a two-bladed upwind teetering rotor on "soft" tower, has proven to be sound. Also, the use of a flexible "quill" shaft to take the torque of the rotor to the gear box (the low speed shaft takes the weight of the rotor) proved to be successful. This information has contributed to the improvement of wind generating technology as a whole, making wind power a more cost-effective power generation alternative (Smith et al., 1989).

### 1.5.3A.3 Coincidence of Wind Plant Output with System Demand

Another aspect of the Altamont wind power plants' performance studied by PG&E was the coincidence of their output to PG&E's needs. Would the wind power plants deliver energy on hot summer afternoons when PG&E's loads are the greatest, allowing PG&E to avoid construction of new conventional power plants? This is a critical question for a utility in that it determines how much capacity credit a utility can assign to a generation source. In order to assign a substantial capacity credit, utilities need a high level of confidence, on the order of 95%, that the amount of capacity accredited to a resource will be available at the time of the utility's peak load period.

PG&E examined the effect of the wind power plants' output on the PG&E loss-of-load probability (LOLP) during the month of June, 1986. The LOLP is the statistical probability that, due to high load levels and/or the failure of sources of supply, the utility will not be able to meet its customers' power requirements. The PG&E system maximum hourly load during that month occurred between 3:00 to 4:00 p.m. on the 26th day, and averaged 15,151 MW. During that hour, the Altamont wind power plants were producing 82.7 MW, or 0.55 percent of the rated capacity (Steeley et al., 1986).

Further analysis of the June 1986 Altamont wind power plant performance showed that the maximum output of 285 MW, which occurred at 10:00 p.m. on the 6th day of the month, represented 2.8 percent of PG&E's load during that hour. On the first day of the month at 3:00 a.m., the wind power plant output of 259 MW was 3.9 percent of PG&E's load while during 91 other hours the wind power plants produced over 3 percent of the load. During only 7 hours did the wind power plants produce no output at all. Overall, during the month the wind power plants produced 1.4 percent of PG&E's energy requirements (Steeley et al., 1986).

PG&E performed a similar analysis for every hour in 1987 and 1988 for both Altamont Pass and the MOD-2 wind turbine located in Solano County, another major wind resource area in PG&E's service territory. The wind at the MOD-2 site was found to have a much better fit to PG&E's hourly loads. Wind speeds at the MOD-2 site were high enough to produce the rated wind turbine output of 2.5 MW during the peak load hours in 1987 and in 1988. Compared to the average 1987 output of the MOD-2 of 413 kW, the output during PG&E's peak 1987 load hours was 5.8 times as much. The wind at Altamont Pass did not fit PG&E's needs nearly as well in either year, especially in 1988. At the time of the 1987 PG&E system peak load, the hour ending at 4:00 p.m. on July 15, the Altamont wind plants were producing 243 MW, which was 2.4 times their average 1987 output. However, on the day of the 1988 PG&E system peak load, the Altamont wind plants were producing only 88 MW, which was less than the 117 MW average 1988 output. The 88 MW output during the 1988 PG&E system peak load was only 11% of the total

**Table 1.19:** Load Carrying Capability (LCC) Of Wind For PG&E (LCC Divided by Maximum Output)

|                     | 1985   | 1986   | 1987 | 1988 |
|---------------------|--------|--------|------|------|
| Actual Solano MOD-2 | --     | --     | 0.74 | --   |
| Potential Solano    | --     | --     | 0.80 | 0.83 |
| Actual Altamont     | 0.30*  | 0.30*  | 0.40 | 0.20 |

\* Approximated from output on peak load day only.

Altamont rated capacity at that time of 708 MW and 18% of the maximum 1988 Altamont output (486 MW) (Smith et al., 1989).

To better understand the potential capacity value of the Altamont wind power plants and the MOD-2, PG&E calculated the load carrying capability (LCC) of the wind turbines in these two locations. The LCC or "capacity credit" was computed for the MOD-2 in 1987 and found to be 1.84 MW, or an impressively high 74% of its rating. If the MOD-2 had been available all year, its LCC would have been 1.99 MW in 1987, or 80% of its rating. Similar computations were made for the winds at the MOD-2 site in 1988, and the potential MOD-2 LCC was an even higher 83% of the rated capacity. According to PG&E, these values are nearly as good as those for conventional peaking power plants. However, since the MOD-2 was not operated during the summer of 1988, it had no actual LCC that year (Smith et al., 1989).

The LCC was also computed for the Altamont wind plants and found to be 154 MW in 1987 and 98 MW for 1988. The 1987 LCC was about 40% of the plants' peak output during that year. The 1988 LCC of 98 MW was 14% of the Altamont rated capacity of 708 MW and 20% of the maximum 1988 Altamont output of 486 MW. The LCC averaged 30% of the maximum output for these two years. A cursory examination of the Altamont output on the peak load days for 1985 and 1986 shows that during those years the LCC was also about 30% of the maximum output for each year (Smith et al., 1989). The results of PG&E's LCC calculations are summarized in Table 1.19.

These findings led PG&E to two significant conclusions. One is that the fit of the wind to PG&E loads was excellent in the windy areas of Solano County during 1987 and 1988. This shows that wind turbines in these areas could have a significant capacity value to PG&E. Another important finding is the great difference in capacity value between the Altamont and Solano County areas (Smith et al., 1989).

Similar computations were made for each of the individual Altamont wind plants for 1988. The ratio of LCC to average 1988 output for the plants varied from 0.54

to 2.3, with an overall value of 0.9. Thus, there is a clear difference in LCC based on location within a resource area. Wind plants in the southwestern part of the Pass produced energy with more value to PG&E (i.e., with a higher ratio of LCC to average output), presumably because the winds come earlier in the afternoon there. Wind plants using turbines with low cut-in wind speeds, such as the Danish two-speed turbines, may also produce energy at times of greater value to PG&E, but this trend was found to be much less strong than the trend based on location. These results indicate the extremely site specific nature of the fit of wind to utility loads (Smith et al., 1989).

### 1.5.3A.4 The Wind Resource in PG&E's Service Territory

PG&E has also assessed the wind energy resource potential in its service territory. This potential was estimated based on the *California Wind Atlas* issued by the California Department of Water Resources for the California Energy Commission. The *California Wind Atlas* indicated a total resource of about 13,500 MW in PG&E's service area. After PG&E eliminated heavily populated areas, Air Force bases, and national and state forests and parks, 6,200 MW of potentially developable resource remained. The best areas for initial development were selected from the 6,200 MW of potential resource areas using the following criteria (Ilyin et al., 1987):

- Annual wind speed: greater than or equal to 17.7 kmph (11 mph)
- Land: sparsely or moderately populated, or Bureau of Land Management property
- Transmission access: transmission line less than 1.6 km (one mile) away
- Terrain: flat or rolling (not steep or mountainous)
- Surface roughness: smooth or moderate (few trees or buildings)
- Towns and roads nearby for wind power plant support

The areas which met the above criteria are in Solano, Napa and Contra Costa counties, and in Altamont Pass. These areas total about 1,340 MW of which 500 MW had already been developed at Altamont Pass by the end of 1986. (There is another 200 MW developable in Altamont Pass that does not meet all the above criteria.) Therefore, PG&E concluded that about 840 MW of easily-developable wind resource potential remained in the PG&E service territory at the end of 1986 (Ilyin et al., 1987).

PG&E's study concluded that future wind power development will depend on the cost of wind energy. The levelized cost of energy was calculated in constant 1986 dollars for the 6,200 MW resource potential in PG&E's service territory using a wind power plant capital cost of $1,100/kW. PG&E estimated that this wind power potential might be developed for the cost ranges shown in Table 1.20 (Ilyin et al., 1987).

According to PG&E, a more detailed study would be needed to identify the best sites if PG&E were to decide to build or buy wind power plants. The wind resource for the best sites would have to be measured thoroughly over a period of at least a year, and the actual wind power plant site-specific costs estimated in detail. Also, the value of the energy would need to be determined based on the diurnal and seasonal variations in the wind and the degree to which these variations match PG&E's needs (Ilyin et al., 1987).

**Table 1.20:** Wind Power Potential vs. Development Cost in PG&E Service Territory

| Portion of 6,200 MW Potential | Levelized Cost of Energy 1986$ (Cents/kWh) | Approximate Amount (MW) | Cumulative Amount (MW) |
|:---:|:---:|:---:|:---:|
| 18 % | 5 - 7 | 1,100 | 1,100 |
| 10 % | 7 - 9 | 600 | 1,700 |
| 24 % | 9 - 11 | 1,500 | 3,200 |
| 43 % | 11 - 14 | 2,700 | 5,900 |
| 5 % | > 14ᐧ | 300 | 6,200 |

Source: Ilyin et al, 1987.

### 1.5.3A.5  Wind Turbine Array Effects Project

The spacing between turbines in wind turbine arrays significantly affects output. The array effects were studied by PG&E at an operational wind power plant that has its turbines spaced two rotor diameters apart along the rows, with 7 rotor diameter spacing between rows. The rows are perpendicular to the prevailing wind. The site chosen for the study is fairly flat and small, so any reduction in wind turbine output in the downwind rows was thought to be due mainly to array effects (Steeley et al., 1986).

Over 750 hours of tests were conducted and computer analyses performed to predict wake effects for the same operational wind power plant. Results indicated significant energy losses due to array effects in both the computer model predictions and the field data for the 2 by 7 spacing. The results indicated that a greater array spacing at this site would be required to achieve less variability in wind turbine output (Steeley et al., 1986).

### 1.5.3B  Southern California Edison

Next to PG&E, SCE purchases more energy from wind power projects than any other utility in the world. There are more than 7,900 wind turbines in SCE's service territory at Tehachapi and San Gorgonio Pass with a total wind turbine capacity of about 670 MW. SCE purchased over one billion kWh from these projects in 1989. In 1980, SCE established a Wind Energy Test Center in the San Gorgonio Pass near Palm Springs California and tested a number of machines. The test center was decommissioned in 1989, and no further wind turbine testing is being conducted by SCE.

One of the experiments conducted while the test center was in operation involved a two-year evaluation of a 500 kW vertical-axis wind turbine, reported in EPRI

AP-5044. On the basis of experience gained in operating a 50 kW vertical-axis wind turbine earlier in the 1980s, SCE became interested in operating a 500 kW vertical-axis turbine manufactured by the same company – DAF Indal, Ltd., of Canada. In 1983 EPRI and SCE arranged a two-year joint evaluation of the larger machine, shown in Figure 1.30, and on-line operation at the Wind Energy Test Center began in August of the same year. The purpose of the project was to develop a better understanding of vertical-axis wind turbine technology, to develop methods for testing and evaluating such wind turbines, and to collect field test data. EPRI report AP-4258 documents the first year results of the project.

The test program covered a number of areas: energy capture, wake characterization, operation and maintenance characteristics and problems, structural loads, electrical behavior, environmental impact, and site meteorology. The project team used two data acquisition systems to capture information for the test, one developed by SCE and the other system loaned from Sandia National Laboratories.

The turbine produced its rated power in winds up to and above 46 mph, and its projected annual energy production was within 3 to 4% of the manufacturer's estimates. Tests indicated that the rotor wake could extend from 8 to 16 rotor diameters. Three design modifications improvised during the project corrected problems with the guying cables, the capacitor switching circuits, and the hydraulic/lubrication pump. In addition to these repairs, the project team experimented with the rotor brake system.

The overall appraisal by SCE, as reported in EPRI AP-5044, is that relative to horizontal axis machines, the generic advantages of vertical-axis wind turbine designs lie in their simplicity, their ability to accept winds from any direction without the need for a yaw mechanism, and the location of their power train and generator at ground level for ease of maintenance. Their disadvantages are associated with a pulsating power output, their inability to drive synchronous generators, relatively lower hub heights (where winds are lower) and the complex rotor and guy cable dynamics. Their lack of aerodynamics controls places greater reliance on mechanical braking for safe shutdowns.

The DAF Indal 500 kW wind turbine was found to be a generally reliable machine with good performance. The turbine noise levels were moderate with no detrimental characteristics. The wind turbine had no detectable effect on TV reception and did not appreciably contribute to bird mortality. Public reaction to the wind turbine was generally favorable. SCE also concluded that in order to be commercially ready and competitive, the turbine needed design improvements and simplifications to guarantee an adequate rotor life and to decrease manufacturing costs.

Overall, SCE concluded that vertical-axis wind turbine designs have not received the same amount of research and development work as horizontal-axis designs. They have not been widely commercialized and their place in wind energy remains to be firmly established.

### 1.5.3C Hawaiian Electric Company

Hawaiian Electric Company (HECO) has also had extensive experience with wind power. HECO's involvement with alternative energy began in the late 1970s at about the time of the oil crisis of 1979. With no domestic fossil fuel resources,

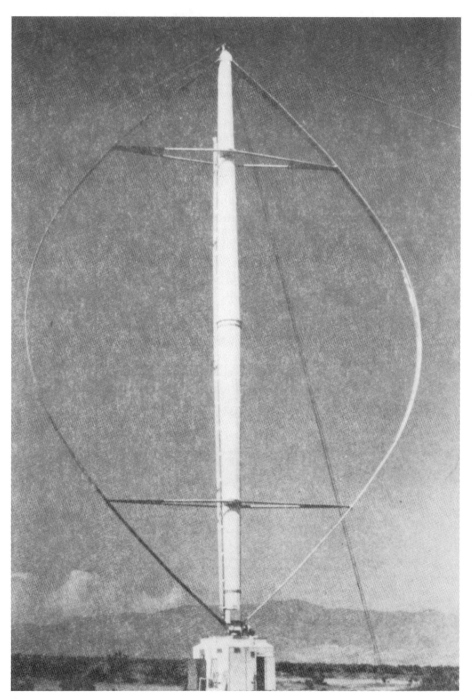

**Figure 1.30:** DAF Indal 500 kW vertical-axis wind turbine Installation at San Gorgonio Pass. Source: Southern California Edison Company.

renewables were Hawaii's only alternative to oil. The activities of HECO in the area of wind have included:

- Operation of a MOD-OA wind turbine
- Assessments of the wind resource in Hawaii
- Assessment of the feasibility of constructing wind power stations
- Purchase and operation of fifteen 600 kW Westinghouse wind turbines
- Purchase and operation of a MOD-5B Wind Turbine
- Establishment of a subsidiary, Hawaiian Electric Renewable Systems, to develop, construct and operate renewable energy systems.

### 1.5.3C.1 MOD-OA Experiment Results

DOE's experimental MOD-OA wind turbine program, discussed earlier, was the first major domestic implementation of wind turbines in a utility application. The four 200 kW wind turbines were experimental in that their purpose was to provide information regarding the operating characteristics, performance, and maintenance requirements of a wind power generating station. To this end, the MOD-OA wind turbines were located at four different utilities of widely differing size, geographic location, electric system configuration, operation and maintenance staffing, and wind resources.

The last of the MOD-OA wind turbines was hosted by HECO and was installed at Kahuku Point on the Island of Oahu. It commenced operation in early June 1980. The Kahuku wind turbine had a high energy production compared to the other three sites. For example, more energy was generated by the Kahuku MOD-OA than by the Clayton, New Mexico MOD-OA, which was in operation more than twice as long. The primary reason for the high energy production at Kahuku was the very good wind regime at the site. The wind speed at Kahuku was above the wind turbine cut-in speed more than 90 percent of the time. Whereas the capacity factors for the other three MOD-OA's fell in the 0.20 to 0.24 range, the Hawaii MOD-OA had a capacity factor of 0.51 – over twice that of the other machines. This is attributable mainly to four factors (AP-2796):

- High wind availability in the region;
- Excellent performance, reliability, and durability of the wood composite blades in contrast to metal or fiberglass blades used on the other machines;
- Improvements in the system control functions; and
- Higher reliability of the drive train due to the use of an in-line alternator.

After the MOD-OA wind turbines had completed their assigned tests, they were disassembled. NASA technicians inspected the Clayton machine at teardown, and EPRI joined in a detailed inspection of the major Kahuku machine components during disassembly in October 1983. The major conclusions from the operating experience with the Kahuku MOD-OA and from the Kahuku disassembly/inspection were reported in EPRI AP-3813. Energy output and average power output were found to be highly dependent on the wind regime at a particular wind turbine site, as evidenced by the operating results for the Kahuku site. Operation of the MOD-OA wind turbines caused little adverse electrical interaction on the utility systems. It should be pointed out, however, that the MOD-OA generators were of the synchronous type with the capability to rapidly control generator excitation.

High penetration of wind turbines with induction generators might result in more noticeable utility interaction than that observed with the MOD-OA wind turbines.

As would be expected with any prototype machine, the later MOD-OA wind turbines tended to be more reliable than the earlier units as the design matured. The number of discrepancy reports generally decreased from the first installation at Clayton to the last installation at Kahuku. From a structural standpoint, the laminated wood-epoxy blades appeared to be more successful than the fabricated aluminum blades. Also, the aluminum blades were difficult and costly to repair in the field compared to the wood-epoxy blades. The wood-epoxy blades were considered by NASA to be satisfactory for long-term operation.

The MOD-OA was designed for operation during a short experimental period. Protection of the wind turbine components from the environment was not sufficient for some of the sites. However, there is a lesson to be learned from the insufficient protection provided. Very significant corrosion damage was noted during inspection of the Kahuku machine, particularly on the yaw support cone. Corrosion was also evident on the machines at the Puerto Rico site. While some of the deterioration occurred after cessation of operation of the machines, the corrosion problem would have had a severe impact on the life of many of the components, including structural members. With continued operation, this deterioration would have resulted in higher maintenance costs and reduce availability. It was concluded that future wind turbine designs must place more emphasis on material selection and protection, particularly in coastal or other adverse environmental areas.

Also, as an experimental machine, maintenance access was not a priority for the MOD-OA designers. Many components of the wind turbine were quite difficult to repair or replace in the event of failure. The severe wear noted on some major components at Kahuku and Clayton would have required eventual replacement on a long-term commercial wind turbine. In some cases this replacement would have required major disassembly of the wind turbine resulting in costly repairs and reduced availability. Future wind turbine designs should include adequate provisions for repair/replacement of major components that cannot reasonably be expected to endure 20–30 years of normal operation.

The HECO experience with the Kahuku MOD-OA also indicated some difficulty with routine maintenance of the equipment since the wind turbine was located remotely from the base of the assigned service personnel. This resulted in longer downtimes than necessary for minor problems. It was concluded that future wind turbines will achieve higher availability if located closer to existing maintenance centers. Also, future wind power generating stations should contain sufficient numbers of wind turbines to justify an on-site maintenance staff.

It was concluded that additional operation of the Kahuku MOD-OA would have been hampered by a number of factors. These factors include the corrosion of many components, the difficulty in performing major maintenance, and the high wear rates of the low-speed shaft coupling and turntable bearing.

### 1.5.3C.2 Resource and Feasibility Assessments

By the early 1980s, HECO, working with the University of Hawaii, had completed an assessment of the possibility of constructing a wind power station consisting of a cluster of 25–75 machines each rated at 2 MW or larger (nominally the MOD-2 machine). A conceptual design was developed for an 80 MW wind power station, and costs for siting at Kahuku or on the island of Molokai were compared. The analysis showed that it would be less costly to site the station at Kahuku, even though Molokai has a large wind resource, due to the cost of the dc cable system, including inverters, which would have been required to interconnect a station on Molokai with HECO's Oahu Island system.

HECO also performed wind resource assessments which resulted in the identification of 15 sites at Kahuku Point in Oahu which exhibited potential for wind power station development based upon available wind data. HECO performed detailed evaluations of these sites, including assessments of turbulence, to define the different wind regimes. Some of the sites were eliminated from further consideration based upon these detailed studies plus consideration of such factors as communication interference.

### 1.5.3C.3 Wind Power Station Development Program

Over the past decade HECO has worked with manufacturers and independent power producers on the development of wind power stations. In 1983, HECO established Hawaiian Electric Industries which, in turn, founded Hawaiian Electric Renewable Systems (HERS) in 1985 to pursue such activities. In 1985, HERS constructed a 9 MW wind power plant consisting of fifteen 600 kW Westinghouse wind turbines at Kahuku Point (Figure 1.31). In 1988, HERS acquired the $55 million, 3.2 MW MOD-5B, also located at Kahuku Point (Figure 1.32). The resulting 12.2 MW rated wind plant delivers power directly into the HECO grid at 46 kV on Oahu's North Shore. In 1989, HERS formed a wholly-owned subsidiary, Lalamilo Ventures, Inc., which operates two wind power plants purchased in 1987. In all, HERS operates three separate wind power plants totaling about 16.4 MW on the islands of Oahu and Hawaii.

The 600 kW Westinghouse units at Kahuku Point are upwind, two-bladed, teetered-hub, variable-pitch, constant-speed wind turbines with synchronous generators. A well conceived control system enables the unattended operation of the units and allows flexible real and reactive power control settings. A modem in the central monitoring computer allows remote operation of the units using telephone lines for communications (it even worked from Pittsburgh). The remote operation feature includes capabilities for power control, failure analysis, startup and shutdown. According to HERS, the control and monitoring systems work well, but the overall wind turbine reliability (80 percent availability) has been below expectations. This is attributed to the immaturity of the wind turbine design.

The 3.2 MW MOD-5B is now the world's largest operating wind turbine. (The Canadian Eole, designed for a 4 MW capacity, will only be operated up to 3 MW.) The MOD-5B has a 320-ft blade diameter and is an upwind, two-bladed, teetered-hub, variable-pitch and variable-speed wind turbine. Boeing Aerospace Company completed its test program in late 1987, and HERS began the MOD-5B's commercial operation in January, 1988. The variable speed generating system controls generator speed, power output and reactive power output to deliver smoothly-changing power to the HECO grid. The variable speed system has significantly improved the production capability of the MOD-5B while reducing drive train loads and power fluctuations under gusty wind conditions. Salt air caused some degradation of the

**Figure 1.31:** Hawaiian Electric Renewable Systems 600 kW Westinghouse wind turbines at Kahuku Point.

unit's generator and the air reaching the generator is now being filtered. The generator has been repaired. As of January, 1990, the MOD-5B had accumulated 10,860 operating hours and produced 13,700 MWh of energy.

### 1.5.3C.4  Utility Interface Operating Experience

On the Island of Oahu, HECO's main 138 kV transmission system supplies power to Honolulu's urban areas and a lower voltage 46 kV transmission system is used to supply the rural areas along the perimeter of the island. The HERS Kahuku wind plant delivers power into the tail-end of a segment of this 46 kV transmission system, 20 miles from the nearest 138 kV substation. This segment has a 15 MW on-peak load and 5 MW off-peak load, with the majority of this load consisting of residential users plus a large resort and golf course development.

On January 6, 1989, under an EPRI funded effort, HECO and HERS engineers experimented with and monitored the wind plant's energy and reactive power output to assess its effect on the utility system. HECO recorded the wind plant power, reactive power, voltage harmonic content, system frequency, system voltage, system tap changer activity and other information during this test. Gusty wind conditions of 24 to 72 kmph (15–45 mph) coincident with the off-peak hours were selected for the test to record the impact of wind turbine fluctuations under worst case conditions. Power fluctuations at the Kahuku wind plant were a major concern for HECO, but the test results indicated the concerns to be unfounded. For example, during the system minimum load, a 77 kW/sec load swing in a three-second interval was recorded for a single Westinghouse unit. Due to the averaging effect of twelve

**Figure 1.32:** 3.2 MW MOD-5B at Kahuku Point, owned and operated by Hawaiian Electric Renewable Systems.

other units delivering power into a common substation at the same time, the utility system absorbed only a 217 kW/sec load swing during this three-second period, out of an average facility output of 8,360 kW. The HECO system load at the time was 511 MW. HECO's report to EPRI (1989) concluded:

- Megawatt load swings created by wind gusts were normally within the control capability of the utility's automatic load dispatch system, due to the wind turbine array averaging effect.
- The averaging effect of the wind plant turbines was found to be similar for reactive power (vars) except that due to the var control programming of the wind turbines, var production generally is inversely proportional to megawatt output. This relationship is actually beneficial in terms of helping to maintain a constant utility interface voltage.

- Variations in the utility's 46 kV interface voltage caused by gusty winds were well within the utility specifications for all time spans tested.
- The harmonic content of the wind plant's output, which was measured to range from 0.5% to 1.1% is considered not at significant variance from the baseline values at 0.8%.

During a large portion on the test period, the wind plant was providing power for the entire 20-mile long segment of Oahu's north shore without causing problems. The 46 kV substation closest to the 138 kV system registered negative power, signifying power being supplied by the wind plant rather than the utility. The engineers also varied the reactive power recorded at this substation from positive to negative and were able to move the 46 kV system voltage in doing so. Even in the worst case scenario, the wind plant had no significant detrimental effect on the utility grid.

HECO concluded that the HERS 12.2 MW Kahuku wind plant continues to deliver power to the 5 to 15 MW load on Oahu's north shore without detriment to the utility's power quality. Hawaii's experience demonstrates the ability to provide utility grade power to lower voltage grids from wind plants.

### 1.5.3D  Bonneville Power Administration

The Bonneville Power Administration (BPA) is a federal power marketing agency with extensive interest and some experience in wind power. Key activities of BPA have included various wind resource and siting assessments; a field test and demonstration of three 2.5 MW MOD-2 wind turbines; the performance of a wind-plant feasibility study; and development of advanced power electronics for wind power applications. These activities, their key results, lessons learned, and conclusions are summarized below:

### 1.5.3D.1  Wind Resource and Siting Assessments

The wind resource and siting assessment activities of BPA have included several wind research projects involving the Oregon State University (OSU) Department of Atmospheric Sciences and a regional wind energy assessment program. Data was obtained at BPA and OSU wind recording stations, from existing records and from other meteorological stations. The results of this work were used to assess the potential for wind energy within a portion of the BPA service area and to evaluate the usability of wind energy as a supplemental resource. The results of this effort were also used to improve data acquisition techniques, select data gathering equipment, and provide early indications of general siting areas for possible future installation of research or commercial wind turbine generators.

The regional wind energy assessment program (BPA, 1985) was a five-year project which enabled BPA to systematically assess all of its 300,000 square mile service area for wind resource potential. Benefits derived included the establishment of a reliable data base for assessment purposes, identification of additional locations in the BPA service area which could support wind power plants or clusters, detailed analysis of the diversity of the wind resources, improved understanding of the impacts of wind and other intermittent generation on the operation of the BPA hydro-electric system, and improved capabilities to predict winds on a short-term basis for generation scheduling and dispatch purposes.

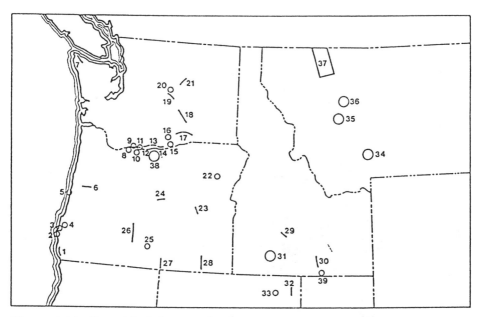

**Figure 1.33:** Potential wind energy development areas in Bonneville Power Administrations service territory. Circled locations indicate general areas with significant wind and development potential. Those locations noted by lines indicate ridgelines. The sizes of the circles, rectangles, and lines indicate relative extent (not absolute magnitude). Source: Oregon State University.

The results of BPA's wind resource assessments (Figure 1.33) indicate a potential for thousands of MW of wind power generation in the BPA service territory, which includes Washington, Oregon, Idaho, Western Montana and parts of Wyoming, Utah, Nevada and Northern California. The principal sites where wind power could be developed include several locations along the Oregon coast, several in the Columbia Gorge, one in Nevada, and several large areas in Montana. The potential in Montana is characterized by BPA as "equivalent to many central stations" but would probably require construction of long distance direct-current (dc) transmission lines to move energy produced to load centers.

### 1.5.3D.2 MOD-2 Demonstration at Goodnoe Hills

BPA's site at Goodnoe Hills, located about 15 miles from Goldendale, Washington, was chosen by DOE for the installation of three prototype (experimental) MOD-2 large wind turbines, pictured in Figure 1.34. The three unit configuration was established primarily to support field research of large scale wind turbines, with research programs sponsored by DOE, NASA, SERI, Pacific Northwest Laboratories, Boeing, EPRI, BPA and others. The first unit was operational in December 1980 and the second and third in the Spring of 1981. The installation included one 60 m (198-ft) and one 107 m (350-ft) meteorologic tower, both instrumented at three levels. As discussed earlier, shortly after commencing operation the three turbines were temporarily shut down due to an overspeed incident in Unit 1 in June 1981. Units 2 and 3 returned to service in September 1981 and Unit 1 early in 1982. All

**Figure 1.34:** MOD-2 wind turbines at Goodnoe Hills.

three units were shut down again in November 1982 due to cracks in the low-speed shafts. According to BPA, debugging of design problems was completed in 1984 and beginning in 1985 all three machines were fully functional. At the conclusion of the test program in June 1986, the machines were shut down. Beginning in the Fall of 1987, the turbines were dismantled and removed from the site.

BPA considered the MOD-2 demonstration as an opportunity to get in on the ground floor in wind power, to have hands-on involvement in the operation and maintenance of "utility-size" machines. It also viewed the demonstration as an opportunity to obtain technical and cost information, and to perform research that BPA considered desirable from a utility perspective. BPA wanted to examine aspects of the project concerned with electrical integration, power quality and system performance. By the time the MOD-2 project was terminated in 1986, BPA and the other research test participants had obtained most of the information they wanted from the project. It should be noted that BPA observed no adverse impacts to utility customers in power quality, and voltage levels only varied 1–2% at synchronization.

The project is viewed as having provided several important benefits. One important benefit, according to BPA, is that the physical size of the machines, coupled with the stresses experienced during operation, helped to surface problems more quickly than with smaller machines. The MOD-2 problems focused attention on those aspects of the wind turbine drive train and airfoils that require special consideration in design. Because of the timing of the MOD-2 experience and the expansive California wind power development, lessons learned could be applied during the development-and-installation stage of the smaller units at Altamont Pass and elsewhere before large numbers of these smaller units were deployed. Many of the computer codes that are now used to design smaller wind turbines benefited from the DOE large wind-turbine program. With the computer design codes, it is possible to locate the high-stress points and identify desirable changes in the design. Prior to the DOE large wind turbine program, such mathematical design tools were either not available or of limited use because of the absence of operational verification.

While the MOD-2 helped to identify the generic weaknesses of early wind machines, it also pointed out to BPA the relatively greater cost of operation and maintenance for large machines compared to small machines. The effort required to repair a large machine, if it does have a failure, is substantial. Most of the working parts needing repair are more than one hundred feet in the air, and thus it involves a considerably greater effort to effect maintenance than for a small machine, which can be lowered relatively easily to the ground or repaired in place. More energy may be lost when a large machine is out of service because more time is needed to accomplish major repairs. Thus, while the MOD-2 helped to identify early generic problems shared with small turbines, it also identified the relatively higher cost of maintenance of large wind turbines.

One of the research objectives of the MOD-2 installation involved the positioning of the three wind turbines at Goodnoe Hills. The machine placement was configured specifically to demonstrate wind turbine interaction during operation and the effects on a down-wind machine from the operation of an up-wind machine. The spatial separations were about 5, 7, and 10 diameters, i.e., 460, 640 and 900 m (1,500, 2,100, and 3,000-ft) – the sides of the triangle defined by the three turbines. A strong influence was expected at 5 diameters, and to some degree, at 7 diameters. No significant interaction at 10 diameters was anticipated. However, BPA's data and analyses showed that even at a 10 diameter separation, a 10–12% power reduction was observed in the down-wind machine. This is manifested by a reduction in energy of the incoming wind and increase in the amount of turbulence. In the very early morning hours, when the wind is steady, the wake can extend much farther downstream because there is little turbulence in the atmosphere to cause the wake to decay. The turbulence may also influence machine life, however, no life cycle analysis was performed to investigate that issue.

Extensive evaluations of field performance for the DOE/BPA MOD-2 machines and other megawatt-scale wind turbines, including loads analysis and operating experience, were supported by EPRI and are summarized in a series of reports issued in the mid 1980s (EPRI AP-4054; AP-4060, 3 Vols.; AP-4239; AP-4288; AP-4319; and AP-4335).

### 1.5.3D.3 Cape Blanco Wind Farm Feasibility Study

The Cape Blanco Wind Farm Feasibility Study (BPA, 1987) investigated the engineering, economic, environmental and meteorological aspects of a proposed 80 MW wind energy project on a 650 hectare (1,600-acre) site southeast of Cape

Blanco, Oregon. The study evaluated the potential impacts of large-size (2.5 MW) MOD-2 and intermediate-size (170 kW) vertical-axis FloWind turbines. When BPA initiated this study in 1983, power shortages were anticipated in the Northwest. Generic and site-specific information on wind energy projects was needed to augment regional planning efforts. While a subsequent forecast showed a power surplus, BPA decided to finish the nearly completed study because it offered valuable generic information on wind power plants. The study was co-funded by BPA, Pacific Power and Light Company, Portland General Electric Company, and Windfarms Northwest and was completed in 1986.

The study found no unusual engineering problems that would preclude development of a wind power plant at the Cape Blanco site. The proposal called for standard engineering and construction practices, with particular attention to coastal environmental conditions and soil characteristics at the site. It was estimated that normal coastal bluff erosion would eliminate 8 to 22 percent of the usable sites over the 30-year lifetime of the project.

Costs of wind energy from such a project were estimated to range from 4.5 to 8.7 cents per kWh (1985 dollars). A regional power surplus, constraints on the ability to market and transmit power outside the region, and the Northwest utilities' low avoided costs led BPA to conclude that the energy from the proposed wind power plant would not be competitive with the avoided fuel cost of a coal-fired power plant, estimated at 1.8 to 3.0 cents per kWh (1985 dollars).

From an environmental perspective, the study found that less than 2 to 4 percent of the site would be used for all of the wind power plant facilities. The disturbed land could be returned to pasture at the end of the project. Less than 0.1 percent (one in a thousand) birds flying low enough to strike turbines would be injured or killed, according to projections. This was considered biologically insignificant. Careful observation, however, was recommended during phased-in construction of turbines. Compliance with state noise regulations would reduce usable MOD-2 wind turbine sites from 31 to 11 (50 dBA limit) or 2 (43 dBA limit), and FloWind sites from 433 to 428 (50 dBA limit) or 418 (43 dBA limit). Eliminating 11 of the FloWind turbine sites would mitigate visual impacts of the FloWind, particularly to the culturally significant Cape Blanco Lighthouse and Cape Blanco Beach areas. Eliminating 8 of the 31 MOD-2 sites would still leave a high-to-moderate visual impact. Hazards to aircraft from the 90-foot FloWind turbines were considered negligible; however, the Oregon State Aeronautics Department and the Federal Aviation Administration would need to evaluate specific impacts of each 350-foot MOD-2 machine. Wind turbine vibrations were expected to be absorbed by the soil and not affect fish populations in the nearby Elk or Sixes Rivers. Television-signal interference could be mitigated by installing a cable system. In the case of the MOD-2 alternative, satellite-dish interference near the site boundary could be reduced through a cable hookup.

The study concluded that a wind power plant at the Cape Blanco site, using commercially available turbines with hub heights of at least 100 feet, is possible given engineering, environmental, and meteorological considerations, but that the regional utility avoided costs, turbine installed cost, and demand for power would need to change significantly before development can be considered.

**Figure 1.35:** SCE Test Site at Tehachapi with adaptive power factor controller in test trailer.

### 1.5.3D.4 Power Electronic Applications to Wind Power

BPA has also been supporting research involving induction motors and generators in conjunction with the University of Washington (U of W) Electrical Engineering Department. Out of this work the U of W developed a concept for an adaptive power factor controller to help deal with the widely fluctuating reactive power requirements of induction generators. The result was the development by the U of W of an "adaptive power factor controller" (EPRI AP-5210).

The controller automatically senses the amount of reactive power being consumed by the generator, selects the proper number of compensating capacitors, and connects them to the system through solid-state switches. It can monitor each phase independently and connect capacitors as needed to provide continuous and adaptive compensation. Southern California Edison, which was experiencing reactive power problems with the wind power plants at Tehachapi, tested the device on a 50 kW vertical axis wind turbine at its Wind Energy Test Center. The controller worked so well that SCE ordered a 300-kVAR unit to test at a wind plant at Tehachapi (Figure 1.35). The device was successfully operated for two years and now is at the U of W. It was refurbished and installed at the U of W hospital to compensate for the rapidly fluctuating reactive power requirements of the building's elevators.

According to BPA, the controller is applicable to needs beyond wind power generators. It can be applied to any induction generator, particularly a prime mover that has a variable source, such as a variable stream flow. It also applies to other variable power factor or motor loads, such as would be found in a sawmill. Hence, it has many applications, although first developed for an induction wind turbine. Although the device is not yet commercial, BPA and SCE plan to look for a firm to commercialize the device.

Another device being studied by BPA, in conjunction with the Oregon State University Electrical Engineering Department, is the series-resonant converter. It would allow operation of a wind turbine at variable speeds while supplying a 60 Hz output to the utility system. A 15 kW series resonant converter is currently under construction and will be tested in a variable speed pumped-hydroelectric storage demonstration on the Island of Hawaii – using surplus energy from a small wind power plant.

### 1.5.3D.5 Lessons Learned

The following are among the lessons learned by BPA as a result of its wind power activities:

- Substantial wind resources exist and, under appropriate circumstances, it is feasible to develop them.
- Television interference is a significant concern in siting wind turbines, but one that can be mitigated with good antenna systems.
- Climate effects must be considered. Wind turbines designed for inland use can corrode and deteriorate quickly in the harsh salt-air atmosphere of a coastal site. Units sited in coastal regions must be designed with provisions for salt corrosion protection.

### 1.5.3E Northern States Power Company

Northern States Power (NSP) has also been involved with wind power development in a variety of ways. NSP initiated its activities in wind because of the energy crisis and concluded that there was a need to examine alternatives to conventional energy generation. There were also external pressures from customers and regulators.

Activities reported for NSP in the 1983 EPRI survey include a detailed regional wind resource assessment study involving 15 monitoring towers located in five states, a grant to the University of Minnesota for research on small-scale wind power feasibility, a field test and demonstration of a 40 to 50 kW wind turbine at the University of North Dakota, monitoring of a small 10 kW wind turbine at the University of Minnesota, and providing low-cost wind-monitoring hardware and consultation on siting to the Minnesota Department of Energy. NSP's most substantial wind effort to date has been the Holland Wind Farm, a project which NSP undertook subsequent to the 1983 EPRI survey and one which grew out of an agreement involving the certification for construction of NSP's Sherburne County 3 coal-fired unit.

### 1.5.3E.1 Wind Resource Assessment Program

In the early stages of its wind assessment efforts, NSP decided that its money would be better spent on evaluating the wind resource in its service area rather than on installing and testing wind machines.

NSP initiated its Wind Resource Assessment Program in 1981. As part of this effort, NSP performed a siting study, established siting criteria, and looked for sites with potential for wind generation. In order to keep the effort unbiased, criteria were established. One criterion was to have one tower in each service division. Consequently, monitoring towers were located in all parts of the service territory, which includes portions of Minnesota, Wisconsin, Upper Michigan, North and South Dakota. The area covers a diversity of topography – from the forests of Wisconsin to rolling prairie to flat plains. Each monitoring site included measurements at 3 elevations 10, 20 and 30 m (33, 66 and 100-ft) on a 30 m (100-ft) tower, wind speed at all elevations, direction at top elevation, and temperature at the lower elevation. Data from this study indicated the Holland site in southwest Minnesota with an average wind speed of 26.1 kmph (16.2 mph) had the highest wind speed of any site monitored.

The Holland site is approximately 32 km (20 miles) north of the City of Pipestone in southwestern Minnesota. The site was named after the nearby town of Holland, Minnesota. The site has an elevation of 590 m (1,930 feet) above mean sea level and is located at the southeast end of the northwest-southeast running Buffalo ridge. The average drop in elevation 64 km (40 miles) from the site at the 8 standard compass points is approximately 150 m (500 feet). Turbines at this location take advantage of a topography that is somewhat unusual in Minnesota (Halet, 1988).

### 1.5.3E.2 Holland Wind Power Plant

After determining the Holland site to be preferred location for the wind power plant, NSP requested bids and selected three 65 kW Bonus wind turbines, which were installed at the Holland site in 1986 (Figure 1.36). They have been operating since December 1, 1986. The Holland Wind Power Plant was operated as an unattended generation facility after May 22, 1987, when a remote monitoring system became functional.

NSP's first annual report on the Holland wind plant describes the installation and summarizes the first year's operating experience. The gross generation of the wind power plant for the period from December 1, 1986 through November 30, 1987 was 324, 480 kWh. This was 131, 274 kWh (28.8%) lower than the projected gross generation of 455, 754 kWh. A computer model estimated that 41.1% of the generation loss could be attributed to lower than expected wind speeds and 58.6% to lower than expected turbine availablity. The average turbine availability for the period was 83.4%. The average capacity factor for the period was 19.2%.

In the three months of March, April and May, the wind power plant generated 39% of the total annual output. In the three months of July, August and September, NSP's peak load period, the wind power plant generated only 14.5% of the total annual output. March was the highest generation month with 16% of total generation. August was the lowest generation month with only 3.9% of total generation.

Most of the operation and maintenance outages that occurred during 1987 were associated with the microprocessor-based turbine controller and its inputs. Problems were also experienced with the braking systems, yaw transmissions, and wind speed sensors. Icing of the wind speed sensors can prevent the turbines from starting when the wind blows. Some outages were extended by maintenance delays. NSP initially had a maintenance agreement with a company which was to stock spares and come to the site to repair the machines when necessary. Unfortunately, the firm went out of business by the time the project commenced operation. This left NSP without spare parts or personnel familiar with repair of the machines. NSP undertook a

**Figure 1.36:** Northern States Power's Holland Wind Power Plant.

crash program to acquire spares and assigned maintenance crews from other NSP plants to become familiar with the system.

Personnel from the Minnesota Valley Plant in Granite Falls, Minnesota, about a 1-hour drive from the Holland Plant, were assigned in December 1986 to maintain the wind machines. Instrumentation support is provided from a plant in the southern part of Minneapolis, about a 3.5-hour drive from the wind power plant. An on-site computer interrogates the wind-turbine controllers and the wind monitoring tower every 5 minutes and contacts an alarm printer at the Minnesota Valley Plant whenever there is a status change. The computer can also be interrogated remotely, and there is a real-time display for each machine to show output and alarm conditions.

During the second year of operation (December 1, 1987 through December 31, 1988), the gross generation of the Holland Plant was 270,253 kWh. The average capacity factor was 14.4%. The average turbine availability was 55.8% compared to 83.4% in 1987. The largest contributors to unavailability during 1988 were wind speed sensor failures, phase-current-imbalance relay trips, yaw transmission failures, and brake release failures. Even so, NSP estimates that turbine availability would have exceeded 93.8% if the turbines at Holland had been part of a large commercial wind power plant with on-site maintenance staffing.

A number of enhancements made in 1988 were expected to improve availability in 1989. The wind speed sensors were moved to a lower vibration area of the turbine and have operated for at least five months with no signs of failure. In addition, control system modifications designed by NSP to reduce the number of false phase-current-imbalance relay trips were implemented. Also, turbine controller settings were modified to reduce the number of yaw operations, and the brake release sensor bolt threads were taped to prevent the bolt from moving out of alignment. These improvements, made in December 1988, have eliminated most of the problems that occurred during the first two years of operation. However, during 1989 NSP replaced a generator on one of the machines and was reporting a blade root problem.

Major maintenance at Holland is conducted annually and is scheduled before winter begins. After a winter storm, the site becomes inaccessible until it can be plowed out. Winter is the windiest season and is also the period when maintenance is most difficult. The wind speed sensors frequently ice up. The ice also tends to coat the blades and the support structure, making maintenance difficult. Another problem has been the need for maintenance personnel to climb exposed ladders in sub-zero temperatures to perform maintenance on the wind machines.

The turbine manufacturer, Bonus, a Danish firm operating out of Long Beach, California, has been very cooperative according to NSP. Bonus has been working to improve the turbine documentation and has developed an improved maintenance manual. Bonus had a representative on-site for startup and during the first major maintenance. Bonus also helped train NSP crews to maintain the equipment.

NSP sponsored a study by the South Dakota State University (SDSU) to determine the effects of the wind turbines on the 23 kV rural distribution line serving the Holland site and other area customers. The study investigated harmonics, voltage and current fluctuations, and power quality. No significant adverse impacts were detected. However, it was observed that more than 50% of the line current variations at the substation in Pipestone could be attributed to changes in the wind farm output. The SDSU report suggests that NSP could begin to see spurious trips of overload and directional power flow relays if the wind power plant generation approached 20% of the line load.

The Holland Plant is to be operated for at least 3 years, according to NSP's agreement with the Minnesota PUC. After that, NSP will review the project and decide whether to continue operation of the plant.

### 1.5.3E.3 Small Wind Turbine Assessment

From 1981–1985, NSP operated a 10 kW wind turbine in St. Paul. The University of Minnesota worked with NSP to evaluate the machine. The wind resource was poor, and the machine had a capacity factor of only about 9%. The average wind speed at the site is only about 8.7 mph. Despite the minimal operating hours due to the low wind speeds, many failures were experienced. Four different inverters were used, and governor and alternator problems were experienced. NSP could not keep the machine running for more than 6 months without a major problem. Eventually it was dismantled and sold.

### 1.5.3F Santa Clara Electric Department

The City of Santa Clara, California Electric Department's approach to wind power development has been radically different from that of other utilities.

Santa Clara's interest in wind power dates back to about 1981. The City wanted to develop renewable resources to supplement its conventional supply sources. As part of its efforts to achieve this objective, plans were prepared for development of regional wind power resources for the City's own use.  The City's Electric Department is a relatively small utility (1987 peak demand of 367.4 MW), and wind power resource increments on the order of 20 MW seemed appropriate for its needs.

An early step was to characterize the wind resource on 4,000 hectares (10,000 acres) acquired by the City in the Sierra Nevada Mountains. In 1981, some wind monitoring towers were erected on that property. That year, the City also purchased 837 hectares (2,068 acres) at Benecia in Solano County, a site close to PG&E's MOD-2 machine, specifically for wind power development. In 1983, Santa Clara also purchased 237 hectares (586 acres) in Altamont Pass mainly as a future waste disposal site. Both of these properties were shown as high wind resource areas in a report published by the California Energy Commission (1980).

By 1981, Santa Clara was involved in EPRI's renewable energy technologies activities and became particularly interested in the small wind machines. Santa Clara also had an interest in photovoltaics and followed developments in photovoltaics technology.

Wind power technology, however, had evolved to the point that small and medium-sized machines were available. This created an opportunity for the City to proceed. Perhaps the leading consideration in the selection of this renewable technology was that it was available and could be used immediately.

The approach followed by Santa Clara was to purchase the property and to seek developers. Although wind machines had advanced significantly, they still represented too great a risk for the City to invest the large amounts of capital required for the development of a wind power station. To avoid taking this risk, the City decided to enter into lease-purchase agreements with third-party wind developers on its properties. Under this arrangement, a third-party developer leases the site from Santa Clara and provides both financial and project management. The developer hires subcontractors to design, build and operate the project and receives the profits from the project. The developer pays Santa Clara an annual lease fee for the use of the site and provides Santa Clara an option to purchase the project facilities and equipment after a fixed period of time. Santa Clara shares in the project profits via the lease fee (which can be a percentage of the project revenue), but has no direct control over project expenditures nor any direct financial liability for the project.

### 1.5.3F.1  Altamont Wind Power Plant

The 237 hectares (586 acres) of land purchased by Santa Clara in Alameda County featured three distinct ridges suitable for wind turbine siting and two canyons usable for landfill development. On March 6, 1985, the City entered into a Ground Lease and Third Party Development Agreement with Wind Developers, Inc., of Sacramento (Wind Developers), for the construction, operation and maintenance of Danish Wind Technology "Windane 31" horizontal-axis wind turbines on the property. Three units with a capacity of 340 kW per unit or 1.02 MW, total, were installed and subsequently subleased to Atkinson Mechanical Contracting Company of Sacramento, California (Atkinson). All three units were placed in commercial operation on July 31, 1985. The turbine-generators in these units are housed atop 100-ft stepped, tubular steel towers.

Table 1.21: Santa Clara's Altamont Wind Power Plant Results

| Lease Year | Calendar Year | Unit Availability | MWh Produced | Income To City |
|------------|---------------|-------------------|--------------|----------------|
| 0 | 1985 | 0 % | - 0.0 - | $15,200 |
| 1 | 1986 | 89 % | 26,538 | $119,124 |
| 2 | 1987 | 98.5 % | 36,285 | $150,069 |
| 3 | 1988 | 98.5 % | 36,539 | $150,449 |

On October 29, 1985, the City approved the assignment of the Wind Developers, Inc., lease to Zond Systems, Inc. (Zond) of Palm Springs, California. The project currently incorporates 200 new "Vestas V17" units installed by Zond and the three existing Windane 31 units installed by Wind Developers and subleased earlier to Atkinson. Zond currently has all responsibilities for project operation and management of the sublease to Atkinson. The Altamont project is designed to produce 21.02 MW of power under ideal conditions. The output is sold to PG&E. The agreement with Zond provides the City with options to buy the wind power plant for its own use starting in 1998.

The Vestas V17 units are 100 kW, horizontal-axis wind turbines manufactured by Vestas North America, Ltd. of Tehachapi, California. The turbine-generators are housed atop four-leg steel lattice towers 45-ft high and, alternatively, 80-ft high.

Other terms and conditions of the agreement with Zond include the following:

- In return for the right to use the property for 20 years, Santa Clara receives a minimum rent.
- Santa Clara receives a portion of the gross revenues (a royalty) from the electricity sales to PG&E.
- Santa Clara receives technical data on the wind resource and wind turbine performance.

Zond came into the project late in 1985 before the tax incentives for wind power development were to expire. Zond had only 2–3 months to install 200 machines, but managed to accomplish the work within that time.

The estimated annual generation at the Altamont site is 44,000 MWh/year. The City's capital investment includes $824,000 in 1983 for property and $285,783 in 1984 to PG&E for transmission line upgrading. Project results are shown in Table 1.21.

### 1.5.3F.2 Benicia Ranch Wind Power Development

The City is also pursuing third-party wind power station development on its Benicia property in Solano County at sites considered prime wind resource land owing to strong air mass movement over the property during all but a few days each year. The proposed development features 215 vertical-axis wind turbines, projected to produce 63,240 MWh per year at a 90% turbine availability. Additional sites are under review and, if proven feasible, the number of sites may be increased to 273 for an estimated energy generation capacity of 72,360 MWh per year at 90% turbine availability. The City applied the same approach to the development of this property as to that at Altamont Pass. The City has signed an agreement with FloWind Corporation of Pleasanton, California to develop the property for wind power. Terms of the development include an option plan for City buyout of the project and payment of rents/royalties to the City during years of FloWind operation.

The proposed turbines are 19 meter "F19" vertical-axis wind turbines. The turbines have a generator capacity of 250 kW per unit or 53.75 MW for 215 units. Ultimately the site was expected to produce about 50–60 MW. About 15–20 MW were planned for the first phase, and FloWind was preparing to implement that plan.

At that juncture, difficulties were encountered in obtaining approval from Solano County for construction of the project. In addition to PG&E's MOD-2, there were another 6 or so wind machines in the area at that time. These were small, stand-alone machines, not part of any large developments. The Solano County Board of Supervisors enacted a zoning ordinance restricting the installation of wind turbines. This ordinance prohibits wind turbines from being installed if they are visible from a nearby freeway or housing development. This situation arose at the last moment with the FloWind ready to proceed and attempting to obtain a permit for construction. The City's site clearly fell within the scope of the restrictions, and no permit was issued. Although there are several ridge lines on the property, Santa Clara considers them unusable for wind machines unless the zoning is changed. As it turns out, this has not posed a great problem for Santa Clara since the value of the property has been appreciating, except that Santa Clara believes that it could generate a substantial amount of energy at this site. (Although Santa Clara's efforts to develop a wind power station in Solano County were not successful, U.S. Windpower is constructing a 60 MW wind power station in the County, as previously noted.)

### 1.5.3F.3 Sierra Nevada Site

Santa Clara also has 10,000 acres near Loyalton, California, about 20 miles north of Reno, Nevada. The land is at high elevation, a base level of 5,000-ft or more, with peaks rising to 10,000-ft. Santa Clara initially purchased the property for its geothermal energy potential. Because of the visible evidence of strong winds, some wind speed monitors were installed by the City. The average wind speeds were found to be excellent, approximately in the 18–22 mph range. Further research indicated, however, that the wind velocity is extremely high when it blows, and low at other times. The average thus resulted from a combination of too much wind for wind machines part of the time and insufficient wind at other times. Because no wind turbine technology designed for such conditions was available, Santa Clara has put aside any plans for development in that area.

Another concern was the high elevation and the potential for ice loading and other problems associated with a high-altitude climate. The site is still owned by

Santa Clara, and the City plans to revisit the site for future wind power development as the wind turbine technology evolves.

### 1.5.3F.4 Lessons Learned

Three key lessons about wind power development have been learned by Santa Clara:

- There is a need to be sensitive to the locale for a project. A good wind regime (resource) is not enough.
- It is necessary to know the wind resource at a site very completely. The City spent three years mapping the Loyalton resource in the Sierras before concluding that it was impractical to install wind machines there.
- There is a continuing need for technology development to evolve wind machines suitable for a variety of climates. Machines that work well in some wind regimes may not work well in others.

### 1.5.3F.5 Future Directions

The interest in renewable energy technology at the Santa Clara Electric Department is more internally developed rather than a response to public or political interest or pressure. The Electric Department has pursued renewable resource options it considers to make economic sense. The City's policy is that the preferred resources are renewable – hydro, wind, etc. The City also owns fossil-fueled combustion turbines and cogeneration projects. Thus, there is a policy level guidance to support this work. The policy is shaped by both the Department and the City Council. The Council can authorize the Department to undertake any type of resource development project.

One goal of the Department is to develop the Benicia site and the developers for Benicia continue to express interest in proceeding with the installation. Together, the Alamont and proposed Benicia site development would provide about 75 MW of community wind resource according to the City's estimates. Santa Clara currently has a peak demand of about 400 MW. In ten years this might grow to 750 MW, and the 75 MW of wind power would be sufficient for the City's purposes. Also, Santa Clara is a summer peaking utility, with peaks at the times of the best winds from Altamont and Benicia.

### 1.5.3G Green Mountain Power

Green Mountain Power (GMP) is a private utility located in Burlington, Vermont. GMP's involvement with wind energy dates back to 1978 when the utility installed several wind monitoring stations in the Champlain Valley at locations that appeared to have potential for wind power development. In 1979–1980, GMP responded to a DOE wind energy program opportunity notice, and its proposed site was selected as one of 35 in the U.S. at which meterological towers were to be placed for wind resource measurements. DOE's plan was to collect two years worth of data and select several sites for placement of large wind turbines. Initial public reaction to GMP's participation in this program was favorable and encouraging, but before it was over GMP encountered substantial opposition to its proposed wind monitoring site, as discussed below.

Despite the adverse public reaction to GMP's proposed participation in the DOE wind monitoring program, GMP continued its wind power efforts on a lower-key and less visible basis. The company was one of three utilities to participate in an EPRI assessment of wind power systems completed in March 1983 (AP-2882). And, in the early 1980s GMP instituted a comprehensive wind turbine site identification and evaluation program. This program was directed toward identifying windy sites which could potentially be developed in terms of environmental acceptability and proximity to existing roads and transmission facilities. As a result of this program, development rights were secured at the most attractive sites and wind measurement equipment was installed at about ten of these sites.

By the mid 1980s the company had accumulated substantial data and knowledge about the quality of the wind resource and practical development constraints at sites in its service territory. It had entertained development proposals from Hamilton Standard, Boeing, General Electric, and other wind turbine manufacturers of the era. However, several factors encouraged GMP to scale down its wind development efforts: load forecasts indicated new generation sources were not needed in the near-term; wind turbine technology was not yet sufficiently mature; and the technology was still too expensive to compete economically with GMP's other supply options.

GMP's interest in wind power rebounded in the late 1980s as wind turbine prices declined and the reliability of the hardware improved significantly. Also, the company could see the need for additional sources of power by the mid 1990s. Given its knowledge of the resource and site availabity, it appeared that wind could become economically competitive with other sources by that time. The major remaining uncertainty was whether or not the technology, which was proven in California, could be made to work in the New England mountaintop environment. To address this question in the most expeditous and practical way, GMP purchased and installed two U.S. Windpower turbines at its Mount Equinox installation. Unlike the DOE monitoring program experience, public reaction to the Mount Equinox project, GMP's latest wind power effort, has been favorable and supportive.

The following paragraphs discuss GMP's experience with the DOE wind monitoring project, the results of the EPRI assessment of distributed wind power systems, and GMP's Equinox project.

### 1.5.3G.1 Effort to Participate in DOE Wind Monitoring Program

The site on which Green Mountain Power proposed to place a meteorological (met) tower is in the Green Mountains of Vermont, on a mountain known as Lincoln Ridge, located in the small rural town of Lincoln. The site is very close to a ski resort and is alongside the famous Long Trail, which winds along this mountaintop. The proximity to the ski resort was seen as a desirable feature of the site. There was already adequate access to the site, and the met tower would not have been visually intrusive because the site was already developed. However, serious opposition surfaced to the placement of a met tower on Lincoln Ridge. The opposition came from people in the town of Lincoln who feared that subsequent placement of a MOD-2 or other large wind system near the site would have adverse visual impacts. An active and effective opposition group, known as the "Save Lincoln Mountain Committee" emerged to protest the proposed development. Little concern was expressed by other groups or communities that might be affected by the decision (Lotker, 1981).

When the opposition to the met tower first surfaced, GMP held a public meeting to explain the proposal. The public reaction at the May 1980 meeting was sufficiently adamant that the Forest Supervisor for the Green Mountain National Forest requested DOE to perform an environmental assessment and impact statement. DOE agreed and prepared an environmental assessment for the Lincoln Ridge site. In January 1981, DOE announced that it had determined that the proposed met tower at Lincoln Ridge would not constitute a major federal action significantly affecting the quality of the human environment. However, due to decreased funding, DOE subsequently made a programmatic decision not to install a met tower at Lincoln Ridge (Lotker, 1981).

### 1.5.3G.2 Assessment of Distributed Wind Power Systems

GMP participated in an EPRI-funded assessment performed by General Electric Company of distributed wind power systems (AP-2882). The project was a two-year effort to develop methods for evaluating the performance, economics, utility system impacts, and penetration limits for distributed wind power generation relative to central-station wind power generation and other possible future forms of distributed power generation. The distributed wind turbines studied range in size from 1.5 MW, which might be employed in a utility substation application, down to 2 kW, which might be employed in a small application such as a single-family residence. Hypothetical future costs for wind turbines were projected down to levels where meaningful studies would be conducted. Both utility and nonutility ownership were considered for several utility case studies. In addition to Green Mountain Power, the utilities studied include Pacific Gas and Electric Company and Southwestern Public Service Company.

The studies concluded that the potential value of distributed wind power systems stems primarily from fuel savings rather than from deferral of the expansion of conventional generating capacity. In other words, little or no capacity credit can be ascribed to this generation form. Also, the influence of transmission and distribution system expansion requirements on distributed wind power system economics relative to central-station economics is case-specific and can be either favorable or unfavorable; however, this influence is generally small relative to that stemming from generation planning considerations. In any case, for safety and system integrity, widespread deployment of distributed generation forms would require costly modifications of existing transmission and distribution networks and changes in the design criteria for future networks. Furthermore, the impracticality of operating and maintaining distributed wind power generation, with only a few machines at individual locations, is likely to introduce significant additional cost penalties relative to central-station wind power generation.

The study also found that the economics of wind power systems owned by utility customers are highly dependent on product marketing and distribution costs; as individual wind power systems tend to smaller ratings, these costs increase rapidly. Small wind turbines in residential applications do not appear to have the potential for making a significant contribution to overall national energy requirements. In summary, centralized deployment of wind power systems appears to have significant advantages relative to distributed deployment.

### 1.5.3G.3 Mount Equinox Project

In its most recent wind power development effort, GMP acquired a site on Mount Equinox in Manchester, Vermont, and in November 1989, the company installed two new 100-kilowatt turbines, which began generating electricity by year-end. If these two new turbines perform well, GMP plans to install two more turbines by the end of 1992.

The company expects wind energy technology will prove to be a cost-effective and environmentally attractive method of producing electricity in Vermont. Toward that goal, the company has compiled one of the most comprehensive wind energy data bases in New England. By installing and operating the new wind turbines on Mount Equinox, GMP will understand better the feasibility of generating electricity with wind in Vermont.

### 1.5.3H  Experiences of Other Utilities

### 1.5.3H.1  Carolina Power & Light

CP&L jointly evaluated wind resources in North Carolina with the North Carolina Alternative Energy Corporation (AEC) to assess the potential feasibility of wind power. CP&L wanted a base of actual data and supporting analyses to be able to respond to inquiries made about wind energy. CP&L had the perception that the wind resource in North Carolina is insufficient for energy production in competition with other sources, but CP&L wanted a more definitive basis for its planning with regard to the wind energy in its service territory.

The assessment was conducted during the mid-1980's in two phases. Phase I concerned the characterization and quantification of the wind energy resource on the coast of North Carolina. CP&L then simulated use of medium-sized machines in wind regimes found along the coast. Studies concluded that significant improvements in wind turbines plus significant reductions in their costs would be required before wind power could be considered as an alternative to conventional power generation along the North Carolina coast. Further, the high value of other uses for coastal land (recreational) plus effects on aesthetics constituted additional barriers to coastal-zone wind power. The sites evaluated were at shore line or just off shore.

Even at the most favorable sites (located at Cape Hatteras), the capacity factors of machines would be low, only 13–14% on an annual basis. To increase capacity factors, machines with different characteristics than those commercially available would be needed. Average wind speeds off the coast were lower than those for which the machines were designed. Important considerations were seasonal, and especially diurnal, fluctuations of the wind as well as weather changes.

The second phase of the effort addressed sites in western North Carolina. CP&L found some sites with significant wind resources but again determined a need for substantial improvements in performance and reductions in cost for wind to be competitive with conventional (coal-fired) generation.

### 1.5.3H.2  Detroit Edison

Detroit Edison jointly funded a study of wind turbine performance in Michigan which was completed by Michigan State University in November 1984 (Michigan State University, 1984). Wind data was collected at 20 sites in lower Michigan from 1982 to 1984. Customer-owned, utility-interconnected wind turbines were located at 16 of those sites. Wind speed and ac output were recorded at all of the sites. At a few sites, power quality parameters such as reactive demand, output fluctuations, and harmonics were measured. Based on data collected, the study found that winds at most lower-Michigan sites are inadequate to produce energy at competitive costs unless wind turbine installed costs are significantly reduced and wind turbine reliability increased. Other findings include (Parks et al., 1984):

- There is a high correlation between monthly wind speed averages at many sites.
- Wind speed and direction are highly variable – especially at altitudes under 30 m (100 feet).
- Induction-generator wind turbines operated at average power factors near 0.5.
- Excessive time-to-repair significantly reduced annual energy output.
- Even during very windy weeks, there are long periods (typically one day) where the power output of widely-dispersed wind turbines summed to less than 5% of capacity. Thus, wind turbines should not receive "capacity credit" in lower Michigan.

### 1.5.3H.3 Northeast Utilities Service Company

Northeast Utilities Service Company (NUSCo) conducted a survey of wind resources in its service area, collecting about two years' of data. NUSCo worked with the University of Massachusetts, which surveyed topography, vegetation, and other characteristics. Most locations with good wind resources were found to be remote. NUSCo selected two sites, one each in Massachusetts and Connecticut, that were expected to have the best wind and that were also sufficiently accessible to make a survey. NUSCo collected 15-minute data at all locations – average wind speed plus direction. NUSCo also flew kites to evaluate wind shear at many locations.

At one location in Massachusetts, NUSCo did a significant amount of work on wind shear and developed a good understanding of the change in wind shear with elevation. However, at the Massachusetts location with the best average wind speed, NUSCo was unable to obtain wind-shear data because the presence of trees on the ridge prevented the launching of kites. The average annual wind speed was about 16 mph at 60 feet above ground on this north-south ridge, but NUSCo wanted wind shear data to evaluate this site for possible wind turbine placement. Obtaining this data would have required clearing the site. At the other location, where the average wind speed was found to be 13.5 mph, open fields enabled measurement of wind shear. NUSCo found that wind speed increased more rapidly with elevation than expected. NUSCo collected detailed data from various locations and found the wind resources to be very site specific. A shift of 200 yards can significantly alter the resource availability. In Connecticut, winds were found to be much lower, averaging less than 10 mph at the sites surveyed.

A developer attempted to install a wind power plant at the site for which the detailed data were available. Although accessible by road, the existing power line's capacity is limited. Hence, it would have been necessary to rebuild the power line to service the proposed wind power plant. This added cost made the project economically unfeasible. Also, there were objections to the concept. It appears that zoning approval would have been resisted. NUSCo concluded that in wooded areas, clearing the land for wind power development is likely to be opposed.

### 1.5.3H.4  Long Island Lighting Company

LILCO's initial interest in wind began in the 1970's due to concerns about fuel cost and availability. Although only 120 miles long, Long Island has up to 1,500 miles of coastline (the result of numerous indentations), and a good wind resource relative to New York State and adjacent areas, especially offshore. Because of this potential resource, LILCO wanted to stay informed about developments in wind technology. Wind resource assessments were conducted and in some locations the average wind speeds matched the requirements for existing wind turbine technology.

LILCO was one of the original utilities in the DOE wind resource monitoring program. LILCO also submitted a proposal in 1979 to DOE to be a host for the demonstration of MOD-2 wind machines at Montauk, a site which LILCO had been monitoring. LILCO was the only east coast utility to reach the final selection round (one of seven). LILCO had the cooperation of much of the community – park officials, town government, etc. There was also some opposition. After losing the MOD-2 award, LILCO received an instrumentation package from DOE. The package was installed at a site closer to the ocean and windier than Montauk. Data were recorded for several years. However, no large wind turbines have been installed on Long Island. LILCO is aware of the problems experienced with large wind turbines. Also, the decline of oil prices and elimination of tax incentives have led LILCO to conclude that the large wind turbine technology is uneconomical at this time. LILCO has a small wind turbine, less than 1 kW, operating at a fuel oil unloading platform offshore from Northport Power Station (perhaps the largest oil-fired station in the U.S. – a 1,500 MW baseload plant). The unloading facility is two miles offshore and has no electrical connection to shore for lights, fog horn, etc. (ships provide their own pumping). Formerly, the platform was battery powered with a diesel generator to charge batteries. To improve reliability, this was supplemented about five years ago with a small wind turbine. This system has worked very well, according to LILCO, even during Hurricane Gloria, which produced winds near 160 kmph (100 mph) across Long Island in 1985.

In LILCO's service territory there are eight customer-owned wind turbines. The largest one is about 10 kW. These are grid-connected with two-way meters. LILCO purchases the surplus energy from these units at the New York state legal minimum of 6 cents/kWh, which includes a subsidy because it exceeds LILCO's avoided cost. Even with this incentive, only 60 kW of customer-owned capacity is installed.

LILCO also cooperated with Grumman Aircraft in the installation of a 15 kW unit in the town of Islip. However, Grumman abandoned the project after two years of monitoring. Grumman had expected a market for wind power to develop in remote areas, such as Alaska.

### 1.5.3H.5  Public Service Electric & Gas

Wind speeds in New Jersey are quite low. As part of its solar heating and cooling experiments, PSE&G installed four meteorological towers and has been collecting weather data since 1977. The average wind speed in New Jersey was found to be less than 5 mph on an average annual basis for typical locations. Wind speeds increase during the winter and decrease during summer. Since PSE&G is a summer-peaking utility, the winds are out of phase with its load.

In cooperation with Westinghouse, PSE&G performed a study in 1975–76 to assess off-shore wind energy. The study examined conceptual designs for machines that would be positioned in the ocean. The designs were for multi-megawatt machines. PSE&G had some offshore meteorological towers that had been installed as part of a plan for a floating nuclear generating station which did not materialize. Reasonably steady winds on the order of 15 mph, annual average, were found offshore. The wind speeds decreased to the 11–13 mph range at the shoreline proper around Atlantic City. Five to ten miles inland, however the average drops to 5 mph or less.

During the early 1980s, PSE&G began modeling the performance of various customer-owned turbines in its service area due to requests from customers and some regulatory pressure to evaluate wind energy as an alternative energy technology. A small computer program was developed to model wind turbine performance with a given wind pattern. Using wind data from various stations, the performance of various wind turbines was predicted at different locations. PSE&G found that the wind in New Jersey could not be used to generate much energy. The capacity factors were very low precluding cost-effective use. PSE&G also examined the northwest corner of the state where the terrain includes some mountains. From temporary instrumentation, it was determined that the average annual wind speeds are 10–11 mph at the peaks of the ridges, considered to be insufficient to justify installation of wind turbines. It was felt that at least 15 mph would be needed to justify an installation.

PSE&G then turned again to the coastline. From 1983–86, the company worked with the Stevens Institute of Technology. The Institute was challenged to conceptually design an offshore 100 MW photovoltaic power plant, floating or on an artificial island, and also a 100 MW wind power station. During the next three years, conceptual designs were developed and evaluated. The results confirmed PSE&G's earlier conclusion that off-shore wind power would be very expensive. On shore, there is insufficient wind to make wind power feasible. In New Jersey, about one square mile of land would be required for 1 MW of wind turbine capacity. Since PSE&G has over 10,000 MW of capacity, enormous amounts of land would be required for the resource to be significant. The environmental effects would be adverse in such a small, well populated state.

### 1.5.3H.6 Virginia Power

An extensive alternative energy study performed by Virginia Power (VP) between 1981 and 1984 concluded that there are potentially good wind sites in the VP service territory and surrounding area but that wind power was not economically feasible at the time. However, VP decided to proceed with some site evaluation studies to better characterize the wind resource. The studies concluded that the best potential wind turbine sites are in the Appalachian Mountains with one of the best sites being at VP's Mount Storm Power Station in West Virginia. Mount Storm Power Station is a coal-fired, minemouth station located outside VP's service area but operated by VP. Coastal sites were also examined but were found to be inferior to the mountain sites in terms of wind resource and very expensive to develop in terms of land cost.

The data used for the site evaluations were obtained over a period of years from ambient air monitoring observations at power stations. Data for the coastal sites were also obtained from federal wind resource assessments. Also, some wind modeling was performed to better define the potential at specific locations.

The variability of the storm driven wind resource at Mount Storm suggested the possibility of using a variable-speed wind turbine. Therefore, a joint program was developed with U.S. Windpower (USW), which was developing an interest in variable-speed wind turbine operation, to further invesigate the potential of the wind resource, design a test wind power station, and determine the cost/benefit of a commercial-size wind power station at Mount Storm. The North Carolina Alternative Energy Corporation also provided funds to the project.

The major purpose of the study was to evaluate the wind energy resource at Mount Storm and to determine the value of wind as an energy source to the Company. Concurrent with this study, USW began developing an advanced, variable speed, wind turbine as part of a joint program with EPRI (discussed later in this section). This new wind turbine, labeled the USW Model 33–300, is a 300 kW machine with a nominal rotor diameter of 33 meters (106-ft). This machine is now undergoing pre-prototype testing in constant speed operation in the Altamont Pass area of California and was slated to be tested at Mount Storm.

Four monitoring stations were installed in October, 1987 along a mountain ridge approximately two miles west of the power station. An ice storm in November 1987 destroyed two of the towers, which were replaced by December of the same year. Three additional towers were installed in March 1988 within a 500-ft radius of the most promising site to determine the variations in wind speed along that section of the ridge.

Energy projections were made using the available wind data and the proposed USW 33-300 power curve for Mount Storm. Assuming 95% availability and no system losses, the estimated energy output of each wind turbine for the Mount Storm site is 910 MWh/yr, calculated based on an expected mean wind speed of 27 kmph (16.9 mph). This estimated output is close to the average for very windy sites in Altamont Pass and sites in San Gorgonio Pass in California. A total of 63 possible turbine sites were identified on VP's Mount Storm property and another 400 to 500 sites are possible along various ridges near the power station.

Based on the data collected, five turbine test sites were selected. A site map showing the five sites, the transmission line, and other terrain features was created. A preliminary one-line diagram showing the features of the electrical interconnection and an estimate of the installation costs were prepared. Combining the cost estimate of the hardware with the estimated energy production, the estimated cost of energy was determined. A break even third-party investor analysis, assuming a present value discount rate of 12.5% and an annual inflation rate of 6%, yielded a life-cycle cost of energy of 4.00 cents per kWh. An economic analysis using the revenue requirement method described in the EPRI Technical Assessment Guide (EPRI, 1986), resulted in a constant dollar cost of energy of 3.98 cents per kWh (Virginia Power, 1989).

The wind turbine program was terminated by VP in mid-1989 due to budget restrictions which allowed funding for only low-risk projects.

### 1.5.3H.7 Wisconsin Power & Light Company

In 1978–79, several of WP&L's customers purchased 1.5 kW wind machines with the aim of becoming more independent from the utility. The prospect of independence from the utility was part of the wind turbine vendors' marketing strategy. In response to customer inquiries and its own concerns about problems which might be created by the installation of many small wind turbines on its system, WP&L decided to conduct its own experiments by installing and monitoring wind machines on customers' premises.

In 1980, WP&L installed its first wind machine at a dairy farm. It was part of a federal program to build ten 20 kW, 10 m (33-ft) machines. Two weeks later the machine fell over, an event reported in the Wall Street Journal and by radio commentator Paul Harvey. Subsequently, WP&L installed and tested three more 20 kW machines and four 10 kW machines. In all, WP&L purchased and tested machines from six manufacturers (DeWinkel, 1983).

The sites for the wind machine installations were selected to be distributed throughout the WP&L service territory, so that field personnel could take customers to a nearby site to demonstrate a machine. WP&L also wanted to be able to show the machines to its own personnel. In addition, WP&L wanted to learn about the wind regime in different areas. Customer cooperation was excellent, in part because the customers received all the electricity produced. Any excess was fed back to WP&L (without credit to the customer). Meters were installed to show when energy was produced, when it was used, and when it was fed back to WP&L (DeWinkel, 1983).

WP&L was also interested in the wind-turbine market itself. As one result, WP&L purchased the Wind Works Company in 1982. The concept was to expand WP&L's activities in wind power through this subsidiary. WP&L thought there would be synergism in promoting small renewable technology for its customers with WP&L's efforts to integrate the technology into its system. Unfortunately for WP&L, the situation did not work out as anticipated, and Wind Works was sold back to its founder in 1984.

WP&L performed many experiments at its wind power test sites. A March, 1988 report by WP&L (Pigg, 1988), summarizes the results of performance analyses on four of WP&L's residential wind-turbine projects. The analyses were based on approximately 15 site-years of 15-minute wind speed and energy production data collected from the four sites. The primary focus of the analyses was to examine the power output curves of the wind turbines to see how they compared with the manufacturers' specifications and to compare actual energy production with what would be predicted using techniques that have become standard for the wind energy industry. In addition, the study examined the effect of the wind machines on the customers' demand profiles and analyzed the economic performance of the systems, based on their energy production and repair histories.

The study found that the use of the manufacturers' power curves led to overestimation of the energy production for the machines. In the case of two of the machines, the error in the estimated annual capacity factor was more than 30%. Also, the daily power production profile for the wind systems tended to be relatively flat, except for some tendency toward greater power output during the afternoon. Seasonally, the machines all followed the expected trend of highest average output during the late winter and spring, and lowest output during the summer. Only one machine returned significant amounts of energy to the WP&L distribution system. This machine met or exceeded the average base demand for the dairy farm on which it was located except during the low-wind summer months.

The estimated cost of power for the four wind machines ranged from about 26 to 50 cents/kWh based upon the initial investment cost of the machines, which ranged from $1,300 to $3,250/kW of machine rating. According to the report, if industry target levels of availability and operating and maintenance expense could be met by these machines at their actual installed cost, the cost of power would drop to a range of 12–30 cents per kWh. Even at these lower costs, the installed cost of the machines would have to be greatly reduced in order for these wind systems to produce economic power in Wisconsin (Pigg, 1988).

On the basis of its experience with small machines, WP&L concludes that the small, customer-size wind machines are unlikely to return to Wisconsin. No wind machines have been sold in WP&L's service territory during the past 5 years. Wind machines are not as economically attractive as energy conservation measures, and the wind regime in the State is marginal. If wind machines return to Wisconsin, it will be in the form of larger machines deployed in groups to facilitate maintenance. They will probably be operated by utilities or independent power producers.

Also, wind power stations would most likely be located near the Great Lakes (Michigan and Superior) and on some of the ridges located in the State. WP&L plans to conduct, in conjunction with other Wisconsin utilities, a detailed survey of the wind resource and potential sites in these areas.

### 1.5.4 EPRI Wind Program

EPRI's efforts in wind power began in the late 1970s. The overall objectives of the EPRI wind power program have been to assess and report the state of the art in wind power technology and to promote the evolution and appropriate use of utility-grade wind power stations. The program has sought to accomplish these objectives by conducting appropriate research activities, by providing information to the utility and wind power industries about the requirements for commercial wind power stations, and by fostering mutually beneficial interaction between the utility and wind power industries. EPRI has conducted numerous wind power projects including workshops on wind power, surveys of utility activity and experience, assessments of wind turbine performance and operating experience, detailed wind turbine test programs, economic and technical requirements assessments for utility applications, the preparation of cost estimates and siting requirements, assessment of the requirements for utility interconnection, market assessments, and various technical studies. About 50 technical reports on wind power have been published by EPRI. These are listed in the bibliography of EPRI reports provided at the end of this chapter.

EPRI's latest initiative in the area of wind power involves a joint program with PG&E and U.S. Windpower (USW) to develop, test and commercialize an advanced variable-speed wind turbine. This is a five-year program to develop and test an economic, utility class machine that will meet utility requirements and can be readily integrated into utility power systems worldwide. The joint program is the outgrowth of a feasibility assessment which was conducted by USW between April 1987 and April 1988 with cofunding from EPRI. PG&E, Virginia Power, and Bonneville Power Administration also participated in the assessment. The objective of the assessment was to determine the technical and economic feasibility of proceeding with a development program for a power-electronic variable-speed wind turbine. The assessment indicated that a variable-speed machine capable of producing energy at 5 cents/kWh is achievable. It also indicated that the power electronics could be used to increase the energy productivity of wind turbines, reduce drive train loads, and improve the power quality of a wind power station.

The five-year joint program now includes USW, EPRI, and PG&E, and other utility participants are likely. The program is to be conducted by USW in three sections. First is the development of the USW model 33-300 Constant Speed Wind Turbine (CSWT). Second is the Technology Development Section in which the power electronic converter will be developed and tested. The third is the Utility Test Stations Section in which prototypes of the USW model 33-300 Variable Speed Wind Turbine (VSWT) will be installed at selected utility locations for testing and evaluation of performance relative to design goals. The program is now underway. The first constant speed prototype has been installed and is being tested.

The results of the feasibility assessment study indicated that a VSWT using a power electronic converter would initially have a cost of energy no higher than a corresponding CSWT. The increase in energy capture of a VSWT will offset the additional cost of a specially designed converter. In addition, there are other benefits of utilizing a power electronic converter. For example, the power factor can be controlled and varied according to the host utility needs. The power fluctuation experienced with a CSWT is greatly reduced with a VSWT because of the inertial smoothing of the variable speed rotor operation. These and other considerations make it preferable for a wind turbine to utilize power electronics. In addition, it is expected that the cost of power electronic converters will decline steadily as new power-electronic components and technologies become available in the market place. The converter development (Section 2) is now underway and is expected to result in the first fully functional converter in the latter part of 1990 (Lucas et al., 1989).

### 1.5.5 Utility Industry Wind Interest Group

Based on a conviction that electric utilities must be directly involved if wind power is to be a success over the long term, DOE, EPRI and a group of utilities in May of 1989 formed the Utility Industry Wind Interest Group (Interest Group). The nine utilities represented in this group include:

- Bonneville Power Administration
- Green Mountain Power Company
- Hawaiian Electric Company
- Hawaiian Electric Renewable Systems
- Niagara Mohawk Power Company
- Northern States Power Company
- Pacific Gas & Electric Company
- Southwestern Public Service Company

- Western Area Power Administration

It is the mission of the Interest Group to expedite the appropriate integration of wind power for utility applications. In keeping with this purpose, the following objectives have been established:

- Provide a forum for the advocacy and critical analysis of wind technology for utility applications.
- Articulate the needs and requirements of electric utilities for wind power to be considered as a viable generating option.
- Enhance the credibility of and identify opportunities for wind power in the electric utility sector.
- Provide guidance to industry, state, and national wind development programs, including those of DOE and EPRI.
- Encourage support for the national effort.

The following strategies have been adopted in support of the above objectives.

1. Publish periodic reports and brochures on wind power aimed primarily at utility management.
2. Highlight the positive trends in wind power development from a balanced perspective, and provide realistic projections for the future.
3. Provide a utility perspective and philosophy to national wind energy programs. Review and comment on state and national government activities related to wind energy, including program plans.
4. Provide input to the federal and state regulatory and legislative program and budget planning process, to the extent allowed by individual corporate charters, in support of wind power development.

The premise for the Interest Group's formation is that wind energy technology is the most promising near term renewable energy option to come out of the energy crisis of the 1970's. A concerted, cooperative effort between DOE, EPRI, and the utility industry can help to ensure that wind energy will become a serious option for the future utility generation mix.

### 1.5.6 Summary and Conclusions About Wind Power

Mid-size wind turbines have been developed to the point that the utility industry can factor wind power into its long-range generation expansion plans and take the early steps necessary to prepare for wind power. Those early steps include intensive resource measurement programs and installation of pilot wind power stations. Although some refinements, such as variable-speed generation through power electronic control, will be forthcoming in future years, wind turbines are already sufficiently attractive for use by utilities that have suitable wind resources and sites and also moderate to high conventional energy costs.

Because of the inherent modularity of wind power stations, a utility can build a wind power station in increments and use the best turbines available at the time any given increment is built.

A significant parallel can be drawn between the progress with wind power and progress with solar thermal parabolic trough technology. In both cases, the most notable success has occurred through entrepreneurially motivated private-sector activities. Government tax credits played a strong and crucial role in reducing financial risks, but the development effort has been directed primarily by the reality that success is measured only in the commercial marketplace. Also, in both cases, a good technology base existed at the outset, in part as a result of government-funded research and development programs; and significant progress has been made from year to year through engineering development programs that did not require technological breakthroughs or long gestation periods to achieve results and allow sufficient evaluation. And in both cases, development programs involving larger-sized variants of similar technology were conducted at substantial public expense through directed government efforts. The latter programs included far less marketplace-oriented entrepreneurial motivation, and met with far less commercial success. There is an important lesson here for those involved in the development of new energy technologies and new technologies in general.

There are a number of areas throughout the United States with wind conditions favorable for generating electricity. The correlation between the wind resource availability and the local utility system demand varies from area to area. In some locales the correlation is good and in other areas it is not. Given the variability of wind, its value in most cases will be primarily as a source of energy. In some cases, such as Solano County in PG&E's service territory, the wind may be sufficiently reliable that a wind power station may be a source of a significant amount of dependable capacity as well.

The largest wind resource area is in the Great Plains, reaching from the Texas Panhandle all the way to the Canadian border. The wind resource here is enormous (thousands of MW). However, this region also has or is close to substantial coal deposits. The variable cost (cost of fuel and variable operation and maintenance) of producing energy from coal-fired generation in the region is relatively low, on the order of 1 to 2 cents/kWh. By contrast, wind power at good sites costs about 8 cents/kWh (Schaefer, 1989). Unless a favorable correlation between the wind resource and utility capacity needs can be established, this will make it difficult for wind generation, which is primarily an energy resource as opposed to a capacity resource, to compete in the Great Plains in the absence of other incentives, such as special environmental credits. Also, the wind sites in the Great Plains are remote from large load centers and would necessitate the development of substantial transmission facilities for delivery of the energy produced to population centers. Finally, due to the variability of the wind resource, large quantities of wind power produced in remote locations may require special measures, such as the incorporation of energy storage, to protect utility system reliability and stability. Thus, although an enormous wind generation potential exists in the Great Plains, it will be expensive to tap in large quantities compared to conventional options.

Good wind resources also exist in a number of other areas of the United States. In fact, the areas of California in which wind turbines have been installed contain a very small portion of the overall national wind resource. Determining the full potential within areas that have been developed will require quantification of localized wind variations, determination of site accessibility, and identification of institutional restrictions on wind power development.

A few utilities have already begun to develop small wind-turbine power projects. This utility use of wind power is likely to show gradual growth over the next 15 years irrespective of advances in wind power technology. However, there are some key areas in which potential improvements in wind-turbine technology could significantly enhance the attractiveness of wind power to the utility industry.

Although no technology barriers remain to the production of electric energy from the wind, the long-term reliability of wind turbines remains an uncertainty which will require additional years of operating experience to be resolved, as evidenced by the fatigue-related problems being experienced by foreign machines. Meanwhile, refinement of existing wind turbine products is expected to continue, although no major technological breakthroughs are foreseen. The prospects for fundamental advances in wind turbine technology are limited due to the relative simplicity of the machine. The primary prospect for a fundamental advance lies in the application of power electronics to enable variable speed wind turbine operation (for prospective gains in turbine dynamics and energy capture), reactive power control and other benefits. Significant opportunities also exist in research and development leading to new airfoils designed specifically for wind turbines and improved analytical tools to help designers deal with the complex aerodynamic and structural dynamic considerations in wind turbine and wind power station design. Also, increased emphasis of wind power research on wind power station technology advancement appears desirable. These improvements could reduce costs by 20–35% over the next decade.

An additional specific conclusion that can be drawn from the aggregate activities of a number of utilities is that, except in very highly specialized circumstances, the installation and operation of a single machine or a very small number of machines does not make economic sense. This conclusion holds regardless of machine size, and is borne out by utility experience with machines ranging in size from a few kW to a few MW. The primary reason is that an appropriate maintenance program cannot be justified on the basis of one or a few machines. In sharp contrast, however, installations that include tens or hundreds of machines and that enjoy a well-conceived maintenance effort have demonstrated equipment availabilities above 95% with operating and maintenance costs near one cent per kWh.

In summary, tremendous strides have been made since 1973 in the advancement of wind power technology, especially with the smaller machines. The electric utilities have played a key role in helping to attain the successes achieved with wind power at Altamont Pass and elsewhere. In spite of some continuing technical and economic uncertainties, there is reason for optimism about the prospects for future expansion in the use of wind power by utilities due to the degree of technology advancement and cost reduction achieved so far, the non-polluting nature of this resource and the continued improvements and cost reductions being achieved. The technology has advanced to the point where wind power costs are approaching par with more conventional power generation sources in some parts of the U.S.

# 1.6 Photovoltaics

As a result of progress achieved in photovoltaic (PV) technology over the past decade, U.S. electric utility interest and involvement in photovoltaics has seen substantial growth. Important advances in cell efficiency have been demonstrated, both for flat-plate cells and modules and for cells using high sunlight concentration. And, during the past five years, PV power technology has progressed considerably toward the goal of acceptance for bulk power generation. Field test experience from megawatt-scale installations virtually assures this acceptance from a power-plant operational standpoint. As a result, a strong sense of optimism about PV's prospects for both utility system and customer (demand-side) applications has emerged among utilities. Key reasons for the growth of electric utility interest and optimism include:

- PV is highly modular, allowing systems to be deployed in a wide range of sizes; allowing relevant, scalable experience to be gained with small, affordable systems; allowing short construction times, and revenue generation as modular segments come on line; in sum, reducing technical and financial risks associated with testing, precommercial deployment, and commercial use.
- PV systems accomplish sunlight-to-electricity and dc-to-ac conversion entirely with solid-state electronic components. Very few, if any, moving parts are used. These features promise high equipment availability and very low operating and maintenance costs.
- With the possible exception of land utilization, environmental impacts are expected to be minimal. No pollutants are released during power plant operation, since no fuel is consumed. Even the potential carbon-dioxide-related global warming problem is completely side-stepped.
- A sizable international R&D community is actively pursuing PV development. In addition, extensive activity in other semiconductor electronic materials and products may benefit PV technology. In sum, an above-critical-mass effort is underway worldwide.
- Applicability of PV to early markets, such as remote power and consumer products, suggests viable paths for orderly business development.
- A steady stream of PV technology advances has been occurring, leading to progress toward achieving cost and performance goals for PV components in prospective bulk power applications.

In contrast with solar thermal and wind power technologies, by the early 1980's no specific PV technology had reached the threshold of economic feasibility for grid-connected applications. Breakthroughs were still required, and incremental progress through orderly engineering development would not be sufficient by itself to cross that threshold. Hence PV development has been largely unaffected by the availability of governmental tax credits and favorable, PURPA-related electricity purchase arrangements. Far more significant in facilitating development of the technology has been research support provided by government and EPRI programs, and of even greater importance, by large corporate sponsors that have applied more than $1 billion of industrial research funds. Principal among these have been several U.S. oil industry firms. Up to now, the promise and appeal of PV have been sufficient to keep these funds flowing, even though the PV industry as a whole has yet to become profitable. As one result of all these factors, the powerful motivations of marketplace success and failure and the associated entrepreneurial drive have not yet reached full force.

Several PV technologies are now approaching the threshold of economic feasibility, and business development in the sense of commercial, market-oriented products is now becoming of primary importance. This is the case all the more because numerous corporate investors have become impatient after many years of negative cash flow. Indeed some have already left the PV arena, but others have joined. Still, major technological advances are required in some PV technologies, particularly in thin-film flat plate approaches that show low cost potential in large-scale manufacturing but have relatively low sunlight-to-electricity conversion efficiencies. Hence the next several years will be a crucial period for the PV industry as it attempts to mature in the commercial marketplace with less and less government and corporate underwriting, while simultaneously maintaining sufficient activity in basic research to allow advancement from near-term high value applications to energy significance in the bulk power arena.

All of these factors have affected the manner in which utilities have participated in PV development. And they will continue to influence strongly the evolution of utility involvement over the next decade.

This section provides an overview of electric utility activities in PV, summarizes the results of key utility field experiences with PV systems, discusses the major issues associated with utility system applications of PV, reviews the purpose and scope of current initiatives to advance PV technology for utility applications, and assesses prospects for the future.

### 1.6.1 Overview of Electric Utility Activities in PV

The growth of electric utility interest in PV can be seen by examining the results of the EPRI surveys in Table 1.1 and Table 1.2. In 1976, only five utilities and five activities were reported. By 1983, this number had grown to 41 utilities and 74 activities. Table 1.3 and Table 1.4 show that a variety of activities, cosponsorships and use sectors were involved, and Figure 1.4 indicates that the activities reported in 1983 were occurring in most regions of the country, especially in the southeast and western United States.

Although the level of utility activity in the other solar technologies has declined over the past decade, attention to PV has increased significantly. This increase is reflected in both the number of activities and the number of utilities involved, but far more importantly, in the type of activity and level of effort. Of all the emerging renewable power technologies, PV is perceived by most utilities as having by far the greatest potential.

A typical early utility activity was monitoring the performance of a small, customer-owned PV installation. While such activities continue, typical activities today include operation of major, utility-owned and -installed test facilities; evaluation of field performance for installations in the 100 kW to 5 MW range that have been financed by others; evaluation of distributed PV systems, either individual or multiple units, that have been financed by the utilities themselves; and internal evaluations of a broad range of business opportunities, both in extending the traditional function of electricity supply to customers, and in exploring other non-traditional avenues (DeMeo, 1988).

In addition, for several years, a small number of utilities have cofunded PV device research and development programs. PG&E and SCE, for example, have helped to support the development of dendritic-web silicon sheet technology through a program being cosponsored by Westinghouse, DOE, and EPRI. As another example, these same two utilities have been joined by the Arizona Public Service Company, the Georgia Power Company and the Los Angeles Department of Water and Power in utility cosponsorship of EPRI's development program for the single-crystal-silicon, high-concentration cell invented at Stanford University (Moore, 1987).

This type of involvement is now expanding and is taking on an important new dimension. In the past, the primary motivation for such cofunding appeared to be the desire to participate in bringing new PV technology along. Today, however, many utilities are attracted to potential business positions involving PV technology. Both near-term and longer-term opportunities are of interest. This trend toward business investment, in some cases through unregulated subsidiaries, may represent a substantial new source of funds for PV technology development.

Alabama Power Company has invested in PV module manufacturing with Chronar, a supplier of amorphous-silicon thin-film modules. At least in the near term, production from this venture is slated for consumer product and non-utility lighting applications, indicating a substantial broadening of utility interests beyond traditional bulk power. Also, several utilities, PG&E for example, (Hoff and Iannucci, 1988) are expressing interest in PV for off-grid power supply in areas where grid extension is currently uneconomical.

It is difficult to quantify the aggregate of electric utility expenditures in PV, since some unregulated activity is not publicly reported and many other activities are not rigorously accounted. Nevertheless, it is likely that, through field test activities, direct support of PV industry development programs, support of the EPRI PV program, internal assessments, and investments and joint ventures, U.S. electric utility industry PV-related expenditures now exceed those of the U.S. DOE program. This level of expenditure, along with the much larger level of the PV supplier industry, suggests that a healthy transition from public to private sector is underway.

### 1.6.2 Field Experience with Bulk Power System Installations

Over the past decade, a number of field demonstrations of PV bulk power systems involving government, industrial and utility organizations have been conducted. Key characteristics of several of these are summarized in Table 1.22, operational results in the form of capacity factors and operating availabilities are summarized in Table 1.23, and maintenance costs in Table 1.24. All systems described in these tables employ single-crystal-silicon cell technology with flat-plate modules, with the exceptions that a portion of the Sacramento Municipal Utility District (SMUD) PV2 project includes polycrystalline silicon modules, and Arizona Public Service Company's Sky Harbor system employed Fresnel lens concentrating modules. Although these PV power installations are not cost competitive with conventional bulk power systems, they have provided a great deal of information on the performance of PV generation within electrical utility networks with profound implications for the acceptance of PV in bulk power use.

**Table 1.22:** Major Experimental PV Power Plants

| Station | Location | Owner/ Operator | In Service Date | Station Capacity Rating (kWac) | Photovoltaic Modules | Tracking | Local Utility |
|---------|----------|-----------------|-----------------|-------------------------------|---------------------|----------|---------------|
| Sky Harbor[a] | Phoenix, AZ | Arizona Public Service (APS) | May 1982 | 200[b] | Martin Marietta Point Focus Fresnel | Two Axis | APS |
| Lugo | Hesperia, CA | ARCO Solar | Dec 1982 | 729[c] | ARCO Solar Flat Plate | Two Axis | SCE |
| Carrisa Plains | San Luis Obispo County, CA | ARCO Solar | Dec 1983 | 5,120[c] | ARCO Solar Flat Plate | Two Axis | PG&E |
| SMUD PV1 | Sacramento County, CA | Sacramento Municipal Utility | Aug 1984 | 932[c] | ARCO Solar, | One Axis | SMUD |
| SMUD PV2 | Sacramento County, CA | District (SMUD) | Feb 1986 | 875[c] | Flat Plate SOLAREX Mobil | One Axis | SMUD |
| Austin PV300 | Austin, TX | Austin Electric Utility Department | Dec 1986 | 250[d] | ARCO Solar Flat Plate | One Axis (Passive) | Austin |

[a] Plant shut down October 1, 1987 and dismantled.

[b] Represents original rating under standard test conditions (850 W/m direct beam insolation, airmass 1.5 solar spectrum, and 53°C cell temperature). Because of open interconnect straps, cannibalization of some modules, and moisture-related grounding of modules, actual power level decreased by about 50 percent after 5 years of operation.

[c] Except for Carrisa Plains, measured at standard operating conditions (1,000 W/square meter insolation, air mass 1.5 solar spectrum, and 20°C ambient temperature). Carrisa Plains calculated from observed ac outputs and inverter efficiencies. Sources: Schaefer, 1990.

[d] Hoffner, 1989.

**Table 1.23**: Capacity Factors and Availabilities Of PV Plants (Percent)

| Plant | Measure | 1983 | 1984 | 1985 | 1986 | 1987 | 1988 | 1989 * |
|---|---|---|---|---|---|---|---|---|
| Phoenix-Sky Harbor | CF | 16 | 10 | 22 | 22 | 29 * | out of service | |
| | Avail. | 58 | 53 | n.a. | 87 | 92 | | |
| Hesperia-Lugo 1 | CF | 25 | 32 | 35 | 23 | 28 | 33 | 35 |
| | Avail. | n.a. | 92 | 96 | 96 | 97 | 98 | n.a. |
| Hesperia-Lugo 2 | CF | 25 | 34 | 35 | 24 | 30 | 33 | 34 |
| | Avail. | n.a. | 94 | 96 | 98 | 97 | 98 | n.a. |
| Carrisa Plains | CF | | 29 | 30 | 29 | 25 | 24 | 22 |
| | Avail. | | n.a. | n.a. | n.a. | n.a. | n.a. | n.a. |
| SMUD PV1 | CF | | n.a. | 23 | 25 | 7 | 0 | n.a. |
| | Avail. | | n.a. | n.a. | 91 | 24 | 0 | n.a. |
| SMUD PV2 | CF | | | | 23 * | 23 | 19 | n.a. |
| | Avail. | | | | 91 | 89 | 83 | n.a. |
| Austin PV300 | CF | | | | | 22 * | 22 ** | 22 |
| | Avail. | | | | | 97 | 100 | n.a. |

Notes:
n.a.   Not available.
*        Six months of data.
**      Hoffner, 1989.

Source: Schaefer, 1990.

The most significant PV field tests to date have been constructed and are being operated by ARCO Solar, Inc., at Hesperia (Lugo Substation) and Carrisa Plains in California, in cooperation with Southern California Edison Company (SCE) and PG&E, respectively. Key field tests listed in Table 1.22 to Table 1.24 in which utilities have been the owners and operators include the Sky Harbor Project of Arizona Public Service Company (APS), the two PV plants of Sacramento Municipal Utility District (SMUD), and the City of Austin, Texas (Austin) PV300 project. These PV systems and the results of the field tests are described below.

### 1.6.2A  Lugo PV Power Plant

The Lugo Photovoltaic Power Plant (Lugo), Figure 1.37, is located approximately 52 miles northeast of Los Angeles, California, near Hesperia in San Bernardino County, adjacent to the SCE's Lugo Substation.

**Table 1.24:** Maintenance Costs for PV Power Plants*

|  | Period Covered | Average Costs ( cents / kWh ) | Potential Cost ** ( cents / kWh) |
|---|---|---|---|
| Phoenix- Sky Harbor | Apr 85 - May 87 | 4.8 | 0.3 to 1.12 |
| Hesperia- Lugo | Jul 85 - Sep 88 | 1.1 | 0.2 to 0.65 |
| Carrisa Plains | Jul 86 - Sep 88 | 0.8 | 0.2 to 0.5 |
| SMUD PV1 and PV2 | Jan 86 - Jun 88 | 0.6 | 0.13 to 0.22 |
| Austin | Jan 88 - Dec 88 | 0.9 *** | 0.4 *** |

\*      Sources: EPRI GS-6625 and GS-6696, and Hoffner, 1989.

\*\*     Potential cost range, estimated in GS-6625, varies from low cost based upon use of high efficiency cells to higher cost based on resolving known problems at sites.

\*\*\*    According to Hoffner (1989), for commercial operation -- assuming weekly site visits and completion of all other scheduled and unscheduled maintenance -- estimated maintenance cost would be 0.4 cents/kWh.

Lugo commenced power production in December 1982. Its peak-power output is 729 kW at standard operating conditions while it nominal ratng is 1,000 kW. The power conditioning unit-utility system interface is at 480 Vac. Two redundant power conditioning systems, one 1,000 kVA unit and two 500 kVA units, were installed. After initial operation, only the two 500 kVA units were used. SCE owns the 12.5 kV/480 Vac step-up transformer and 12.5 kV switchgear used to interconnect the station with the SCE system. A block diagram of the station electrical system is presented in Figure 1.38.

The PV array field consists of 108 two-axis solar trackers. Each tracker has 256 flat-plate modules. The nominal peak power output per tracker is 9.6 kW. The trackers are wired in parallel to form the array field. The trackers are electrically connected (current summed) in groups of six, with main feeder cables connecting the power collection box of each six-tracker group to the control building; which is located near the center of the array field to minimize dc power cabling losses and wiring costs. The control building houses the switchgear, inverters, and control computers.

**Figure 1.37:** Lugo Photovoltaic Power Station.

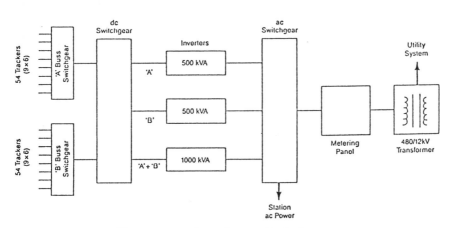

**Figure 1.38:** Lugo System block diagram

The Lugo plant has been operating since early 1983. Both EPRI and SCE have tracked the performance of this plant for over four years (EPRI AP-5229, AP-5762, AP-6251, GS-6696, and Patapoff, 1985). The experience is typified by the 1985 annual performance summary, which showed an annual 24-hour capacity factor of 35%, estimated to be 96% of that achievable with perfect system operation (EPRI AP-5229). Operating and maintenance (O&M) costs for this plant have also been estimated; with results to date, after compensating for non-routine expenses associated with the experimental nature of the plant, of about 1.1 cents per kWh (Schaefer, 1990).

Overall, the Lugo plant has performed very well. However, as can be seen in Table 1.23, energy production declined markedly in 1986 and 1987. Wind loads at the site turned out to be greater than anticipated, and several tracker drive units failed. During 1986 and 1987, all tracker drives were upgraded. Replacement was carried out on several trackers at a time, and those trackers were out of service during the repair. However, the remaining trackers continued to generate power, demonstrating the ability of a modular system to produce power output even if a portion of the PV field is out of service. Based on the data available and the assumption of no further tracker failures, a plant like Lugo can be expected to operate with about a 35% annual capacity factor (Schaefer, 1990). This is shown by the recovery of the plant's capacity factor in 1988 and 1989 after the completion of the tracker drive replacements.

### 1.6.2B  Carrisa Plains PV Plant

The second ARCO Solar PV plant is located at Carrisa Plains in Central California and is interconnected with the PG&E system.

### 1.6.2B.1  Plant Description

Carrisa Plains is the world's largest photovoltaic power plant. A picture of the plant is shown in Figure 1.39 and a block diagram in Figure 1.40. The plant is rated at 5.1 MWac. It was installed in two phases. The initial phase (Phase 1A) became operational in December 1983 and includes 756 dual-axis trackers arranged in 9 segments. Each segment consists of 84 trackers, a power collection system and a 650 kW Helionetics power conditioning unit (PCU). Each of the trackers has a $36 \times 32$ feet array arranged in eight panels of ARCO Solar flat-plate PV modules. Each tracker array also uses 16 laminated reflected glass (LRG) panels to increase the sunlight striking the photovoltaic panels in order to increase their electrical output. Each tracker is mounted on a dual-axis gearbox which in turn is mounted on a 24-inch diameter pipe pedestal. The plant is designed as an unmanned facility and is controlled and monitored by computers housed in the control room.

The second project phase (Phase 1B) utilizes a tracker design that is different from Phase 1. It was brought on line in October 1984. This segment contains 43 trackers and one 750 kW Toshiba PCU. The Phase 1B trackers contain no LRG's, only photovoltaic modules. The array size is $40 \times 40$ feet and has a mounting similar to that of the Phase 1A tracker. The array design was modified for Phase 1B to maximize the photovoltaic active area per tracker and, thereby, minimize the area-related balance-of-system cost for the installation.

The Phase 1A trackers are wired in series to produce the plant dc operating voltage of 480 Vdc at 46 amperes (see Figure 1.40). The output of two (2) groups of three trackers each is collected in a junction box (fourteen per segment) and the dc power is sent to the PCU in each segment. The dc power is then converted to 480 Vac with an average efficiency of 97 percent. The power then goes to a transformer, which steps up the 480 Vac to 12 kV and then to the 12 kV switch gear which is centrally located near the Control Room. The power from all segments is then collected and sent to the substation. At the substation, the voltage is stepped-up to 115 kV and

**Figure 1.39:** Carrisa Plains Photovoltaic Power Plant.

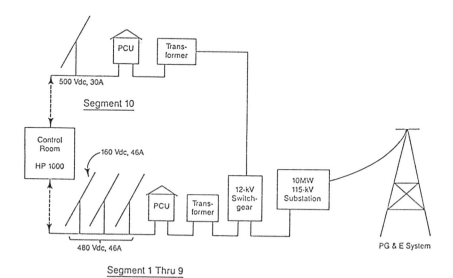

**Figure 1.40:** Schematic of Carrisa Plains Photovoltaic Power Plant.

fed into the PG&E power grid. The electrical interconnection scheme of Phase 1B is almost identical to that of Phase 1A.

Additional equipment in the facility includes the weather monitoring system with five data collection stations, a 750 kW diesel generator for use when utility system power fails, and an uninterruptible power supply for the computers.

### 1.6.2B.2 Plant Performance

Figure 1.41 shows the monthly capacity factors for Carrisa for the years 1984 through 1987 (Sumner et al., 1988). Capacity factors for the Carrisa plant have been somewhat less than those of the Lugo Plant, although they have been quite high. A primary contributor to the lower output at Carrisa has been module temperature elevation resulting from the reflector augmentation in Phase 1A. Module temperatures greater than 90 °C have been observed for extended periods during the summer months (Sumner et al., 1988). Beginning in 1987, significant reductions in energy produced and capacity factor were observed at Carrisa. This is attributed to module discoloration (a browning of the cell encapsulant, the EVA polymer in front of the cell, changing the cell's appearance from blue to brown) that has begun to occur in Phase 1A. The elevated temperatures and elevated ultraviolet irradiance are thought to be responsible for this discoloration also (Sumner et al., 1988). The problem has continued with further declines in output in 1988 and 1989. Some discoloration is also beginning to appear in Phase 1B, but the problem does not appear to be as severe as that in Phase 1A. No such effects have been noted at Lugo, and the problem appears to be unique to the Carrisa Plains modules. One lesson is clear; elevated module temperatures must be accounted for when mirror-enhanced designs are considered (Schaefer, 1990).

PG&E has analyzed the Carrisa Plains plant capacity factors during its peak load hours in the summer months; i.e., 12 noon to 7:00 p.m., for the years 1984 to 1986. Figure 1.42 shows the peak-period capacity factors for the same three-year period. The high values observed (70–80%) indicate a very high degree of correlation, in the Carrisa location, between solar energy production and PG&E's system peak needs. This in turn indicates that a substantial capacity credit can be ascribed to this plant by PG&E. Subsequent analyses by PG&E indicate similarly high correlations between insolation availability and system power demand throughout its service territory (Hoff and Iannucci, 1988).

### 1.6.2B.3 Operation & Maintenance

The most dependable information on PV station operation and maintenance is that compiled by ARCO Solar. This is because of the relatively large size of the ARCO Solar plants and the extensive photovoltaic experience the company has already accumulated. The Carrisa plant is operated and maintained by a field operations group that consists of three people who also operate and maintain the Lugo plant and other ARCO Solar test facilities. Among the major lessons learned from the experiences at Carrisa are (Schlueter, 1987):

- Photovoltaic central power stations can be safely and reliably operated as unmanned sites.
- Photovoltaic station maintenance is uncomplicated relative to that required at conventional plants, reflecting the simplicity of photovoltaic plant designs.

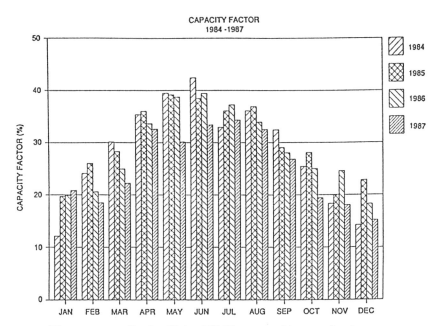

**Figure 1.41:** Carrisa Plains PV Plant monthly capacity factors.

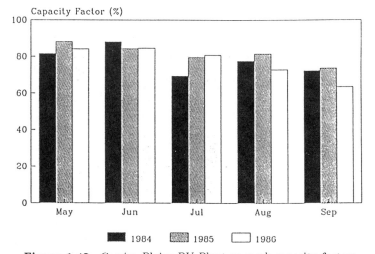

**Figure 1.42:** Carrisa Plains PV Plant on-peak capacity factors.

- Maintenance costs at large photovoltaic plants should be $0.008/kWh or less.

To a considerable extent, the success achieved to date by ARCO Solar in operating Carrisa Plains can be attributed to the plant maintenance philosophy, which emphasizes preventive maintenance.

About 65% of maintenance labor is devoted to preventive maintenance while only 35% is required for corrective maintenance. According to ARCO Solar, an important aspect of PV station maintenance is the simplicity of scheduled maintenance tasks as compared to those necessary in more conventional power plants. This is a reflection of both the relative simplicity of photovoltaic system designs, and, owing to the modular nature of such plants, the small probability of serious consequences should a malfunction occur. This results in lower maintenance costs, reduced administrative needs, and less extensive training requirements for personnel.

### 1.6.2C SMUD PV1 and PV2

The Sacramento Municipal Utility District (SMUD) Photovoltaic Power Station is a 2 MW installation comprised of two phases of nominally 1 MW each, referred to, respectively, as SMUD PV1 and SMUD PV2. These are the first two phases of what was to have been a five-phase development resulting in a 100 MW photovoltaic power station. The project was built by SMUD and cofunded by DOE and the California Energy Commission. Phase 3, which was to have gone into commercial operation in 1987 and would have increased the total plant rating to 7 MW, has been indefinitely delayed by SMUD until the economics for this phase of the project improve.

PV1 began operating in August 1984 and PV2 began operating in February 1986. Both PV1 and PV2 use single-axis tracking. PV1 and 765 kW of PV2 use flat-plate, single-crystal-silicon solar cell modules manufactured by ARCO Solar. Two hundred kW of PV2 are polycrystalline silicon modules manufactured by Solarex and the remaining 35 kW are polycrystalline ribbon silicon modules manufactured by Mobil Solar Energy Corporation.

The PV1 facility, shown in Figure 1.43, consists of 112 trackers, each containing eight panels. Each panel holds 32 modules. The trackers are electrically connected in four subfields of seven series strings each. PV1 uses a line-commutated Omnion PCU to convert dc power to ac power. As discussed below, the original PV1 PCU, also manufactured by Omnion, Figure 1.44, was destroyed by a fire in April 1987, and was replaced by a new Omnion PCU in late 1988.

The PV2 field consists of 92 trackers in an overall design similar to PV1. A self-commutated PCU manufactured by Toshiba converts the dc array output to ac for delivery to the grid.

The independent PV1 and PV2 PCUs, control computers, and data acquisition systems are housed in control buildings centered within each field. Uninterruptible power supplies provide backup power to the computers in each field. The performance of the array, inverter, and weather station are monitored by the computers. The plant is also linked with the SMUD dispatch office through an on-site remote terminal unit (RTU) and a dedicated telephone line. The SMUD dispatcher has a remote trip/permissive interlock, which allows isolation of the PCU output from the grid via a 12 kV fault interrupter. Selected plant performance data is available to the dispatcher from the RTU (EPRI GS-6625).

The operating voltage window of SMUD PV1 is 575 to 700 Vdc and the current supplied by the 112 arrays ranges up to about 1,700 amps. The 28 source circuits are isolated by blocking diodes and are collected in positive and negative dc buses, first at the subfield level and then in two main buses (positive and negative). The main dc fuses are separated from the PCU by the main dc contactors (see Figure 1.45).

**Figure 1.43:** SMUD PV1 Photovoltaic Power Plant.

Table 1.23 shows the SMUD plant's capacity factor for 1986 through 1988. The 1986 energy production was 94% of that achievable with perfect system operation (EPRI AP-5762). SMUD's reduced output for 1987 and 1988 relative to 1986 reflects the outage of PV1 for the latter eight months of 1987 and most of 1988 due to the fire which destroyed the PV1 power conditioning unit on April 30, 1987. The cause of the fire and lessons learned are discussed in a paper presented at the 20th IEEE Photovoltaic Specialists Conference (Collier and Key, 1988). These lessons are of substantial significance for PV power plant design.

The suspected cause of the fire is a positive to negative fault in a congested area of cabling leading to the dc collection bus in one of the PCU cabinets. The cable insulation failure apparently occurred about 3 inches below the point where the power cables connect to the main dc contactors (Collier and Key, 1988). This fault apparently caused power to be fed back into the inverter from the ac system, and it is believed that the surge of power and resulting arcing destroyed the two main dc contactors connecting the positive and negative dc circuits to the PCU. A reverse power relay then tripped and caused the PCU's ac contactors to open and separate the PCU from the utility system. Thus, the PCU was electrically isolated only from the utility system. The subfield dc contactors in the array field did not open, and the array field continued to supply current at about 600 amperes to the fault for approximately 45 minutes before the array field was deactivated and the

**Figure 1.44:** 1 MW Omnion Power Conditioner at SMUD PV1.

resulting fire extinguished. The fire destroyed the main cabinets of the Omnion PCU including the inverter, dc collection and controls cabinets.

A number of important design lessons were learned from the SMUD PV1 fire. Most important are those that relate to fault protection. The experience at SMUD highlighted the very special fault protection problems unique to PV array power sources. It is the open-circuit voltage and short-circuit current characteristics of a power system that determine the appropriate protection method. The current-voltage (I-V) characteristic of a PV array is completely different from that of a conventional ac power source, and understanding this difference is fundamental to PV array protection design. Contributing to the problems of array fault protection in PV1 was the fault-prone mounting of the blocking diodes. Also the rough installation of wiring, both direct burial and that pulled in conduit, apparently damaged the wiring insulation (Collier and Key, 1988).

To prevent a recurrence of the fire, SMUD has modified the plant fault-protection design including extension of the zone of protection, fusing all source circuits, increasing the sensitivity of and reducing the time delays to actuation of contactors, and uprating the inverter rating and operating voltage. SMUD paid about $300,000 to replace the PCU. The replacement PCU was supplied by Omnion and installed in the fall of 1988. Including the value of the energy lost during the outage, the estimated total cost of the accident to SMUD was about $500,000.

The SMUD PV1 fire provides another example of the advantages of modularity. Even though the PV1 array field was out of service until the PCU was replaced, half

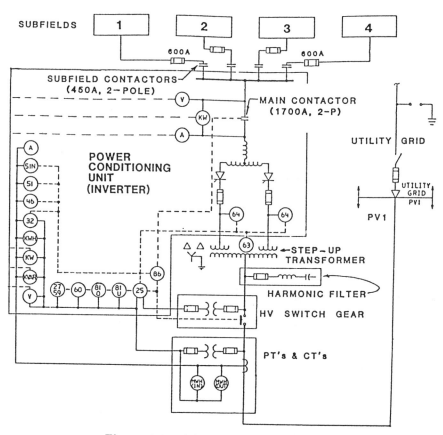

**Figure 1.45:** Schematic of SMUD PV1.

of the entire SMUD power plant still operated normally. Furthermore, if this were a larger PV plant, say nominally 20 MW, loss of the PCU for a 1 MW subfield would cause a reduction in plant output of only 5%. This can be contrasted with many conventional electric generating units where damage to a key component, such as a combustor or turbine generator, can render the entire unit inoperative.

### 1.6.2D Austin Photovoltaic Plant

The Austin PV300 facility, shown in Figure 1.46, is a nominal 300 kW plant constructed in 1986. The total cost of the facility, which is located on 2.5 acres adjacent to Austin's Decker Creek Power Plant, was approximately $3.0 million. Construction was performed under a turnkey contract with ARCO Solar, Inc. Construction began in August 1986, and the plant was operational at dedication on December 5, 1986 (a period of less than 5 months). The plant was accepted from ARCO by the City on July 29, 1987, when the plant was considered fully operational.

**Figure 1.46:** Austin PV300 Photovoltaic Power Plant (Austin's Decker Creek Power Plant in background).

Detailed operating data were collected and research tests conducted during the first full year of operation. The operation of the plant has been highly successful, as summarized below.

The key plant components include 154 flat-plate ARCO solar panels ($2,600$ m$^2$/28,000-ft$^2$), single-axis tracking using Robbins Engineering passive trackers, and a Toshiba self-commutated inverter. There are a total of 7 source circuits (22 panels per source circuit) connected to common positive and negative buses through disconnect switches. The two buses are separated from the inverter by dc contactors. The array field is not center-tapped and operates with a floating ground. The plant is connected to the City's electric distribution system at 12.5 kV. The following is a summary of Austin's experience with the plant (Hoffner, 1989. See also EPRI AP-6251 and GS-6696):

### 1.6.2D.1 Plant Performance

The plant generated $522,000$ kWh on a gross basis and delivered $485,000$ kWh (net of station service) to the Austin system during calendar year 1988. The gross output was within 6% of the projected output of $550,000$ kWh. The average gross energy output during the six-month summer peak period was $51,000$ kWh per month, while the average output during the winter period was $36,000$ kWh per month. The 1988 energy output, shown in Figure 1.47, was near design level projections on both a monthly and yearly basis. The 6% lower yearly output relative predictions can partially be attributed to the fact that the unit was shut down manually at times througout the year for testing and experimentation.

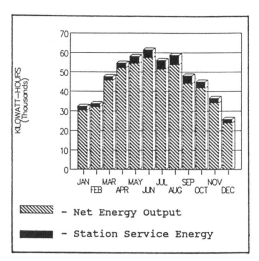

**Figure 1.47:** PV300 1988 monthly energy production.

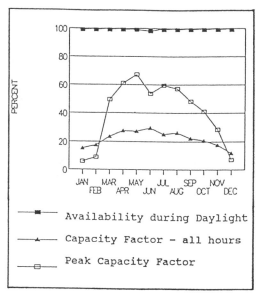

**Figure 1.48:** Austin PV300 1988 availability and capacity factors.

## 1.6.2D.2 Reliability

The plant performed very reliably during 1988 with an operating availability during the daylight averaging greater than 99% for the year. See Figure 1.48.

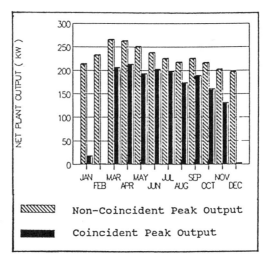

**Figure 1.49:** Austin PV300 1988 monthly peak plant output.

### 1.6.2D.3  Capacity Factor

The plant's 1988 annualized capacity factor based upon a plant rating of 250 kW under standard operating conditions (See Table 1.22) and a net output of 485,000 kWh, was 22 percent, comparable to that of a traditional peaking plant. See Figure 1.48.

### 1.6.2D.4  System Coincidence

The plant's output (net) was 265 kilowatts in March 1987. The coincident peak output (plant peak at time of City's peak demand, which occurred on August 23, 1988 at 5:00 p.m.) was 173 kilowatts. During the summer months, the net plant output consistently averaged approximately 185 kilowatts during the City's peak demand period (Figure 1.49). This indicates that the City could rely on 50 to 60% of rated power, or 125 to 150 kW, net, from the system with a high level of confidence during summer peak periods.

### 1.6.2D.5  Maintenance Costs

Austin is maintaining careful records of maintenance costs for the plant. Even though actual maintenance costs averaged $.009/kWh, as shown in Table 1.24, the City estimates that yearly maintenance costs, after the one-year warranty period and when site visits are reduced to once per week, will be on the order of $.004/kWh.

### 1.6.2E  Arizona Public Service Company's Sky Harbor Plant

The Sky Harbor Photovoltaic Power Plant (Figure 1.50) was originally funded by DOE. It was owned and operated by APS and located at the Sky Harbor Airport in Phoenix, Arizona. The project commenced operation in May of 1982 and continued to operate in parallel with the APS utility system until October 1987, when the plant was shut down and dismantled. Although APS would rather have kept the plant operating, the land had been leased from the City of Phoenix, and the City did not want to renew the lease because of other needs for the land.

**Figure 1.50:** Arizona Public Service Company's Sky Harbor Photovoltaic Power Station.

The system initially included 80 Martin-Marietta Corporation two-axis-tracking arrays, along with the requisite electrical wiring, power conditioning, controls, and other balance-of-station items. The PV arrays were a first-generation Martin-Marietta design. Point-focus Fresnel lenses were used to concentrate sunlight to approximately 36 times its normal intensity and focus it onto 2.25 inch diameter solar cells. Two and one-half arrays were wired in series to provide a 300 Vdc (normal operating voltage), 750 A source circuit. Thirty-two source circuits, called branches, were connected in parallel to supply the input to the power conditioning unit. The output of the PCU was 480 Vac and after site loads had been tapped off, the voltage was transformed to 12.5 kV and fed into the APS distribution system (Lepley, 1986).

The Sky Harbor plant has been the only major test in the U.S. of two-axis-tracking, point-focus, sunlight-concentrating PV technology. The system is of an early design relative to those currently receiving attention. During its early years, operation was plagued by inverter problems; these were reduced substantially by an inverter replacement in 1984. Other difficulties surfaced, however. Ground faults within the modules were attributed to water entry and the deterioration of Kapton insulation between the bypass diodes at high voltage and the grounded heat sinks. One ground fault led to a module fire, and the realization in the PV community that protection against double contingency failures is sometimes appropriate. A number of other difficulties plagued the Phoenix-Sky Harbor system: tracker control boxes, cable failures, and gearbox slippage in one axis because of lubricant leakage (Schaefer, 1990).

Because of the difficulties noted, this system was cannibalized over the years to provide properly functioning components; as a result the kW rating declined commensurately. Ratings in kW from 1982 to 1987 were 173, 167, 151, 151, 131, and 114, respectively. In spite of these difficulties, the Sky Harbor system demonstrated a generally rising capacity factor over its five-year life, as can be seen in Table 1.23, because APS made the engineering changes necessary to improve its reliability. The system's best performance was achieved during the six-month period prior to its decommissioning in 1987 (Schaefer, 1990).

Operation and maintenance costs per kWh for 1985 to 1987 were substantially higher for Sky Harbor than for the other systems discussed, as can be seen in Table 1.24. This was due primarily to the tracker problems (EPRI AP-5762 and AP-6251) and the reduced output due to declining ratings. It is clear that, from a system operation standpoint, concentrating technology is not as advanced as flat-plate technology.

### 1.6.3 Other Utility-Scale Experimental Installations

A number of other demonstration and test facilities that are also important to the utility industry's understanding of utility-scale applications of PV have been installed and operated by utilities. For example, PG&E is spearheading a major effort called Photovoltaics for Utility-Scale Applications that is evolving into the major U.S. test and demonstration activity of the 1990s for utility-oriented PV applications; APS has established a Solar Test and Research Center; Alabama Power is operating a 75 kW experimental installation employing thin-film PV technology; Florida Power Company has installed and is operating a 15 kW system using state-of-the-art commercially available amorphous-silicon thin-film modules; Florida Power and Light Company has designed, built and is operating a 10 kW array integrated into a Miami substation; and Platte River Power Authority has installed and is now operating a 10 kW pilot plant. These experimental installations are described below.

### 1.6.3A PVUSA

The Photovoltaics for Utility Scale Applications (PVUSA) project was initiated in 1987, following nearly two years of planning discussions among the founding cosponsors. This project is being cosponsored by federal and state government, EPRI, PG&E and several other utility organizations. The project is intended as a proving ground for advanced PV module technologies; and, equally important, for advanced balance-of-system (BOS) technologies in utility-scale applications (Hester, 1988).

#### 1.6.3A.1 PVUSA Objectives and Scope

The main objectives of PVUSA are:

- To directly compare and evaluate the electrical performance and reliability of promising current and emerging photovoltaic modules and balance-of-system components side by side at a single location.
- To conduct assessments of the operation and maintenance costs of photovoltaic systems within an electric utility.

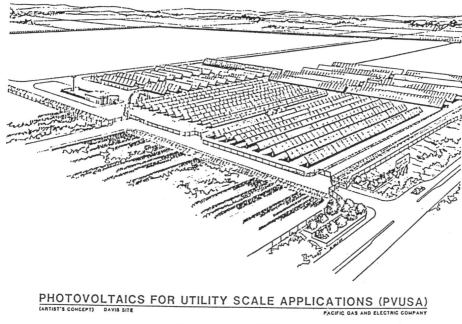

**PHOTOVOLTAICS FOR UTILITY SCALE APPLICATIONS (PVUSA)**
(ARTIST'S CONCEPT)   DAVIS SITE                                    PACIFIC GAS AND ELECTRIC COMPANY

**Figure 1.51:** PVUSA.

- To test and compare promising technologies in differing geographic areas with diverse environmental characteristics.
- To provide electric utilities with hands-on experience in the installation and operation of utility-scalable central station PV power generation systems.
- To document and disseminate the knowledge gained from the project.

To reach these objectives, the PVUSA project (Figure 1.51) will monitor the performance, reliability, operation, maintenance, and related technical and economic characteristics of many promising PV technologies/systems at a single location near Davis, California, and individual PV systems at other utility locations in the nation. PVUSA provides a utility-scalable test bed for system-level design, installation, operation, and evaluation of emerging photovoltaic module technologies and innovative, complete PV systems. The design, procurement, and installation will follow utility standards and specifications and the systems will be operated by utility personnel. The information derived from these efforts will be documented and presented to various audiences. The aim is that these systems become the building blocks for future utility PV power plants.

The project uses information developed by the California Energy Commission, EPRI, DOE and its national laboratories. The project also capitalizes on the knowledge and experience gained from PG&E's PV module test facility in San Ramon, California, and from other recent operating photovoltaic installations (e.g.,

SMUD PV1 and PV2, ARCO Solar's Carrisa Plains and Lugo PV stations, and other PV plants).

### 1.6.3A.2 Participants

PVUSA, as a cooperative research project, maintains broad based support from industry, utilities, and government agencies. The organizations formally participating in PVUSA as of early 1990 include:

- U.S. Department of Energy (DOE)
  - Sandia National Laboratories
  - Solar Energy Research Institute (SERI)
  - Jet Propulsion Laboratory (JPL)
- Electric Power Research Institute (EPRI)
- California Energy Commission (CEC)
- Pacific Gas and Electric Company (PG&E)
- Virginia Power/Commonwealth of Virginia
- State of Hawaii/Maui Electric
- San Diego Gas and Electric Company
- Salt River Project (Arizona)
- Niagara Mohawk Power Corporation (New York)
- New York State Energy Research and Development Authority (NYSERDA)

Of this group, PG&E, DOE, EPRI and CEC are founding cosponsors. Other organizations may join as project activities evolve.

### 1.6.3A.3 Schedule

Phase One of PVUSA is being developed around three major procurement stages. They are:

- Emerging Module Technologies 1 (EMT-1)
- Emerging Module Technologies 2 (EMT-2)
- Utility Scalable 1 (US-1)

The PVUSA Phase One schedule has been structured, as much as possible, to match the on-going technical progress within the U.S. photovoltaics industry. These procurement stages incorporate as many new and promising PV technologies into the project as is practical. The estimated cost of Phase One, from initiation of the construction activities to completion, is $18.5 million.

The Emerging Technology Segment procurements (EMT-1 and EMT-2) consist of 20 kW (nominal) arrays installed on PVUSA-supplied BOS (support structures, dc/ac inverters, and utility grid interconnection equipment) and are designed to demonstrate state-of-the-art PV technologies which show promise, but have not yet been field tested. The utility scalable procurements (US-1) (nominal 200 kW and 400 kW complete turnkey systems) will provide realistic operation and maintenance experience with utility-scalable PV systems. These procurements also expose both the project participants and the PV industry to typical commercial procurement and construction practices.

Contracts for EMT-1 were awarded in the spring of 1988. EMT-1 includes four 20 kW (nominal) fixed flat-plate arrays and one 20 kW (nominal) 2-axis tracking concentrator array, installed at the Davis site. Contracts for both EMT-2 and US-1 were awarded during June 1989. EMT-2 consists of three 20 kW (nominal) fixed flat-plate arrays. US-1 includes two 200 kW and one 400 kW complete turnkey PV systems. Additionally, there is a potential opportunity for PVUSA to evaluate an EPRI-developed point-contact concentrator array at the Davis site. Other array technologies may also be evaluated as opportunities arise. Table 1.25 lists the companies and PV technologies/systems to be tested at the Davis site.

Utility participants in the project have the opportunity under certain conditions to install a similar 20 kW EMT array or downsized (25 kW or 50 kW) US-1 system in their own service territory. This "hosting" of PVUSA systems will insure that those utilities that have sufficient interest will also obtain hands-on experience. Also, the project will provide an opportunity to compare the performance of similar systems under different environmental conditions. As of the end of 1989, Virginia Power, Hawaii/Maui Electric, and NYSERDA have agreed to host PV systems as part of PVUSA.

### 1.6.3B  APS Solar Test and Research Center

Arizona Public Service Company's PV experience includes, in addition to the Sky Harbor project, a 4 kW system installed at a residence in Yuma, Arizona. This was a retrofit project to study the operation of customer-sized PV systems connected to utility distribution feeders. It was performed in cooperation with the State of Arizona and Arizona State University. According to APS, the projects at Sky Harbor and in Yuma have shown that photovoltaics can produce electricity with the quality and reliability required for utility applications. However, APS agrees with most members of the PV community that many issues remain to be resolved. For example: a valid method of rating PV systems to reflect a realistic expectation of the power and energy they can produce is needed; a direct comparison of the various methods of tracking should be made; different types of modules and systems need to be compared and tested, and any design flaws or special utility requirements should be identified and the information conveyed to manufacturers to foster product improvements.

To address some of these issues in its service territory, APS has constructed a Solar Test and Research (STAR) Center on an acre of land bordering APS's Ocotillo Power Plant in Tempe, Arizona. (Figure 1.52.) About 3 more acres are available for future expansion. The purpose of the center is to test and evaluate solar technologies from a utility perspective. A primary criterion used in the design of the center is that it provide flexibility for future expansion. The test facility has been fully operational since January, 1988.

Although APS funded the initial installation, additional funding has been provided by EPRI, Salt River Project, and WEST Associates. In addition, in cooperation with the Arizona Solar Energy Office, the facility is also testing equipment owned by the State of Arizona.

### 1.6.3B.1  Facility Description

The facility includes a 1,900 square foot control building, which houses the inverters, switchgear, instrumentation and control systems, and computers. Seven of the Martin Marietta two-axis trackers which were originally installed at the Sky Harbor project have been erected at STAR. Identical 2.1 kW flat plate arrays have been mounted on three of the trackers, and 2 kW concentrator arrays have been placed on another two of the trackers. The sixth array is used to test a variety of

**Table 1.25:** Technologies and Systems to be Tested at PVUSA Davis Site

| Emerging Technologies 1 | |
| --- | --- |
| Company | Technology |
| ARCO Solar (Installed) | Microgridded Crystalline Silicon |
| Sovonics (Installed) | Tandem Junction Amorphous Silicon |
| Solarex | Poly-Silicon Bifacial |
| Utility Power Group (Installed) | Tandem Junction Amorphous Silicon |
| ENTECH | 22X Linear Focus Concentrator, 2-Axis Tracking Silicon Cells |

| Emerging Technologies 2 | |
| --- | --- |
| Company | Technology |
| ARCO Solar | Copper Indium Diselenide (CIS) |
| AstroPower | Polycrystalline Silicon film on ceramic substrate |
| Photon Energy | CadmiumSulfide / CadmiumTelluride (CdS/CdTe) |

| Utility Scalable System 1 | |
| --- | --- |
| Company | Technology |
| ARCO Solar | 1-Axis Passive Tracking Flat-Plate, Crystalline Silicon Modules, Dickerson Inverter |
| Chronar | Fixed 30° Flat-Plate, Amorphous Silicon Modules, Magna Power Inverter |
| Integrated Power Corporation | 1-Axis Active Tracking Flat-Plate, Mobil Solar Modules, Omnion Inverter |

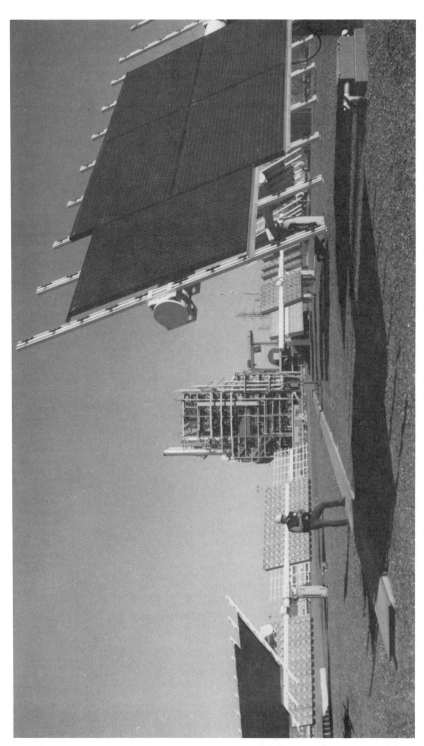

**Figure 1.52:** Arizona Public Service Company's STAR Center (Ocotillo Power Plant in background).

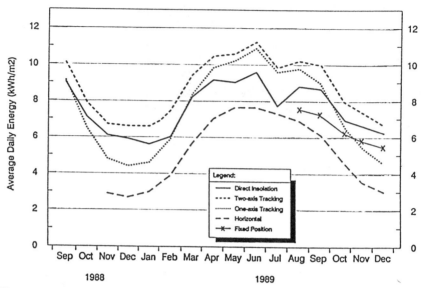

**Figure 1.53:** Arizona Public Service Company STAR Center insolation summary.

individual flat plate modules, and the seventh array is used primarily for testing of high-concentration PV modules for EPRI.

Seventeen individual flat plate and concentrator modules are also under test. They are loaded at their peak power point, and current-voltage (I-V curve) data are being collected five times daily.

### 1.6.3B.2 Preliminary Results

Insolation data recorded at STAR, Figure 1.53, show that two-axis-tracking total insolation exceeded one-axis-tracking total insolation by 14%. However, in the critical summer months (June–August) the difference was only 3.3%. This implies to APS that in Arizona very little additional expense will be justified in going to two-axis tracking for flat plate PV systems.

Of the five 2 kW systems operating at STAR, three are identical flat-plate systems each comprising 10 Mobil Solar modules and an Omnion inverter. These systems have been operating very reliably. One of these flat-plate systems is under two-axis tracking, another is programmed for one-axis tracking (N-S axis, horizontal), and the third is in a fixed position (south facing, and tilted at the latitude angle of 33 °). Energy production data from these systems have been used to calculate capacity factors. During winter months, the monthly capacity factor for all three arrays is in the range of 12–18%, and during summer months 20–35%. Capacity factors have also been calculated for APS's peak load period, and according to APS, the resulting values, shown in Figure 1.54, are significantly lower than values which have been reported by California utilities. This is primarily due to the fact that APS's peak load period extends later into the evening. After

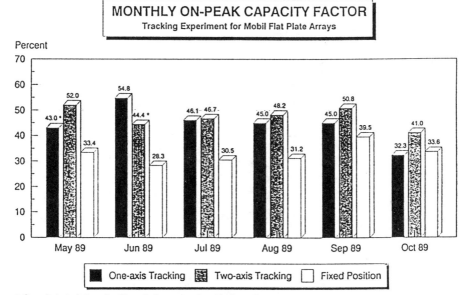

**Figure 1.54:** Arizona Public Service Company STAR Center capacity factor results.

approximately two years of testing are complete, the arrays will be reprogrammed to test different tracking options or fixed positions.

### 1.6.3C  Alabama Power Company

Alabama Power Company's involvement with PV, which began about 1984, was motivated by two considerations. One was that Alabama Power wanted to be recognized as an innovative, progressive utility. The other is that the Company viewed PV as potential competition for conventional utility generation if installed as customer-owned systems. To offset the potential impact of loss of load (market share), Alabama Power wanted to explore the possibility of becoming an installer and provider of PV systems to its customers. Alabama Power's effort began with a study which concluded that any PV installations in Alabama would most likely be distributed systems, rather than large central station systems, because the state has no desert-like areas that could be utilized for a large PV station.

### 1.6.3C.1  Resource Monitoring

As part of its efforts to evaluate PV, Alabama Power installed three solar insolation monitoring stations in different parts of its service territory. These stations collect solar radiation and weather data and complement similar stations installed by other companies in the Southern Company System (Georgia Power, Gulf Power, Mississippi Power, and Southern Company Services). Southern Company is

**Figure 1.55:** Alabama Power Company 75 kW photovoltaic demonstration project.

a utility holding company with five operating companies, a service company, an international engineering marketing company (Southern Electric International) and an investment company (Southern Investment Group).

### 1.6.3C.2  PV Manufacturing Facility

In 1985, Southern Investment Group entered into a Joint Venture and Formation Agreement with Chronar Corporation of New Jersey under the terms of which the Southern Company gained access to Chronar's technology, and a PV cell manufacturing facility was constructed in Birmingham in Alabama Power's service territory. This facility, which is managed by Alabama Power, produced its first modules in 1987. The plant has a capacity of about 1,000 kW per year of amorphous-silicon thin-film PV modules. As of the end of 1989, future operation of this production facility is uncertain, owing to corporate restructuring underway at Chronar Corporation.

### 1.6.3C.3  PV Demonstration Plant Construction

In the negotiation to obtain access to Chronar's technology, it was agreed that Alabama Power Company would install a demonstration PV generation facility in its service territory utilizing Chronar's amorphous-silicon thin-film photovoltaic modules. This resulted in the construction of a 75 kW photovoltaic array at Varnons, Alabama (Figure 1.55).

Alabama Power Company constructed the PV generation facility, including site preparation, grounding and foundation installation. Chronar was responsible for

**Table 1.26:** Alabama Power Company 75 KW Photovoltaic Demonstration Project Cost Data

| Item | Total Cost |
|---|---|
| Site Preparation & Installation | $ 343,000 |
| Chronar Contract | |
|     PV Array | 660,000 |
|     100-kW Inverter | 90,000 |
|     System Design and Engineering | 90,000 |
|     Test and Monitoring Equipment | 80,000 |
| Total Cost | $1,263,000 |
| $ / kW | $16,800 |

supplying all of the material above the foundation posts necessary to install the arrays, the arrays themselves and the equipment necessary to connect the arrays to the Alabama Power distribution grid. Construction began in late 1985 and was completed in early June 1986.

During the construction period, Alabama Power personnel worked closely with Chronar's personnel in making modifications to the photovoltaic array design and to the dc/ac inverter design. The project construction cost was $1.26 million or about $17/W (See Table 1.26). Because the project was intended to be presented as a "showcase," many features were included that would not be required for a competitive installation. Examples are the equipment building, restrooms, two HVAC systems, weather station and heavy duty PV panel support steel and posts.

The site was constructed for operation as an unmanned facility. Special features were added to allow early detection of problems by Alabama Power operating personnel. Also, special accounting techniques were employed to determine the operating and maintenance expenses associated with the photovoltaic array.

### 1.6.3C.4 Project Results

The array at Varnons has yielded much information of use in the development of Chronar's modules and of PV plant designs. A number of problems arose during construction and initial operation. The major construction problems encountered were late PV panel deliveries, resulting in demobilization of the project for about four months; and problems with the inverter and elsewhere in the system after energizing, resulting in several design changes to the inverter and system.

As is typical with amorphous-silicon thin-film modules, efficiency degradation occurred upon initial illumination of the cells. Plant performance is typified by results for August 1986 to September 1987 presented in a paper (Schaefer et al., 1988) summarizing experiences with several utility-operated thin-film PV test installations. Overall array efficiency at Varnons was about 2%. Also, as has been found with other experimental amorphous-silicon modules, corrosion of the cell material was observed. Ground faults at the project began to occur in early 1987, often initiating in the lower end of the modules where water collected. Usually these faults burned themselves clean, leaving a charred hole in the module. Alabama Power began corrective action by replacing modules in December 1987. The experience with the modules, which included problems with encapsulation and electrical isolation, has provided Chronar with important information leading to improvements in module design.

Alabama Power's experience indicates that future installations should be constructed with a more austere equipment building, and the PV panel supports should be sized to be more compatible with the strength of the panels themselves. The use of driven pipe supports could also be investigated to replace the poured concrete posts used at Varnons. Measurements have shown that harmonics are not a significant problem with the inverter, which was custom designed for this installation. In the future, inverters preferably should be standardized, shelf-item equipment. Overall, the installation has given Alabama Power an opportunity to work with the system, observe its characteristics, learn how the components fit together, and determine the safety problems associated with it.

### 1.6.3D Florida Power Corporation's "Solar Progress" PV Array

The Florida Power Corporation's (FPC's) "Solar Progress" 15 kW (nominal) photovoltaic system was installed in August 1988. The "Solar Progress" PV array is a joint project involving FPC, EPRI, Sandia National Laboratories (Sandia), and the Florida Solar Energy Center. The Florida Solar Energy Center was responsible for the array design, and is also monitoring system performance with support from EPRI. Sandia provided the inverters, and FPC was responsible for the system construction.

The array has a total area of 217.1 m$^2$ and faces true south at a tilt of 25 °. The array consists of 624 ARCO Solar, Inc. amorphous-silicon modules divided into three subfields, each designed to produce 5 kW of dc power. FPC procured and installed 640 modules; however, 16 were not connected until a new three-phase power conditioning unit (PCU) became available in 1989. During 1988, each subfield was connected to a separate DECC PCU with a single-phase 240 volt ac output. All three outputs were combined in a three-phase connection with the utility grid. Performance data are available from August 23, 1988. Some key results (EPRI GS-6696) for the last four months of 1988 are:

- The total energy production was 8.76 MWh dc and 7.79 MWh ac.
- Total solar insolation recorded on the array was 625.4 kWh/m$^2$.
- The plant availability averaged over 97 percent.
- There were no system failures during operation.

• The only system maintenance consisted of I-V curve testing and visual array inspection.

These results yield an array dc efficiency, based on total electrical energy production over the period, of 6.45%, and a system ac efficiency of 5.74%.

### 1.6.3E Florida Power & Light's Flagami Substation Experiment

Florida Power and Light Company (FPL) first became involved in photovoltaic power in 1975 during the consideration of a possible business venture with a major photovoltaic manufacturing company. Although these efforts did not bear fruit, they heightened the utility's awareness of PV, and FPL has since continued to pursue various assessments and investigations of PV energy systems.

FPL has been a supporter of the Florida Solar Energy Center since its origin in 1977, and was among the first supporters of the Center's successful effort to obtain a $2 million DOE grant for the Southeast Residential Photovoltaic Experiment Station. The objective of the Southeast Residential Photovoltaic Experiment Station is to perform integrated research, development and testing, and to evaluate experiences with grid-connected PV systems in the Southeastern United States. FP&L has also conducted or participated in numerous other photovoltaic projects including DOE and EPRI funded activities. The most significant of its own activities to date is the Flagami Substation PV system experiment.

The Flagami Substation PV system (Figure 1.56) is rated to produce 10.5 kW of dc and 9.7 kW of ac electricity. The PV system is comprised of 256 ARCO Solar single crystal silicon modules. The system was designed for South Florida building codes, which require 125 mph wind speed design (the maximum wind speed experienced to date has not exceeded about 60 mph, according to FPL). It was also designed for minimum installation time. Once the foundations were installed, the FPL crew installed the eight prewired panels, pivots, posts and drive motors in half a day.

The system was designed to require a minimum of maintenance. In the event of a PV module failure, a unique detection system locates the malfunctioning module so that it can be replaced. The system operates unattended. The panels are rotated east to west 90 degrees each day by a computer which is programmed to derive the maximum power output based on the position of the sun. A DECC PCU converts the direct current into alternating current for use in the substation. This alternating current supplements the utility supply to substation transformer fans, control-house lights, air conditioning, and circuit breakers. Testing of the project began in June 1984 and has been ongoing since that time. The total project budget was $280,000.

### 1.6.3F Platte River Power Authority

Platte River Power Authority supplies electricity to four municipalities in northern Colorado: Estes Park, Fort Collins, Longmont and Loveland. In 1986, Platte River began to examine the feasibility of constructing a 10 kW photovoltaic pilot plant. Objectives of this project were to demonstrate the performance of full-scale modules, determine the feasibility of a utility-scale array, determine the reliability and efficiency of different photovoltaic technologies, and determine the combination of cell type and tracking mode best suited for the climate of northern Colorado.

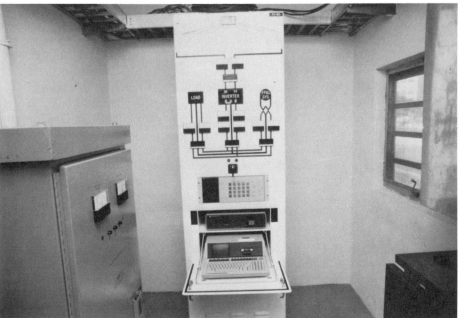

**Figure 1.56:** Florida Power & Light's 10 kW Flagami Substation PV system.

### 1.6.3F.1 Assessment of Colorado's Solar Resource

The installation of this project followed a multi-year solar resource assessment begun by Platte River in 1985. Although this study is still in progress, a report on the first two years of data collection has been published (EPRI AP-5883). The study found average daily insolation totals in kWh/m$^2$/day for the two-year monitoring period beginning June 1985 to be as follows:

|                       | Actual | Normalized |
| --------------------- | ------ | ---------- |
| Direct Normal         | 4.78   | 0.93       |
| Two-Axis Global       | 7.05   | 1.37       |
| Adjustable-Tilt Global | 5.43  | 1.06       |
| Fixed-Tilt Global     | 5.13   | 1.00       |
| Horizontal            | 4.52   | 0.88       |

While these results are specific to the northern Colorado area studied, they are important because, unlike the preponderance of insolation information, they are actual rather than estimated or projected data; and because they are very probably representative of the central part of the U.S.

### 1.6.3F.2 Project Feasibility

In January 1988, Platte River completed a detailed cost estimate and feasibility study for a 10 kW PV pilot plant (Platte River, 1988). This study examined the preliminary design and cost of three commercially available cell/tracking configurations. The study indicated that a 10 kW pilot plant could be constructed to demonstrate the three systems, each rated approximately 3.3 kW. Ultimately, four systems (A, B, C and D) were constructed:

- Two-axis tracking concentrator, System A
- Two-axis flat plate, System B
- One-axis tracking flat plate, System C
- Adjustable-tilt south facing flat plate, System D

Systems B and C are shown in Figure 1.57 and Figure 1.58, respectively. A two-year performance monitoring period of all four systems began in July 1988 (Emslie and Dollard, 1988).

It would have been difficult for Platte River to justify the cost of such systems without assistance. Therefore, project sponsors were sought. Ultimately six sponsors were obtained including:

| | |
| --- | ---: |
| American Public Power Association's Demonstration of Energy Efficient Developments (DEED) Program | $35,000 |
| City of Colorado Springs, Colorado | 6,250 |
| State of Colorado Office of Energy Conservation | 100,000 |
| Platte River Power Authority | 93,750 |
| Western Area Power Administration | 45,000 |
| Western Energy Supply and Transmission Associates | 20,000 |
| Total | $300,000 |

**Figure 1.57:** Platte River two-axis tracker with flat plate modules (System B).

## 1.6.3F.3 System Costs

The installed costs of the project are summarized in Table 1.27. These costs do not include the six additional concentrator modules provided at no cost by the manufacturer.

## 1.6.3F.4 Preliminary Results

Platte River's preliminary results (based on 15 months of monitoring) show that:

- The single axis tracking system (System C) produced the most energy for all months monitored.
- The single axis tracking system (System C) produced 5% more energy per unit of rating per operating day than the south facing adjustable tilt system (System D).
- The two axis tracking concentrator system (System A) produced only 54% of the energy per unit of rating per operating day of the single axis tracking system (System C). (However, as indicated below, results suggest that the nominal rating of System A was unreasonably high.)
- There needs to be a more realistic rating system, particularly for the concentrating systems. The concentrating PV modules never exceeded 67% of rating. This in part explains why the per unit energy production of the concentrator was so low. The flat plate modules achieved and even exceeded their ratings on the days with low temperatures and enhanced insolation (1000 W/m$^2$) due to ground reflection from snow cover.
- During a period of extreme haziness in August 1988 (caused by smoke from the fires in Yellowstone National Park), the efficiency of the concentrating system

**Figure 1.58:** Platte River one-axis passive tracker with flat plate modules (System C).

(System A) was cut in half while the flat plate systems did not drop in efficiency. The output of all the systems dropped because of the reduced insolation, but the concentrating system was more adversely affected because efficiency dropped also. This finding is especially important for urban locations, because the level of haziness was similar to that caused by persistent levels of air pollution that exist in some of these locations.

The Pilot Plant will continue to be monitored through 1990, and a final report will be prepared in early 1991.

**Table 1.27:** Installed Costs of the 10 KW Pilot Plant at Platte River Power Authority

| | |
|---|---|
| Site Work (Fencing, Grading, etc.) | $ 12,000 |
| PV Arrays | 69,000 |
| Trackers and Foundations | 33,000 |
| Inverter (three 3.5-Kw Units) | 19,000 |
| Cable and Conduits, Other Electrical | 9,000 |
| Indirect Costs | 13,000 |
| TOTAL | $155,000 |

Source: Emslie and Dollard, 1988.

### 1.6.4 Utility Customer-Size PV Installations

Another avenue through which utilities are becoming more involved with PV is customer-sized, line-connected systems. To date, these cannot be justified on economic grounds; nontheless, such systems are being installed. A recent survey conducted by EPRI identified 219 line-connected PV systems in the U.S. (EPRI GS-6306), totaling some 11 MW in capacity. A handful of these, including some of the megawatt-scale systems discussed earlier, account for the great majority of this capacity; but the remainder, some 200 systems, are residential-size systems, ranging from a few kW to a few tens of kW, either customer or utility owned. Many customers have installed such systems for non-economic reasons. This trend is likely to continue, particularly as costs are reduced and reliability improves.

Electric utilities have also been actively involved in tests and experiments involving customer-size PV systems. The following are descriptions of several of these activities, including residential PV arrays installed and tested in Gardner, Massachusetts by New England Electric; San Diego Gas & Electric's Laguna Del Mar project; Detroit Edison's 4 kW Oakland Community College PV demonstration; Los Angeles Department of Water and Power's Optimum Energy House; Madison Gas and Electric's Goodwill Project, Philadelphia Electric Company's PV test site, and Virginia Power's Integrated Solar Test Arrays.

### 1.6.4A New England Electric Residential PV

The companies comprising New England Electric System (NEES) have an ambitious research effort underway to study various aspects of the interaction of residential photovoltaic generation with utility distribution feeders. This effort includes the

**Figure 1.59:** Aerial view showing 6 PV homes in New England Electric System's Gardner, Massachusetts model PV community.

installation of a major cluster of residential PV arrays in Gardner, Massachusetts and three associated research projects. Selection of the houses comprising the Gardner Model PV Community was predicated on establishing a high saturation of inverters as may become typical on New England distribution feeders in the next century. The total cost of this program including installations and the associated research is about $2 million.

On September 12, 1985, Massachusetts Electric Company, a retail subsidiary of New England Electric System, signed a contract with Mobil Solar Energy Corporation to install residential photovoltaic arrays and inverters on a cluster of thirty homes in a neighborhood in Gardner. The roof mounted 2.2 kW PV installations (Figure 1.59) were completed in the fall of 1986. Each of the systems is comprised of a grid-connected PV array mounted parallel to the plane of the roof and occupies approximately 250 square feet. Direct-current source circuits are fed through the roof to a junction box in which the dc circuits are connected in parallel. The dc circuit is then connected to a dc to ac power inverter and connected directly to a circuit breaker on the electric power distribution panel.

All of the homes with these single-phase units are located on adjacent streets and are electrically connected at the very end of the same phase of a 13.8 kV distribution feeder. This arrangement was designed to model the saturation levels of photovoltaics on distribution feeders which might be expected in the twenty-first century. The three research projects conducted as part of this program are described below.

### 1.6.4A.1  Photovoltaic Inverter Voltage Excursion Research Project

Before energizing the photovoltaic cluster at Gardner, a research project was completed that analyzed the possibility of voltage excursions on isolated distribution feeders caused by photovoltaic inverters. This research was conducted by Electric Research Management, Inc. for NEES and three other industry cofunders.

The objective of this project was to study whether a distribution feeder could be kept energized by a photovoltaic inverter if isolated from the rest of the utility system. A hybrid analog/digital power system simulator located at Purdue University was used to model the residential inverters connected to a representative distribution feeder. The Gardner inverters were modeled and analyzed to determine if they could keep an isolated feeder energized.

The results indicate that in the event of separation from the utility system, inverters without any protective schemes could keep a distribution feeder energized. It was also concluded that the inverters to be used at Gardner had adequate integrated protective schemes to prevent this from happening. Recommendations were made on the types of protective schemes that could be employed in inverters to prevent continued generation or voltage excursions after isolation from the utility system. It was recognized that most reputable commercial inverters have these protective schemes incorporated within their designs (Bower et al., 1985).

### 1.6.4A.2  PV Inverter Harmonic Baseline Monitoring Research Project

During the installation of the photovoltaic arrays in Gardner, NEES funded a research project by the Worcester Polytechnic Institute (WPI) to study the voltage and current harmonic distortions produced by a photovoltaic system. This study was done on one of the PV homes with and without operation of th PV system. A further objective was to measure the level of harmonic activity recorded at the distribution substation supplying the area before all the installations were complete. The final objective was to field test a prototype harmonic monitoring system designed by WPI to efficiently and accurately store harmonic data over a lengthy testing period.

The results of this research indicated that the inverters used at Gardner are of a high quality design and inject currents low in harmonic distortion. The monitoring system was very successful in capturing a baseline harmonic recording of both a PV residence and the distribution feeder for comparison with the scenario with the PV cluster in total operation. A computer program was developed as part of this project to display and summarize the harmonic data collected into an easily understandable format (Orr et al., 1986).

### 1.6.4A.3  PV Generation Effects Research Project

At the end of 1986, New England Power Service Company (NEPSCo), a subsidiary of NEES, was awarded a $550,000 research contract by EPRI to study the wide range of mutual interactions between a high saturation of residential photovoltaic inverters and an average utility distribution feeder. NEPSCo subcontracted with Ascension Technology, Inc., Electric Research and Management, Inc., and Worcester Polytechnic Institute to conduct this research.

Ascension Technology studied the 60 Hz steady-state and slow transient response of the local distribution system with its relatively high saturation of inverters. This included modeling and confirmation of cloud effects on PV output and voltage regulation of the feeder. The goal was to develop a means to study these effects and incorporate them into distribution system planning and operating procedures. Electric Research and Management studied the fast transient effects of the inverters on the feeder. This included investigating the effects of islanding the PV cluster from the utility feeder as well as studying the effects of electrical transients caused by lightning. Worcester Polytechnic Institute investigated the harmonic distortion of the distribution feeder. This included studying harmonic interactions of the inverters with home appliances, other inverters and utility distribution equipment.

The study's conclusion regarding the impact of cloud-induced effects on distribution circuits with customer-size PV generation is that conventional feeder designs and voltage regulation techniques deal adequately with the PV-induced load flow fluctuations. Also, it was found that the inverters did not create problems for either the utility or the customers during utility system faults, lightning-induced surges, capacitor switching, or large load changes. And, well-designed PV inverters result in very small amounts of harmonic voltage and current distortion on the distribution system, even when operated in clusters or at high levels of penetration (New England Power Service Company et al., 1989).

### 1.6.4A.4 Overall Project Results
The results of the Gardner Project show that PV systems can contribute strongly and frequently to reducing residential demand during summer peak hours. The monitoring of the PV cluster showed that cloud-induced net load fluctuations caused only minimal effects on feeder voltage regulation and that the forced commutated inverter with its internal sine wave generator and protection package performed flawlessly and contributed very little additional harmonic voltage distortion to the distribution feeder. The monitoring is continuing with the objective of developing analytical tools to predict the impacts of large customer-size PV penetrations on distribution feeder loading, design, protection, and control (Gulachenski et al., 1988).

### 1.6.4B SDG&E Laguna Del Mar Project
San Diego Gas & Electric Company is participating in a residential PV demonstration involving 36 townhouses at Laguna Del Mar, Carlsbad, California. Each PV system consists of a PV array rated at 1.1 kWdc and an inverter. The following information about the project is based upon a report prepared by SDG&E (Mishler, 1989).

### 1.6.4B.1 Project Description
The Laguna Del Mar project was sponsored by SDG&E, Native Sun/Carew Development Company, Photocomm, Inc., ARCO Solar, Inc., Pacific Inverter, Sandia National Laboratories, and EPRI. SDG&E's objectives for its participation in this project were to determine the electric system impacts of residential PV and to evaluate the potential of PV as a future energy resource. This was accomplished by recording and evaluating operating performance data for the rooftop PV systems. The project afforded SDG&E the opportunity to investigate PV's impact on residential load profiles and the issues associated with the interconnection of residential PV systems.

The first phase of a three-phase subdivision development was targeted for this project. Each of the 36 townhomes in the first phase was equipped with a PV system. The PV systems were optional for the latter two phases. A standard PV array consisted of 27 single-crystal ARCO Solar modules totaling 1.1 KWdc. The aperture area of each of the arrays was 10 m², and they were flush-mounted to the roofs of the townhomes. The dc energy produced was converted to ac using a self-commutated inverter rated at 2.1 kW manufactured by Pacific Inverter. This oversized inverter allowed for future expansion of the arrays.

Native Sun/Carew and Photocomm worked together to incorporate PV in the subdivision. Photocomm was able to install the PV systems quickly and inexpensively by hiring a solar water heater contractor to preassemble the panels and install them as a single, 27-module assembly.

### 1.6.4B.2  Project Results

Based on 22 months of testing during the period 1987–1988, the PV systems were found to be performing adequately, in spite of some sub-optimal installations. Many systems are facing east or west or south or have minor shading from nearby obstructions.

The PV electricity production pattern, as expected, corresponded to the level of insolation throughout the day. A monthly average of 115 kWh was produced by a typical PV system. This PV output displaced energy normally provided by the utility and enabled sale of some excess energy back to the utility. However, the residences remained net purchasers of utility energy. The majority (81%) of the total household requirements were supplied by the utility and the balance from the PV systems. On average, the household PV systems produced an amount of energy equal to 35% of the average household needs. Approximately 45% of the energy generated was sold back to SDG&E.

PV reduced the household energy costs by reducing the amount of energy purchased from SDG&E and by enabling the households to obtain credits for excess energy sold to the utility. SDG&E, on the other hand, lost sales revenue throughout the year. SDG&E found that its 1987 summer system peak demand (which occurred at 3:00 p.m.) was reduced while the 1987 winter peak (which occurred at 5:30 p.m.) was unaffected by the output of the PV installations. Without storage capacity, the PV systems were unable to impact SDG&E's secondary residential evening peak (which occurred at 8:00 p.m.). The utility benefits from the reduction in peak demand, but is adversely impacted by the reduction in overall sales revenue.

Currently these customers take service under a standard residential rate schedule and are billed for the net energy consumed at the household. The average PV system saved the customer $171 per year in electric service costs compared to a townhome without PV.

In spite of these savings, SDG&E's economic analyses results indicate that the PV systems are not cost effective. This is attributed to the high capital cost of the PV systems relative to the value of the reduction in system coincident peak demand. The high capital cost of the PV systems substantially outweighs any benefits from avoided energy and capacity costs. SDG&E estimates that a reduction to one-sixth of the current PV capital cost is required before the economics begin to favor PV,

**Figure 1.60:** Detroit Edison's 4 kW amorphous-silicon PV array installed at Oakland Community College in Southeast Michigan.

all else being held constant. Other events that could shift the economics in favor of PV are PV performance improvements and escalating electric rates.

Limited testing of harmonics generated by the inverters showed no significant problems.

### 1.6.4C Detroit Edison PV Demonstration Project

Detroit Edison is leading a group of corporate participants in demonstrating and evaluating an amorphous-silicon, thin-film photovoltaic power production facility in southeast Michigan. Other participants in the project include Consumers Power Company, the Lansing Board of Water and Light; Energy Conversion Devices/Sovonics Solar Systems; Oakland Community College (Auburn Hills, MI); EPRI; and the Michigan Energy and Resources Research Association.

A 4 kW array of amorphous-silicon photovoltaic panels has been installed on the Auburn Hills Campus of Oakland Community College near Rochester, Michigan (Figure 1.60). The dc power produced by the array is delivered to a nearby classroom building where it is converted to ac and used for lighting, air conditioning and other electrical equipment within the campus.

The 4 kW array became operational on May 6, 1987. Since startup, the system has performed reliably and consistently. Generation outages have been caused primarily by power line outages and power surges during electrical storms. While

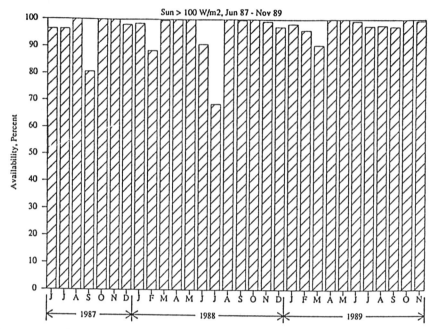

**Figure 1.61:** Detroit Edison's Oakland Community College 4 kW array monthly PV system availability.

the inverter should automatically restart after such occurrences, it has failed to do so on several occasions and manual resetting has been required. In over two and one-half years of continuous operation, very few components or parts have been replaced. Detroit Edison considers it a remarkable achievement to build a power generating facility that is essentially maintenance free (Pratt, 1990).

Figure 1.61 shows the percentage of time positive ac power was available during times of insolation greater than 100 watts/m² over a 31 month period following startup in 1987. In Figure 1.62, dc capacity factors are shown for 30 months of operation. The average yearly capacity factor over that time period was 14.3 percent. Monthly energy output from the array has varied considerably, mostly as a result of seasonal and daily insolation variation. The lowest monthly energy delivery (124 kWh dc, 56 kWh ac) occurred in December, 1987, an uncommonly dark and cloudy month in southeast Michigan. The best month was July, 1987 when 652 kWh dc were produced by the array and 507 kWh ac were delivered to the campus (Pratt, 1990). The 1988 performance is summarized in a recent EPRI report (GS-6696).

Conversion efficiency since startup has decreased from in excess of 4% to about 3.5%. Changes in efficiency have been shown to be related to ambient temperature, air mass and duration of exposure at high summer temperatures, the latter likely contributing to an annealing effect which raises efficiency. In addition, a gradual continuing efficiency decline has been identified, although its cause is still unknown (Pratt, 1990).

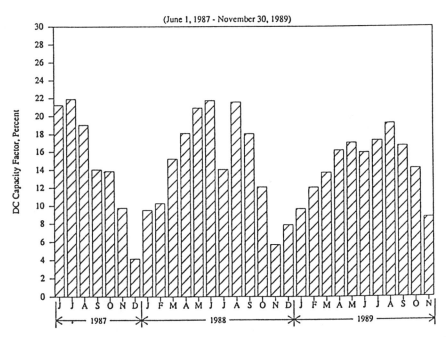

**Figure 1.62:** Detroit Edison's Oakland Community College 4 kW array monthly PV system capacity factor.

### 1.6.4D LADWP Optimum Energy House

A 2 kW demonstration PV system was installed at the Los Angeles Department of Water and Power's (LADWP's) Optimum Energy House (Figure 1.63) during the summer of 1984. The purpose of this project was to provide consumers and LADWP personnel with an opportunity to see and analyze a solar-electric technology. The Optimum Energy House is a single-family residential unit in which various energy and water saving ideas are showcased.

The addition of the PV system provided the house with the capability of generating some of the electricity needed for cooking, lighting, appliances, and air conditioning. The 2 kW size was believed large enough for test purposes. Based on preliminary calculations (before construction), it was anticipated that a typical fixed flat-plate PV system in the Los Angeles area would produce about 3,500 kWh net annually (20% capacity factor). LADWP collected data on the system for three years and reports that it performed well.

As with many of the other PV installations, the LADWP system's actual output has been within a few percent of predictions (McAvoy, 1986). The house and PV system have been turned over to another City department and the system is still operating.

**Figure 1.63:** Los Angeles Department of Water and Power's Optimum Energy House.

**Figure 1.64:** Madison Gas & Electric's Goodwill Photovoltaic System installed in an eight-unit apartment building.

### 1.6.4E  MG&E's Goodwill PV Project

The Goodwill Photovoltaic Project is one of a number of demonstration projects that Madison Gas and Electric Company (MG&E) is investigating as part of its Innovative Energy Program. This program has a goal of obtaining experience in the innovative and alternative energy areas that will allow MG&E to provide customers and employees with information on the effectiveness of new energy technologies. The Goodwill project, Figure 1.64, is a 2 kW PV system that serves an eight-unit apartment building. MG&E funded the design and installation of the PV system. The installation was completed in 1983.

The system consists of two rows of 27 PV panels (54 total) mounted on the roof at a 45 ° angle facing south. All 54 PV panels consist of polycrystalline silicon cells and were manufactured by Solarex. Auxiliary equipment for the project includes an inverter, an isolation transfomer, a voltage/frequency protection relay, a monitoring system and miscellaneous disconnect switches, meters and circuit breakers.

The PV system performs automatically without any special operating procedures. Maintenance has been limited to the power conditioning equipment. Data collected shows panel efficiency ranging between 4 and 8%, with an average of 6.4% (MG&E, 1988).

### 1.6.4F  Philadelphia Electric PV Test Site

Philadelphia Electric Company (PECo) has been operating and monitoring the performance of PV systems for several years. PECo has been studying

**Figure 1.65:** Philadelphia Electric Company's PV Test Site. Photo by Charles Peatross, Philadelphia Electric Company.

residential-size systems because it believes that these installations are more likely to appear in its service territory than large-scale central station PV. In order to continue these experiments and acquire field experience with a new technology, PECo joined with Solarex Thin Film Division (Solarex) in a research project with the following objectives:

- to investigate the integration of thin-film amorphous-silicon PV modules into the residential and commercial load sectors,
- to obtain "real world" experience with the environmental effects of first generation amorphous-silicon modules,
- to examine the results of series and parallel connection of small 5-watt modules into panels and arrays, and
- to gain utility system experience in a residential-sized PV experiment.

### 1.6.4F.1 Project Description

For the first phase of this ongoing project, PECo purchased 4.0 kW of amorphous-silicon thin film modules. The modules are connected to two residential-sized inverters at a new PV test site at the Pottstown-Limerick Airport in Limerick Township, Pennsylvania. A picture of this site is shown in Figure 1.65. The site was completed and became fully operational in August 1986.

The PV system consists of two arrays of approximately 2.2 kW and 1.8 kW (peak dc power) output. The physical support structure and installation of the arrays, along with the electrical interface, were designed by Solarex to meet site and PECo system requirements.

The PV array output is interconnected with the PECo distribution system through both line- and self-commutated single-phase inverters. The line-commutated inverter is a 1.6 kW unit which operates at 120 Vac. The self-commutated inverter is a 2 kW unit which operates at 240 Vac. The inverters are connected in parallel on the ac side. Because one inverter operates at 120 Vac and the other at 240 Vac, they only share one common leg of a 240 V, 3-wire electric service. According to PECo, this mixed arrangement could be typical of residential systems in the future. The inverters are connected into a breaker panel which is also equipped to supply an assortment of ac loads. This panel with its inverters and loads can be isolated from the rest of the building for islanding studies (Fagnan et al., 1987).

### 1.6.4F.2 Experience with Inverters

The 2 kW self-commutated inverter has achieved a peak output of 2.1 kW while the 1.6 kW line-commutated inverter has only attained a peak output of 1.36 kW. The line-commutated inverter trips and shuts down at power levels that exceed this 1.36 kW level. Test results suggest that these premature trippings are caused by dc ripple on the array. When the inverter is reconnected to a dc power supply, it is able to obtain a 1.6 kW output without tripping. As a result, the line-commutated inverter must be operated below its peak rating (Fagnan et al., 1987).

A series of islanding tests were performed to determine if the inverters would continue to operate when separated from the utility system. The results of these islanding tests show that the line-commutated inverter did run on (continue to operate) when running in parallel with the self-commutated inverter. The self-commutated inverter did not run on. Voltage surges were also noted under some conditions (Fagnan et al., 1987).

Both of the inverters at the PECo site required modification to operate properly with amorphous-silicon's relatively high ratio of open circuit voltage to operating voltage. Leakage current to ground was detected, particularly during wet weather, and because one inverter employs a ground current detection circuit, it sometimes failed to operate with wet modules. This failure of inverters to operate in the presence of ground leakage current has been noted with other modules, where the solution sometimes has been to disable the inverter's ground current sensor (Schaefer et al., 1988).

### 1.6.4F.3 Experience with Module Array

Data on the performance of selected modules and strings were collected and comparisons to the overall array performance were made. Observations of the effects of the environment on the modules indicated that most defects contained in the modules did not change in appearance, size or shape. This was of interest since some of these defects (such as large pinholes, small cracks or missing sections of amorphous-silicon near scribe lines) were thought to be weak areas which would grow or be sites for accelerated corrosion.

However, as discussed earlier with regard to the Alabama Power PV installation, the amorphous-silicon modules at PECo did suffer corrosion damage. The visible manifestation was a delamination of the material within individual cells, starting at the bottom and working its way up the vertically oriented cells. These were called "bar graphs." The 2.0 kW system, with a solidly grounded positive, showed worse damage than the 1.6 kW system, in which neither polarity was grounded (Schaefer et al., 1988).

Ten open circuited strings were identified in the Philadelphia system in December, 1987, and new modules were installed (Schaefer et al., 1988).

### 1.6.4F.4  Phase Two Activities

Site modification for Phase 2 was conducted during the winter of 1988, and inverter installation and system check-out were completed during the summer. The PV array size was increased with the purchase by PECo of 512, second generation Solarex modules. The array field was partitioned into four arrays of about 2 kW each. Three of these will feed a 6 kW "mini central" inverter. The other array, which is bipolar ($+/-$ 250 Vdc), will feed a transformerless 2 kW inverter. PECo planned to test the operation of this system for at least a year. The program will include islanding tests, power quality measurements, and an assessment of connecting several PV sources into a common inverter.

Solarex plans to continue to monitor the module and total dc performance of both the older and new arrays. The new modules are expected to show improved environmental durability owing to a non-conducting polymer frame and improved high voltage isolation. After these modules had been in the field for 8 months, no bar-graph or other corrosion effects had been noted. Also, preliminary measurements indicate a much lower rate of light-induced degradation (D'Aiello et al., 1988; Fagnan and D'Aiello, 1990).

As with the Alabama Power Company installation, the PECo test program is providing a great deal of practical operating experience of importance to a particular manufacturer of thin-film PV modules, and to the flat plate community at large.

### 1.6.4G  Virginia Power Integrated Solar Test Arrays

The Virginia Integrated Solar Test Arrays (VISTA), located at Virginia Power's North Anna Power Station, near Richmond, Virginia and shown in Figure 1.66, consist of three different and independent 25 kW PV power systems designed to provide information on the performance of prospective commercial-size PV systems in the Virginia Power service area. The three segments produced over 97 MWh during the twelve month test period from October, 1986 to September, 1987. Two different operating modes and two different PV technologies (single and polycrystalline materials) were tested during this period. By using 2-axis tracking, annual output increased nearly 14% compared with stationary PV operation. However, more data is needed to fully determine the economics of tracking compared to fixed angle systems. Annual capacity factors for the three 25 kW segments ranged from 14% to nearly 17%, which is typical of PV systems on the east coast where the available solar energy is substantially less than that in the Southwest. Annual average availability (the percentage of daylight hours during which the PV system was able to produce energy) for the three systems ranged from 86% to nearly 96% in the first year of testing (Green, 1988).

During the second test year (October 1987 - September 1988) one system (A) was operated in the 2-axis tracking mode, one (B) was set at a fixed angle adjusted at the solar equinoxes, and one (C) was set horizontally to simulate the simplest

**Figure 1.66:** Virginia Power's integrated solar test arrays.

operating mode possible. The differences in performance of the systems and from the previous test year were analyzed. In addition, power quality and other grid interconnection issues were evaluated.

### 1.6.4G.1 Operating Comparisons

Measured total global insolation for the second test year was $2,214$ kWh/m$^2$, similar to the first test year and greater than the typical year data used for predictions. After adjusting to account for differences in availability, System A produced only about 18% more energy than System B, and System C produced about 42% less energy than System B due to its horizontal orientation. However, System C also showed a 20% decrease in efficiency of the single-crystal cells in the system. This result indicates that the efficiency of cells in System C may decrease more rapidly under dispersed or off-optimum insolation conditions than the cells in the other two systems. This finding may be significant since a large portion of the Virginia area's solar resource is dispersed (Virginia Power, 1989).

### 1.6.4G.2 Reliability

The three systems continued to demonstrate that availabilities greater than 90% are achievable for unmanned commercial PV systems. Power conditioning unit trips due to small random dc ground faults and low utility voltage continue to be the major causes of downtime. Virginia Power believes this downtime can easily be reduced in commercial systems by remote reset capabilities. The tracking devices continued to be high maintenance items and, late in the test period, were found to have some previously damaged azimuth gears due to high winds and snow. Some blown fuses and polycrystalline module delaminations were also experienced during the period. Power quality from the systems continues to be satisfactory with voltage THD of less than 5% (Virginia Power, 1989).

### 1.6.4G.3  Conclusions

VISTA's results indicate that PV promises to be a reliable energy technology in the Virginia area. The project is providing information necessary to evaluate and maximize the performance of PV in the Virginia Power service area. In the next test year, repairs to the damaged panels and trackers will be made and testing continued. In addition, in 1990, a new thin-film system will be installed in cooperation with the Commonwealth of Virginia and DOE as part of the PVUSA project (Virginia Power, 1989).

### 1.6.5  Utility Application and Integration Assessment

In addition to tests such as those described above, utilities have conducted numerous studies and assessments involving PV technology. Two such efforts are summarized here. One is a very substantial effort undertaken by PG&E to evaluate the potential for utility-owned distributed PV. The other is the power quality testing work conducted by Salt River Project at a privately-owned housing development.

### 1.6.5A  PG&E Evaluation of Utility-Owned Distributed PV

One of PG&E's several PV projects involves investigation of prospective non-centralized electric utility applications of PV. The ultimate goal of this project is to demonstrate the economic viability of PV for a range of utility applications and to consider use of PV in cases where it is more cost effective than the alternatives (assuming comparable reliability and functionality). Where possible, PG&E characterized key economic considerations and other factors which substantially influence the feasibility of using utility-owned PV systems to serve customers in remote locations or to meet the utility's own internal operations needs. The results will be used to assist PG&E in evaluating the merits of PV for various applications.

Based on PG&E's preliminary findings, there are many remote sites, perhaps several thousand, where stand-alone photovoltaic power generation may be the most economical power source. The findings also indicate that PV could be used for many applications within the utility's electrical distribution system. Overall, three broad categories of application types have been identified:

- Utility-system infrastructure applications. These are distributed, grid-connected PV systems which are usually connected directly to distribution lines or equipment.
- Internal/operations applications. These are utility applications which are not grid-connected and non-power service related.
- Customer service applications. These are primarily remote, stand-alone PV systems with capacities ranging from 0.5 to 5 kW to provide power to individual customers.

To further pursue these options, PG&E's R&D Department initiated an ongoing "technology transfer" activity within the Company. The Department's Advanced Energy Systems Program staff met with marketing representatives, job cost estimators, new business representatives, land management persons, civil engineers, distribution system engineers, and others throughout the PG&E system. The purpose of these meetings was to seek additional input concerning the potential for PV as a service option or for PG&E's own utility system or internal operations applications. To quantify the potential for PG&E's internal uses for PV, PG&E surveyed all of its gas and electric operating departments. By the end of 1989, PG&E had identified over 400 cost-effective applications ranging from 10 to 7,200 watts in size, for a total of 32 kW. A list of these applications is provided in Table 1.28 (Jennings, 1989).

Key factors that will influence PG&E's decision to install PV for distributed applications include the comparative cost of alternatives, peak and average power requirements of the load, specific reliability requirements of a customer or project, and the cost of connecting to the grid (i.e., labor costs, distance, local terrain, local geology, land costs, easement costs, etc). The opinions and suggestions of PG&E engineering and field personnel who are familiar with the applications and who would implement any company-wide effort to use PV extensively are important to this process (Eyer et al., 1988).

PG&E's preliminary conclusion is that in specific cases, PV power can provide the most cost-effective power supply for a range of utility-related applications. Some preliminary analysis results comparing utility-owned PV system costs to utility line extensions and also to private diesels are shown in Figure 1.67 and Figure 1.68, respectively. However, further study is required to quantify the benefits and impacts that utility-owned PV systems could have on utility operations. Additional study is also needed to develop a more definitive characterization of the conditions that will make PV the power source of choice for various applications. The evaluations should adapt accepted methodologies for assessing the feasibility of various projects so that they include provisions to compare PV with conventional alternatives (Eyer et al., 1988).

### 1.6.5B Salt River Project Power Quality Testing

The Salt River Project (SRP) has also participated in a variety of PV activities and projects. One of these involved power quality testing at John F. Long's "Solar 1" subdivision, a privately-owned 24-home development with 200 kW of generation from an ARCO Solar PV array. The power from the PV array feeds a 150 kW Toshiba inverter. The power flows through a 600/12.47 kV power/isolation transformer and is distributed at 12.5 kV to three 50 kVA pad-mount transformers serving the homes. Any excess power is purchased by SRP. The PV system started producing power on November 19, 1985.

The study measured electrical noise and power quality at various points on the system with particular emphasis on harmonic content. To minimize the harmonics on its utility system, SRP requires that interconnected customer or third-party facilities keep harmonic distortion under 5 percent total harmonic distortion (THD) for current with a maximum of 3 percent for any component and 2 percent THD for voltage with a maximum of 1 percent for any component.

The power quality testing at John F. Long's "Solar 1" photovoltaic system revealed that SRP's specified limits for current and voltage harmonic distortion

**Table 1.28:** PG&E's Cost-Effective PV Installations by Application

| Application | April 1989 | | December 1989 | |
| | Number of Systems | Total PV Capacity (watts) | Number of Systems | Total PV Capacity (watts) |
|---|---|---|---|---|
| Water Level Sensors | 108 | 5,210 | 128 | 5,710 |
| Automated Gas Meters | 35 | 390 | 124 | 1,410 |
| Gas Flow Computers | 19 | 190 | 104 | 1,040 |
| Gas SCADA RTUs | 16 | 4,550 | 16 | 4,550 |
| Cloud Seeders | 8 | 480 | 8 | 480 |
| Meteorological Towers | 6 | 60 | 8 | 80 |
| Microwave Repeaters | 5 | 3,110 | 5 | 3,110 |
| Warning Sirens | 5 | 50 | 5 | 50 |
| Aircraft Warning Beacons | 4 | 1,520 | 4 | 1,520 |
| Gas Samplers | 4 | 40 | 4 | 40 |
| Cathodic Protection | 2 | 9,800 | 4 | 10,800 |
| Rupture Control Valve | 1 | 60 | 1 | 60 |
| Lights | 1 | 40 | 1 | 40 |
| Automatic Gate Opener | 1 | 40 | 1 | 40 |
| Backup Generator Starter | 1 | 20 | 1 | 20 |
| Water Temperature Sensor | 1 | 10 | 1 | 10 |
| Generator Replacement | 0 | 0 | 1 | 3,000 |
| Total | 217 | 25,570 | 416 | 31,960 |

Source: Jennings, 1989.

were met by the project. These limits are about twice the average level of distortion from other sources on Salt River Project's system (Morris, 1987).

## 1.6.6  Additional Input From Utilities

The following selected summaries of information provided by the utilities indicated offer some further insights into the nature and extent of utility industry involvement in photovoltaics.

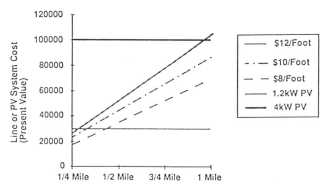

**Figure 1.67:** Present value cost comparison for line extensions and PV systems (sample calculation by PG&E. Source: Eyer et al., 1989).

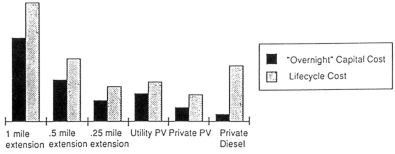

**Figure 1.68:** Relative costs of power source alternatives for typical residential customer (analyses by PG&E. Source: Eyer et al., 1988).

### 1.6.6A  Public Service Electric and Gas

PSE&G's involvement with PV began in the early 1980s with testing of several PV panels on the roof of PSE&G's testing laboratory. The objective was to monitor the technical performance of each panel. Subsequently, during 1981–1982, PSE&G worked with the RCA David Sarnoff Laboratories (now a part of SRI International) in a joint effort to develop a conceptual design for a 50 MW PV power plant sited in central New Jersey. The projected capacity factor of the conceptual design was about 19%. An analysis of the value of PV to PSE&G was also performed. The study conclusion was that PV power plants would not be economically justified unless their cost is reduced to the range of $1 to $1.50 per installed peak watt, well below the then current cost of about $10 per installed peak watt. These values were derived on the basis of PSE&G's load and generation mix. They assumed a directly-connected system without storage backup. PSE&G still considers these results to be valid.

Also, PSE&G determined that large amounts of land would be required for a central PV plant (about one square mile/100 MW). The high cost of central-station PV and the problem that would arise with land utilization led PSE&G to conclude that central-station PV is not likely to become competitive in New Jersey. PSE&G

also evaluated an offshore 100 MW photovoltaic installation with Stevens Institute. The results were similar to the conclusions on land-based PV. The option is too expensive and space consuming.

PSE&G also studied distributed PV through efforts involving several other utilities under a contract with Sandia National Laboratories. Other utilities involved in the study were Georgia Power, Salt River Project, and Southern California Edison. The objective was to examine the effects of interconnected PV on the utility system. The project involved a number of tasks. As part of the effort, PSE&G devised various tests and arranged to have them performed by MIT. Also, some conceptual studies were conducted to assess the effect on utility planning, from the transmission level down to the customer feeder level.

Sandia also loaned PSE&G a 3 kW array, which PSE&G installed and is still operating at its Battery Energy Storage Test (BEST) Facility for evaluation purposes. As part of this effort, PSE&G has evaluated the quality of the power produced by the system. Various commercially available inverters were installed and tested under different load conditions. PSE&G found that the harmonic content of the output depended on the characteristics of system load at the time of measurement. Some of the results were used to develop specifications that PSE&G has established for any grid-connected sources owned by customers, such as PV, wind, motor-generation, etc. These specifications address such requirements as grounding, safety, interconnection, and inspection by PSE&G.

In addition to PV, PSE&G has pursued solar heating and cooling and wind, as previously discussed. PSE&G believes that if there are any major developments in renewables, they will most likely occur in the area of PV. For the other technologies, they feel, there can be only marginal improvements, not breakthroughs, as can occur with PV.

### 1.6.6B  Philadelphia Electric Company

Philadelphia Electric Company's experience with PV dates back to the mid-1970s when PECo acquired some aerospace solar cells and used them to charge batteries. This was done principally to acquire some hands-on experience with PV. The system cost about $500/watt. PECo believes that residential customers will be the first to use PV. Hence, the PECo program has concentrated on smaller, residential-sized installatic 's rather than central stations.

Around 1980, PECo began working with General Electric (GE) which had developed a type of PV roof shingle. GE installed a 1.2 kW test module using the shingle at its King of Prussia, Pennsylvania facility and invited PECo to use the output. PECo purchased a 2 kW residential-sized inverter and connected the array into its grid in August 1982. The inverter worked quite well. Later, another inverter was also purchased and tested. Subsequent to this project, PECo installed its amorphous-silicon test site described earlier.

PECo has discussed several potential scenarios for customer applications of PV. One scenario is that builders will install PV and add the cost to the price of new homes. PECo is already receiving inquiries from customers who consider PV to be affordable; i.e., not necessarily economic, but within their range of acceptable cost. PECo believes that eventually PV manufacturers will produce a shingle with solar cells or other roof-top PV array at $4–$5 per watt and encourage builders to install them. In a home that costs $200,000 or more, an additional $10,000–$15,000 for a PV system may be viewed as affordable. PECo considers it likely that market penetration would come about in this way before PV is economical in an absolute sense. As PV becomes cost-effective, customers may even retrofit PV to existing homes. The first penetration, however, will most likely come from renovation or new construction, for which the cost of the PV can be included in the price for the entire project.

Looking to the future, PECo has determined that it may be advantageous for its customers and the company for PECo to provide the inverter as an item of distribution equipment, as it does now in the case of the distribution transformer. Typically, a customer does not worry about a distribution transformer's size, voltage ratio, or any other characteristic. Likewise, if the utility supplies the inverter, the customer would not worry about its size or specifications.

Assuming that the utility will provide the inverter, the question arises if there should be one at each residence or a central inverter at the distribution transformer? PECo is considering both alternatives. The central inverter, called a mini-central inverter by PECo, might be a 15 kVA inverter serving about six homes. However, a drawback of this approach is that it requires a separate dc distribution system. This raises a number of safety and other technical issues that would need to be addressed, such as the possibility of electrical interaction among houses connected to a common line.

Because of these potential problems, PECo has proposed another approach, which is to build the inverter into the customer's meter box. The inverter might, for example, be incorporated into an intelligent electronic package inserted into a replacement meter panel. In addition to the inverter, the package might include such features as remote switching and communications capabilities for remote meter reading and load management functions. The inverter might also be controlled to regulate harmonics and control power factor. The potential exists to have one microprocessor control all these functions.

### 1.6.6C Wisconsin Electric Power Company

Wisconsin Electric Power Company (WEPCO) has three 1 kW PV arrays located at Appleton, Waukesha, and the Point Beach Nuclear Plant Information Center (Figure 1.69). WEPCO personnel handled all of the wiring, trenching, digging, etc. as well as assembling the arrays along with the vendors. Two of the systems were supplied by Solarex and one by ARCO Solar.

The arrays have been operating since 1982. WEPCO has not washed the panels; normally there is sufficient rain to provide adequate washing. The Solarex systems are semi-crystalline, and the ARCO system is single-crystal silicon. All are flat-plate, fixed arrays. WEPCO's perception is that if customers are going to install PV systems, which it considers to be the more likely renewable energy option to be implemented in its service area, they are going to install roof-mounted fixed arrays. Therefore, the test arrays were placed in a fixed configuration to determine how they would perform.

For each system, WEPCO monitored output, power consumption by the inverter, power output from the inverter, and solar insolation at the plane of the array. There has been one major problem with the array at Appleton, which is located 40 miles

**Figure 1.69:** Wisconsin Electric Power Company 1 kW PV array. Photograph by Steven T. Jensen, Wisconsin Electric Power Company.

away from and at the same latitude as the Point Beach Plant array. Insolation is equal at the two sites, but the Appleton array has experienced significantly lower energy production.

WEPCO and Solarex have worked at length to determine the reasons for the difference in output at Appleton and Point Beach, including having Solarex measure voltage and current characteristics on each panel of both arrays. Solarex replaced the Appleton panels and the array still did not produce as much energy as the Point Beach array. The inverters were interchanged and also dismantled in an attempt to identify any differences, but none were found. The reasons for the difference have so far eluded both Solarex and WEPCO.

WEPCO now has over 7 years of data on insolation and output. WEPCO projects little PV capacity to be installed in its service territory until PV system prices decline. With a retail cost of about 6.5¢/kWh for utility-produced firm energy, there is currently little cost incentive for a customer to install a PV system.

### 1.6.7 Key Issues in the Development, Introduction, and Use of PV Power Systems

Although the technical feasibility of PV power plants has been convincingly demonstrated, several key issues concerning the development, introduction and use of such plants have not yet been resolved. A number of the ongoing utility tests and activities are aimed at the resolution of these issues over the coming years.

These involve both business and technical issues pertaining to module cost and performance targets, potential markets for PV, land requirements, and PV array, power conditioning, and balance of plant issues.

### 1.6.7A  PV Module Cost/Efficiency Targets

With respect to prospective bulk power applications, the key need in photovoltaics is to demonstrate that PV modules–flat-plate, concentrating, or both–can achieve the module efficiencies and costs required to achieve busbar energy costs of 6 to 8 cents per kWh in levelized, constant 1982 dollars (EPRI AP-3351, and Moore, 1985). These energy cost goals and corresponding performance and cost targets for modules were originally derived at EPRI in 1982 (EPRI AP-3176-SR) based on an approach developed at the Institute in the mid 1970's (EPRI ER-539-SR). Subsequently, they were adopted by the U.S. government program (DOE, 1987) and by other programs overseas. These goals are still valid today, even though oil and other fuel costs have dropped substantially since 1982. However, in part to compensate for these effects, we express the goals now in 1989 dollars. We have taken a middle ground between ignoring the effect of general inflation from 1982 to 1989 (which would, by itself, increase a 6 cent goal to over 10 cents) and including the full effect; i.e., we have used a cost-of-energy goal of 8 cents per kWh, in 1989 dollars. We believe the resulting goal to be valid, based on: (a) our assumption that oil prices will not remain depressed indefinitely; (b) the realization that the goals were originally based more on power costs from advanced coal-based power plants than on oil-fired generation; and (c) the expectation that PV power technology will begin to penetrate the bulk power market in the mid 1990's rather than today. Our conclusion is that 8 cents per kWh for PV power plants represents a meaningful goal in the sense that its achievement would qualify PV to service a substantial multimegawatt market in peaking- and intermediate-load bulk electric power supply. The updated cost and efficiency targets, which apply to areas with abundant sunlight like the Southwestern U.S., are given in Table 1.29.

The cost and efficiency targets in Table 1.29 represent combinations of PV module costs and stable efficiencies that would enable PV systems to achieve annual average power costs of 8 cents per kWh in electric utility bulk power applications. It was recognized, however, that early PV power applications in the electric utility market can occur in some situations at significantly higher costs (perhaps 12–15 cents per kWh). The corresponding area-related balance-of-system (BOS) cost target was assumed to be $50/m$^2$ for flat-plate systems, and $100/m$^2$ for concentrator systems. The methodology for deriving these targets and the financial parameters used are essentially those employed in DOE's Five Year Photovoltaic Research Plan, 1987–1991 (DOE, 1987). The cost and efficiency targets in Table 1.29 correspond to an assumed 20- to 30-year module life.

Although lower efficiencies could lead to equivalent energy costs if corresponding module (per-unit-area) costs were sufficiently low to compensate for the resulting increase in module area and BOS area-related plant hardware; EPRI views the achievement of per-unit-area module costs below the goals of $85/m$^2$ for flat plate and $95/m$^2$ for concentrator modules as extremely difficult. Hence, the EPRI program continues to strive for flat-plate and concentrating module efficiencies

**Table 1.29:** EPRI PV Module Cost and Efficiency Targets (1989 \$, Southwest U.S. Site)

| | Module Efficiency * (Percent) | Module Cost ** ( \$ / m² ) |
|---|---|---|
| **Flat-Plate** | | |
| | 10 | 40 |
| | 15 | 95 |
| | 20 | 155 |
| **Concentrator** | | |
| | 15 | 75 |
| | 20 | 150 |
| | 25 | 225 |

* Stable efficiency.

** Area-related balance-of-systems cost target assumed to be \$50 / m².

of 15% and 20%, respectively, the latter implying a need for concentrator cell efficiencies of at least 25%.

### 1.6.7B  U.S. Bulk Power Markets

While EPRI has not performed any comprehensive projection of U.S. bulk power market size and potential penetration for PV power technology, it has conducted sufficient informal analyses to know that the market opportunity for PV through the 1990's exceeds, by a wide margin, the PV industry's supply and capability in any reasonable business expansion scenario. For example, analysis of published data on generation-addition plans and load forecasts for electric utilities in the Southwestern United States indicates a market of some 40 gigawatts (GW) for new generating capacity of all types through the year 2000. If one assumes that 10 to 20% of this total is peaking capacity and that PV can capture one-fifth of the new peaking capacity market, then the resulting opportunity for PV is in the neighborhood of 1 GW. An annual breakdown suggests a corresponding market of 100 to 200 MW per year by the late 1990's in the Southwestern United States. This potential market estimate is judged to be conservative since (a) it is based on modest load growth projections that many feel will soon be exceeded; (b) no real escalation of fuel costs is assumed; and (c) no disruptions in fuel supply are assumed.

Hence, assuming that PV cost and efficiency goals are achieved and that user confidence is established through orderly programs of PV testing and demonstration, PV penetration into bulk power markets in the 1990's is not likely to be limited by market size, but rather by the ability of the PV industry to expand its production capability.

PV penetration beyond the year 2000 will be determined by factors that cannot be foreseen at this time. Assuming continued successful development, however, it is likely that achievable levels are far greater than those estimated above for the late 1990's.

### 1.6.7C PV Land Requirements

The requirements for large areas has often been raised as a constraint on penetration of PV and other solar power plants. In heavily urban locations, available area, including rooftops, is insufficient to allow all but a minor contribution to local energy needs. But in most U.S. locations with good solar resources, land availability is not a major issue. Mature solar power technology is expected to generate approximately 100 MW per square mile, an amount comparable to coal power plants when coal mining area is included. By comparison, the total installed capacity in the U.S. in the year 2000 is projected to be some 900 GW. The average load factor or capacity factor for the aggregate national system is estimated at about 60%. Hence, if the PV plant average capacity factor is 25%, the amount of PV generation required to generate all of the electricity needed in the U.S., ignoring for the moment the matters of time of delivery and sunlight intermittency, would be $(60/25) \times 900$ GW or some $2,200$ GW. This in turn would require some 22,000 square miles, or 20% of the area of the state of Nevada. Vast areas of this order are available in much of the Southwestern U.S. and other reasonably sunny areas of the country. So land requirements will not be the factor that limits PV penetration in the U.S. It is also worthwhile to note that land cost does not significantly affect the total cost of solar plants until the price of land approaches about $50,000/acre. At this price, one square mile (640 acres) would cost $32 million or about 15% of the expected cost of a 100 MW plant using mature PV technology (DeMeo, 1989).

### 1.6.7D Intermittency of Insolation Availability

The intermittent nature of sunlight sets a far more significant constraint on penetration of PV into the power supply network. Even in areas of high correlation between utility peak loads and sunlight availability, such as much of California, PV cannot support the baseload requirements that remain after peak loads are served. Further penetration will require available storage and/or fuel production from the solar electricity. Storage availability is a major uncertainty in itself, and its status and outlook are reviewed in a recent *EPRI Journal* article (Boutacoff, 1989). There is great incentive to develop economical storage for utility systems in general, both for operating flexibility and for peak load shaving. But to the same extent storage becomes available for use with baseload power plants to increase their capacity factors, it will also be available for use with intermittent renewable power plants to serve load during periods of resource intermittency.

### 1.6.7E Early Cost-Effective PV Applications

The commercial use of PV in remote power applications has been well established for many years, and accounts for a significant portion of the roughly 30 MW of module sales in 1988. Less well known, however, is that U.S. utility companies are beginning to use PV in a variety of such applications, ranging from a few watts for automated meter power supplies to a few hundred watts for transmission tower aircraft warning beacon lighting to a few thousand watts for cathodic protection.

Although total capacity for these applications is relatively small, their significance should not be underestimated. First, PV is the system of economic choice for these applications. Even at current prices, PV is less costly than available alternatives. In some cases, such as perimeter lighting and transmission tower beacons, great quantities of grid power are available nearby; but the cost of cable trenches or step-down transformers far exceeds the cost of the PV system.

Of even greater significance, these systems provide an important means for PV technology to achieve credibility in the eyes of utility engineers for whom reliable service is paramount. Positive PV experience within the ranks of utility organizations will facilitate the introduction of PV into larger, more energy-significant applications as costs come down.

EPRI recently initiated a project to assist a number of utilities in identifying early, cost-effective PV applications such as those mentioned above. The intended approach is to work closely with distribution engineers within participating utility organizations to understand their needs, and then to ascertain to what extent PV can help meet these needs. PG&E initiatives in this area, discussed earlier, make that utility a leader among U.S. utilities in identifying and supporting early cost-effective uses of PV.

### 1.6.7F   Technical Issues

#### 1.6.7F.1   Array-Related Issues

Almost all PV systems currently deployed show observed ratings substantially below the nominal plant ratings. This indicates that optimistic values were assumed for module-mismatch and system-electrical losses, and that nameplate ratings are based on "laboratory" rather than actual ambient temperatures. Second, for systems employing sunlight tracking, the electromechanical tracking components are still evolving. Stemming initially from experience with high wind loading at Lugo, ARCO Solar carried out several redesigns of tracker drives. During 1986 and 1987, these were replaced on most of the Hesperia arrays. However, there is strong evidence that the Carrisa phase 1B trackers, enjoying the benefit of earlier experience, are highly reliable. Continued operation over extended periods will be needed before resolution can be assured.

#### 1.6.7F.2   Power Conditioning

The PCUs at PV stations have also been a source of downtime, although experience varies substantially from site to site. To investigate this matter further, EPRI and Sandia National Laboratories recently completed a joint assessment of the state of the art in PV power conditioning. The issues addressed in the study include performance, reliability, and operation and maintenance experience for existing power conditioners; power conditioner procurement experience; technology, cost, and market trends; integration of conditioner design with PV system design; and the utility system interface.

The key findings of the study were:

- Power conditioning technology needs to be improved in the areas of cost, performance, reliability, and maintainability.
- Power conditioner designs need to be made more integral to overall PV system designs.

• Utilities desire appropriate information and assistance to address problems with power conditioners, PV electrical system design, and utility interface issues for both large and small PV systems.

The current cost of custom-built photovoltaic PCU's appears to be in the neighborhood of $800 per kW for systems in the several hundred-kW size range and about $500 per kW for 1 MW systems, (based on the latest utility-scale units installed at Austin and SMUD PV2). These costs are assumed to include the vendor contract price, including transformer and switchgear, housing if housed, delivery to the site, warranty, and any vendor provided assistance with installation, check out, and startup. Advances in power electronic devices, circuit topologies, and packaging, combined with volume production, are factors which have the potential to significantly reduce PCU costs in the future.

As the cost of the photovoltaic array field and balance-of-station decline due to advances in photovoltaic technology, it is important that the PCU cost also decline proportionately to minimize the overall station cost. It would appear from the information accumulated that a reasonable cost target for photovoltaic PCU's is $100 to $150 per kW or $0.10 to $0.15 per watt in 1987 dollars, corresponding to some 5 to 7% of a mature PV power plant total installed cost. If anticipated technology advances, volume production and optimization of other key parameters can be achieved, this cost target appears attainable and might even be exceeded in large-scale applications.

### 1.6.7F.3 Balance-of-System Costs

While the acceptance of PV from a power-plant operational standpoint is now a virtual certainty, the high capital costs of PV systems still preclude their use for economically competitive bulk power applications. Today's system costs of approximately $10 per watt installed are generally divided evenly between PV module and balance-of-system (BOS) costs. BOS costs are expected to yield to value engineering, quantity production, installation ingenuity, and carefully tested relaxation of standards and practices for power plant construction.

However much work needs to be done, much experience needs to be gained, many mistakes need to be made, and many systems need to be built on the way down the learning curve toward acceptable BOS costs. But the weight of informed opinion argues that affordable BOS costs of about one dollar per watt can be achieved by orderly engineering development without the need to strike down any fundamental barriers (see, e.g., EPRI AP-2474 and 3264).

### 1.6.8 Small System BOS Issues and Outlook

A significant amount of operational experience has been gained from experiments involving small PV systems such as those at Gardner, Massachusetts (New England Electric Service) and at Laguna Del Mar (San Diego Gas & Electric). While the performance of these systems has met expectations, there has been no real progress in the advancement of small systems technology in recent years. It can even be argued that setbacks have occurred with the loss of valuable small businesses that supplied designs and products and formed a small PV infrastructure during the early 1980s.

The most promising aspect of the outlook for small systems is the growth in interest expressed by utilities. Real opportunities exist for the utility industry to enhance the outlook for small PV systems by marketing or helping to market such systems to their customers. However, three areas need attention if these opportunities are to be pursued. First, business strategies need to be identified under which utilities can effectively market cost-effective small systems to their customers. Second, the demand side management aspects of small PV need to be evaluated. Smart power conditioning units incorporating metering, communications, and direct load control features might be integrated into PV systems to add value. Third, as small PV system technology evolves, individual utilities will need to know specifically how small PV systems will affect conventional generation requirements in their service territories. Networks of low-cost utility solar resource instruments and standards for their use may provide the most cost-effective means to study these effects (Kern, 1988).

### 1.6.9  EPRI PV Program

Owing to the steady stream of progress from R&D programs around the world with multiple sources of support and to the positive experience from major field test installations, EPRI has developed a growing sense of optimism that PV power technology will emerge as an option for bulk power applications in the U.S. by the mid-to-late 1990's. One indicator of PV's emergence as a future electricity supply option for utilities is the level of emphasis on PV within the Solar Power Program at EPRI. In the Program's early years, PV represented about 20% of annual expenditures. By 1980, the level had risen to about $800,000, or 35% of the program's total. And in 1987, EPRI PV expenditures totaled some $6 million, or 85% of the Program's total. This level of funding and emphasis has been maintained during 1988 and 1989.

The EPRI PV program contains four primary thrusts: (a) high-concentration silicon cells and systems, (b) amorphous-silicon materials research aimed at high-efficiency ($> 15\%$) tandem flat-plate modules, (c) flat-plate silicon sheet technologies, and (d) system-level evaluations, including requirements assessments and field experience evaluations. The foundation for these thrusts stems from the major, in-depth evaluation of PV status and research needs conducted in the 1982–1983 time frame, mentioned previously (AP-3351).

Since its inception in the mid 1970's, the program has been guided by the belief that the need for high cell/module efficiency is paramount. Consequently, the program has pursued PV technologies and system concepts that offer the hope of required high efficiencies. The program's underlying management philosophy has been to seek fundamental scientific understanding as a precursor to high performance, rather than to apply an empirical approach. Consistent with this philosophy, a high premium has been placed on long-term, multi-year commitments to the needed basic research efforts.

During 1988–1989, EPRI updated its earlier evaluation of PV status and research needs. Although no significant redirection of EPRI's PV program was indicated, an expansion of basic materials research efforts was recommended. An additional recommendation favored greater emphasis on the development of near-term PV markets. This finding was a primary impetus for EPRI's initiation of the project, discussed previously, to assist a number of utilities in identifying early, cost-effective PV applications.

Over the years, EPRI has sponsored or supported numerous studies, assessments, workshops, projects and field tests dealing with photovoltaic technologies. EPRI has also published approximately 50 reports as well as numerous articles and papers describing these activities. A listing of EPRI's reports on PV is presented in the bibliography to this chapter.

### 1.6.10 Summary and Prospects

As described above, electric utilities have been involved in a broad spectrum of efforts to advance the evolution and application of PV technology, from research and development aimed at higher efficiency cells to demonstrations of customer-size and central-station systems. Most important, electric utility interest in PV is growing and has reached the point where some utilities are viewing PV as a very real near-term business opportunity and not just an R&D curiosity with possible long-term potential.

The experience with utility-scale PV plants has demonstrated that PV power systems using state-of-the-art, single-crystal-silicon PV technology can operate reliably with performance close to predictions. These plants have also provided conclusive demonstrations of modularity in design and construction (most consist of two or more identical subfields) and of short construction times. Furthermore, the demonstrated capability for unattended operation, along with favorable maintenance experience, bodes very well for attractively low operating and maintenance costs. The significance of these experimental plants in assessing prospects for PV power systems should not be underestimated; for they have shown beyond a reasonable doubt that, from a system operation standpoint, today's PV power technology and the associated power conditioning apparatus are technically viable in electric utility applications.

These plants, as well as all of the others mentioned in Table 1.22, have also shown minimal environmental impact. Even land utilization, at 15 to 20 acres per megawatt, has not posed a problem; since the plants have been constructed in desert or other areas where competition for land use is not great. However, it should be recognized that utilities in more urban areas such as the Eastern United States do not foresee large PV installations (i.e., more than several hundred kilowatts at one site) in their service territories, largely because of land-availability restrictions.

From an electrical interface standpoint, another key finding of efforts to date, both with centralized and customer-sized systems, is that power quality impacts from properly designed systems, including issues such as power factor, harmonics and voltage transients, are acceptable to utilities even in cases of high penetration on utility network distribution feeders.

Our overall assessment from the recent developments discussed above is that PV's prospects for success have advanced substantially over the past several years. Field test experience has been very positive; substantial technology advances have occurred; and business developments, although mixed, indicate continued industrial commitments and growing interest on the part of U.S. electric utility organizations.

The prevailing view among the utilities interviewed in the course of the preparation for this chapter is that the prospects for widespread, energy significant applications by utilities and/or their customers are better for PV than for any other emerging renewable energy technology. However, for widespread use of PV to occur, further cost reductions and performance improvements must be achieved. Key needs now are to demonstrate (a) reliable operation over extended operating periods for thin-film and concentrating PV technologies; (b) sufficiently high conversion efficiencies for thin-film PV modules; and, ultimately, (c) acceptable system costs for any or all of the PV technologies with good performance prospects.

Another critical need, especially in the near term, is the development of early markets for PV products in high-value applications to provide the market pull needed for the evolution and maintenance of a healthy and growing PV industry. It is likely that many of these applications, including distributed systems that may be either customer or utility-owned, will be of greater significance than large bulk power applications for the next decade. Efforts such as PVUSA, PG&E's assessment of distributed PV applications, APS's STAR Center, and various utility-scale and customer-size system demonstrations can help to initiate the necessary market pull.

If current programs can be maintained over the next several years through continued support from industrial, utility and government sources, the role of PV in U.S. power applications should be well understood within the next decade.

## 1.7 Other Solar Technologies

In addition to the solar and wind power technologies discussed previously, electric utilities have pursued activities in the area of biofuels, solar thermal process heat, and ocean thermal energy conversion (OTEC). As noted in the introduction, the category of biofuels has not been included within the scope of this paper, which addresses only the developing solar technologies.

### 1.7.1 Utility Activity in Other Solar Technologies

As can be seen in Table 1.1 and Table 1.2 and Figure 1.2 and Figure 1.3, the EPRI surveys reported significantly fewer electric utility activities in process heat and OTEC than in the other solar technology areas. Activity in the area of process heat was first reported in 1978 and peaked in 1982 at 13 utilities and 10 activities. By 1983, the number had declined to nine utilities and nine activities. Likewise, activity in the area of OTEC got off to a relatively slow start with the first reported activity in 1978 and the peak at five utilities and 10 activities in 1982. By 1983, according to the EPRI survey, the level of involvement had declined to four utilities and five activities. Today, the level of electric utility activity in these two solar technology areas is very limited. The electric utility industry experience in these two areas is described below.

### 1.7.2 Process Heat

Industry requires large amounts of process heat, normally in the form of hot water and/or steam, to perform various production and manufacturing operations. The concept of solar-thermal process heat is typically envisioned as involving the use of solar-thermal technology, central receivers, dishes, and troughs, to concentrate sunlight to produce steam for use as industrial process heat (IPH). DOE sponsored a number of IPH demonstration projects around the U.S. in the late 1970's and early 1980's. However, these projects involved little electric utility participation.

Of the nine process heat activities described in EPRI's 1983 survey, seven related to applications of solar process heat to agricultural uses, mostly warm air for grain drying or hot water for other farm uses. Only two of the reported utility activities could be construed to relate to industrial applications and these involved hot water production and not process steam.

It is apparent that electric utility activity in the area of process heat has been limited because this is a non-power application of solar thermal technology. Even more important, the production of process heat requires integral involvement with the manufacturing or production facilities and processes of industrial companies. Even if utilities were interested in pursuing IPH applications of solar energy, it is very difficult to find companies with both (1) land and facilities suitable for assessment and technology demonstrations and (2) a willingness to cooperate with utilities in such efforts. Thus, most of the activity which did not occur was undertaken by DOE and a fledgling solar IPH industry.

By the early 1980's, however, the relatively low price and availability of gas and oil and competition from cogeneration had effectively driven solar-thermal systems from the IPH market. Industries, which in the late 1970's had some interest in solar thermal energy, were by then exhibiting very little interest in purchasing solar-thermal industrial process heat systems, even with the availability of tax incentives. Many reasons for this can be cited including the favorable fossil-fuel supply situation and drawbacks associated with the use of solar-thermal IPH systems. Reasons often given by companies include (Brown, 1983):

- Major R&D expenditures are devoted to new products, not energy supply. For example, in the highly competitive chemicals industry, resources cannot be allocated to activities that do not directly contribute to productivity or market share.
- Land is not available; plant personnel are generally disinterested.
- Recent applications of conservation through heat recuperation make solar impractical and infeasible.
- Time and talent of plant personnel is much more profitably invested in new process development; involvement in a technology (such as solar) that has such obviously high risks and small returns is of little interest.

The above reasons, coupled with the high initial cost of solar-thermal IPH systems and the desire of industry for payback within a 5-year or shorter time frame, made it very difficult for solar thermal to penetrate the IPH market. In fact, the value of the solar-thermal energy produced by these systems was generally so small that companies could not afford to maintain the systems.

The barriers encountered were heightened further by developments in cogeneration technology and energy conservation measures. These provided industry with more cost-effective means to control energy costs.

### 1.7.3 OTEC

Ocean thermal energy conversion is a concept of electric power production using the temperature differences between seawater close to the surface and the much colder deep water in tropical seas. It is an established thermodynamic principal that power can be generated using two reservoirs at different temperatures. Even though the temperature differences between surface and deep water at the best sites is only about 20 °C (36 °F), using very large water flows, it is theoretically possible to generate megawatt levels of power.

Several U.S. utilities have investigated the potential of using OTEC for power generation. The possibility of using tidal power has also been investigated by U.S. utilities. A 1986 EPRI Report (AP-4921) assessed the state of the art in ocean energy technologies including tidal power, OTEC, wave power and other ocean energy technologies.

#### 1.7.3A Utility Involvement in OTEC

The U.S. electric utilities involved in OTEC, as reported in the 1983 EPRI survey, include Florida Power Corporation, Hawaiian Electric Company, and the Puerto Rico Power Authority. The activities of these utilities were described as follows:

#### 1.7.3A.1 Florida Power Corporation

Florida Power Corporation's activities included an integration and assessment study and component development and testing. A Florida Power Corporation (FPC)/DOE task force conducted a planning study using FPC's planning models and systems to evaluate OTEC opportunities, capabilities, and parameters in the 1985–2000 time frame. The final report was completed in 1980. Also, in conjunction with Alfa Laval, Inc., Florida Power Corporation equipped a trailer with prototype heat exchangers and a data analyzer to test biofouling rates on new types of plate heat exchangers. The equipment was installed at an existing power plant site and removed in November 1981. The data were to be used in the design of OTEC systems.

#### 1.7.3A.2 Hawaiian Electric Company

HECO cooperated in two DOE funded studies to develop Phase 1 conceptual designs for a 40 MW OTEC pilot plant to be located off Oahu. General Electric and Ocean Thermal Corporation performed the two studies, which were awarded in February 1980. The Ocean Thermal Corporation proposal was selected for further federal funding through preliminary design. Ocean Thermal Corporation, along with TRW, Hawaiian Dredging, Burns and Roe, R. J. Brown Associates, and Science Applications Inc., completed preliminary design for a 40 $MW_e$ plant on an artificial island off Kahe, Hawaii. The power plant would use the cooling water from the Hawaiian Electric Company's oil-fired power plant to enhance the warm water temperature by 1.7 °C (3.1 °F). The proposed power plant would have four modules each composed of shell and tube condensers and evaporators with titanium tubes, and a four-stage double-flow turbine. The preliminary design phase was completed during 1984. At that time it was decided by DOE that industry and not the Government should fund the remaining phases of the program (AP-4921).

### 1.7.3A.3 Puerto Rico Electric Power Authority

The Puerto Rico Electric Power Authority performed a study assessing the integration of an OTEC plant into its electric system.

### 1.7.3B State-of-the-Art Study Results

EPRI's state of the art study on ocean energy technologies (AP-4921) was performed by the Massachusetts Institute of Technology (MIT). The objectives were to document the state of the art for various ocean energy technologies and to evaluate the potential of these technologies for utility power generation.

Investigators gathered information on tidal, OTEC, wave, ocean currents, ocean winds, and salinity-gradient power generation technologies from their own research, from the literature, and from developers. For each technology, they analyzed the historical development, theoretical considerations, resource potential, environmental effects, and research needs. For tidal power, OTEC, and wave power, the most developed technologies, they then calculated levelized busbar energy costs for representative base case systems and a few variations. To do this, they combined predicted capital costs, operation and maintenance costs, performance figures, and financial information. The results are summarized in Table 1.30.

Ocean energy technologies are in different stages of development. Tidal power is a proven technology, with an 18 MW plant in Canada and a 240 MW plant in France. OTEC and wave power systems are still undergoing development and demonstration. A 50 kW demonstration of OTEC technology has taken place near the Hawaiian Islands. And in Japan and Norway, wave energy technology systems ranging from 100 kW to 500 kW have been demonstrated. Ocean currents, ocean winds, and salinity gradient technologies, however, are in the early stages of research. The estimated levelized busbar costs in constant 1985 dollars for the base case systems were 113 mills/kWh for tidal power, 217 mills/kWh for OTEC, and 274 mills/kWh for wave energy. The estimated energy efficiencies were 20% for tidal power, 2% for OTEC, and 21% for some wave energy devices.

The study results indicate that utility-owned tidal, OTEC, and wave power plants are unlikely to be economical in the United States in the near term, except in very special situations. With the possible exception of tidal power, these technologies would require further research, development, and demonstration before utilities could consider them as serious power generation options. Substantially more work would be involved in developing ocean current, ocean wind, and salinity gradient technologies into practical utility alternatives.

## 1.8 Summary and Conclusions

This chapter has sought to provide a comprehensive overview of the nature and extent of the electric utility industry's activities in solar and wind energy technologies. This section provides an overall summary of the information presented in the chapter and its implications in terms of the prospects for the energy-significant application of these technologies by electric utilities and their customers.

### 1.8.1 Overall Summary

The electric utility industry's involvement with solar and wind energy systems has been varied and extensive. Hundreds of utilities from all parts of the U.S. have

**Table 1.30:** Ocean Energy Technologies

|  | Tidal Energy | OTEC | Wave Energy |
|---|---|---|---|
| **Status:** | | | |
| Installations | Full Scale | Demonstration | Demonstration |
| Sites | Canada, France | Pacific Islands | Japan, Norway |
| Rated Power | 18 - 240 MWe | 20 - 50 kWe | 100 - 500 kWe |
| **Base Case** | | | |
| Site | Maine | Hawaii | East Coast |
| $ / kW | 3,700 | 12,200 | 4,026 |
| Power Rating | 12 MWe | 46 MWe | 115 MWe |
| Capacity Factor (average / rated power) | 35.5 % | 68 % | 17 % |
| Busbar Cost, mills/kWh | | | |
| Current $ | 192 | 370 | 466 |
| Constant $ | 113 | 217 | 274 |

Source: EPRI AP-4921.

conducted or participated in literally thousands of different activities and projects including resource assessments, technical and economic analyses, basic research and development, component and systems experiments and field tests, customer assistance and information programs, and various business ventures. These utilities became involved in solar and wind for various reasons, some as a result of their own initiative and curiosity and others in response to pressures from consumers and government. Many pursued these technologies even in situations where neither technical nor economic feasibility prospects were encouraging.

The utilities embarked on these activities and projects independently, in cooperation with other utilities and with EPRI, and with other sectors including the solar and wind power industries, government, universities, and their own customers. A major part of the aggregate utility experience with the solar and wind technologies has been gained through the EPRI Solar Power Program. This experience was obtained through EPRI funded activities and projects documented in approximately 170 technical reports and numerous papers and articles (See bibiliography of EPRI Solar Power Program Publications at the end of this chapter).

Over the 17-year period since the 1973 oil embargo, during which utilities have been actively involved with solar and wind technologies, many changes have occurred in a number of key external factors influencing the utilities' activities. Among these are changes in fuel prices and supply availability, load growth rates, the economic climate, federal and state tax credit policies for renewable energy systems, the level of federal spending on renewable energy systems, the roles of conservation and least-cost planning in the formulation of utility strategies for meeting customer electricity needs, the mix of commercially-available power generation technologies, and public perceptions about the environmental impacts of various energy technologies. In addition, a trend toward deregulation of the power generation industry is evolving that is likely to have far-reaching consequences for the entire electric utility industry.

Overall, the changes in external factors, especially during the 1980s, have tended to discourage utility involvement in solar and wind energy. This appears to be a primary reason why utility activity in solar and wind technologies trended downward in the early 1980s. Another key reason is that in the aftermath of the 1973 oil embargo, a great deal of activity was initiated by utilities throughout the country in response to the "crisis" atmosphere of the time. Much of this activity was not thoroughly planned and lacked a clear purpose and logic. Many utilities had high hopes for these early efforts, but progress was often slow and disappointments were encountered. By the early 1980s, perspective had begun to emerge, and the better-conceived activities and programs have endured while others fell by the wayside or were channeled toward more meaningful directions. As one might expect, the best planned activities were usually the most successful and will have the most lasting value.

Solar thermal systems (heating and cooling, electric power, process heat) were the initial renewable technology choices of most utilities. The technologies seemed relatively simple and straight-forward, and activity was encouraged by federal programs providing cost-sharing opportunities and tax incentives. However, by the early 1980s, after understanding began to emerge that solar heating and cooling's main impacts were to increase awareness of energy conservation and energy efficient design practices, and, after the technical and economic risks associated with most solar thermal electric systems had become apparent, utilities shifted their attention more to wind power and photovoltaics. In the last several years, the interest of utilities in PV and, more recently, wind has grown as technical advances have increased confidence in the suitability of these technologies for utility system applications, particularly in locations where it can be shown that the solar and wind resources are such that a significant capacity value, in addition to an energy value, can be derived from a plant's output.

Also, as information and experience have been gained, utilities have become more selective and sophisticated in their solar and wind activities. During the latter part of the 1980s, a few utilities began a transition from involvement in solar and wind power on an assessment- and experimentation-level basis toward involvement with the more promising of these technologies through serious business ventures. Most telling, however, utilities in areas with good solar and wind resources have begun to view solar and wind technologies as viable future power generation options, some to the point where they are factoring these technologies into their generation planning. This extensive and continuing involvement of U.S. utilities in solar and wind energy is all the more remarkable upon realization that most of the incentives afforded to developers and independent power producers, such as renewable energy tax credits and favorable energy sales conditions, have not been available to the utilities.

The following is a summary of the electric utility experience with each of the solar and wind technologies covered in the chapter.

### 1.8.1A  Solar Heating and Cooling

The earliest and, for several years, most extensive of the electric utilities' activities in solar energy were in the area of solar heating and cooling. The utility SHAC activities ranged from resource assessments to system demonstrations and customer information programs. The key result of these efforts was the development of a base of information and understanding that utilities could use to advise their customers concerning such systems. The electric utility activity in the area of SHAC continues, but is now confined mainly to customer information services in conjunction with demand-side programs because SHAC, especially the passive variety, is really a conservation measure and not an electric energy supply option. The technology of SHAC systems is fairly well established and the advancement of this technology involves primarily applications engineering rather than research and development.

### 1.8.1B  Solar-Thermal Electric Power

To be cost effective, solar-thermal power systems require a relatively high level of direct solar radiation. For this reason, electric utility activity in solar-thermal electric power has been confined mainly to the Southwest.  The electric utility experience with solar-thermal power systems is that they are technically feasible but costly to construct, operate and maintain.  The principal reason is that the conversion of sunlight to thermal energy and then to electricity requires significant amounts of hardware and relatively complex systems.  In the case of the central receiver, it is clear that at least one or two (probably several) more iterations of systems development and demonstration would be required to evolve a system suitable for utility system applications.  However, the cost of each such iteration is very high, and the outcome in terms of energy production is highly uncertain in light of the experience with Solar One and other central-receiver projects. Prospects for dish-engine systems appear more promising due to their high energy conversion efficiencies and rapid response times, but the development of an engine with sufficient reliability and longevity to lead to cost-effective systems again involves a very costly undertaking with very uncertain prospects.

On the other hand, the experience of Luz with its SEGS plants suggests that the simplest of all solar thermal electric technologies, parabolic troughs, is already commercially viable or nearly so in solar-fossil hybrid applications. The SEGS Plants are a private development benefiting from favorable avoided-cost energy contracts with Southern California Edison, from substantial federal and state tax credits, from an experience base developed with federal R&D support, and most importantly, from entrepreneurial drive motivated by the desire for marketplace success.  The continuing enhancements and cost reductions being achieved by Luz with these plants suggests that this technology potentially could be cost-effectively used to produce power in the Southwest on a large scale, at least in the near term and perhaps in the longer term as well. The successes achieved at the SEGS Plants also suggest that future development efforts involving solar-thermal energy systems need to revisit the benefits of solar-fossil hybrid operation. Such operation enhances equipment availability, reduces thermal cycling, and eliminates the need for costly backup generation whenever the solar resource is unavailable.

### 1.8.1C  Wind Power

Among the most substantial utility activities and achievements in renewables have been those in the area of wind power. In large measure, this is a reflection of the relatively advanced state of the art in this technology when compared to the solar technologies. In spite of the disappointing experience with megawatt-scale machines, as well as adverse publicity associated with tax credits, wind power has made rapid advances in the past decade. These advances were influenced by the federal wind power program, but came primarily from the activities of entrepreneurially driven wind power developers, who benefited from the tax credits and the provisions of PURPA. The developers installed thousands of wind turbines, mainly in California, and sold their output to electric utilities at the electric utilities' avoided costs. The electric utilities purchased the output of these projects and also installed and operated their own wind power projects. In addition, electric utilities performed extensive wind resource assessments to identify locations where wind power stations might be sited. The results indicate that wind power is technically feasible and that wind plants can also be economically feasible using available technology in locations with good wind resources (nominally 15 mph or more average wind speed) and moderate-to-high alternative energy costs. A key requirement is the location of a suitable site; i.e., one with adequate wind resources, adequate land, minimal environmental impacts, and access to the regional transmission system. Even so, the wind turbine technology is still at an early stage in terms of maturity relative to conventional power generation technologies.

As with any generation technology, wind power requires a well organized program of planning, siting, machine selection, construction, and operation and maintenance to be cost effective. Wind power is most likely to be feasible for wind plants consisting of many machines in the several hundred kW size range and having a well organized and administered maintenance program. Smaller numbers of wind turbines, such as one or two or even 10 machines at a site, do not afford the ability to establish a sufficiently cost-effective maintenance program. Without the ability to establish such a program, the cost of maintaining wind turbines can exceed the value of the energy produced. Also, large wind machines thus far have been shown to be too costly to operate and maintain due to the large size of the structures and the excessive downtime when repairs are required. However, as the wind turbine technology evolves, it is possible that the optimum wind turbine size will increase to the point where the optimal wind machine is a larger machine.

### 1.8.1D  Photovoltaics

Photovoltaic power systems have been demonstrated to be technically feasible for electric utility system applications, both in central station and distributed systems. Even though photovoltaic power technologies are not yet cost effective except in certain specialized, high-value applications, electric utilities are attracted to photovoltaics by the relative simplicity and low maintenance requirements of such systems, their modularity which allows them to be utilized in virtually any size range, and the potential for breakthroughs in cell efficiencies and module cost reduction. For these reasons, electric utility interest in photovoltaics has been growing, with many utilities installing demonstration systems or participating in other research activities. It was the consensus of the electric utilities interviewed in the process of preparing this chapter that PV shows the greatest promise of any emerging renewable technology in terms of the potential for wide-spread and energy-significant electric utility system (central station and distributed) applications. PV was also considered to be the only technology with good potential for customer applications as well.

Photovoltaics is seen as either a central station or distributed option in the western United States and as primarily a distributed option in the eastern United States. Central stations would typically be sited remotely from load centers and could be from tens to hundreds of megawatts in size while distributed systems would range in size from a few kW in residential applications to the megawatt level for use in industrial or utility transmission and distribution system applications.

### 1.8.1E  Solar Thermal Process Heat

The electric utility involvement in solar-thermal process heat has been very limited. Most of the activity in the area of process heat involved a number of DOE demonstration projects in the late 1970s and early 1980s. The stabilization of oil prices in the 1980s, the elimination of tax credits, and the problems which emerged in attempting to integrate solar process heat systems with manufacturing operations resulted in the demise of the small industry that had emerged. Among the principal reasons that industries shied away from the installation of solar-thermal process heat systems was the effort required to operate and maintain such systems, the land requirements, and the complications introduced to their manufacturing operations. In most cases, the costs of operating and maintaining the system were found to exceed the value of the energy produced. Unless significant benefits could be shown, manufacturers preferred to focus their attention and resources on their ongoing operations and avoid the complications associated with operating solar-thermal process heat systems. Also, it was difficult to find industrial plants with suitable locations for the installation of such systems.

### 1.8.1F  OTEC

Ocean-thermal energy conversion systems are technically feasible only in tropical seas. This limits their application in the United States to the Gulf Coast and Caribbean Ocean areas and to Hawaii. Although several electric utilities investigated ocean-thermal energy systems, it does not appear that such systems will be cost effective for the foreseeable future.

### 1.8.2  Overall Conclusions

Based upon our review of the activities and experiences of electric utilities in solar and wind energy, the following overall conclusions can be drawn:

- The utilities' efforts in solar and wind power development after the onset of the energy crisis were undertaken for various reasons. Many pursued these technologies in response to public pressures from consumers, renewable energy activists, regulatory commissions, and government agencies. Some began to pursue these technologies on their own initiative. A few utilities had investigated these technologies on a limited basis in decades prior to the 1970s.

- The development and application of solar and wind power on a cost-effective basis presented a formidable technological challenge to utilities because of the variability and diffuse nature of the solar and wind resources, the inherent complexity and immaturity of the technologies, the very limited prospects for near-term economic feasibility, and the competition for utility financial and human resources from other areas. At the same time, proponents of solar and wind power tended to overestimate the energy production potential from such sources while significantly underestimating the technological impediments and the time and effort needed to evolve systems suitable for commercial use.

- In spite of the problems, utilities, by themselves and through EPRI programs, pursued solar and wind power on a broad basis in a great variety of research projects and activities including studies and assessments; basic research and development; component and systems experiments and field tests; cooperative efforts with government, industry, and customers; and commercial ventures to develop and market solar and wind energy systems.

- Of the solar technologies investigated by the utilities, PV appears most promising in terms of potential for widespread, energy-significant applications. This is due to the modularity and simplicity of the PV systems, their capability for unattended operation, and the lack of need for water or other working fluid for cooling or thermal transport purposes. These attributes result in low operation and maintenance costs and a high degree of reliability. Although the technical feasibility of PV for utility bulk power applications has been demonstrated, it is clear that much R&D is still needed to make this technology cost effective for widespread use.

- Based on their experiences with wind power, a growing number of utilities view this technology as promising for utility application in areas with good wind resources (nominally 15 mph or more average wind speed) and assuming the use of intermediate-sized machines in well-maintained wind plants. The importance of careful siting has become clear. In many areas of the country the most attractive places for wind power generation are on ridges or near coastal areas. Siting wind power plants may prove difficult in some areas due to environmental objections, from an aesthetic viewpoint, to using such sites.

- Based on their experiences with solar-thermal power systems, utilities are generally less optimistic concerning the prospects for such systems. Capital costs and costs of operation and maintenance are high in relation to the economic benefit such systems can provide. The prospects for technology advances sufficient to make these technologies suitable for widespread application are limited. The exception to date is the solar-fossil-hybrid trough technology employed by Luz in the SEGS plants. It appears to have good prospects for more widespread application in the Southwest.

- For the solar and wind power technologies in general, economic feasibility rather than technical feasibility remains as the primary hurdle to widespread use. The solar and wind power technologies are not yet cost competitive with other utility power supply and demand-side (conservation and load management) options except in special cases. Feasibility prospects are considerably enhanced where a utility can obtain a significant capacity value in addition to an energy value for a plant's output. The primary drawbacks associated with the solar and wind technologies are high capital costs and uncertainties with regard to energy availability due to solar and wind resource variability.
- On the other hand, growing concerns about air pollution, global warming, and the expected return of higher oil and gas prices are seen by utilities as factors favoring the emergence of some of the solar and wind technologies as economic options for utilities in the near future. And, it is probable that some of these technologies will become economically competitive even without the need for incentives that might arise from these concerns.

### 1.8.3 Acknowledgements

Many people from electric utilities and other organizations contributed to the preparation of this chapter on the electric utility industry's activities in solar and wind energy. The authors want to first acknowledge, posthumously, the considerable contribution of Mr. Neil F. Lansing, who was centrally involved in the early stages of the effort but passed away unexpectedly after having completed in-depth interviews with some 35 utilities and affiliated organizations. The notes left by Mr. Lansing of his interviews were a primary reference source for this Chapter.

The authors also wish to acknowledge and extend their deep appreciation to the following individuals who contributed to this effort by submitting to interviews, providing information and pictures, reviewing draft text and providing other forms of support:

Mr. Robert S. Allan, Florida Power and Light Company
Dr. Gobind H. Atmaran, Florida Solar Energy Center
Mr. Ronald Barlowe, San Diego Gas and Electric
Dr. Philip Baumann, TU Electric
Mr. James Beck, City of Santa Clara, California Electric Department
Mr. Timothy Bernadowski, Virginia Power Company
Mr. Herbert Boyd, Alabama Power Company
Dr. A.G. Bullard, Carolina Power and Light Company
Mr. Nicholas Butler, Bonneville Power Administration
Mr. John Bzura, New England Power Service Company
Mr. Richard H. Chastain, Southern Company Services, Inc.
Mr. David E. Collier, Sacramento Municipal Utility District
Mr. James E. Crews, Florida Power Corporation
Dr. Carel C. DeWinkel, Wisconsin Power and Light Company
Mr. David Docter, Seattle City Light
Mr. Raymond J. Dracker, Pacific Gas and Electric Company
Mr. Timothy Driscoll, Long Island Lighting Company

Mr. Charles B. Duckworth, Salt River Project
Mr. William Emslie, Platte River Power Authority
Mr. William Esler, Southwestern Public Service Company
Mr. Donald Fagnan, Philadelphia Electric Company
Mr. Brian Farmer, Pacific Gas and Electric Company
Dr. Roosevelt Fernandes, Niagara Mohawk Power Corporation
Mr. Brian Firtion, Florida Power and Light Company
Mr. Don E. Fralick, San Diego Gas and Electric Company
Mr. Mark Freeman, Southwestern Public Service Company
Dr. Robert W. Goodrich, Northeast Utilities Service Company
Mr. Robert Gowing, New England Power Service Company
Mr. D. Jack Groves, Jr., Public Service Company of New Mexico
Mr. Edward M. Gulachenski, New England Power Service Company
Mr. Richard Halet, Northern States Power Company
Mr. Stephen Hester, Pacific Gas and Electric Company
Mr. Thomas Hillesland, Pacific Gas and Electric Company
Mr. John Hoffner, City of Austin, Texas Electric Department
Mr. Ronald H. Holeman, Bonneville Power Administration
Dr. Richard J. Holl, HGH Enterprises
Mr. Joseph Iannucci, Pacific Gas and Electric Company
Ms. Mary Ilyin, Pacific Gas and Electric Company
Ms. Christina Jennings, Pacific Gas and Electric Company
Mr. Walter Johnson, Hawaiian Electric Company
Dr. David Kearney, Luz Development and Finance
Mr. J. Willard King, Georgia Power Company
Mr. Michael Lechner, Public Service Company of New Mexico
Mr. C.W. Lopez, Southern California Edison Company
Mr. Robert Lynette, R. Lynette & Associates, Inc.
Mr. Kevin McAvoy, Los Angeles Department of Water and Power
Mr. William J. McGuirk, Arizona Public Service Company
Mr. Michael McNeil, Wisconsin Electric Power Company
Ms. Marlene I. Mishler, San Diego Gas and Electric Company
Mr. James A. Morris, Salt River Project
Mr. William E. Muston, TU Electric
Mr. Walter Myers, Bonneville Power Administration
Mr. Edward J. Ney, Georgia Power Company
Mr. George C. Penn, Wisconsin Power and Light Company
Mr. J. Timothy Petty, Southern Company Services
Mr. Stephen R. Pitts, Madison Gas and Electric Company
Mr. Robert G. Pratt, Detroit Edison Company
Mr. Derek Price, Wisconsin Electric Power Company
Mr. James N. Prothero, Wisconsin Electric Power Company
Mr. Joseph N. Reeves, Southern California Edison Company
Mr. William J. Reichmann, City of Santa Clara, California Electric Department
Mr. F. Russell Robertson, Tennessee Valley Authority
Mr. Mark D. Rogers, Northern States Power Company
Mr. Christopher C. Rolfe, Duke Power Company

Mr. Harry Roman, Public Service Electric and Gas Research Company
Mr. John C. Roukema, City of Santa Clara, California Electric Deparment
Dr. Henry Sanematsu, Los Angeles Department of Water and Power
Dr. John C. Schaefer, Electric Power Research Institute
Mr. Laurence E. Schlueter, ARCO Solar, Inc.
Mr. Mark Shank, Nevada Power Company
Mr. Graham Siegel, Wisconsin Electric Power Company
Ms. Peggy Sheldon, Luz International
Mr. Donald Smith, Pacific Gas and Electric Company
Mr. Dan Suehiro, Hawaiian Electric Renewable Systems
Mr. Roi    Terrell, Wisconsin Power and Light Company
Mr. A. Norman Terreri, Green Mountain Power
Dr. Robert Thresher, Solar Energy Research Institute
Mr. J.W. Waddill, Virginia Power
Mr. Howard Wenger, Pacific Gas and Electric Company
Mr. Michael Wehrey, Southern California Edison Company
Mr. Carl Weinberg, Pacific Gas and Electric Company
Ms. Tonawanda Woodall, Detroit Edison Company
Mr. Alan T. Yamagiwa, Seattle City Light
Mr. John Zimmerman, Consultant to Green Mountain Power

The Authors also wish to express their appreciation to: Wisconsin Public Power, Inc. SYSTEM for the administrative support provided by that organization to this effort; to Betty L. Menick for her enthusiastic and professional word processing support; and to Thomas A. Thrun for his assistance in coordinating the figures and tables used throughout the chapter.

# References

Arizona Public Service Company (1988), *Alternate Utility Team Utility Solar Central Receiver Study* Volumes 1 and 2, DOE/AL/38741-2.

Arizona Public Service Company (1988), *Arizona Public Service Utility Solar Central Receiver Study* Volumes 1 and 2, DOE/AL/38741-1.

Arthur D. Little, Inc. (1977), System Definition Study: Solar Heating and Cooling Residential Project, EPRI ER-467-SY, *Summary Report.*

Bonneville Power Administration, U.S. Department of Energy (1985), *Pacific Northwest Regional Wind Energy Assessment Program* (Wind REAP) Summary, Report No. BPA 85-19.

Bonneville Power Administration, U.S. Department of Energy (1987), *Cape Blanco Wind Farm Feasibility Study*, Summary, DOE/BP-1191-13.

Boutacoff, D. (1989), *Emerging Strategies For Energy Storage*, EPRI Journal.

Bower, W., F.B. Brumley, and B. Petterson (1985), *Engineering Evaluation Summary Report for American Power Conversion Corporation Model UI-4000 Utility-Interactive Residential Photovoltaic Power Conditioning System*, Sandia Report, SAND 83-2601.

Brown, K.C. (1983), *Reexamining the Prospects for Solar Industrial Process Heat*, Annual Review Energy, **8**, pp. 509–530.

Burkhart, L.A. (1989), *External Social Costs as a Factor in Least-Cost Planning – An Emerging Concept*, Public Utilities Fortnightly, pp. 43–45.

California Energy Commission (1980), *Wind Prospecting in Alameda and Solano Counties*.

Collier, D.E. and T.S. Key (1988), *Electrical Fault Protection For A Large Photovoltaic Power Plant Inverter*, The Conference Record of the Twentieth IEEE Photovoltaic Specialists Conference – 1988, Volume II, IEEE 88CH2527-0, pp. 1035–1042.

D'Aiello, R.V., E.N. Tsesme, and D.A. Fagnan (1988), *Performance of Solarex/Philadelphia Electric Co. Amorphous Silicon PV Test Site*, The Conference Record of the Twentieth IEEE Photovoltaic Specialists Conference – 1988, Volume II, IEEE 88CH2527-0, pp. 1092–1097.

DeMeo, E.A. (1988), *Solar Photovoltaic Power: Evolution of U.S. Electric Utility Activities Over the Past Decade*, Presented at Euroforum New Energies Congress and Exhibition, Saarbrücken, FRG, 7 pp.

DeMeo, E.A. (1989), *Solar Photovoltaic Power for Electric Utilities: Status, Outlook, and Impacts*, 1989 Conference on Technologies for Producing Electricity in the Twenty-First Century, San Francisco.

DeWinkel, C.C. (1983), *Wisconsin Power and Light Wind Energy Research and Demonstration Program*, Presented at Energy Sources Technology Conference & Exhibition, Houston, Texas.

DiGiovacchino, D.J. (1983), *Wind Power Status and Outlook*, EPRI Conference Proceedings: Solar and Wind Power – 1982 Status and Outlook, EPRI AP-2884-SR, pp. 1–25 to 1–30.

Eckert, P. (1987), *Lessons Learned and Issues Raised at Sky Harbor – A Utility-Interactive Concentrator Photovoltaic Project*, 1987 ASME Solar Energy Conference, Honolulu, Hawaii.

Electric Power Research Institute (1986), *Technical Assessment Guide*, EPRI P-4463-SR.

Eldridge, F.R. (1975), *Wind Machines*, prepared for National Science Foundation with Cooperation of Energy Research and Development Administration, NSF-RA-N-75-051.

Emslie, W.A. and C.J. Dollard (1988), *Photovoltaic Pilot Plant*, The Conference Record of the Twentieth IEEE Photovoltaic Specialists Conference – 1988, Volume II, IEEE 88CH2527-0, pp. 1283–1288.

Eyer, J.M., K. Firor, and D.S. Shugar (1988), *Utility-Owned Distributed Photovoltaic Systems, Pacific Gas and Electric Company*, The Conference Record of the Twentieth IEEE Photovoltaic Specialists Conference – 1988, Volume II, IEEE 88-CH2527-0, pp. 1051–1055.

Fagnan, D.A., R.V. D'Aiello, and J. Mongon (1987), *SOLAREX/Philadelphia Electric Amorphous Silicon PV Test Site*, *Proceedings of the 19th IEEE Photovoltaic Specialists Conference*, New Orleans, LA.

Fagnan, D.A. and R.V. D'Aiello (1990), *Update on the PECo/Solarex Thin Film Test Site*, Solar Cells **28**, No. 2, p. 151.

Fair, David (1985), *Georgia Power Company Solar Total Energy Project Test Report for Thirty Consecutive Day Test*, Shenandoah, Georgia: Georgia Power Company.

Georgia Power Company (1988), *Solar Total Energy Project Summary Report*, Sandia Contractor Report SAND87-7108.

Green, J.C. (1988), *Virginia Power Photovoltaic Field Test Experience*, *Proceedings 1988 Energy Technology Conference*.

Gulachenski et. al. (1988), *The Gardner, Massachusetts, 21st Century Residential Photovoltaic Project*, presented at Solar 88, MIT.

Halet, R.M. (1987), *NSP Holland Wind Farm Annual Report for the Period December 1, 1986 to November 30, 1987*, Northern States Power Research Department.

Halet, R.M. (1988), *NPS Holland Wind Farm Annual Report for the Period December 1, 1987 to December 31, 1988*, Northern States Power Research Department.

Hallet, R.W., R.J. Holl, and C. Bratt (1985), *The Dish Stirling System for Solar Electric Power Production*, Huntington Beach, California: McDonnell Douglas Astronautics Company.

Hawaiian Electric Company (1989), *Performance Assessment of Kahuku Point Wind Turbines*, report to EPRI, RP1590-11.

Hester, S. (1988), PVUSA - *Advancing the State of the Art in Photovoltaics for Utility-Scale Applications*, Pacific Gas and Electric Company, The Conference Record of the Twentieth IEEE Photovoltaic Specialists Conference – 1988, Volume II, IEEE 88CH2527-0, pp. 1068–1074.

Hicks, T.H. (1985), *Georgia Power Company Solar Total Energy Project Test Report for Continuous Fourteen Day Commercial Operations, Shenandoah, Georgia*: Georgia Power Company.

Hoff, T. and J.J. Iannucci (1988), *Siting PV Plants: A Value Based Approach*, Pacific Gas and Electric Company, The Conference Record of the Twentieth IEEE Photovoltaic Specialists Conference – 1988, Volume II, IEEE 88CH2527-0, pp. 1056–1068.

Hoffner, J.E. (1989), *Analysis of the 1988 Performance of Austin's 300-Kilowatt Photovoltaic Plant, Proceedings of the 1989 Conference of the American Solar Energy Society '89*, Denver, CO.

Holgersson, S. (1984), *United Stirling's Solar Engine Development - The Background for the Vanguard Engine, Proceedings Fifth Parabolic Dish Solar Thermal Power Program Annual Review December 6–9, 1983*, Indian Wells, CA, Pasadena, CA: Jet Propulsion Laboratory, DOE/JPL 1060-69, pp. 95–101.

Ilyin, M.A., W.J. Steeley, and D.R. Smith (1987), *Pacific Gas and Electric Department of Engineering Research's Wind Energy Activities*, presented at Windpower '87, SERI/CP-217-3315, pp. 127–141.

Jennings, C. (1989), *PG&E's Cost-Effective Photovoltaic Installations*, Pacific Gas and Electric Company, Department of Research and Development, San Ramon, CA.

Jennings, C., C. Whitaker, and D. Sumner (1990), *Thin-film PV Performance at PG&E*, Solar Cells **28**, p. 145.

Kearney, D. (1989), *Solar Electric Generating Stations (SEGS)*, IEEE Power Engineering Review **9**, No. 8, pp. 4–8.

Kearney, D. (1990), verbal communication.

Kern, E.C. (1988), *Small PV System BOS: Issues and Outlook*, Presented at EPRI Workshop on Evaluation of Photovoltaic Power Technologies and Applications, Palo Alto, CA.

Lepley, T. (1986), *Sky Harbor Photovoltaic Concentrator Project, Phase III: First Forty-Two Months of Operation*, Arizona Public Service Company.

Lopez, C.W. (1988), *Solar One - Five-Year Operating Result*, Southern California Edison Company, SCE 14-201.

Lotker, M. (1981), SERI Report TR-778, p. 49–50.

Lucas, E.J., G.M. McNerney, E.A. DeMeo, and W.J. Steeley (1989), *The EPRI-Utility-USW Advanced Wind Turbine Program – Status and Plans*, presented at Windpower '89.

Lynette, R. (1988), *Status of the U.S. Wind Power Industry*, Journal of Wind Engineering and Industrial Aerodynamics, **27**, pp. 327–336.

Lynette, R. (1989), written communication.

Lynette, R. (1990), written communication.

Madison Gas & Electric (1988), *Goodwill Photovoltaic Project*, Innovative Energy Program Project Report No. 30.011.

McAvoy, K. (1986), *Optimum Energy House, 2 KW Photovoltaic Demonstration Project*, Performance Report, September 1984–August 1986.

McDonnell Douglas Astronautics Company (1985), *Pilot Plant Station Manual* (RADL Item 2-1, Volume 1, System Description, Revised, prepared under Department of Energy Contract DE-AC03-79SF10499.

McGlaun, M.A. (1987), *LaJet Energy Company Update of Solar Plant 1, Proceedings of the Solar Thermal Technology Conference*, Albuquerque, New Mexico, SAND87-1258, pp. 142–150.

McGraw-Hill, Inc. (1989), *Electrical World Directory of Electric Utilities*, 97th Edition.

Michigan State University (1984), *Windmill Performance in Michigan*, Division of Engineering Research, East Lansing, MI.

Mishler, M.I. (1989), *Laguna Del Mar Photovoltaic Project*, San Diego Gas & Electric.

Moore, T. (1984), *Wind Power: A Question of Scale*, EPRI Journal.

Moore, T. (1985), *Opening the Door to Utility Photovoltaics*, EPRI Journal, pp. 2–19.

Moore, T. (1987), *Opening the Door For Utility Photovoltaics*, EPRI Journal, pp. 4–15.

Morris, J.A. (1987), *Power Quality Testing at John F. Long's 'Solar 1,' Salt River Project*, presented at 1987 American Solar Energy Society Annual Meeting, Solar 87.

New England Power Service Company, *Ascension Technology, Electric Research and Management*, and Worcester Polytechnic Institute (1989), *Photovoltaic Generation Effects on Distribution Feeders*, EPRI RP 2838-1, Final Report.

Northern States Power Company (1987), *Solar Demonstration Program Evaluation Report*.

Orr, J.A., D. Cyganski, A.E. Emanuel, and R.F. Saleh (1986), *Design of a System for Automated Measurement and Statistics Calculation of Voltage and Current Harmonics*, IEEE Transactions on Power Delivery, Vol. PWRD-1, pp. 23–30.

Pacific Gas and Electric Company (1988), *Solar Central Receiver Technology Advancement for Electric Utility Applications Phase I* Topical Report Volumes 1 and 2.

Pacific Gas and Electric Company (1988), *Utility Solar Central Receiver Study Phase IIB* Review Meeting, San Ramon, California.

Patapoff, N.W., Jr. (1985), *Two Years of Interconnection Experience with the 1 MW Plant at Lugo, Proceedings of the 18th IEEE Photovoltaic Specialists Conference.*

Pigg, S.K. (1988), *A Performance Analysis of Four Wind Energy Systems*, prepared for WP&L's Wind Energy Test Program.

Platte River Power Authority (1988), *Preliminary Design and Feasibility Study 10 Kilowatt Photovoltaic Pilot Plant.*

Pratt, R.G. (1976), *Solar and Wind Power*, Detroit Edison Power Club Meeting, p. 21–25.

Pratt, R.G. (1990), *Two Year Performance Evaluation of a 4 kW Amorphous-Silicon Photovoltaic System in Michigan*, Solar Cells **28**, No. 2, p. 163.

Public Service Company of New Mexico (1976), *Technical and Economic Assessment of Solar Hybrid Repowering*, proposal submitted to Energy Research and Development Administration.

Public Service Company of New Mexico (1978), *Solar Hybrid Repowering Project*, submitted to the U.S. Department of Energy, Sunday, May 3, 1978.

Public Service Company of New Mexico (1978a), *Technical and Economic Assessment of Solar Hybrid Repowering*, Final Report, prepared for U.S. Department of Energy.

Putnam, P.C. (1948), *Power From the Wind*, Van Nostrand Reinhold Company, New York.

Reid, R.L. and A.H.P. Swift (1989), *Solar Ponds: A Potential Low Cost Option for Providing Energy for Wisconsin*, Wisconsin Professional Engineer, p. 20–21, 25.

Roman, H.T. (1988), *Public Service Electric and Gas Company*, verbal communication.

Schaefer, J., H. Boyd, C. DeWinkel, D. Fagnan, and R. Pratt (1988), *Experiences with U.S. Line-Connected Amorphous Silicon Systems, Proceedings 9th European Photovoltaic Solar Energy Conference*, Florence, Italy.

Schaefer, J. (1989), *Wind Systems*, EPRI Journal, pp. 49–52.

Schaefer, J. (1990), *Review of Photovoltaic Power Plant Performance and Economics*, presented at IEEE Winter Power Meeting.

Schlueter, L.E. (1987), *Maintenance Requirements and Costs at the Carrisa Plains Photovoltaic Plant*, ARCO Solar, Inc., presented at 19th IEEE Photovoltaic Specialists Conference, New Orleans, LA.

Smith, D.R., M.A. Ilyin, and W.J. Steeley (1989), *PG&E's Evaluation of Wind Energy*, presented at Windpower '89.

Southern California Edison Company, McDonnell Douglas Corporation, and Bechtel Power Corporation (1982), *Solar 100 Conceptual Design* Final Report.

Steeley, W.J., M.A. Ilyin, and D.R. Smith (1987), *Summary of Wind Energy Activities at Pacific Gas and Electric's Department of Engineering Research*, Sixth ASME Wind Energy Symposium **3**, pp. 107–111.

Sumner, D.D., C.M. Whitaker, and L.E. Schlueter (1988), *Carrisa Plains Photovoltaic Power Plant 1984–1987 Performance*, The Conference Record of the Twentieth IEEE Photovoltaic Specialists Conference – 1988, Volume II, IEEE 88CH2527-0, pp. 1289–1292.

The SOLAR THERMAL Report (1981), *Owens Lake Selected as Site for 210 kWe Solar Pond* **2**, No. 5.

U.S. Department of Energy (1989), *Annual Outlook for U.S. Electric Power*, 1989, DOE/EIA-0474.

U.S. Department of Energy (1987), *National Photovoltaics Program*, Five Year Research Plan, 1987–1991, DOE/CH10093-7.

U.S. Department of Energy, Conservation and Renewable Energy (1987), *Solar Thermal Technology Annual Report, Fiscal Year 1986*, prepared by Sandia National Laboratories and Solar Energy Research Institute, DOE/CH10093-12, DE87001171.

U.S. Department of Energy, Office of Solar Technologies, Wind Energy Technology Division (1985), *Five Year Research Plan 1985–1990, Wind Energy Technology: Generating from the Wind*, DOE/CE-T11.

Virginia Power (1989), NEW TECHNOLOGY EVALUATIONS, *Summary Report* MT Storm Wind Energy Program.

Virginia Power (1989), NEW TECHNOLOGY EVALUATIONS, VISTA *Solar Facility Second Annual Report*.

Wahrenbrock, H.E. (1979), *Can We Afford Not to Develop the High Tower Windmill Now?*, Public Utilities Fortnightly, pp. 42–47.

Washom, B.J. (1984), *Vanguard I, Solar Parabolic Dish Stirling Engine Module* Final Report (May 28, 1982–September 30, 1984), El Segundo, California: Advanco Corporation, DOE-AL-16333-2.

Weinberg, C.J., W.J. Steeley, M.A. Ilyin, and D.R. Smith (1988), *Wind Energy Research and Development Activities at PG&E*, presented at Windpower '88.

Zaininger, H.W. (1980), *Review of Economics for Seven DOE Solar Repowering Studies*, Draft Final Report, EPRI RP1348-10 (Unpublished).

## 1.8.4 Publications of the Solar Power Program, Storage and Renewables Department, Generation and Storage Division, Electric Power Research Institute

### Solar Thermal

**1989**

**GS-6577 Volumes 1 & 2:** "Molten Salt Solar Electric Experiment" - McDonnell Douglas Astronautics Company, December 1989 (RP2302-2)

**GS-6573** "Status of Solar Thermal Electric Technology" - HGH Enterprises, December 1989 (RP2003-9)

**1986**

**AP-4608** "Performance of the Vanguard Solar Dish-Stirling Engine Module" - Energy Technology Engineering Center, July 1986 (RP2003-5 Final Report)

**1984**

**AP-328 Volume 2: Project Documentation** "10 MW$_e$ Solar-Thermal Central Receiver Pilot Plant" - Burns & McDonnell Engineering Company, March 1984 (RP2003-2 Final Report)

**AP-328 Volume 1: Report on Lessons Learned** "10 MW$_e$ Solar-Thermal Central Receiver Pilot Plant" - Burns & McDonnell Engineering Company, March 1984 (RP2003-2 Final Report)

**1983**

**AP-2852** "Large Gas Turbine Modifications for Solar-Fossil Hybrid Operation" - United Technologies Research Center, March 1983 (RP1348-8 Final Report)

**Published Prior to 1983**

**AP-2550** "Centaur Gas Turbine Modification and Development for Solar-Fossil Hybrid Operation" - Solar Turbines Incorporated, September 1982 (RP1270-1 Final Report)

**AP-2435-SY** "1 MW$_{th}$ Solar-Thermal Conversion Full-System Experiment" - Boeing Engineering & Construction Company, August 1982 (RP1509-1 Summary Project)

**AP-2398-SY** "Design and Fabrication of a 1 MW$_{th}$ Ceramic Tube Bench-Model Solar Receiver" - Black & Veatch Consulting Engineers, May 1982 (Summary Report)

**AP-2267** "Analysis of Thermal and Mechanical Stresses in the Ceramic Seal of the 1 MW$_{th}$ Bench-Model Solar Receiver" - AiResearch Manufacturing Company, February 1982 (RP475-9 Final Report)

**ER-1101-SY** "Design and Fabrication of a 1 MW$_t$ Bench Model Solar Receiver" - Boeing Engineering and Construction, August 1979 (RP377-2 Interim Summary Report)

**ER-869** "A Methodology for Solar-Thermal Power Plant Evaluation" - Westinghouse Electric Corporation, August 1978 (RP648-1 Final Report)

**ER-652** "Solar-Thermal Conversion to Electricity Utilizing a Central Receiver, Open-Cycle Gas Turbine Design" - Black & Veatch Consulting Engineers, March 1978 (RP475-1 Final Report)

**ER-651** "Physiological Effects of Redirected Solar Radiation" - Black & Veatch Consulting Engineers, August 1978 (RP955-1 Topical Report)

**ER-629** "Closed-Cycle, High-Temperature Central Receiver Concept for Solar Electric Power" - Boeing Engineering and Construction, January 1978 (RP377-1 Final Report)

**ER-434** "Preliminary System Analysis of a Fixed Mirror Solar Power Central Station" - General Atomic Company, June 1977 (RP739-1 Final Report)

· **ER-403-SY** "Closed Cycle High-Temperature Central Receiver Concept for Solar Electric Power" - Boeing Engineering and Construction, August 1976 (RP377-1 Summary Report)

**ER-387-SY** "Solar Thermal Conversion to Electricity Utilizing a Central Receiver, Open Cycle Gas Turbine Design" - Black & Veatch Consulting Engineers, March 1977 (RP475-1 Summary Report)

## Wind Power

**1989**

**GS-6567** "Solano MOD-2 Wind Turbine Operating Experience Through 1988" - Pacific Gas and Electric Company, November 1989 (RP-1590-6)

**GS-6256** "Assessment of Wind Power Station Performance and Reliability" - R. Lynette and Associates, Inc., March 1989 (RP1590-10)

**GS-6245 Volume 2: Appendixes** "Experiences with Commercial Wind Turbine Design" - R. Lynette & Associates, Inc., April 1989 (RP1590-12 Final Report)

**GS6245 Volume 1: Main Report** "Experiences with Commercial Wind Turbine Design" - R. Lynette & Associates, Inc., April 1989 (RP1590-12 Final Report)

**1988**

**AP-5824** "Altamont Wind Power Plant Evaluation for 1986" - Pacific Gas and Electric Company, May 1988 (RP1590-6 Final Report)

**1987**

**AP-5220** "Electrical Behavior of Wind Power Stations" - Zaininger Engineering Company, Inc., June 1987 (RP1996-24 Final Report)

**AP-5084** "Status of Commercial Wind Power: 1986 Survey" - Strategies Unlimited, March 1987 (RP1590-9 Final Report)

**AP-5044** "Test and Evaluation of a 500 kW Vertical-Axis Wind Turbine: Final Report" - Southern Company Edison Company, March 1987 (RP1590-3 Final Report)

**1986**

**AP-4794** "Proceedings: Workshop on Prospects and Requirements for Geographic Expansion of Wind Power Usage" - Steitz and Associates, November 1986 (RP1996-20 Proceedings)

**AP-4682** "Guidelines for Testing Wind Turbines" - Southern California Edison Company, August 1986 (RP-1996-25 Final Report)

**AP-4639** "Wind Power Stations: 1985 Performance and Reliability" - R. Lynette and Associates,Inc., June 1986 (RP1996-2 Final Report)

**AP-4638** "The Solano MOD-2 Wind Turbine: Operating Experience September 1984-August 1985" - Pacific Gas and Electric Company, June 1986 (RP1590-6 Final Report)

**AP-4591** "Utility Considerations for Wind Energy Purchase Agreements" - R. Lynette and Associates, Inc., June 1986 (RP1996-2 Final Report)

**AP-4590** "Testing Requirements for Variable-Speed Generating Technology for Wind Turbine Applications" - Electrotek Concepts Inc., May 1986 (RP1996-22 Final Report)

**AP-4586** "Wind Power Instrumentation Directory" - R. Lynette and Associates, Inc., May 1986 (RP1996-2 Final Report)

**AP-4528** "Cost Estimates for Vertical Axis Wind Turbines" - Bechtel Group, Inc., May 1986 (RP1989-2 Final Report)

**1985**

**AP-4348** "Proceedings: Utility-Wind Turbine Industry Interaction Workshop" - Science Applications International Corporation, December 1985 (RP1996-17 Proceedings)

**AP-4335** "Rotationally Sampled Wind and MOD-2 Wind Turbine Response" - Battelle, Pacific Northwest Laboratories, November 1985 (RP1996-12 Final Report)

**AP-4319** "Fatigue-Life Assessment Methods and Application to WTS-4 Wind Turbine" - Hamilton Standard, October 1985 (RP1996-4 Final Report)

**AP-4288** "EPRI Wind Turbine Test Activities, 1983–84" - Burns and McDonnell Engineering Company, October 1985 (RP1996-11 Final Report)

**AP-4261** "Variable Rotor Speed for Wind Turbines: Objectives and Issues" - Power Technologies, September 1985 (RP1996-9 Final Report)

**AP-4258** "Testing and Evaluation of a 500 kW Vertical-Axis Wind Turbine" - Southern California Edison Company, September 1985 (RP1590-3 Interim Report)

**AP-4239** "MOD-2 Wind Turbine Field Experience in Solano County, California" - Pacific Gas and Electric Company, September 1985 (RP1996-3 Final Report)

**AP-4199** "Wind Power Stations: 1984 Performance and Reliability" - R. Lynette and Associates Inc., August 1985 (RP1996-2 Interim Report)

**AP-4089** "Wind Turbine Structural Loads Resulting from Wind Excitation" - Oregon State University, June 1985 (RP1996-13 Final Report)

**AP-4077** "Early Market Potential for Utility Applications of Wind Turbines" - Science Applications International Corporation, August 1985 (RP1976-1 Final Report)

**AP-4060 Volume 3: Rotational Wind-sampling Tests and Analyses** "Goodnoe Hills MOD-2 Cluster Test Program" - Boeing Aerospace Company, May 1985 (RP1996-6 Final Report)

**AP-4060 Volume 2: Structural and Mechanical Loads Tests and Analyses** "Goodnoe Hills MOD-2 Cluster Test Program" - Boeing Aerospace Company, May 1985 (RP1996-6 Final Report)

**AP-4060 Volume 1: Energy Production Performance, Availability, and Reliability** "Goodnoe Hills MOD-2 Cluster Test Program" - Boeing Aerospace Company, May 1985 (RP1996-6 Final Report)

**AP-4054** "WTS-4 Wind Turbine Test Program" - Hamilton Standard, June 1985 (RP1996-4 Final Report)

**AP-3963** "Wind Power Stations: 1984 Survey" - Strategies Unlimited, June 1985 (RP1348-17 Final Report)

**AP-3735** "Wind Turbine Operation and Maintenance Experience" - R. Lynette and Associates Inc., February 1985 (RP1996-2 Interim Report)

**1984**

**AP-3896** "Solano County MOD-2 Wind Turbine Field Experience" - Pacific Gas and Electric Co., December 1984 (RP1996-3 Interim Report)

**AP-3813** "Wear and Environmental Effects on the Kahuku MOD-OA Wind Turbine" - Burns and McDonnell Engineering Company, December 1984 (RP1196-11 Topical Report)

**AP-3578** "Wind Power Parks: 1983 Survey" - Strategies Unlimited, August 1984 (RP1348-17 Interim Report)

**AP-3447** "Wind Turbine Performance Assessment: Technology Status Report No. 7" - Arthur D. Little, Inc., April 1984 (RP1996-1 Final Report)

**AP-3233 Volume 3: Bonneville Power Administration Goodnoe Hills Project** "Early Utility Experience with Wind Power Generation" - JBF Scientific Corporation, January 1984 (RP1590-1 Final Report)

**AP-3233 Volume 2: Pacific Gas and Electric Company Solano County Project** "Early Utility Experience with Wind Power Generation" - JBF Scientific Corporation, January 1984 (RP1590-1 Final Report)

**AP-3233 Volume 1: Summary Report** "Early Utility Experience with Wind Power Generation" - JBF Scientific Corporation, January 1984 (RP1590-1 Final Report)

**1983**

**AP-3284** "Wind Turbine Performance Assessment: Technology Status Report No. 6" - Arthur D. Little, Inc., November 1983 (RP1996-1 Interim Report)

**AP-3276** "Cost Estimates for Large Wind Turbines" - Bechtel Group, Inc., November 1983 (RP1989-1 Final Report)

**AP-3259** "Methods for Wind Turbine Dynamic Analysis" - Systems Control, Inc., October 1983 (1977-1 Final Report)

**AP-2906** "Application Examples for Wind Turbine Siting Guidelines" - Battelle Pacific Northwest Laboratories, March 1983 (RP1520-1 Final Report)

**AP-2882-SY** "Assessment of Distributed Wind Power Systems" - General Electric Company, February 1983 (RP1271-1 Summary Report)

**AP-2882** "Assessment of Distributed Wind Power Systems" - General Electric Company, February 1983 (RP1271-1 Final Report)

**AP-2796** "Wind Turbine Performance Assessment: Technology Status Report No. 5" - Arthur D. Little, Inc., January 1983 (RP1996-1 Interim Report)

**AP-2795** "Siting Guidelines for Utility Application of Wind Turbines" - Battelle Pacific Northwest Laboratories, January 1983 (RP1520-1 Final Report)

**Published Prior to 1983**

**AP-2456** "Wind Turbine Performance Assessment: Technology Status Report No. 4" - Arthur D. Little, Inc., June 1982 (RP1996-1 Interim Report)

**AP-1959** "Large Wind Generator Performance Assessment: Technology Status Report No. 3" - Arthur D. Little, Inc., July 1981 (RP1348-1)

**AP-1641** "Large Wind Turbine Generator Performance Assessment: Technology Status Report No. 2" Arthur D. Little, Inc., December 1980 (RP1348-1 Interim Report)

**AP-1614** "Wind Power Generation Dynamic Impacts on Electric Utility Systems" - Zainginer Engineering Company, November 1980 (TPS 79-775 Final Report)

**AP-1317** "Large Wind Turbine Generator Performance Assessment: Technology Status Report No. 1" - Arthur D. Little, Inc., January 1980 (RP1348-1 Technical Report)

**ER-1110-SR** "Proceedings of the Workshop on Ecomomic and Operational Requirements and Status of Large Scale Wind Systems" - Altas Corporation, July 1979 (Special Report)

**ER-978 Volume 3: Appendixes** "Requirements Assessment of Wind Power Plants in Electric Utility Systems" - General Electric Company, January 1979 (RP740-1 Final Report)

**ER-978 Volume 2:** "Requirements Assessment of Wind Power Plants in Electric Utility Systems" - General Electric Company, January 1979 (RP740-1 Final Report)

**ER-978 Volume 1: Summary Report** "Requirements Assessment of Wind Power Plants in Electric Utility Systems" - General Electric Company, January 1979 (RP740-1 Final Report)

### Photovoltaics

**1990**

**GS-6696** "Photovoltaic System Performance Assessment for 1988" - Southwest Technology Development Institute, January 1990 (RP 1607-6 Final Report)

**GS-6689** "Carrisa Plains Photovoltaic Power Plant: 1984–1987 Performance" - Pacific Gas & Electric Company, January 1990 (RP1607-2 Final Report)

**1989**

**GS-6625** "Photovoltaic Operation and Maintenance Evaluation" - R. Lynette & Associates, Inc., December 1989 (RP1607-5 Final Report)

**GS-6306** "Survey of U.S. Line Connected Photovoltaic Systems" - EPRI, March 1989 (Special Report)

**GS-6251** "Photovoltaic Field Test Performance Assessment: 1987" - Southwest Technology Development Institute, March 1989 (RP1607-6 Final Report)

**1988**

**AP-6006** "Models of Photovoltaic Module Performance" - Pacific Gas and Electric, September 1988 (RP1607 Final Report)

**AP-5883** "Photovoltaic Resources Assessment in Colorado" - Platte River Power Authority, July 1988 (RP1607-4 Final Report)

**AP-5762** "Photovoltaic Field Test Performance Assessment: 1986" - New Mexico Solar Energy Institute, March 1988 (RP1607-6 Final Report)

**1987**

**AP-5229** "Hesperia Photovoltaic Power Plant: 1985 Performance Assessment" - New Mexico Solar Energy Institute, July 1987 (RP1607-6 Final Report)

**AP-5166** "Hydrogenated Amorphous Silicon Films Produced by Chemical Vapor Deposition" - Poly Solar, Inc., April 1987 (RP1193-2 Final Report)

**1986**

**AP-4821-SR** "Joint Technical Workshop: Amorphous Silicon Alloys for Photovoltaic Power" - EPRI, MITI, NEDO, September 1986 (Special Report)

**AP-4752** "High Concentration Photovoltaic Module Design" - Black & Veatch Engineers-Architects, August 1986 (RP1415-7 Interim Report)

**AP-4466** "Photovoltaic Field Test Performance Assessment: Technology Status Report Number 4" - Boeing Computer Services Company, March 1986 (RP1607-1 Final Report)

**AP-4464** "PG&E Photovoltaic Module Performance Assessment" - Pacific Gas and Electric, March 1986 (RP1607-2 Final Report)

**AP-4369 Volume 2: Appendixes** "Photovoltaic Manufacturing Cost Analysis: A Required-Price Approach" - Research Triangle Institute, September 1986 (RP1975-1 Final Report)

**AP-4369 Volume 1: Preliminary Assessment** "Photovoltaic Manufacturing Cost Analysis: A Required-Price Approach" - Research Triangle Institute, September 1986 (RP1975-1 Final Report)

**1985**

**AP-3533** "Laser Recrystallization of Compound Semiconductor Films" - Poly Solar, Inc., January 1985 (RP1193-2 Final Report)

**1984**

**AP-3792** "Photovoltaic Field Test Performance Assessment: Technology Status Report Number 3" - Boeing Computer Services Company, November 1984 (RP1607-1 Interim Report)

**AP-3532** "Thin Film Indium Phosphide Photovoltaic Device Research and Development" - Poly Solar, Inc., May 1984 (RP1193-2 Final Report)

**AP-3285 Volume 2:** "Project Documentation 10 MW$_e$ Solar-Thermal Central-Receiver Pilot Plant" - Burns & McDonnell Engineering Company, March 1984 (RP2003-2 Final Report)

**AP-3264** "Integrated Photovoltaic Central Station Conceptual Designs" - Black & Veatch Engineers-Architects, June 1984 (RP2197-1 Final Report)

**AP-3263** "Conceptual Design for a High-Concentration (500X) Photovoltaic Array" - Black & Veatch Engineers-Architects, December 1984 (RP1415-7 Interim Report)

**AP-3124** "Interconnecting DC Energy Systems: Responses to Technical Issues" - Systems Control, Inc. and James B. Patton, June 1983 (SIA82-412 Final Report)

**1983**

**AP-3351** "Photovoltaic Power Systems Research Evaluation" - Strategies Unlimited, December 1983 (RP1348-15 Final Report)

**AP-3244** "Photovoltaic Field-Test Performance Assessment: Technology Status Report Number 2" - Boeing Computer Services Company, September 1983 (RP1607-1 Interim Report)

**AP-3176-SR** "Photovoltaic Systems Assessment: An Integrated Perspective" - EPRI, September 1983 (Special Report)

**AP-2859** "Point Contact Silicon Solar Cells" - Stanford University, May 1983 (RP790-2 Interim Report)

**AP-2475** "Photovoltaic Requirements Estimation-A Simplified Method" - Science Applications, Inc., February 1983 (RP1975-3 Final Report)

**1982**

**AP-2687-SY** "Assessment of Distributed Photovoltaic Electric Power Systems" - JBF Scientific Corporation, October 1982 (RP1192-1 Summary Report)

**AP-2687** "Assessment of Distributed Photovoltaic Electric Power Systems" - JBF Scientific Corporation, October 1982 (RP1192-1 Final Report)

**AP-2636** "Assessment of Distributed Solar Power Systems: Issues and Impact" - Science Applications, Inc. and Systems Control, Inc., November 1982 (RP1995-1 Final Report)

**AP-2544** "Photovoltaic Field-Test Performance Assessment: Technology Status Report Number 1" - Boeing Computer Services Company (RP1607-1 Interim Report)

**AP-2474** "Photovoltaic Balance-of-System Assessment" - Bechtel Group, Inc., June 1982 (RP1975-2 Final Report)

**AP-2473 Volume 2: Technology Basis** "Photovoltaic Cell and Module Status Assessment" - Research Triangle Institute, October 1982 (RP1975-1 Final Report)

**AP-2473 Volume 1: Technology Overview** "Photovoltaic Cell and Module Status Assessment" - Research Triangle Institute, October 1982 (RP1975-1 Final Report)

**AP-2408** "Summary Status of Conceptual High-Performance, High-Specific-Output Silicon Photovoltaic Systems" - Black & Veatch Consulting Engineers, May 1982 (RP1348-3 Final Report)

**Published Prior to 1982**

**AP-1940** "Technical and Economic Assessment of Solar Thermophotovoltaic Conversion" - Science Applications, Inc., July 1981 (RP1415-1 Final Report)

**ER-1272** "Silicon Photovoltaic Cells in Thermophotovoltaic Conversion" - Stanford University, December 1979 (RP790-2 Interim Report)

**ER-1262** "Thermophotovoltaic Conversion from Conventional Heat Sources" - Black & Veatch Consulting Engineers, December 1979 (RP1348-3 Final Report)

**ER-685 Volume 3** "Requirements Assessments of Photovoltaic Power Plants in Electric Utility Systems" - General Electric Company, June 1978 (RP651-3 Appendices)

**ER-685 Volume 2** "Requirements Assessments of Photovoltaic Power Plants in Electric Utility Systems" - General Electric Company, June 1978 (RP651-1 Technical Report)

**ER-685 Volume 1** "Requirements Assessments of Photovoltaic Power Plants in Electric Utility Systems" - General Electric Company, June 1978 (RP651-1 Summary Report)

**ER-633** "Silicon Photovoltaic Cells in TPV Conversion" - Stanford University, February 1978 (RP790-1 Interim Report)

**ER-589-SR** "Perspectives on Utility Central Stations Photovoltaic Applications" - EPRI, January 1978 (Special Report)

**ER-478** "Silicon Photovoltaic Cells in Thermophotovoltaic Conversion" - Stanford University, February 1977 (RP790-1 Progress Report)

**ER-258** "Solar Data Verification Project" - Science Applications, September 1976 (Final Report)

**ER-198** "Penetration Analysis and Margin Requirements Associated with Large-Scale Utilization of Solar Power Plants" - The Aerospace Corp., August 1976 (TPS 75-611 Final Report)

**ER-188** "Assessment of Cadmium Sulfide Photovoltaic Arrays for Large Scale Electric Utility Applications" - Brown University, February 1976 (WS72-21 Final Report)

### Surveys & Conference Proceedings

**1984**

**AP-3665-SR** "Electric Utility Solar Energy Activities: 1983 Survey" - EPRI, September 1984 (Special Report)

**1983**

**AP-2884-SR** "EPRI Conference Proceedings: Solar and Wind Power-1982 Status and Outlook" - The University of Kansas Center for Research, Inc., February 1983 (Special Report)

**Published Prior to 1983**

**AP-2850-SR** "Electric Utility Solar Energy Activities: 1982 Survey Update" - Electric Power Research Institute, December 1982 (Special Report)

**AP-2516-SR** "Electric Utility Solar Energy Activities: 1981 Survey" - Electric Power Research Institute, July 1982 (Special Report)

**AP-1713-SR** "Electric Utility Solar Energy Activities: 1980 Survey" - December 1980 (Special Report)

**ER-1299-SR** "Electric Utility Solar Energy Activities: 1979 Survey" - Electric Power Research Institute, December 1979 (Special Report)

**ER-966-SR** "Electric Utility Solar Energy Activities, 1978 Survey" - Electric Power Research Institute, May 1979 (Special Report)

**ER-649-SR** "Electric Utility Solar Energy Activities, 1977 Survey" - Electric Power Research Institute, February 1979 (Special Report)

**ER-515-SR** "Proceedings of Semiannual EPRI Solar Program Review Meeting and Workshop" - Altas Corporation, August 1977 (Special Report)

**ER-371-SR** "Proceedings of Semiannual EPRI Solar Program Review Meeting and Workshop" - Altas Corporation, February 1977 (Special Report)

**ER-321-SR** "Electric Utility Solar Energy Activities 1976 Survey" - EPRI, January 1977 (Special Report)

**ER-283-SR Volume II** "Proceedings of First Semiannual EPRI Solar Program Review Meeting and Workshop" - Altas Corporation, May 8–12, 1976 (Special Report)

**ER-283-SR Volume I** "Proceedings of First Semiannual EPRI Solar Program Review Meeting and Workshop" - Altas Corporation, March 8–12, 1976 (Special Report)

**Other**

**1988**

**AP/EM/EL-5470** "Materials and Devices for Power Electronic Applications" - Steitz and Associates, March 1988 (RP1996-20 Final Report)

**1987**

**AP-5219** "Specification of Test Equipment for Power Electronic Drive Research" - Omnion Power Engineering Corporation, June 1987 (RP2790-1 Final Report)

**AP-5210** "Reactive Power Management Device Assessment" - University of Washington, August 1987 (RP1590-8 Final Report)

**1986**

**AP-4921** "Ocean Energy Technologies: The State of the Art" - Massachusetts Institute of Technology, November 1986 (RP1348-28 Final Report)

**1985**

**AP-3842** "A State-of-the-Art Study of Nonconvective Solar Ponds for Power Generation" - Massachusetts Institute of Technology, January 1985 (RP1348-18 Final Report)

**Published Prior to 1983**

**AP-2265** "Evaluation of Biomass Systems for Electricity Generation" - Battelle, Columbus Laboratories, February 1982 (RP1348-7 Final Report)

**AP-1548** "Satellite Power System: Utility Impact Study" - Systems Control, Inc., September 1980 (TPS79-752 Final Report)

**AP-1169** "Comparative Assessment of Marine Biomass Materials" - Science Applications, Inc., September 1979 (TPS77-735 Final Report)

**AF-974** "Evaluation of the Potential for Producing Liquid Fuels From Biomaterials" - University of Oklahoma, January 1979 (TPS77-716 Final Report)

**ER-1113-SR** "Ocean Thermal Energy Conversion: A State-of-the-Art Study" - Massachusetts Institute of Technology, July 1979 (SOA78-411 Special Report)

**ER-1070 Volume 3: Environmental Impact Assessment Application:** "Solar Power Plant Applications" - Woodward-Clyde Consultants, May 1979 (RP551 Final Report)

**ER-1070 Volume 2: Industrial Implications and Secondary Impacts Methodology:** "Solar Power Plant Applications" - Woodward-Clyde Consultants, May 1979 (RP551 Final Report)

**ER-1070 Volume 1: Environmental Impact Assessment Methodology Enviremental Assessment Methodology:** "Solar Power Plant Applications" - Woodward-Clyde Consultants, May 1979 (RP551 Final Report)

**ER-746-SR** "Biofuels: A Survey - EPRI, June 1978 (Special Report)

**EPRI-550-1** "Solar Materials and Components Test Program" - Black & Veatch, April 1976 (Final Report)

**ER-258** "Solar Data Verification Project" - Science Applications, Inc., September 1976 (RP552 Final Report)

**ER-198** "Penetration Analysis and Margin Requirements Associated with Large-Scale Utililization of Solar Power Plants" - The Aerospace Corporation, August 1976 (TPS75-611 Final Report)

## 1.8.5 List of Acronyms

AEC - Alternative Energy Corporation
APS - Arizona Public Service Company
BEST - Battery Energy Storage Test
BPA - Bonneville Power Administration
CEC - California Energy Commission
CP&L - Carolina Power & Light
CRTF - Central Receiver Test Facility
CSWT - Constant Speed Wind Turbine
DHW - Domestic Hot Water
DOE - Department of Energy
EEI - Edison Electric Institute
EPRI - Electric Power Research Institute
ERDA - Energy Research and Development Administration
ESI - Energy Sciences Inc.
ESI - Energy Sciences Institute
FERC - Federal Energy Regulatory Commission

FPC - Florida Power Corporation
FPL - Florida Power & Light Company
GE - General Electric Company
GMP - Green Mountain Power Company
GPC - Georgia Power Company
HCE - Heat Collection Element
HECO - Hawaiian Electric Company
HERS - Hawaiian Electric Renewable Systems
HTF - Heat Transfer Fluid
IPH - Industrial Process Heat
IPP - Independent Power Producer
ISES - International Solar Energy Society
JPL - Jet Propulsion Laboratory
LADWP - Los Angeles Department of Water & Power
LCC - Load Carrying Capability
LILCO - Long Island Lighting Company
LOLP - Loss of Load Probability
LRG - Laminated Reflector Glass
MDC - McDonnell Douglas Corporation
MSEE - Molten-Salt Electric Experiment
MSS/CTE - Molten Salt Subsystem/Component Test Experiment
NASA - National Aeronautics and Space Administration
NEES - New England Electric System
NSP - Northern States Power Company
NUSCo - Northeast Utilities Service Company
NYSERDA - New York State Energy Research & Development Authority
OSU - Oregon State University
OTEC - Ocean-thermal Energy Conversion
PCU - Power Conditioning Unit
PECo - Philadelphia Electric Company
PG&E - Pacific Gas & Electric Company
PNM - Public Service Company of New Mexico
PSE&G - Public Service Electric & Gas of New Jersey
PUC - Public Utility Commission
PURPA - Public Utility Regulatory Policies Act of 1978
PV - Photovoltaic
PVUSA - Photovoltaics for Utility Scale Applications
R&D - Research and Development
REA - Rural Electrification Administration
RTU - Remote Terminal Unit
SCA - Solar Collector Assembly
SCE - Southern California Edison Company
SDG&E - San Diego Gas & Electric Company
SDSU - South Dakota State University
SEGS - Solar Electric Generating System
SERI - Solar Energy Research Institute

SHAC - Solar Heating and Cooling
SMUD - Sacramento Municipal Utility District
SRP - Salt River Project
STAR - Solar Test and Research Center
STEP - Solar Total Energy Project
THD - Total Harmonic Distortion
TUE - Texas Utilities Electric Company
TVA - Tennessee Valley Authority
U of W - University of Washington
UCB - Utility Coordination Board
USAB - United Stirling, AB
USW - U.S. Windpower
VISTA - Virginia Integrated Solar Test Arrays
VP - Virginia Power
VSWT - Variable Speed Wind Turbine
WEPCO - Wisconsin Electric Power Company
WPI - Worcester Polytechnic Institute
WP&L - Wisconsin Power & Light Company

CHAPTER 2

# The Status of Solar Thermal Electric Technology

Richard J. Holl and Edgar A. DeMeo

## 2.1 Abstract

Solar thermal electric technology was evaluated as a future source of power for United States utilities. The technology status was developed using an experimental data base derived from full system experiments and component development programs. A representative configuration was selected for each of the major solar collector concepts: the central receiver, the parabolic dish and the parabolic trough. The ultimate energy costs that could be expected from these three concepts as mature, commercial systems were estimated. The estimates were in the form of upper and lower bounds to the expected energy costs, where the upper bound could be supported by the existing experimental data base and the lower bound represented the potential possible for the system if present uncertainties in costs and performance were to be resolved favorably. The development status for each of the three concepts was reviewed, and estimates were made of the costs and time that would be required to complete the development of commercial versions of the systems. Central receiver and dish collector system technology are both at an early stage of development. The potential energy costs for these systems could be as low as 5 cents per kilowatt-hour, but the uncertainty in achieving this potential is large. The technology of the trough system is relatively mature. The potential energy costs for this system are about a factor of two higher than the potential of either the central receiver or the dish, but the confidence level of achieving the energy costs estimated for the trough system is high.

## 2.2 Introduction

Solar thermal electric power systems convert solar energy into electricity by first concentrating the incoming sunlight, then converting it to heat and finally converting the heat into electricity. This process distinguishes this technology from photovoltaics which produces electricity from sunlight directly, without the intermediate thermal conversion.

### 2.2.1 Background

While the use of solar energy has been evolving over many years, concentrated research and development was initiated in the early 1970's in response to the first oil embargo. The realization that our fossil fuel resources were finite and that their costs were subject to large escalation motivated a vigorous worldwide response to develop renewable energy alternatives. Solar thermal technology, the production of heat from sunlight, received about 1.2 billion dollars of government funding in the United States alone (Weingart, 1986) and significant support overseas. Much of this was devoted to solar thermal electric technology.

With this vigorous support, many approaches for concentrating and converting sunlight into electricity were examined. Many were carried to test hardware. The more promising components were retained and improved over successive generations of hardware development. A number of full system experiments were built and tested around the world.

This growth reversed in the 1980's coincident with the decline in world oil prices, and, equally important, the reduced projections for fuel prices in the future.

While the present scenario for fuel prices does not encourage continued development of alternative energy sources, this situation will almost certainly change. The only question is when. Depletion and price growth are likely to be seen in the next 20 to 30 years. The development of a new commercial electric power system and an industry capable of meeting any significant fraction of our electricity generating needs is likely to take more than 20 years; an estimate of 50 years for the development and introduction of a new energy technology was made (Kesselring and Winter, 1987).

The extensive development conducted during the past 15 years should be reviewed to assess the potential of this technology and to identify any continued research and development that would be justified through a period of little or no immediate need.

### 2.2.2 Solar Thermal Electric System Concepts

Three basic concepts have emerged as being potentially attractive in collecting and concentrating sunlight for electric power production. They are the central receiver, the parabolic dish and the parabolic trough.

In a central receiver system, a field of computer-guided heliostats (mirrors) focuses sunlight onto a tower-mounted receiver. The concentrated heat energy absorbed by the receiver is transferred to a circulating heat transfer fluid to power an electric generator. Part of the heated fluid may be diverted to an energy storage subsystem that allows the power plant to operate during non-solar hours.

Parabolic dish systems use point-focusing collectors that track the sun in two axes and focus radiant energy onto a receiver at the focal point of the paraboloidal concentrator. Each dish module can be used alone or in multi-module systems. Energy from a heat transfer fluid circulating through the receiver can be converted directly into electrical energy by using a heat engine/generator coupled to the receiver. Alternatively, the total thermal output from a field of collectors can be transported to a central power plant for conversion to electrical energy.

Parabolic troughs are U-shaped collectors lined with highly reflective material that concentrate sunlight onto a linear receiver tube positioned along the focal line of the trough. A fluid in the receiver is heated by the radiant energy and then transported to a central point by means of a piping network. The thermal energy can then be used directly or converted to electrical energy. Troughs are often used for process heat applications because they operate at lower temperatures than central receivers or dishes.

### 2.2.3 Objectives

The overall objective of this review was to assess the prospects for solar thermal electric technology as a future source of power for United States utilities. Specific objectives were to:

1. Evaluate the status of technology,
2. Provide an estimate of the range of future energy costs that could be expected from commercial plants based on this technology, and
3. Scope the development requirements, duration and cost to commercialize the technology.

### 2.2.4 Approach

The approach adopted to accomplish each of the objectives is outlined below.

#### 2.2.4A Technology Status

An experimental data base was collected and utilized for assessment of the technology status. Solar thermal electric full system experiments provided the primary data base. The systems were selected for evaluation according to the following criteria: (1) the ultimate commercial version of the system would be suitable for utility-scale electric power production and (2) the experiment itself was of sufficient size to provide meaningful data. The systems evaluated were limited to the categories of: central receivers, dish systems and trough systems. Data were collected from sufficient system experiments to characterize the technology. Published information on these experiments was augmented by interviewing technically-cognizant project participants to add unpublished data or insights and to aid in interpreting project results. Pertinent component development programs were reviewed to complete the experimental data base.

The status of the three different system categories (central receiver, dish and trough) will be described separately. The approach will be to (1) summarize the status of component development, (2) describe the major system experiments, (3) select a representative candidate for each system category and (4) summarize the status of the technology from the results of the system experiments and the component development programs.

The technical description of the system experiments will cover:

1. a brief physical description of the test hardware,
2. a summary of the system's operation and operating modes,
3. results of performance tests,
4. significant incidents encountered during the experiment and
5. lessons learned from the experiment.

The primary criterion for the information that was selected and presented was to provide a basis for assessment of the technology status. Additionally, the data shown should allow a potential user to gain a familiarity with the technology. Finally, a review of the major incidents encountered in previous system experiments and the

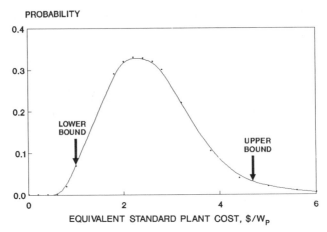

**Figure 2.1:** Probability distribution.

lessons learned from them should allow a system developer to avoid repeating some of the same mistakes.

### 2.2.4B Expected Energy Costs

There are always large uncertainties in the costs for a mature technology when the predictions are made at an early stage of its development. An actual probability distribution for such a prediction might look like the curve shown in Figure 2.1. Any value along the distribution curve could be used as the "prediction", depending upon one's degree of optimism. Recognizing this limitation, the approach adopted for this evaluation was to estimate two limits to the energy costs to be expected from commercial systems employing this technology. These limits would hopefully bound the energy costs from the mature system with a probability of 90 percent (as illustrated in Figure 2.1). The upper bound, the highest expected cost of energy, was estimated based on system performance that could be justified using existing experimental data and system costs that employ all contingencies appropriate to the level of maturity of the technology and the level of detail included in the design of the commercial system. The lower bound to energy costs, the lowest cost with any credibility, was estimated based on performance predictions that, while not confirmed experimentally, at least were not ruled out by existing experimental data. The estimates were based on the expectation that developments already under way would be successful in achieving the low costs predicted.

The cost of energy was estimated for plants operating solely from solar energy. The reason for this selection was that it is not likely for this technology to become fully commercial until its energy costs are somewhat comparable to those from fossil fuels. Accordingly, systems that are well adapted to operation with both solar and fossil energy inputs will be noted for their operating flexibility, but no benefit from the burning of low-cost fossil fuel will be included in the estimates of the expected energy costs from the mature commercial systems.

In order to compare solar concepts with each other independent of economic assumptions that vary so much over time, the energy costs from all concepts will be related to those from a standard solar plant defined as having an annual capacity factor of 0.25 (relative to its net electrical rating) and having operation and maintenance costs of one cent per kilowatt-hour. This, in effect, will be a comparison in terms of $/KWHr-yr which is widely used in evaluating and comparing alternative energy technologies that use no fuel.

The capital costs of the solar thermal electric plants will be expressed as $/m$^2$ of reflective surface area to facilitate comparisons of different concepts. The annual operation and maintenance costs will be expressed as an equivalent capital cost by discounting to present value after subtracting the one cent per kilowatt-hour allowance for the standard plant. An equivalence in energy cost between the solar thermal plant (with equivalent capital costs expressed in $/m$^2$) and the standard plant (with capital costs expressed in dollars per peak watt rating, $/Wp) will be determined for an annual insolation level of 2500 KWHr/m$^2$-yr (representative of Barstow, California) and for various annual net efficiencies of the solar thermal plant.

The following coarse system evaluation guide was employed:

- Energy costs equal to or less than those from the standard plant at $2 per peak watt are very attractive.
- Energy costs greater than those from the standard plant at $8 per peak watt are unattractive.
- Energy costs between these bounds are in a potentially interesting range.

The reason for using these values is that the alternative of photovoltaic solar electric power generation has achieved $8 per peak watt, and $2 per peak watt is a credible goal for this system. Field tests of experimental photovoltaic power plants have demonstrated annual capacity factors of 25% or more and operation and maintenance costs under 1 cent/KWHr (Risser and Stokes, 1987, Risser and Stokes, 1988, and Risser and Stokes, 1989).

Applying these factors for determining the cost equivalence for electricity from solar thermal and the standard plant results in:

$$\text{Standard Plant Cost/Annual Output} = 1000\, C_1/(0.25 \times 8760)$$
$$= 0.4566\, C_1\, (\$/\text{KWHr/yr}) \qquad (2.1)$$

where $C_1$ is the cost of the standard plant in $/Wp, 0.25 is the annual capacity factor and 8760 are the number of hours per year.

$$\text{Solar Thermal Capital Cost/Annual Output} = C_2/(\eta \times 2500)$$
$$(\$/\text{KWHr/yr}) \qquad (2.2)$$

where $C_2$ is the effective cost of the solar thermal plant in $/m$^2$, including the present value of any difference in annual O&M relative to the standard plant, and $\eta$ is the net annual conversion efficiency from solar to electricity.

Equating these yields:

$$C_1 = C_2/1142\eta. \qquad (2.3)$$

Thus, a solar thermal electric power plant with an effective cost of $400/m$^2$ and a net annual efficiency of 10% would produce power at a cost equivalent to the standard plant costing $3.5 per watt peak. The net annual efficiency is defined as

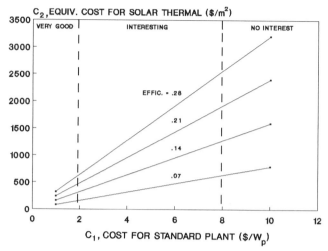

**Figure 2.2:** Thermal electric compared with standard plant.

the net electrical energy produced by the plant during the year divided by the direct normal insolation times the plant's collector area.

Equation (2.1) is plotted in Figure 2.2 with $C_1$ as the abscissa, $C_2$ as the ordinate and various values for the efficiency, $\eta$. The evaluation criteria are shown on the plot.

### 2.2.4C  Development Requirements

Following a summary of the technology status for each of the three system categories, the major technology issues remaining to be resolved will be tabulated. These will be used to prepare an estimate of the development requirements for each system. A representative development program will be identified in order to provide a basis for estimating the duration and cost that would be required to commercialize the candidate selected as representative of each system category.

## 2.3 The Central Receiver System

The central receiver system has been the primary solar thermal concept developed for utility-scale electric power production. The central receiver is so-named because it collects and concentrates solar energy from a large field onto a "central receiver" as reflected sunlight. This feature results in the following advantages for this system concept in utility-scale electric power production:

1. Minimum "plumbing" is required to collect solar energy from the field.
2. A very high concentration of sunlight can be achieved; this allows high temperature energy collection with high efficiency. The high temperature, in turn, allows high efficiency in conversion to electricity.

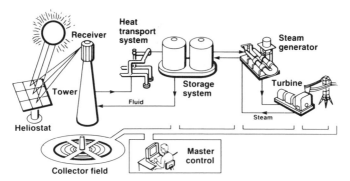

**Figure 2.3:** Solar Central Receiver System (Falcone, 1986).

3. Large size installations are possible with a resulting economy-of-scale.

Solar components have been developed and improved over several successive "generations", and a number of full system experiments have been built and tested to prove the basic system concepts. These developments will be reviewed here to assess the current status of this concept.

### 2.3.1 System Concepts

A typical central receiver system is illustrated in Figure 2.3. The major subsystems are:

- The collector subsystem consists of a field of suntracking mirrors, called heliostats, which collect and concentrate sunlight onto a tower-mounted receiver.
- The receiver absorbs the reflected sunlight, converting it into thermal energy.
- The heat transport subsystem circulates a heat transfer fluid through the receiver to extract the absorbed energy and transport it to either the storage or power conversion subsystems.
- The storage subsystem stores the collected solar energy as sensible heat. It can also decouple energy collection from power generation.
- The power conversion subsystem converts the thermal energy into electricity for dispatch on the grid.
- A master control subsystem is most generally used to coordinate the control of the different subsystems.

The collector field layout and the receiver configuration are interrelated. The two alternatives for this choice are:

- North Field/Cavity Receiver – All heliostats are located to the north of the receiver/tower (in the northern hemisphere), and the receiver absorptive surfaces are located within a cavity enclosure.
- Surround Field/External Receiver – The heliostats completely surround the receiver/tower, and the receiver absorptive surfaces are on the outside of the receiver.

The cavity and external receiver configurations are illustrated in Figure 2.4.

Selection of the heat transport fluid, the thermal storage media and the power conversion cycle and working fluid define the system concept. Table 2.1 lists

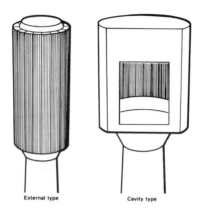

**Figure 2.4:** Cavity and external receiver configurations.

**Table 2.1:** Candidate Central Receiver System Concepts

|  | Water/Steam | Advanced Concepts | | |
|---|---|---|---|---|
| Heat Transport Fluid | Water/Steam | Nitrate Salt* | Liquid Sodium | Liquid Sodium |
| Thermal Storage Media | Hot Water, Oil/Rock or Hitec Salt** | Nitrate Salt* | Liquid Sodium | Nitrate Salt* |
| Power Conversion | Steam Rankine | Steam Rankine | Steam Rankine | Steam Rankine |

*60% NaNO$_3$, 40% KNO$_3$ (Melting Point = 221$^o$ C, 430$^o$ F)
**53% KNO$_3$, 40% NaNO$_2$, 7% NaNO$_3$ (Melting Point = 142$^o$ C, 288$^o$ F)

the major candidates considered for these selections. All plants would employ distributed digital controls coordinated by a master control unit.

## 2.3.2 Component Development

The Central Receiver Test Facility (CRTF), located in Albuquerque, New Mexico, has been the major center for component and system testing for the central receiver (Darsey et al., 1977, Holmes et al., 1980, and Maxwell and Holmes, 1987). The facility is operated by Sandia National Laboratories for the United States Department of Energy. An aerial view is shown in Figure 2.5.

### 2.3.2A Heliostats

The heliostats is a two-axis tracking assembly that reflects sunlight onto the tower-mounted receiver. It is the major component of the collector subsystem and represents the major cost element of the central receiver system. The heliostat assembly is comprised of:

- Reflector Modules (Mirrors)
- Support Structure
- Drive Units (2 axes) – gear boxes and motors

**Figure 2.5:** The Central Receiver Test Facility.

- Controller
- Pedestal
- Foundation

Field wiring for power and control, field controllers for a group of heliostats, an array controller for the entire field and support equipment complete the collector subsystem.

Because the heliostat is a totally new component and because of its major cost impact, substantial effort has been devoted to its development. Most of the development utilized glass mirrors with steel structural support; these are called glass/metal heliostats. A recent innovation is the use of a reflective membrane; these are called stretched-membrane heliostats.

### 2.3.2A.1 Glass/Metal Heliostats

Many alternative designs for glass/metal heliostats were built and tested. Figure 2.6 illustrates the design evolution over successive "generations" (Sandia National Laboratories, 1981, Sandia National Laboratories, 1982, Champion, 1984, and Mavis, 1986). The most striking feature of this evolution is the continual increase in reflective surface area. Heliostats of 150 m$^2$ and 200 m$^2$ were built and tested (Thomas et al., 1987 and Alpert and Houser, 1988). Economics improve with increased area to about 150 m$^2$ because, for a given reflective surface area, there are fewer drives, controllers, structural assemblies, pedestals and foundations; also, installation and maintenance costs are lower per unit reflective area with the larger units.

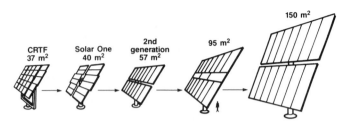

Figure 2.6: Glass/metal heliostat evolution.

Table 2.2: Glass/Metal Heliostat Design

| Feature | Design Specification |
| --- | --- |
| Reflective Area | 150 m$^2$ |
| Reflector | Silvered glass/steel laminate mirrors (curved and canted for focus) |
| Structure | Steel torque tube; trusses and braces |
| Drives-both axes | Worm and spur gearing |
| Support | Steel Pedestal |

Glass/metal heliostats are fully developed technically and have shown good performance and availability under test and in system experiments. The major issue remaining is the achievement of low cost (about $100/m$^2$) under mass production. Major characteristics of the preferred glass/metal heliostat design that has evolved from this extensive development program are given in Table 2.2.

### 2.3.2A.2 Stretched Membrane Heliostats

An innovative design that could be lower in cost than the glass/metal heliostat is the stretched membrane concept (Murphy, 1983, Murphy et al., 1985, Murphy, 1986, Murphy et al., 1986, and Beninga and Butler, 1987). The mirror module consists of a ring, front and back membranes, a tensioning system, and an active focus control system. Together these elements provide a means of achieving a large, focused, reflecting mirror. The radius of curvature can be set at a range of values which allows this single mirror module to effectively focus the sun at exactly its slant range from the tower. By placing a slight vacuum in the plenum between the two membranes, the mirror module can be focused onto the receiver. By placing a slight pressure, with respect to atmosphere, in the plenum, the mirror module can be defocused.

The major cost benefit for the stretched membrane heliostat compared with the glass/metal design are:

- Lower weight.  Mirror modules can be as low as one-fourth the weight of glass/metal designs.
- Fewer parts and pieces.
- Potential for drive and foundation cost reductions with advanced rim drives.

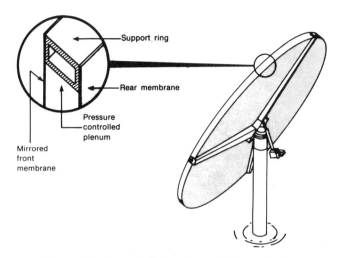

**Figure 2.7:** Stretched membrane heliostat design.

The commerical design of the stretched membrane heliostat also utilizes a reflective surface area of about 150 m$^2$ to preserve the cost advantages of that feature.

Two stretched membrane prototype heliostat assemblies were built and installed on existing pedestal/drive units at the CRTF for testing in 1986. Both units had a reflective surface area of 50 m$^2$. One unit was produced by the Science Application International Corporation (SAIC) and employed thin stainless steel membranes (Butler et al., 1986). The other unit, produced by Solar Kinetics, Incorporated (SKI) used aluminum membranes (Solar Kinetics, 1986). Both used a silvered-polymer as the reflector on the front membrane. Photographs of the SKI unit at the CRTF are shown in Figure 2.8. The SAIC design was not workable and did not produce a viable image; the SKI design was tested over 2 years and the optical performance was at least as good as glass/metal designs (Alpert et al., 1988).

Development of improved designs is continuing by both companies using the lessons learned from the first prototypes (Beninga et al., 1988, and White, 1988).

### 2.3.2B  Receivers

Solar receiver designs and development have progressed through early water/steam concepts to those employing the advanced receiver coolants in a conventional tube/panel configuration and finally to an even more advanced concept in which the reflected solar energy is absorbed directly in the coolant. Receiver panels have been designed, built, tested and evaluated as part of this development.

### 2.3.2B.1  Water/Steam Receivers

Receiver tests were conducted for both a recirculating boiler model and a once-through-to-superheat panel as part of the development program preceding construction of Solar One, the first central receiver plant built in the United States. The once-through-to-superheat design was selected for Solar One. The full-scale panel was tested at the CRTF (Coleman and Friefeld, 1980 and Wolf and Hernandez, 1981).

**Figure 2.8:** Stretched membrane heliostat prototype.

Water/steam receivers were tested as part of the Eurelios, CESA-1 and Solar One system experiments which will be described later in this section.

### 2.3.2B.2 Advanced Heat Transfer Fluid Receivers

The advantages of the advanced receiver coolants (liquid sodium or molten salt) over water/steam are:

- The receiver can be smaller, lighter in weight and potentially cheaper since the coolant is at a much lower pressure and the receiver tubes can accept a higher average heat flux.
- Power production can be decoupled from solar energy availability through storage of the heated fluid until needed for power generation. This allows greater flexibility for receiver operation through cloud transients. It also allows a power production schedule which matches the users' demands and/or generation at full rated output which gives maximum conversion efficiency.
- Thermal energy storage can be incorporated into the plant without reducing the temperature of the steam supplied to the turbine. This is not possible with superheated steam as the coolant leaving the receiver.

A molten salt-cooled receiver was built and tested at the CRFT during 1980-81 (Martin Marietta Denver Aerospace, 1981). This test program provided the basis for development of the molten salt system concept. The receiver was subsequently refurbished and used for the Molten Salt Electric Experiment (MSEE) which will be described later.

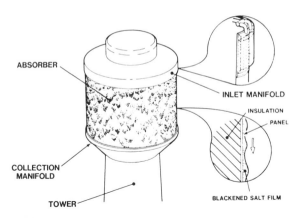

**Figure 2.9:** Direct absorption receiver concept.

A liquid sodium-cooled receiver panel was tested at the CRTF during 1981-82 (Rockwell International, 1983). This test program contributed to the subsequent design of a commercial-type receiver.

Other advanced receivers were tested as part of the THEMIS (molten salt) and the International Energy Agency (liquid sodium) central receiver system experiments.

### 2.3.2B.3 Direct Absorption Receiver

The direct absorption receiver (DAR) concept is a significant departure from tube-type receiver technology. It offers promise for performance improvements and cost reductions in future generations of receivers (Bohn et al., 1986, Anderson et al., 1986, and Tyner and Wu, 1988). The DAR concept involves the absorption of concentrated solar flux directly into a film of molten salt that is flowing over a nearly vertical plate. The concept is illustrated in Figure 2.9.

Because the solar flux is absorbed directly into the coolant, the flux limits associated with tubular receivers can be relaxed substantially. This higher flux density should result in better thermal performance; smaller, lighter and less complex receivers and lower capital costs.

The DAR concept is at a very early stage of development. Water flow tests have been conducted to investigate film characteristics, including waves, in long flowing films, and to develop effective manifolding. A 2 $MW_t$ solar research experiment employing molten salt is planned for future testing at the CRTF (Chavez et al., 1988).

### 2.3.2C Thermal Storage

Although thermal energy storage development is less critical than either heliostats or receivers, subsystem test programs were conducted in support of both the water/steam and molten salt systems.

A 4 $MWhr_t$ storage system employing dual liquid (oil) and solid (rock/sand) media was built and tested to qualify this concept for Solar One (Radosevich and Wyman, 1982). A thermocline was utilized to store both the hot and cold fluid in a single tank. Heating of the storage media was achieved by removing colder oil from the bottom of the tank, heating it with steam in a heat exchanger and returning the oil to the top of the tank. During heat extraction, the process was reversed. The tests were successful and this concept was employed in the Solar One plant.

Operation from storage with this type of system results in reduced power output and lower conversion efficiency because the pressure and temperature of the steam are much lower than that available from the receiver.

Higher efficiency operation from storage can be achieved if fluid at or near the receiver outlet temperature can be stored. This is a major motivation for using the "advanced" heat transport fluids.

A 7 MWhr$_t$ storage system employing molten nitrate salt was built and tested at the CRTF in 1982 (Martin Marietta Aerospace, 1982 and Kolb and Nikolai, 1988). It utilized 2 tanks to store the salt at 566 °C (1050 °F) and 288 °C (550 °F). A fossil-fired heater was employed to simulate a solar receiver, and an air cooler simulated a steam generator. The test program demonstrated that a molten salt thermal storage system could operate efficiently, reliably and safely in all modes expected for a solar power plant.

This storage system was subsequently utilized in the Molten Salt Electric Experiment (MSEE) at the CRTF.

The other full system experiments also employed various types of thermal energy stroage. They will be described later in this section.

### 2.3.2D  Other Subsystems

A molten salt-heated steam generator was designed and constructed at the CRTF (Babcock and Wilcox, 1984). It was incorporated into the Molten Salt Electric Experiment and was tested and evaluated as part of that program.

An on-going program is testing large-scale pumps and valves for use with molten salt (Bator, 1987). Substantial problems have been encountered in this program (Sandia National Laboratories, 1988). It is hoped that the continuing development and test program will resolve these problems and confirm the technical capability of transporting molten salt at the scale required in a commercial plant.

The balance of the central receiver system utilizes conventional technology and equipment, so little component development was conducted apart from the full system experiments.

Many of the system experiments, most notably Solar One, conducted significant development of solar plant controls as part of their total program. In fact, these developments contributed to the application of distributed digital controls to conventional power plants.

### 2.3.3  First Generation System Experiments

The major source of information to be used in assessing the status of central receiver technology will be the experimental results from the system experiment projects. Three water/steam system experiments were evaluated: Solar One, CESA-1 and EURELIOS.

### 2.3.3A  Solar One

Solar One was a 10 MW$_e$ central receiver full system experiment located at Daggett, California (McDonnell Douglas Astronautics Company, 1985, Sandia National Laboratories, 1985, Sandia National Laboratories, 1985a, Sandia National Laboratories, 1985b, McDonnell Douglas Astronautics Company, 1983, McDonnell

**Figure 2.10:** Solar One.

Douglas Astronautics Company, 1982, Burns and McDonnell Engineering Company, 1983, Sandia National Laboratories, 1988a, Sandia National Laboratories, 1988b, Solar One monthly operation and maintenance reports, and Chuck Lopez, 1987 and 1988). It was operated from April 1982 through September 1988. An aerial view of the plant is shown in Figure 2.10.

The principle programmatic objectives of the Solar One project were:

1. To establish the technical feasibility of solar thermal power plants of the central receiver type, including collection of data for retrofit applications of solar boilers to existing power plants fueled by oil or natural gas.
2. To obtain sufficient development, production and operating and maintenance data to identify potential economics of commercial solar plants of similar design, including retrofit applications on a comparable scale.
3. To determine the environmental impact of solar thermal central receiver plants.

Secondary project objectives were:

1. To gather operational data that can be analyzed to determine system operating and safety characteristics.
2. To develop both utility and commercial acceptance of solar thermal central receiver systems.
3. To stimulate industry to develop and manufacture solar energy systems.
4. To enhance public acceptance and familiarity with solar energy systems.

Sponsors of the project were:

- United States Department of Energy
- Southern California Edison Company
- Los Angeles Department of Water and Power
- California Energy Commission

The major participants and their roles during the design/construction phase were:

- McDonnell Douglas Astronautics Co.: Solar facilities design integrator. Also supplied the receiver, thermal storage and control subsystems.
- Martin Marietta Denver Aerospace: Heliostat fabrication and installation.
- Rocketdyne: Furnished receiver and thermal storage subsystems.
- Stearns-Catalytic: Furnished solar related plant support systems.
- Townsend and Bottum: Solar facilities construction manager.
- Southern California Edison: Designed and installed the electric power generation equipment and had responsibility for plant startup and subsequent operations and maintenance.

The major participants and their roles during the operation and evaluation phase were:

- Sandia National Laboratories, Livermore: Analyzed and evaluated data obtained from startup tests, the experimental test and evaluation phase and the power production phase.
- Southern California Edison: Operated and maintained plant and recorded data. Evaluated plant efficiency and accomplished reliability improvements. Provided a utility perspective to plant evaluation.
- Electric Power Research Institute (EPRI): Analyzed experience from design and construction, established a library of documentation and contracted other studies on parasitic power and the control system.
- McDonnell Douglas Astronautics Co.: Operations and performance evaluation.
- University of California, Los Angeles: Environmental assessments.

### 2.3.3A.1 Plant Description

Solar One converted solar energy into electricity using water/steam as the working fluid for both energy collection and power conversion. A schematic of the plant is shown in Figure 2.11; major subsystems are described on Table 2.3. A separate thermal storage system coupled to the water steam loop by heat exchangers provided the capability (although at significantly reduced efficiencies) to operate in various alternate operating modes. The thermal storage system employed a single storage tank which operated on the thermocline principal. The storage media were Caloria HT-43, a heat transfer oil, and a packed bed of rock and sand.

The plant was designed to operate in eight modes. The normal procedure for a mode 1 startup was to command all available east field heliostats and a lesser portion of the north and west field heliostats onto the receiver at sunrise. During warm-up, water circulated through the receiver to the receiver flash tank and back to the deaerator. As the warm-up progressed, boiling began and eventually superheated steam was produced. When all boiler panels reached a superheated steaming condition, the steam was diverted to the downcomer and warming of the downstream

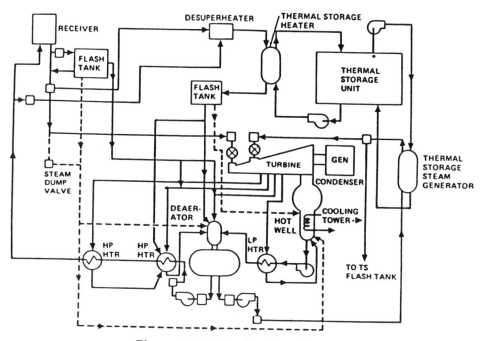

**Figure 2.11:** Solar One flow schematic.

components began. Once superheated steam conditions were established at the turbine main stop valve, the turbine steam chest was warmed followed by turbine roll and generator synchronization. The time required for receiver startup from sunrise until superheated steam flowed in the downcomer was reduced to about 1.2 hours during the course of the project.

For mode 2 operation, turbine direct and charging, a portion of the receiver-generated steam was diverted to charge thermal storage. The remainder of the system was operated as in mode 1. This mode was less desirable than mode 1 because both the turbine/generator and thermal storage charging loop were operated off-rated and at reduced efficiency.

Mode 3, storage-boosted operation, involved discharging the thermal storage system to supplement the steam generated by the receiver. This mode was intended to supplement the receiver-generated steam during periods of low energy collection; however, due to the inefficiencies associated with using stored energy for power generation, operation in this mode was avoided.

Mode 4, in-line flow, provided decoupling of the energy collection process from the power generation process by using all solar-generated steam to charge thermal storage while simultaneously generating steam from storage for plant operation. This mode allowed the system to be operated during partly cloudy conditions when continuous mode 1 operation was not possible. The system was seldom operated in this mode because of its operating complexity and performance penalties.

**Table 2.3:** Solar One Plant

### Collector Subsystem

| | |
|---|---|
| Type | Two-axis tracking heliostats |
| Number | 1818 |
| Total Reflective Area | 71,095 m$^2$ (764,868 ft$^2$) |
| Drives | Pedestal-mounted drive assembly with azimuth/elevation gear motors |
| Power Supply | Commercial grid with back-up feed |

### Receiver Subsystem

| | |
|---|---|
| Rating | 43.4 MW thermal |
| Configuration | Cylindrical, External, 24 flat panels |
| Height | 13.7 m (45 ft) |
| Diameter | 7 m (23 ft) |
| Heat Transport Fluid | Water/Steam |
| Flow | Once-through-to-superheat |
| Operating Temperatures | 204$^o$C (400$^o$F) Inlet, 516$^o$C (960$^o$F) Outlet |
| Operating Pressure | 100 bar (1450 psi) |
| Tower Height | 90.8 m (298 ft) |

### Thermal Storage Subsystem

| | |
|---|---|
| Type | Single tank thermocline |
| Rated Storage Capacity | 182 MWhr thermal |
| Storage Medium | Caloria HT-43 oil/rock/sand |
| Operating Temperatures | 302$^o$C (575$^o$F) upper, 218$^o$C (425$^o$F) lower |

### Steam Generator Subsystem
(Not used for turbine direct operation)

| | |
|---|---|
| Type | Caloria heated preheater, boiler and superheater (twin 50% capacity trains) |
| Rating | 33 MW$_t$ |
| Operating Temperatures | 302$^o$C (575$^o$F) In, 218$^o$C (425$^o$F) out (oil) |
| Feed Water Conditions | 121$^o$C (250$^o$F), 490 psi |
| Main Steam Conditions | 277$^o$C (530$^o$F), 400 psi |

**Table 2.3:** Solar One Plant (con't.)

Power Conversion Subsystem

| | |
|---|---|
| Prime Mover | Single flow, 4 point extraction, condensing steam turbine |
| Inlet Conditions | 510°C (950°F) |
| Outlet Conditions | 45°C (109°F), 6.4 cm (2.5 in) Hg |
| Generator Nameplate Rating | 12.5 MW$_e$ |
| Load | Southern California Edison Grid |

Controls

| | |
|---|---|
| Type | Distributed digital |
| Start-up & Shutdown Sequencing | Automatic or manual |
| Controllers | 5 Minicomputers<br>3 Distributed process controllers<br>21 Digital to analog multi-variable control units<br>6 programmable logic units |
| Data Acquisition | Dedicated computer & peripheral equipment |
| Console Displays | Nine color and 1 B&W CRT's with custom graphics |
| Safety Provisions | 5 of the programmable logic units shutdown the system if unsafe condition detected. Hardwire backup. |

In mode 5 operation, storage charging, all steam generated by the receiver was directed to the thermal storage charging heat exchangers. This was the preferred mode for charging storage.

Mode 6, storage discharging for power production, was demonstrated but was not used during the power production phase because of low efficiency.

Mode 7, dual flow, combined mode 1 and mode 4 operation. this mode also was demonstrated during the test and evaluation phase but was not used during the power production phase because of inefficiencies and operating complexity.

During mode 8, the system was shutdown and the thermal storage system was operated to meet the auxiliary steam needs of the plant.

The plant was operated routinely throughout most of the power production phase. The plant operated almost exclusively in mode 1 except for short periods of mode 5 operation on a regular basis. Mode 5 operation was only used to charge the thermal storage system for the generation of auxiliary steam; it was not used for power production.

### 2.3.3A.2 Performance

Starting August 1, 1984, plant operations emphasized power production. During the first three years of the power production phase, the plant generated 42,780 MWe – hr (gross) and delivered 27,670 MWe – hr to the Southern Cali-

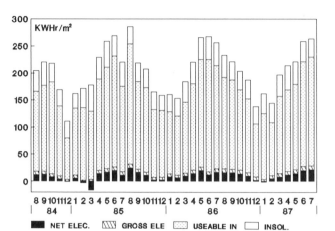

**Figure 2.12:** Solar One monthly power production.

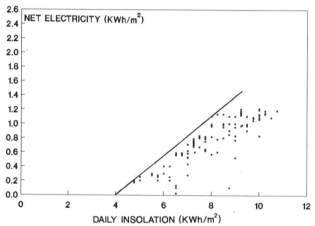

**Figure 2.13:** Solar One input-output (January-June, 1986).

fornia Edison grid based on 24 hr/day generation and consumption measurements. The monthly results normalized to reflector area shown in Figure 2.12.

System design requirements which were met or exceeded included the delivery of 10 MWe net from receiver steam, the delivery of 7 MWe net from thermal storage, and the delivery of 28 MWe – hr from thermal storage. Two design requirements which were not met are the delivery of 10 MWe net for 7.8 hours on the most favorable day of the year (summer solstice) and for 4 hours on the least favorable day of the year. The combined effects of low direct insolation, heliostat outage, mirror soiling, receiver efficiency and turbine efficiency are the major factors which prevented the plant from meeting these 2 design requirements.

Although not specified as a design requirement for the plant, the predictions of annual energy production made during the design phase were not met. Operating characteristics of the plant combined with plant equipment outages and weather-related outages contributed to significantly lower than expected power generation.

Daily output versus insolation input over a period of 6 months is shown in Figure 2.13. This illustrates a major operating characteristic of the plant; there is an insolation threshold (4 KWHr/m$^2$ − day) below which no net generation is possible.

### 2.3.3A.3 Significant Incidents

Significant incidents encountered during Solar One operation are summarized in Table 2.4.

### 2.3.3A.4 Lessons Learned

Major lessons that were learned from the Solar One project are summarized below:

- System design and operation are more complex than in an oil- or gas-fired plant. More instrumentation and controls are required. Greater design margins and tolerances should be considered to reduce this complexity.
- Cyclic operation must be considered in the plant design and equipment specifications. Flanged joints should be avoided.
- Distributed digital controls worked well.
  - Automation reduces operator workload.
  - Well accepted by utility operators.
  - Operators made significant improvements in display graphics.
  - Schematics preferred to tabular data.
  - Greater integration of subsystem controllers is desirable.
  - On-line instrumentation diagnostics should be considered.
  - Fewer brands of hardware should be used.
- Heliostat availability was good and maintenance requirements were low.
- Proper grounding and shielding of the heliostat control wiring against lightning is required.
- Heliostat washing is cost effective.
- An improved Beam Characterization System is desirable.
- The once-through receiver performed well. Control and flow stability were better than expected. With its relatively low thermal inertia, it is probably superior to a recirculating broiler-type of receiver.
- Alloy 800 withstood frequent, rapid thermal cycles and greater than design temperatures and stresses with unexpectedly good reliability.
- Thermal expansion capability of the receiver panels should be provided in all directions and to the temperature limits of the material rather than for the design temperature range.
- Membrane welded receiver tubes were not a good design choice.
- Conservative design margins should be provided for the receiver design temperatures and heat fluxes because of variations in heliostat beam quality and aim point.
- Repainting of receiver panels is required to maintain aborptivity.
- Receiver heat flux measurement capability is desired.
- Although the thermala storage subsystem performed adequately, its value for extended plant operation is questionable.
- Better access for maintenance, including instrumentation should be provided in the receiver.

**Table 2.4:** Significant Incidents - Solar One

| Incident | Cause | Solution/Repairs |
|---|---|---|
| More than 50 receiver panel tube leaks. | Combination of local stresses due to physical restraint of panel expansion & low cycle fatigue. | Leaks repaired by grinding & welding cracks. Installation of shields & modifications to flow rates for panel edge tubes to limit temperature gradients, along with lower temperature operation & modifications to expansion guides, have reduced frequency of failures. |
| Receiver panel expansion guide roller seizures (contributed to panel tube leaks). | Inadequate roller/ shaft clearance & binding between rollers & supporting members. | Roller design modifitions. |
| Out of service heliostats. | Frequently due to drive motor seal wear and malfunction of azimuth & elevation encoders. | Repaired failures. Long term solution requires use of components having higher reliability. |
| Heliostat mirror corrosion of approximately 0.10% of reflective area. | Water seeping into honeycomb mirror support. | Vents installed in mirror modules and modifications to stow position have reduced corrosion rate. Avoid air space in future. |
| Sulfuric acid spill (2500 gal.) from demineralizer acid storage tank. | Plastic pipe failure. | PVC pipe replaced with carbon steel. |
| Caloria oil spill (12,000 gal.) from thermal storage train. | Rupture disk failure at pressure below relief valve setting due to limited fatigue of disk. | Replace rupture disks annually and increase pressure margin between relief valve and rupture disk. |
| TSS flash tank rupture disk failure. | Rupture disk did not meet design specifications. | Installed disk which met design requirements. |
| Caloria oil leaks at heat exchangers and other component flanges. | Thermal cycling. | Flange bolt tightening and gasket material changes reduced frequency of leakages. Not recommended for cyclic operation. |
| Steam leaks at component flanges. | Thermal cycling. | Gaskets replaced and/or flange bolts tightened routinely. |

**Table 2.4:** Significant Incidents - Solar One (con't.)

| Incident | Cause | Solution/Repairs |
|---|---|---|
| Water/steam leaks in heat exchangers. | Thermal cycling caused weld cracks at tube to tube sheet welds. | Leaks seal welded. Double tube sheet not recommended for cyclic operation. |
| Thermal storage tank fire. | Overpressurization due to water in the tank flashing to steam resulted in tank rupture. | Thermal storage system has remained out of service since this August 1986 incident. Provide for water detection & removal. |
| Loss of plant power resulted in heliostats concentrated on uncooled receiver for 15 minutes on multiple occasions. | Substation switching error resulted in loss of transmission line. Other similar incidents. | Only minor damage. Control system interlock logic modified. Backup power grid connection should operate automatically. |
| Receiver panel overheating & subsequent tube yielding. | Leaking receiver drain valve resulted in reduced flow through receiver panel. | Valves repaired and continued operation with deformed panel. |
| Turbine gland seal exhauster pump failure. | Solids in service water plated out on pump rotor causing it to seize. | Pump repaired and changed to condensate water supply in place of service water. |
| Burned off collector field power cables at power center transformers. | Improper installation during original construction. Inadequate quality control. | Undamaged cables and connectors removed, cleaned inspected and reassembled. Damaged components replaced. |
| Heliostat controller failures. | Lightning strikes. | Control system wire shields grounded. Defective power supply capacitors replaced. |
| Loss of receiver control system operating data and unit trips. | Power grid voltage excursion. | All control systems should be connected to an uninterruptable power supply. |
| Heliostat reflectivity degradation. | Long-term accumulation. | Chemical wash. |
| Receiver absorptivity degradation. | Long-term, high temperature operation & cycling. | Repainted receiver. Should be about every 2 years. |
| Many failures causing forced outages. | Inadequate quality control during construction. | Provide adequate quality control. |

- More hand-operated isolation valves should be provided for equipment maintenance.
- Operational outages can reduce plant output up to 50 percent from the nominal prediction. A very detailed operations analysis, including panel warmup, startup and the response to insolation transients, is required to predict the annual output accurately.
- Thermal inertia, allowable equipment warmup rates and the impact of these factors on plant operation and performance should be considered during plant design.
- The low efficiency of east field heliostats in the early morning limits plant startup. This factor should be considered in the layout of the heliostat field.
- The electric boiler was unreliable and was a large parsitic load.
- Parasitic losses were less than predicted.
- All solar energy is not usable for electric power production. Factors to be considered include:
  - Plant startup
  - Wind
  - Passing clouds

### 2.3.3A.5  Summary

Solar One was an effective project. The long-term operation (more than six years) and the size of the plant allowed a thorough assessment of the technology. The results from Solar One would be highly applicable to a commercial plant employing the same design concept. However, a more advanced design employing molten salt is now preferred for a utility-scale plant. Still, the wealth of experience and lessons learned from Solar One are useful as a guide to plant designers. The 10 MWe size of the plant allows a meaningful interpretation of performance data without being masked by the losses and parasitics as in other, smaller experiments. The five years of routine operation by Southern California Edison provide a sound basis for selecting an operations and maintenance approach and predicting many of the costs. A summary of major accomplishments of the Solar One project follows:

- The project met all of the original objectives.
- The technical feasibility of a water/steam central receiver system was demonstrated.
- Extensive operation and performance data were collected over nearly six years of operation.
- Areas requiring additional development were identified.
- Demonstrated that flow stability can be achieved in a once-through water/steam receiver.
- Verified that the system can be effectively operated by utility industry operators.
- Utility operation and maintenance personnel demonstrated aggressive and innovative improvements which maximized net power production and system availability.
- Established a heliostat chemical wash program to restore heliostat reflectivity.
- Repainted receiver to restore design absorptivity.
- The plant was operated with lower than anticipated costs for maintenance of the solar subsystems.

- The plant provided an excellent method for the transfer of advanced digital controls technology to the utility industry.
- Greater than 99% heliostat availability was achieved over extended periods of time.
- Confirmed that ecological consequences of constructing, testing and operating the plant were not significant.
- The plant had high international visibility and was heavily covered in magazines, television anf film.

### 2.3.3B CESA-1

CESA-1 was a 1.2 $MW_e$ central receiver full system experiment located at Tabernas (near Almeria), Spain (Torralbo et al., 1983, Avellaner et al., 1985, Sanchez, 1985, Balanza and Roman, 1985, Martinez, 1985, Sanchez, 1986, Ministerio de Industrio y Energia, 1983, and Avellaner, Sanchez, and Ortiz, 1985). Plant startup began in January 1983. Turbine roll and connection to the grid were achieved in October 1983, and the plant was operated through December 1984. An aerial view of the plant is shown in Figure 2.14.

The overall program objective was to investigate the economic viability of large solar power systems. Specific objectives of the CESA-1 project were:

- To study and demonstrate the feasibility of this type of plant.
- To develop the specific technology.
- To acquire the necessary experience to develop, construct and operate commercial plants that could help to cover the national energy demand.
- To encourage the use of solar energy for electric power generation.

The project was sponsored by the Spanish Ministry of Industry and Energy. The major participants and their roles were:

- Centro de Estudios de lar Energia (CEE): Project management (until 1983).
- Instituto de Energias Renovables (IER): Project management (after 1983).
- INITEC: System Engineering, Detailed System Design.
- SENER: Design and Construction of 1/2 of heliostats (except mirror modules) and controls.
- C.A.S.A.: Design and Construction of 1/2 of heliostats and all mirror modules.
- Technicas Reunidas, S.A.: Receiver design.
- Babcock and Wilcox, Espana: Receiver construction & auxiliaries.
- Siemens: Turbo alternator and instrumentation.
- Sevillana de Electricidad S.A.: Operation after June, 1983.

### 2.3.3B.1 Plant Description

CESA-1 used water/steam as the receiver coolant and as the working fluid in the steam Rankine power conversion cycle. Molten salt was used for thermal storage. A schematic of the plant is shown in Figure 2.15; the major subsystems are described on Table 2.5.

The plant was designed to operate in any of the following modes:

- Mode I: Direct operation. The steam flow produced in the receiver was expanded in the turbine.
- Mode II: Charging storage. The steam generated in the receiver was used to charge thermal storage.

**Figure 2.14:** CESA-1.

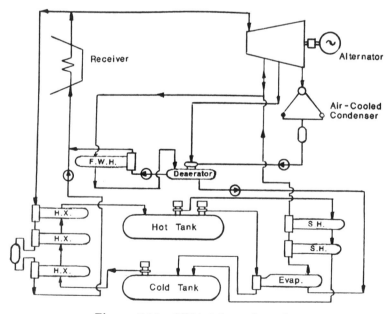

**Figure 2.15:** CESA-1 flow schematic.

**Table 2.5:** CESA-1 Plant

Collector Subsystem

| | |
|---|---|
| Type | Two-axis tracking heliostats |
| Number | 300 |
| Total Reflective Area | 11,880 m$^2$ |
| Mirror Module | 3 mm silvered float glass sand-wich with foam core & steel back sheet |

Receiver Subsystem

| | |
|---|---|
| Rating | 7.7 MW$_t$ |
| Configuration | Cavity with 3.4 m x 3.4 m (11.2 ft x 11.2 ft) Aperture |
| Heat Transport Fluid | Water/Steam |
| Flow | Forced reciculation boiler, steam drum, separate superheater |
| Operating Conditions | Inlet: 185$^o$C(365$^o$F); 132 bar (1914 psi) Outlet: 525$^o$C(977$^o$F); 108 bar (1566 psi) |
| Control | Steam drum level, 3 attemperators |
| Tower Height | 80 m (262 ft); Receiver at 60 m (197 ft) level |

Thermal Storage Subsystem

| | |
|---|---|
| Type | Two tank |
| Rated Storage Capacity | 16 MWHr$_t$ |
| Storage Medium | Hitec salt |
| Operating Temperatures | 220$^o$C (428$^o$F); 340$^o$C (644$^o$F) |

**Table 2.5:**  CESA-1 Plant (con't.)

### Steam Generator Subsystem
(Used only from storage)

| | |
|---|---|
| Type | Shell and tube heat exchangers (charging and discharging) |
| Rating | 4.2 $MW_t$ |
| Operating Temperatures (Salt from storage) | 340°C (644°F) in, 220°C (428°F) out |
| Feed Water Conditions | 185°C (365°F) |
| Main Steam Conditions | 330°C (626°F); 14.7 bar (213 psi) |

### Power Conversion Subsystem

| | |
|---|---|
| Prime Mover | Dual admission, multi-stage, condensing turbine |
| Inlet Conditions | From Receiver: 520°C (968°F); 98 bar (1421 psi) From Storage: 330°C (626°F); 14.7 bar (213 psi) |
| Outlet Conditions | 55°C (131°F); 0.15 bar (2.2 psia) |
| Generator Nameplate Rating | 1500 KVA, P.F.=0.8, 400V, 50Hz |
| Load | Grid |

### Controls

| | |
|---|---|
| Type    Heliostat Field: Plant: | Computer controlled Conventional analog controls |
| Start-up & Shutdown Sequencing | Manual |
| Data Acquisition | Computer-controlled, multiplexer, CRT's, tape & disk storage, printer |
| Safety Provisions | Programmable interlocks, alarms |

- Mode III: Discharging storage. Secondary steam produced by thermal storage discharge was sent to the turbine.
- Mode IV: Direct operation and charging storage. Generated steam was distributed between turbine operation and thermal storage charging.
- Mode V: Direct operation and discharging storage. Primary and secondary steam were both sent to the turbine.
- Mode VI: Buffered operation. Steam generated in the receiver was used to charge storage, and storage discharging was utilized to produce secondary steam for turbine operation.

### 2.3.3B.2 Project Results

The short duration of operation of CESA-1 provided only limited, generally qualitative, information useful for central receiver technology evaluation. Most results are only applicable to the specific plant design. In January 1985, operation of the CESA-1 system experiment was terminated and the facility was utilized for component testing in support of the Spanish/German joint research project on a Brayton cycle solar power plant.

The total number of hours of operation were: receiver, 2120 hr; turbine, 434 hr, and the alternator, 324 hr. The total gross electricity produced by the plant was about 130 MWHr. Data are insufficient to evaluate plant annual efficiencies.

Project Results which are applicable for this technology evaluation are:

- The time required for receiver startup was very lengthy. Much attention was given to this problem and the startup time was reduced to three hours. This result suggests that the "thermal inertia" of a recirculating, steam drum receiver would be higher than for the once-through receiver used for Solar One.
- Recovery from insolation loss due to cloud passage required 30 minutes to one hour. This is another result of the high thermal inertia.
- 25% of the mirror modules showed significant corrosion after 1 1/2 years in the field.
- Heliostat availability was only 90%. The major cause was failure in the power control circuit.
- Many heat trace failures occurred and caused salt to freeze and cause pipe blockage.
- Salt leakage occurred in heat exchanger headers and pipes due to cracks caused by defective materials.
- Two rupture disk failures occurred in heat exchangers.
- The Turbine encountered thrust bearing failure and casing distortion.

The CESA-1 project demonstrated the technical feasibility of the system and determined its serious limitations. Design and development of heliostats was accomplished. The facility is available for future testing.

### 2.3.3C EURELIOS

EURELIOS was a 1 MW$_e$ central receiver full system experiment located near Adrano in Sicily (Gretz, 1981, Borgese et al., 1983, Corvi and Dinelli, 1985, Gretz et al., 1984, and Gretz, 1986). It was the first central receiver plant to be operated. It was completed in December 1980, and was connected to the grid in April 1981. Operation continued through 1984. A view of the plant is shown in Figure 2.16.

**Figure 2.16:** EURELIOS.

The objectives of the EURELIOS project were to:

1. Demonstrate the operation of a solar power plant directly linked to the grid.
2. Evaluate plant performance.
3. Gain experience in operation and improve operating procedures.
4. Obtain data for technical and economic evaluation.

Sponsors of the project were:

- Commission of the European Communities (CEE).
- The Governments of France, Italy and Germany.
- Ente Nazionale per l'Energia Elettrica (ENEL).

The major participants and their roles were:

- Ente Nazionale per l'Energia Elettrica (ENEL): host utility, operator of plant, control system design, civil works and operating procedures.
- Cethel: 1/2 of the Heliostat Field and associated monitoring, controls, the molten salt thermal storage subsystem, electrical system (part), safety control, installation and commissioning.
- Ansaldo ($S_pA$): Receiver, tower, steam cycle, hot water storage, electrical system (part) and installation work.
- Messerschmitt-Boelkow-Blohm GmbH (MBB): 1/2 of the heliostats and associated monitoring & controls and back-up equipment.
- Helioelectric plant consultative Committee (HPCC): Consulting services.
- General Technology Systems Ltd. (GTS SA): Assisted project leader in monitoring the work and acted as commissioning agent during acceptance testing.

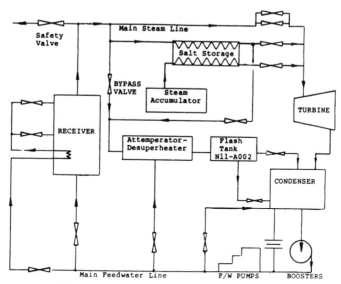

**Figure 2.17:** EURELIOS flow schematic.

## 2.3.3C.1 Plant Description

EURELIOS used water/steam as the receiver coolant and as the working fluid in the steam Rankine power conversion cycle. Only short duration, buffer thermal storage was provided with a water/steam accumulator and molten salt for superheat. Two types of heliostats from two suppliers were employed. A schematic of the plant is shown in Figure 2.17; major subsystems are described on Table 2.6.

Normal startup was accomplished by pressurizing the receiver and allowing water to flow through the receiver and a turbine bypass. After flow was established, the heliostats were moved onto the receiver (group by group at one minute intervals). After about a half-hour of illumination, steam was produced and sent to a flash tank. When the temperature at the receiver outlet reached 340 °C, the turbine main header inlet was preheated. Then the turbine was started and its speed increased under operator control at a standard rate of 200 rpm per minute until synchronous speed was reached. At this time the alternator was coupled to the grid. The total startup time was typically about two hours.

Heliostats were moved to standby if the solar radiation dropped below about 20% of the nominal value for more than three minutes; if charged, the salt storage system was put into operation. Steam flow from the storage system began as soon as the pressure at the turbine dropped below the set point of the regulator. Power production was maintained as long as sufficient steam was available from either the receiver or storage.

## 2.3.3C.2 Project Results

This plant was the first of the central receiver system experiments to go on-line. It pioneered much new ground. Power production was very disappointing. The total gross electricity produced by the plant was about 130 MWHr; the net output was very negative. This has been largely attributed to the conditions required for plant

**Table 2.6:** EURELIOS

Collector Subsystem

| | |
|---|---|
| Type | Two types of 2-axis tracking heliostats |
| Number | 70 Cethel ($52m^2$); 112 MBB ($23m^2$) |
| Total Reflective Area | 6216 $m^2$ |
| Mirror Module | Cethel: 6 mm back silvered float glass |
| | MBB: 3 mm silvered float glass, foam core, galvanized back sheet |

Receiver Subsystem

| | |
|---|---|
| Rating | 4800 $KW_t$ |
| Configuration | Conical cavity, tilted down $20^o$ |
| Heat Transport Fluid | Water/Steam |
| Flow | Once-through-to-superheat |
| Water Inlet Temperature | $36^oC$ ($97^oF$) |
| Outlet Steam Conditions | $512^oC$ ($954^oF$); 65 bar (942 psi) |
| Tower Height | 55 m (180 ft) |
| Control | Outlet steam temperature by control of feedwater flow rate and attemperator |

Thermal Storage Subsystem

| | |
|---|---|
| Type | Water/steam accumulator; 2-tank |
| Storage Media | Water/steam; Hitec |
| Rated Storage Capacity | Press. water: 300 $KWHr_t$; Hitec: 60 $KWHr_t$ |
| Operating Temperatures | Press. water: 19 bar to 7 bar |
| | Hitec: $430^oC$ ($806^oF$) hot, $275^oC$ ($527^oF$) cold |

Steam Generator Subsystem<br>(Not used for turbine-direct operation)

| | |
|---|---|
| Type | Accumulator; concentric pipe, salt-heated superheater |
| Feed Water Temperature | $36^oC$ |
| Main Steam Conditions | $410^oC$ ($770^oF$); 19 to 7 bar (276 to 102 psi) |

**Table 2.6:** EURELIOS (con't.)

Power Conversion Subsystem

| | |
|---|---|
| Prime Mover | Multi-stage, condensing turbine, no extraction |
| Inlet Conditions | 510°C (950°F); 61 bar (885 psi) |
| Outlet Conditions | 36°C (97°F); 0.06 bar (0.9 psi) |
| Heat Rejection | Wet cooling tower |
| Generator Nameplate Rating | 1500 KVA |
| Load | Grid |

Controls

| | |
|---|---|
| Heliostats | Computer controlled |
| Balance of Plant | Conventional panels and boards |

operation; these were: (1) a minimum insolation level of 450 W/m$^2$, (2) cloud cover less than 25%, (3) no haze and (4) greater than 75% of the heliostats available. A solar to net electric conversion efficiency of 16% had been predicted.

The disappointing performance results are more due to the specific experiment design than to inherent limitation of the technology.

### 2.3.3D Summary of First Generation System Experiments

The major system characteristics of the first generation system experiments are summarized in Table 2.7.

It can be seen in the table that these three experiments covered a significant variation of specific design features.

One of the greatest differences was that the 10 MW$_e$ plant rating of Solar One was about 10 times the rating of the other two plants; this made the Solar One experiment approximately 1/10th the scale of a commercial plant while the other two experiments were 1/100th scale. This difference in the scale of the plants may be a major reason for a large difference in the value of the experimental results. The data from Solar One would be applicable to the design of a commercial plant employing the same system configuration. The data available from the other two experiments are much less useful.

Heliostat field configurations included both a surround field with an external receiver and a north field with a cavity receiver. Heliostats were first generation designs from four different vendors. All performed adequately or better for the system experiments; this was one very favorable result of these tests.

Steam conditions from the different receivers were comparable. The once-through-to-superheat flow configuration was used for two of the system experiments.

**Table 2.7:** First Generation Water/Steam System Experiments

| | Solar One | CESA-1 | Eurelios |
|---|---|---|---|
| PLANT RATING, NET | 10 MW$_e$ | 1.2 MW$_e$ | 1 MW$_e$ |
| HELIOSTATS | | | |
| Number | 1818 | 300 | 70/112 |
| Size (m$^2$) | 39.1 | 39.6 | 52/23 |
| Total Area (m$^2$) | 71,084 | 11,880 | 6216 |
| Field Configuration | Surround | North | North |
| RECEIVER | | | |
| Configuration | External | Cavity | Cavity |
| Tower Height, m(ft) | 55(180) | 60(197) | 77(253) |
| Coolant | Water/Steam | Water/Steam | Water/Steam |
| Type | Once-through | Recir. boiler | Once-through |
| Outlet Temp. $^o$C ($^o$F) | 516(961) | 525(977) | 512(954) |
| Outlet Press (bars) | 105 | 108 | 62 |
| THERMAL STORAGE | | | |
| Type | Single Tank Thermocline | 2 Tanks | 3 Tanks |
| Media | Oil/Rocks/ Sand | Hitec Salt | Hot Water Hitec Salt |
| Rating (MWHr$_t$) | 135 | 18 | 0.36 |
| POWER CONVERSION | | | |
| Type | Turbine | Turbine | Turbine |
| Working Fluid | Steam | Steam | Steam |
| Inlet ($^o$C, bar) From Receiver | 510, 100 | 520, 98 | 510, 62 |
| From Storage | 274, 28 | 330, 15 | 410, 19 to 7 |
| Heat Rejection | Wet Cooling | Dry Cooling | Wet Cooling |

This flow configuration performed well and is probably preferred over a recirculating boiler with a separate superheater because of lower thermal inertia.

Three different types of thermal storage system concepts were tested. The substantial reduction in steam conditions required for operation from storage is apparent in the table. It was not established that any of these thermal storage concepts would be cost effective in a commercial plant.

The significant incidents and the lessons learned from the Solar One project provide an excellent data base. Their review will allow the plant designer to avoid many of the problems encountered in these first generation experiments.

### 2.3.4 Advanced System Experiments

Four system experiments employing advanced heat transfer fluids were evaluated: the Molten Salt Electric Experiment (MSEE), the Molten Salt Subsystems/Component Test Experiment (MSS/CTE), THEMIS, and the International Energy Agency-Central Receiver System (IEA-CRS).

### 2.3.4A The Molten Salt Electric Experiment

The Molten Salt Electric Experiment (MSEE) was a 750 $KW_e$ full system experiment conducted at the Central Receiver Test Facility (CRTF) at Albuquerque, New Mexico (Martin Marietta Aerospace, 1985, Babcock and Wilcox, 1986, Holl et al., 1989, and Delameter and Bergen, 1986). The system was operated from May 1984 through July 1985. The CRTF tower and heliostat field were employed for the system tests. Major elements of the MSEE are illustrated in Figure 2.18.

The objectives of the MSEE project were to:

1. Verify that the system works.
2. Obtain performance data for each subsystem and for total system interaction sufficient to verify the design and identify uncertainties.
3. Determine the operating range, flexibility and limitations of the system as installed.
4. Document performance results and evaluations that may be used for scale-up.
5. Identify areas that need additional development.
6. Provide training and hands-on operating experience for utility and industry personnel.

The project was a joint cooperative program between the United States Department of Energy, the Electric Power Research Institute and a consortium of utilities and industry with solar technology experience and interest. The major project participants and their roles were:

- Sandia National Laboratory, Livermore: Project management.
- Sandia National Laboratory, Albuquerque: Test facility (CRTF) management, MSEE construction, test and support personnel.
- Martin Marietta Denver Aerospace: System design, integration and start-up.
- Black and Veatch: Subsystem interface piping, water/steam equipment refurbishment, civil and structural activities.
- Babcock and Wilcox: Design, fabrication, installation and test support of the steam generation subsystem.

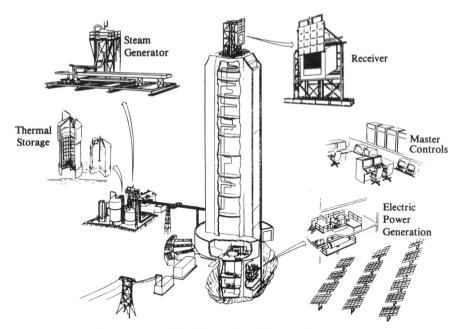

**Figure 2.18:** The Molten Salt Electric Experiment.

- Olin Chemical Group: Salt chemical analysis and materials test loop design, fabrication & maintenance.
- McDonnell Douglas Astronautics Company: Testing, operator training and integrated system operation.
- Public Service Co. of New Mexico: Turbine-generator installation and operation support.
- Arizona Public Service: Engineering and operation services.
- Pacific Gas & Electric: Engineering and operation services.
- Southern California Edison: Engineering and operation services.
- Bechtel Power Corporation: Engineering and operation services.
- Stearns-Catalytic: Engineering and operation services.

### 2.3.4A.1 Plant Description

The MSEE employed molten nitrate salt as both the receiver coolant and the thermal storage medium. The steam Rankine cycle was used for power conversion. A system flow schematic is shown in Figure 2.19; major subsystems are described on Table 2.8.

The salt equipment and piping needed to be preconditioned to about 230 °C prior to startup to prevent salt freezeup. The receiver panel was preconditioned by sunlight from selected heliostats soon after sunrise. Salt was then circulated from the cold salt storage tank through the receiver and back to the cold tank. The balance of the heliostat field was then brought onto the receiver. When the salt temperature reached 400 °C, flow was diverted into the hot storage tank and thermal storage

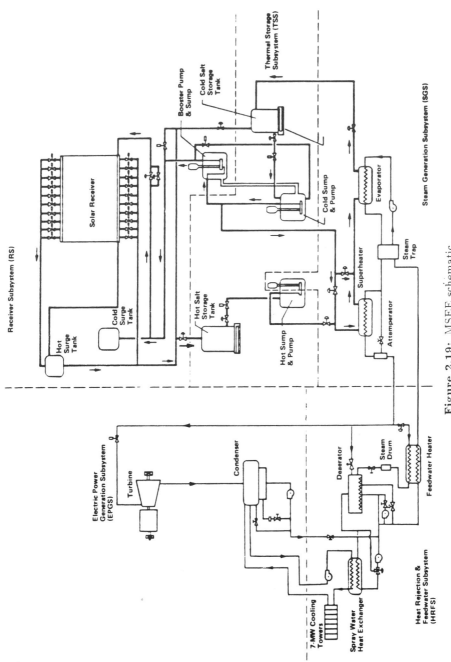

Figure 2.19: MSEE schematic.

**Table 2.8:** MSEE Plant

### Collector Subsystem

| | |
|---|---|
| Type | Two-axis tracking heliostats |
| Number | 211 |
| Total Reflective Area | 7854 $m^2$ (84,500 $ft^2$) |
| Drives | Base-mounted azimuth, yoke-mounted elevation |
| Power Supply | Primary: Diesel generator<br>Back-up: Commercial power |

### Receiver Subsystem

| | |
|---|---|
| Rating | 5 MW thermal |
| Configuration | Flat panel within cavity enclosure |
| Heat Transport Fluid | Molten nitrate salt (60% $NaNO_3$, 40% $KNO_3$) |
| Operating Temperatures | 299$^o$C (570$^o$F) Inlet,<br>566$^o$C (1050$^o$F) Outlet |
| Tower Height | 61 m (200 ft) |
| Control | Distributed digital |

### Thermal Storage Subsystem

| | |
|---|---|
| Type | 2 tank |
| Rated Storage Capacity | 6.5 MWht |
| Storage Medium | Molten nitrate salt |
| Operating Temperatures | 299$^o$C (570$^o$F); 566$^o$C (1050$^o$F) |

### Steam Generator Subsystem

| | |
|---|---|
| Type | Salt-heated evaporator and superheater with steam drum separator and forced recirculation. |
| Rating | 3.13 MWt |
| Operating Temperatures (salt) | 566$^o$C (1050$^o$F) in,<br>299$^o$C (570$^o$F) out |
| Feed Water Conditions | 288$^o$C (550$^o$F), 83 bar (1200 psi) |
| Main Steam Conditions | 510$^o$C (950$^o$F), 76 bar (1100 psi) |

**Table 2.8:** (con't.) MSEE Plant

Fossil Fuel Subsystem

| | |
|---|---|
| Function | Heats salt for operational flexibility; not required for normal operation operation. |
| Fuel | propane |
| Rating | 2.93 MWt |
| Operating Temperatures | 299°C (570°F) inlet, 566°C (1050°F) outlet |

Power Conversion Subsystem

| | |
|---|---|
| Prime Mover | 7 stage, condensing steam turbine (no extraction) |
| Inlet Conditions | 504°C (940°F), 73 bar (1065 psi) |
| Outlet Conditions | 56°C (133°F), 12.5 cm (5 in) Hg |
| Generator Rating | 750 KW, 450 V, 3 phase, 60Hz |
| Load | Local Grid |

Controls

| | |
|---|---|
| Type | Distributed digital process control. |
| Start-up & Shutdown Sequencing | Semi-Automatic |
| Data Acquisition | Separate HP-1000 system |
| Console Displays | Standard formats; custom graphics |
| Safety Provisions | Independent, hard-wired relays |

charging started. To generate electricity, hot salt was pumped through a steam generator to produce steam to drive a conventional turbine generator. The salt was then returned to the cold tank. An alternative of charging storage using the propane heater was provided. Salt was drained into the storage tanks for overnight shutdown, but all salt lines were maintained warm by electric trace heating. The evaporator recirculation pump was operated and the steam generator was maintained warm by using an electric heater.

### 2.3.4A.2 Performance

A 28-day power production test run was conducted to determine the maximum amount of energy that could be generated by this plant. Startup operations were begun before sunrise and shutdown completed after sunset in order to maximize

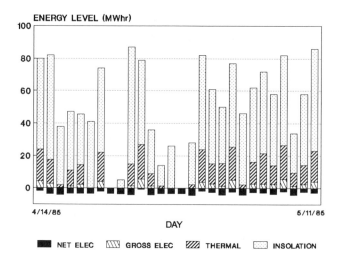

output. The system was refurbished from December 1984 until early April 1985 in order to maximize equipment availability during this test. Although the turbine failed prior to the start of the test run, the test was carried out with the electric power output calculated from the steam flow from the steam generator and the power conversion subsystem efficiency measured earlier in the program. The test run was conducted from mid-April to mid-May 1985. The daily performance during this test is shown in Figure 2.20. It can be seen that the daily net output was never positive.

A composite efficiency chart for the full 28-day test run is shown in Figure 2.21. System availability (0.946) was excellent due to extensive maintenance conducted before this run. Nevertheless, the other operational losses reduced the cumulative efficiency to 0.617 before the subsystem efficiencies were applied. The major contributors were scheduled and unscheduled delays during startup and shutdown, the inability to follow insolation transients and operation at low power levels. Electric parasitic losses were very high and largely accounted for the inability to produce a net positive output. Some of these losses are generic to this system type and some are due to employing non-optimized equipment. All are exaggerated because of the very small scale of the experiment.

Input-output charts for the thermal energy collected and the gross electricity (the net was always negative) for the 28-day test are shown in Figure 2.22. The points represent each of the test days; the line represents the envelope to the system's maximum performance. A threshold for the daily insolation level can be noted on this figure.

### 2.3.4A.3  Significant Incidents

Significant incidents encountered during operation of the MSEE are summarized in Table 2.9.

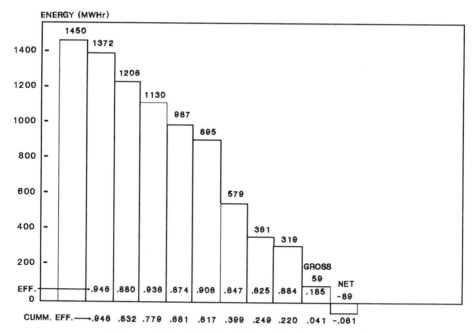

**Figure 2.21:** MSEE efficiency (28 day test run).


## 2.3.4A.4  Lessons Learned

Lessons learned from the MSEE project are summarized below:

- The high melting point of the nitrate salt, 221 °C–238 °C (430 °F–460 °F), created significant problems in maintaining the elevated temperature required for startup.
- The plant design is complex with resulting impact on cost, availability and maintenance requirements. The design should be simplified where possible.
- Parasitic power is extremely high and must be considered as a major factor in plant design.
- Electric trace heating reliability is inadequate for this application, even with multiple redundancy.  Failure rates were excessive, and the cost to repair, involving removal of insulation and lagging from equipment and pipes, is very high. A much more reliable trace heating system, or an alternative approach to maintaining the system warm overnight, must be developed and qualified.
- If reliable electric trace heating elements are developed and used for thermal conditioning, the system must be carefully designed. Since the heaters operate on an on-off control cycle using thermocouples, a good match between heating and cooling rates within a single control loop must be maintained in order to avoid heater shutdown by high temperature at one point while another area is below freezing.  This requires the use of multiple control loops, sufficient thermocouples and a detailed thermal analysis. Also, care must be exercised to avoid any variation in heat loss rates over time due to localized deterioration of insulation, wind effects or other causes.

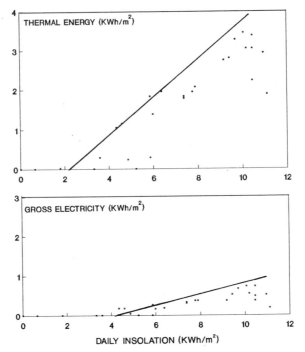

**Figure 2.22:** MSEE input-output (28 day test run).

- Conformal, blanket-type insulation should be used for complex equipment shapes and bends. Calcium silicate crumbled in many locations due to thermal cycling; it also allowed thermal convection to occur within the cutout spaces provided to fit around equipment.
- Daily startup and shutdown caused many problems with commercial water/steam components. Equipment must be designed and qualified for this cyclic operation.
- The electric startup heater for the steam generator was unreliable and a high parasitic load.
- 24 hour recirculation of water in the steam generator was required because the recirculation pump was incapable of daily startup. This caused many problems in maintaining the water level needed in the drum for startup. This mode is undesirable and is a serious parsitic loss. Any alternative would require an adequate heat source to condition the water loop for startup.
- Adequate access should be provided for equipment maintenance.
- Salt pumps in series should not be used.
- Due to salt corrosion, material selection for salt piping and components is limited. Quality control is required to ensure that proper materials are used. This will increase costs over normal commercial practice.
- Distributed digital control is very effective, and a high degree of automated operation can be achieved.

**Table 2.9:** Significant Incidents - MSEE

| INCIDENT | CAUSE | SOLUTION/REPAIRS |
|---|---|---|
| More than 25 failures of electrical trace heaters. | Usually the cold junction overheated. | Repaired/replaced. Did not achieve long term reliability. |
| Calcium silicate insulation crumbled & settled within aluminum lagging. | Thermal Cycling | Reinsulated with blanket material and conformal lagging over complex shapes & bends. |
| Receiver valves & headers cold with trace heat on. | Wind produced variable heat losses. | Added active trace heat control circuits. Added wind shielding. |
| Receiver hot surge tank overflowed. | Surge flow during receiver drain. | Modified procedures. |
| Froze salt in receiver panel. | Filled receiver while below salt freezing temperature. | Modified procedures. Thawed panel with Solar. |
| Receiver boost pump bearing failed and bent shaft. | Wearout due to frequent freezeup of salt in shaft packing. | Replaced. Added protection against salt entering packing. |
| Salt valve bellows failed. | Incorrect material used by manufacturer. | Replaced. Maintain quality control. |
| Steam generator startup electric heater failed 3 times. | Water loss uncovered heating element. | Replaced and relocated protection thermocouple. |
| Steam generator recirculation pump burned out. | Insufficient cooling during startup. | Replaced & operated 24 hours per day. |
| Burst disks on salt side of steam generator burst. | Steam Generator not designed with sufficient margin to accommodate burst disk tolerances over the pump discharge pressure. | Replaced. Added interlocks. |
| Feed water pump failed. | Insufficient cooling. | Rebuilt. Modified loop and procedures. |
| Turbine failed. | Oil pump failure. | Completed test program without turbine. |

### 2.3.4A.5 Summary

The Molten Salt Electric Experiment contributed substantially to our knowledge and understanding of the potential and problems of a molten salt, central receiver solar power plant. An overall technical assessment follows:

1. the system has been demonstrated to be technically feasible and could be operated effectively.
2. The thermal storage configuration effectively decouples solar energy collection from power production. This allows the collection function to follow solar availability and power production to follow user demand.
3. Molten nitrate salt is an effective, low-cost heat transfer fluid for energy collection and storage.
4. Distributed digital controls were very successful. Many operating sequences were automated. Further automation is possible which will minimize operator requirements for future plants.
5. Thermal and hydraulic performance can be predicted with good accuracy.
6. Reasonably rapid startup can be accomplished without excessive loss of collectible solar energy.
7. The high melting point of the salt creates substantial problems in maintaining all receiver loop piping and equipment at the temperature required for startup. The major problems are associated with:
   • Trace heating
   • Insulation
   • Wind protection
   • Extensive instrumentation for temperature monitoring
   • High parasitic power
   • General system complexity

   Improved solutions to these problems are required.
8. Net positive power production has not been demonstrated. Major losses have been identified.
9. Equipment reliability under cyclic operation was poor. This must be considered for future development.
10. A large-scale experiment will be required to confirm technical performance for a commercial plant.

### 2.3.4B Molten Salt Subsystem/Component Test Experiment

The Molten Salt Subsystem/Component Test Experiment (MSS/CTE) had three major parts: (1) a model receiver test, (2) a valve seal bench test and (3) a pump and valve test in molten salt test loops (Sandia National Laboratories, 1987 and Smith and Chavez, 1988). All tests were conducted at the CRTF at Albuquerque, New Mexico. The model receiver test, which will be summarized here, utilized much of the MSEE equipment which was described above. Receiver testing was conducted during 1987. A picture of the MSS/CTE receiver located on top of the tower is shown in Figure 2.23.

The objectives of the receiver test program were to:

1. Demonstrate the features of a commercial type, scale-model receiver.

**Figure 2.23:** The MSS/CTE receiver.

2. Define the operating range, flexibility, and limitations of current molten salt receiver technology.

3. Provide confidence that a scaled-up commercial receiver will meet performance requirements.

The project was cost-shared between the U.S. Department of Energy and a consortium of utilities and industry. The major project participants and their roles were:

- Sandia National Laboratories: Project management; test facility (CRTF); construction; operation; maintenance; and support personnel.
- Babcock & Wilcox Co.: project prime contractor. Responsible for system design, integration, and start-up. Receiver design & fabrication.
- Black and Veatch: Safety analysis and consulting services.
- Foster-Wheeler: Design and fabrication of portions of the receiver.
- McDonnell Douglas Astronautics Company: Receiver requirements analysis, receiver control system design, test plan, testing and test data evaluation.
- Arizona Public Service: Utility consultant.
- Southern California Edison: Utility consultant.

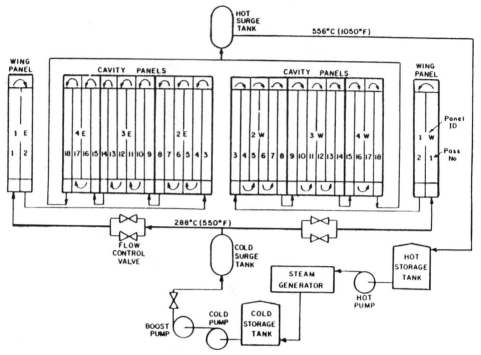

**Figure 2.24:** MSS/CTE receiver loop flow schematic.

### 2.3.4B.1 Description

The MSS/CTE receiver was retrofit to the MSEE system. Most of the subsystems are the same as for that experiment except that the turbine, which had failed during the MSEE project, was bypassed to the heat rejection equipment to condense the steam from the steam generator and reject the heat absorbed in the receiver. A schematic of the receiver loop is shown in Figure 2.24. Major subsystems are described in Table 2.10.

The MSS/CTE receiver test program was directed primarily towards the testing of a scaled-down commercial receiver and was not designed as a full system experiment; there was no electric power production capability. The energy collection process was identical to the MSEE with molten salt providing the energy collection and storage medium. The thermal energy absorbed by the salt was used to generate steam; this energy was rejected to the atmosphere via a separate cooling loop.

From the drained overnight shutdown condition, the MSS/CTE receiver panels were preconditioned to a temperature above the salt freezing point before the receiver was filled. The panels were heated by focusing selected heliostats on the receiver or by using electric cavity heaters which were operated with the cavity door closed to create an "oven" effect. Both of these methods also required the salt piping to be preconditioned prior to filling by heating with electric trace heaters. With the piping, components, and receiver panel preconditioned, the receiver loop was then filled and transitioned to the energy collection mode.

**Table 2.10:** MSS/CTE Plant

### Collector Subsystem

| | |
|---|---|
| Type | Two-axis tracking heliostats |
| Number | 192 used for MSS/CTE (221 available) |
| Total Reflective Area | 7150 m$^2$ (76,900 ft$^2$) for 192 heliostats |
| Drives | Base-mounted azimuth, yoke-mounted elevation |
| Power Supply | Primary: Diesel generator |
| | Back-up: Commercial power |

### Receiver Subsystem

| | |
|---|---|
| Rating | 4.8 MW thermal |
| Configuration | "C" shaped cavity with wing panels |
| Heat Transport Fluid | Molten nitrate salt (60% NaNO$_3$, 40 KNO$_3$) |
| Operating Temperatures | 299°C (570°F) inlet, 566°C (1050°F) outlet |
| Tower Height | 61 m (200 ft) |
| Control | Two receiver control zones. Distributed digital control system. |

### Thermal Storage Subsystem
(existing from MSEE project)

| | |
|---|---|
| Type | 2 tank |
| Rated Storage Capacity | 5.8 MWHth |
| Storage Medium | Molten nitrate salt |
| Operating Temperatures | 299°C (570°F); 566°C (1050°F) |

**Table 2.10:** MSS/CTE Plant (con't.)

### Steam Generator Subsystem
(existing from MSEE)

| | |
|---|---|
| Type | Salt-heated evaporator and superheater with steam drum separator and forced evaporator recirculation. |
| Rating | 3.13 MWth |
| Operating Temperatures (salt) | 566°C (1050°F) inlet 299°C (570°F) outlet |
| Feed Water Conditions | 288°C (550°F), 83 bar (1200 psi) |
| Main Steam Conditions | 510°C (950°F), 76 bar (1100 psi) |

### Fossil Fuel Subsystem
(existing from MSEE)

| | |
|---|---|
| Function | Provided alternate capability to heat salt. Not required for normal operation. |
| Fuel | Propane |
| Rating | 2.93 MWt |
| Operating Temperatures | 299°C (570°F) inlet, 566°C (1050°F) outlet |

### Power Conversion Subsystem

None

### Controls

| | |
|---|---|
| Type | Distributed digital process control (Bailey Network 90) with separate heliostat control system. |
| Start-up & Shutdown Sequencing | System capable of automatic sequencing; however, most procedures were executed manually. |
| Data Acquisition | Separate HP-1000 system |
| Console Displays | Standard formats plus customized control graphics |
| Safety Provisions | Safety interlocks and automatic equipment protection sequences. Uninterruptable power supplies. |

Using solar energy, the receiver panels were warm enough for filling 35 minutes after sunrise. After filling, the remaining heliostats were focused on the receiver to begin collecting energy at approximately 40 minutes after sunrise. Due to minimum flow constraints, the rated outlet temperature was not achieved until 90 minutes after sunrise.

When the cavity heaters were used for preconditioning, the receiver could be filled prior to sunrise. This made it available for energy collection at sunrise. Use of the cavity heaters, however, had the disadvantage of a high electric parasitic load which testing showed could not be offset by early morning gains in additional thermal energy collection. The cavity heaters primary value was in the flexibility provided to the test program; they would not normally be used for optimized operations.

The alternate of circulating cold salt through the receiver loop throughout the night was tested. This method has the advantage of substituting stored thermal energy for electrical energy to provide thermal conditioning. All trace heaters and cavity heaters were off except those on the few lines or components without flow. With the system operating in this mode, the energy collection process could begin immediately after sunrise. Testing of overnight cold salt circulation demonstrated potential advantages in the form of lower total parasitic energy consumption (both electric and thermal). This method would only be applicable to a cavity-type receiver.

### 2.3.4B.2 Performance

Most of the test program relating to performance was devoted to thermal loss testing and receiver control testing to be used in subsequent simulation of receiver operation. Consequently, only limited data were available for solar noon and partial day performance and no data were available for longer-term performance. The test results indicated that the receiver performance was within the expected range. However, no power production test phase was conducted as in the MSEE project. Additional testing would be required in order to determine the receiver's energy collection capability.

### 2.3.4B.3 Significant Incidents

Significant incidents encountered during the MSS/CTE receiver test program are summarized in Table 2.11.

### 2.3.4B.4 Lessons Learned

Major lessons that were learned from the MSS/CTE receiver test are summarized below:

- The MSS/CTE program demonstrated improved trace heater reliability over ealier test programs; however, reliability is still a problem. Further quality control improvements, both in the heater and their installation, is required.
- A single master control system computer to control the subsystem computers (e.g. process control, heliostat control, and trace heater control) should be provided. This would simplify operations and allow integrated sequencing and interlocking.
- The digital process control system was easy to learn, user friendly and well suited for this type of application.
- At least one CRT should be provided for each subsystem being controlled.

**Table 2.11:** Significant Incidents - MSS/CTE Receiver Tests

| INCIDENT | CAUSE | SOLUTION/REPAIRS |
|---|---|---|
| Receiver tube salt leaks. | Poor weld quality. | Seal Weld. Weld quality improvements required during fabrication. |
| Salt leaks at flanged fittings. | Loosened joints and gasket failures. | Replaced gaskets and/or tightened joints. |
| Receiver surge tank salt overflow. | Operating procedure. | Modified procedures. |
| Salt plugs. | Cold lines/components. | Heated lines or dissolved plugs with water. Modify trace heat and/or insulation. |
| Salt valve leaks. | Failed bellows. | Repair or replaced bellows. |
| Cold salt boost pump failures (2). | Mechanical failure. | Pump rebuilt. |
| Boiler recirculation pump failure. | Mechanical failure. | Pump rebuilt. |
| Insulation on salt lines saturated by water. | Facility water line failed due to freezing. | Removed lagging and waited for trace heat to dry insulation (several weeks). |
| Spray water heat exchanger leaks. | Recurring gasket failures. | Replaced heat exchanger. |

- Subsystem computer simulations which reasonably emulated the real system processes were valuable in evaluating control strategies and providing better understanding of the system behavior. These simulations can lead to considerable time savings during plant start-up and check-out.
- Automated heliostat control for receiver panel warm-up would simplify start-up operations and eliminate a significant burden on the operators.
- System performance analyses require more of an overall systems approach rather than subsystem by subsystem approach in order to improve the accuracy of performance predictions. Such analyses must also consider reduced temperature energy collection during early morning, late afternoon and cloud transient operations.
- Future designs require better coordination between designers and controls engineers to resolve strategies for controlling receiver stresses. At issue are major implications for receiver operations, procedures, controls, annual performance and lifetime.

- Flanged pipe fittings should be avoided since they have been prone to leakage.
- Use of cavity heaters to precondition the receiver panel prior to filling is not recommended for daily use. Thermal energy collection gains provided by the earlier start-up do not offset the additional parasitic load of the electric heaters.
- Rapid early morning start-up can be achieved; however, warm-up aiming strategy requires considerable development.
- Air-cooled flux gages performed better and were more reliable than water-cooled gages.

### 2.3.4B.5 Summary

A summary of the major accomplishments of the MSS/CTE receiver test follows:

- The project demonstrated features of a scaled commercial receiver.
- The adequacy of the receiver's structural design, including the design of panel supports, was verified.
- The predicted thermal and hydraulic performance was verified.
- Test data were collected for correlation with thermal models.
- Several methods for receiver thermal conditioning and for early morning start-up were demonstrated.

### 2.3.4C THEMIS

THEMIS was a 2.5 MW$_e$ central receiver full system experiment located near Targasonne in the French Pyrenees (Bonduelle and Rivoire, 1987, Drouot, 1983, Hillairet, 1983, Masele et al., 1985, Bezian and Bonduelle, 1985, Pharabod et al., 1986, Bonduelle and Cazin-Bourguignon, 1986, Boutin and Rivoire, 1986, Rivoire and Bonduelle, 1986, Bonduelle, 1986, Etievant et al., 1987, Etievant et al., 1988, Etievant et al., 1988a, Etievant et al., 1988b, Etievant et al., 1988c, Rivoire and Bonduelle, 1986a, Lopez, 1986, and Faas, 1986). It was tested for three years from mid-1983 until mid-1986. A picture of the plant is shown in Figure 2.25.

The objectives of the THEMIS project were to:

1. Demonstrate the feasibility of a solar thermal electric power plant suitable for tropical and subtropical countries.
2. Define the principles of this type power plant.
3. Be a display window for French manufacturers and develop an export market.
4. Train engineers.
5. Conduct experiments on the full system and, particularly, the solar subsystems.

The sponsors of the project were:

- Centre National de La Recherche Scientifique (CNRS).
- Electricité de France (EDF).
- Solar Energy Authority (COMES), became Agence Francaise pour la Maitrise de l'Energie (AFME) in 1982.
- French Government.
- Public Regional Authority of Languedoc-Roussillon (site).
- General Committee of Pyrenees-Orientales (site).

The major project participants and their roles were:

- EDF: Project manager, system design, prime contracting, operation and maintenance.

**Figure 2.25:** THEMIS.

- CETHEL: Main contractor for the heliostat field.
- C.N.I.M.: Solar receiver.
- C.F.E.M.: Thermal Storage tanks and condenser.
- C.I.T.E.C.: Steam Generator.
- Alsthom Atlantique: Turbo-alternator.
- Groupe D'Evaluation Scientific THEMIS (GEST) (CNRS and AFME personnel): Project evaluation team.

### 2.3.4C.1  Plant Description

THEMIS employed molten salt (Hitec) as the receiver coolant and as the thermal storage medium. The steam Rankine cycle was used for power conversion. A system flow schematic is shown in Figure 2.26; major subsystems are described in Table 2.12.

The operation of the power plant had flexibility because of the separation of the primary and secondary circuits. Three types of operation were employed: (1) thermal energy storage for deferred generation of electrical energy, (2) generation of electrical energy in accordance with sunshine hours and (3) discharge of thermal energy from storage to generate electrical energy at a power level greater than that of the solar receiver. The advantages of this configuration were especially evident during cloudy periods; a temporary drop in the power of the primary circuit had no effect on the operation of the secondary circuit. Since the salt temperature in the thermal storage unit remained constant, the steam characteristics could be kept at their nominal values.

Operation began by placing the heliostats in the stand-by position one hour before sunrise. At sunrise the receiver door was opened and salt circulation started. The heliostats were then focused onto the receiver. Conditioning of the secondary

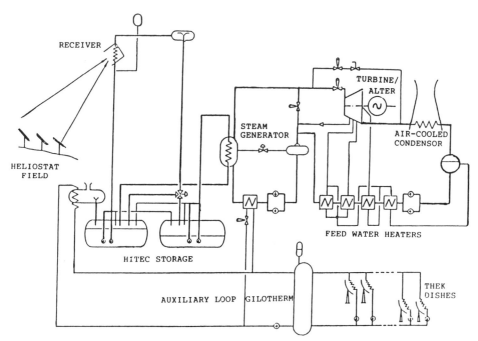

**Figure 2.26:** THEMIS schematic.

circuit was started as soon as a sufficient quantity of hot salt was produced (2 to 3 hours after the start). The turbine generator was started 30 minutes after the start of the conditioning. Power production began 15–20 minutes later. The total startup time was about 3 hours. At sundown, the heliostats were placed in the stow position; the turbine generator was shut down after exhaustion of the available hot salt (or earlier for fast startup on the following day). The temperature of the primary circuit was maintained by periodic circulation of salt until startup the following morning.

### 2.3.4C.2 Performance

THEMIS was operated for a total of 3 years. Power production was substantially lower than predicted. The average was 643 KWHr per year gross (net was negative) versus a prediction of 3000 KWHr per year for the *net* generation of electricity. Much of the performance analyses was based on 43 selected days (out of 3 years) when the plant operated well. Input-output plots of the data from these 43 days are shown in Figure 2.27.

A performance summary for all 3 years of operation is given in Table 2.13. It can be seen that there was a continual improvement in the gross electric power with time, but net power production remained negative. The large parasitic loads were due to the small size of the experiment and non-optimum design.

### 2.3.4C.3 Significant Incidents

Significant incidents encountered during operation of THEMIS are summarized in Table 2.14.

**Table 2.12:** THEMIS Plant

### Collector Subsystem

| | |
|---|---|
| Type | Two-axis tracking heliostats |
| Number | 200 |
| Total Reflective Area | 10,740 $m^2$ |
| Reflective Surface | Silvered-glass laminate |
| Power Supply | Grid; battery at each heliostat |

### Receiver Subsystem

| | |
|---|---|
| Rating | 8.9 $MW_t$ |
| Configuration | Cavity with door, 16 $m^2$ aperture |
| Heat Transport Fluid | Hitec salt (53% $KNO_3$, 40% $N_aNO_2$, 7% $N_aNO_3$) |
| Operating Temperatures | 300°C (572°F) inlet, 450°C (842°F) outlet |
| Tower Height | 100 m (328 ft) |
| Control | Flow control using outlet temperature and flux. |

### Thermal Storage Subsystem

| | |
|---|---|
| Type | 2 tank |
| Rated Storage Capacity | 40 $MWHr_t$ |
| Storage Medium | Hitec Salt |
| Operating Temperatures | 250°C (482°F); 450°C (842°F) |

### Steam Generator Subsystem

| | |
|---|---|
| Type | U-tubes (salt) in shell (water/steam) |
| Rating | 8000 $KW_t$ |
| Operating Temperatures (Salt) | 450°C (842°F) inlet; 250°C (482°F) outlet |
| Outlet Steam Conditions | 430°C (806°F); 42 bar (609 psi) |

**Table 2.12**: THEMIS Plant (con't.)

Power Conversion Subsystem

| | |
|---|---|
| Prime Mover | Ten stage axial-flow condensing turbine with 4 extraction ports |
| Inlet Conditions | 410°C (770°F); 40 bar (580 psi) |
| Outlet Conditions | 89°C (192°F); 0.186 bar (3 psi) |
| Generator Nameplate Rating | 2800 KVA (6.6 KV) |
| Load | Local Grid |

Controls

| | |
|---|---|
| Collector Field | Central computer, through master microprocessors to individual microprocessors at each heliostat |
| Process Plant | Hybrid of microprocessors and conventional control panel (limited use of 2 CRT/keyboard processors at control console). |
| Turbine/Generator | Conventional |
| Safety Provisions | Programmable logic units |

## 2.3.4C.4 Lessons Learned

Major lessons learned from the THEMIS project are:

- The design of the salt loops should be simplified.
- It was concluded that the present technology of electric trace heating was not mature enough for the temperature range of the system.
- The "high temperature" electrical trace heating does not adapt well to the design approximations and construction imperfections normally accepted. For example, pipes were perforated by corrosion due to repeated overheating. The resistance heaters employed must be of top quality. The design must be thoroughly detailed, and construction and installation must be to the highest standards.
- Keep the salt loops warm overnight rather than drain.
- Mount salt valves in a drainable position.
- Avoid shaft vibrations of the cantilever pumps.
- Maintenance operations on salt piping caused by trace heater failures are long, difficult and tedious. These operations should be simplified by using removable heater elements on most salt piping sections.
- Salt pumps were a constant source of trouble and special attention should be given to the salt pump design in future projects.
- Because the loops are most often in the overnight shutdown mode, the design should try to reduce power consumption during this mode.

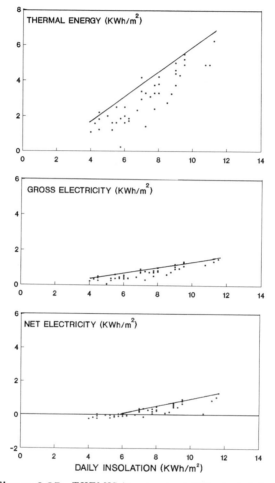

**Figure 2.27:** THEMIS input-output (43 selected days).

- Access for maintenance should be a major design criterion.
- The requirements of thermal cycling should be included in the design and specifications.
- Heliostat mirror reflectivity should be increased by the use of thin glass.
- The design should avoid the use of batteries at each heliostat.
- An isolation valve should be installed between the steam generator and the turbine.
- Plant automation should be completed prior to the start of the test program.

### 2.3.4C.5 Summary

A summary of the major accomplishments of the THEMIS project follows:

- The project confirmed the technical feasiblity of using molten salt for solar energy collection and storage.

**Table 2.13**: THEMIS Performance Summary

|  | 1983 (6 months) | 1984 | 1985 | 1986 (6 months) | Total |
|---|---|---|---|---|---|
| **Plant Output:** | | | | | |
| Gross electric (MWhr) | 45.3 | 573.9 | 765.7 | 543.7 | 1929 |
| Net electric (MWhr) | (743) | (2697) | (2182) | (1131) | (6752) |
| Power Production (hrs) | 51 | 470 | 526 | 388 | 1435 |
| **Annualized & Normalized:** | | | | | |
| Gross elec. (KWHr/m$^2$-yr) | 4.2 | 53.4 | 71.3 | 101 | 59.1 |
| Net elec. (KWHr/m$^2$-yr) | (69.2) | (251) | (203) | (211) | (198) |
| Power Production (% time) | 1.2% | 5.4% | 6.0% | 8.9% | 5.5% |

- The advantages of decoupling solar energy collection from power generation by the use of thermal storage was demonstrated.
- Reliable operation of the collector field was achieved (95% heliostat availability).
- The laminated glass mirrors had no corrosion.
- The full plant was operated for 3 years, and operating procedures were improved.
- Many of the problems associated with the system concept and design were identified.

### 2.3.4D International Energy Agency - Central Receiver System

The International Energy Agency-Central Receiver System (IEA-CRS) was a 500 KW$_e$ full system experiment located at Tabernas, Spain adjacent to the CESA-1 site (Interatom, 1982, Becker, 1982, Becker et al., 1983, Becker, 1983, Bucher, 1984, Hansen, 1984, Grasse, 1985, Sandgren and Andersson, 1985, Grasse and Becker, 1983, Selvage, 1984, Grasse and Ruiz, 1985, Grasse, 1986, Baker, 1986, SSPS Operating Agent IER/CIEMAT, 1987, SSPS Monthly Plant Operation Reports and Daily Evaluation Summaries, 1986, and Baker, 1987). The plant was operated from mid-1981 until August 1986. An aerial view of the plant is shown in Figure 2.28.

The overall objectives of the project were:

1. Determination of design viability
2. Evaluation of operational behavior
3. Extrapolation for advanced commercial designs
4. Comparison of central receiver and distributed collector system performance

Specific objectives of the initial operations phase (1982) were:

1. Maximization of yearly electric output
2. Minimization of plant outages

Objectives of the subsequent test and evaluation phase (1983-1986) were to:

1. Measure heliostat field performance.

**Table 2.14:** Significant Incidents - THEMIS

| Incident | Cause | Solution/Repairs |
|---|---|---|
| Heliostats seriously damaged in 2 bad wind storms. All the concrete supports fractured. | Heavy wind, oblique incidence. | Replaced broken heliostats and reinforced all other heliostats, particularly pedestals. Increased foundation. |
| Heliostats controller cards destroyed. | Lightning. | Replaced. |
| Frequent battery failure. | Batteries unreliable, undersized charger, environment, other. | Replaced. Recommend that batteries not be used in future. |
| Substantial corrosion on back structure of 30 heliostats. | Unsatisfactory welds | Partly resolved. |
| Frequent zero resets required at the heliostats. | Inductive encoder failures. | Replaced with modified encoder. |
| Flux meters difficult to use and accuracy uncertain. | Inherent. | Replaced with optical fiber photon flux meters. |
| Excessive trace heater failure. | Poor installation, hot spots, other. | Replaced. |
| Turbine rotor damaged. | Operator error. | Repaired. |
| Two of 4 salt pumps failed. | Shafts jammed. | Repaired. |
| Frequent salt pump bearing wearout. | Vibration. | Repaired. |
| Steam generator leakage. | Not designed for thermal cycling. | Maintained steam generator warm. |
| Receiver cavity heater failed. | Unknown. | Redesigned. |

2. Determine receiver behavior.
3. Compare cavity and external receiver configurations.
4. Determine plant losses.
5. Investigate system aspects and gain operational experience.

The project was sponsored by the following agencies of the nine participating member countries of the International Energy Agency:

- German Federal Ministry of Research and Technology
- U.S. Department of Energy
- Spanish Ministry of Industry and Energy
- Italian National Research Council
- Belgian Ministry of Scientific Research

**Figure 2.28:** IEA-CRS Plant.

- Swiss Federal Ministry of Science and Research
- Austrian Federal Ministry of Science and Research
- National Swedish Board of Energy Source Development
- Greek National Energy Council of the Ministry of Coordination

  The major project participants and their roles were:

- DFVLR (Germany): Operating Agent (through 1984).
- CIEMAT-IER (Spain): Operating Agent (from January, 1985).
- Compania Sevillana De Electricidad S.A. (Spain): Plant Operation (through 1985).
- MOMPRESA (Spain): Plant operation (from January, 1986).
- Interatom (Germany): Prime Contractor and plant engineering.
- Martin Marietta (U.S.A.): Heliostat field and control.
- MAN (Germany): Energy conversion system design.
- CASA (Spain): Overall system engineering plus civil and site engineering.
- Gebruder Sulzer A.G. (Switzerland): Cavity receiver and steam generator.
- Franco Tosi (Italy): Advanced sodium receiver.
- SAIT (Belgium): Data acquisition system.
- Spilling (Germany and Switzerland): Steam engine and power conversion.
- Voest Alpine A.G. (Austria): Sodium storage tanks.

### 2.3.4D.1 Plant Description

The IEA-CRS plant employed liquid sodium as the receiver coolant and as the thremal storage medium. A reciprocating steam engine was used for power

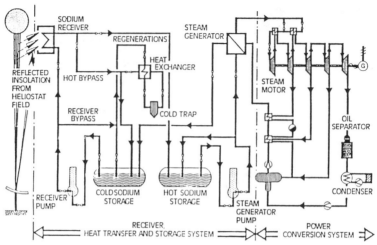

**Figure 2.29:** IEA-CRS schematic.

conversion. A system flow schematic is shown in Figure 2.29; major subsystems are described in Table 2.15.

During normal operating periods, liquid sodium flow was maintained through the receiver and the steam generator throughout the night in order to facilitate faster system startup and to reduce electric trace heating loads. In the morning, when the insolation reached between 250 W/m$^2$ and 400 W/m$^2$ (varied depending on the day of the year), heliostats were focused on the receiver. When the outlet sodium temperature reached 500 °C the sodium was sent to the hot tank to begin hot storage heating which lasted 3 to 4 hours. During this period, heat was rejected from the steam using a steam motor by-pass dump until the hot sodium from storage reached approximately 490 °C. After about 30 minutes of steam motor preheating, the generator was synchronized to the grid and the load increase began. Power production continued until the hot sodium tank was depleted to near its minimum level. Normal startup took approximately 7 hours compared to a planned duration of 30 minutes.

When the insolation fell to the point where the receiver outlet temperature could not be maintained above its inlet temperature, heliostats were defocused, the receiver doors were closed, sodium flow through the receiver was reduced to 10%, and the steam generator was transitioned to cold sodium flow. At this point the system was configured for a "hot start" the next day.

For extended shutdown periods, all sodium in the heat transfer loops was drained to the storage tanks where it was maintained above its freezing point by electric heaters.

Great attention was given to the identification of ways to improve plant operability. It was determined that several design deficiencies existed, which, if eliminated, would have improved plant startup and operational performance. Elimination of these deficiencies would have involved design modifications to reduce the "thermal inertia" of the plant and the addition of heliostats to increase the solar multiple.

**Table 2.15:** IEA-CRS Plant

### Collector Subsystem

| | |
|---|---|
| Type | Two-axis tracking heliostats |
| Number | 93 |
| Total Reflective Area | 3655 m$^2$ |
| Drives | Pedestal-mounted drive assembly with azimuth & elevation gear motors |
| Power Supply | Primary: Utility grid |
| | Back-up: Diesel generator |

### Receiver Subsystem (Advanced sodium receiver)

| | |
|---|---|
| Rating | 2.5 MW thermal |
| Configuration | External flat panel |
| Heat Transport Fluid | Sodium |
| Operating Temperatures | 270$^o$C (518$^o$F) Inlet, 530$^o$C (986$^o$F) Outlet |
| Tower Height | 44 m (144 ft) |

### Thermal Storage Subsystem

| | |
|---|---|
| Type | 2 tank |
| Rated Storage Capacity | 5 MWHr |
| Storage Medium | Sodium |
| Operating Temperatures | 270$^o$C (518$^o$F); 530$^o$C (986$^o$F) |

### Steam Generator Subsystem

| | |
|---|---|
| Type | Once-through sodium/water shell & helical tube heat exchanger |
| Rating | 2.2 MWt |
| Operating Temperatures (sodium) | 525$^o$C (977$^o$F) in, 275$^o$C (518$^o$F) out |
| Feed Water Conditions | 193$^o$C (379$^o$F), 115 bar |
| Main Steam Conditions | 500$^o$C (932$^o$F), 105 bar |

**Table 2.15:** IEA-CRS Plant (con't.)

Power Conversion Subsystem

| | |
|---|---|
| Prime Mover | 5 stage, steam engine |
| Inlet Conditions | 500°C (932°F), 100 bar |
| Outlet Conditions | 193°C (379°F), 0.3 bar abs |
| Generator Nameplate Rating | 600 KW, 400 V, 3-phase, 50 hz |
| Load | Utility grid & substitute load |

Controls

| | |
|---|---|
| Type | Independent, hardwired circuits |
| Start-up & Shutdown Sequencing | Primarily manual |
| Data Acquisition | Dedicated computer & peripherals |
| Console Displays | Color graphics (DAS only) and "conventional" control board |
| Safety Provisions | Automatic interlocking system. Heliostat defocussing provisions. Independent safety Instrumentation. |

**Table 2.16:** IEA-CRS Operation/Performance Summary

| Year | Direct Insolation (KWHr/m²-yr) | Power Production (Hours) | Gross Electric (KWHr/m²-yr) |
|---|---|---|---|
| 1982 | 1730 | 58 | 5.04 |
| 1983 | 2000 | 53 | 4.26 |
| 1984 | 1960 | 127 | 10.60 |
| 1985 | To May only | 10 | 1.09 |

## 2.3.4D.2 Performance

Numerous equipment outages and operating complications limited to the collection of long term performance data. Table 2.16 summarizes the number of power production hours and the gross electricity produced during the project. The total gross electrical output, over a period of nearly 3 1/2 years, was 77 MWHr; this compares with a predicted *net* output of 925.4 MWHr per year. A positive net daily electrical output was not achieved; the predictions for plant net electrical efficiencies were: design point, 16.5% and annual average, 11.4%.

A major contributor to the poor performance was the "thermal inertia" created by a plant design which forced substantial thermal energy to be wasted during startup before the sodium was hot enough to generate steam compatible with the steam motor but too hot to recirculate to the cold tank and the receiver. The high thermal inertia was due to design deficiencies unique to the plant and could be reduced in an optimized plant.

### 2.3.4D.3 Significant Incidents

Significant issues encountered during operation of the IEA-CRS plant are summarized in Table 2.17.

### 2.3.4D.4 Lessons Learned

Lessons learned from the IEA-CRS project are summarized below:

- Design deficiencies in the CRS plant contributed to an extremely high thermal inertia which constrained operations and reduced efficiency. Designers must recognize these deficiencies and ensure that they are not duplicated in future systems.
- Methods for estimating plant equipment and operational availability must be improved in order to avoid overly optimistic performance predictions.
- Electric trace heating was considered to be the weakest part of this system. A more reliable system is required.
- Good quality control of the design, manufacture and construction of sodium components is required in order to avoid costly or hazardous defects such as those experienced with leaking sodium vessels.
- Special consideration is required in designing the sodium system for normal maintenance. The lack of sufficient isolation valves, the danger of sodium fires and the need for cleanliness hindered normal maintenance and repairs of valves and other sodium components.
- Maintenance access, especially for instrumentation and trace heating in the receiver area, needs to be improved
- Receiver preheating and filling procedures need to be simplified and made more efficient.
- Problems were encountered with many "conventional" components as a result of thermal cycling. Components must be designed and qualified for cyclic operation.
- An inadequate inventory of spare parts prolonged plant outages. Components subject to failure need on-site or readily available replacements.
- Heliostat electronics were not adequately protected from lightning strikes. Protection must be provided.
- Plants located in an environment similar to that of the IEA-CRS will require frequent heliostat washing.
- A steam motor is a poor choice for the prime mover. A steam turbine should be used.

### 2.3.4D.5 Summary

A summary of the major accomplishments of the IEA-CRS project follows:

- The project demonstrated the technical feasibility of a high-flux sodium receiver.
- The performance predicted for the receiver was verified.

**Table 2.17:** Significant Incidents - IEA-CRS Plant

| Incident | Cause | Solution/Repairs |
|---|---|---|
| Frequent electric trace heater failures. | Combination of incorrect installation, junction failures and poor heater element reliability. | Repaired/replaced as required. Difficult access in some areas extended repair time. |
| Sodium blockages in pipes & other components. | Inadequate trace heater installation and control. | Added insulation and modified trace heat and control. |
| Steam generator rupture disk leak. | Weld failure. | Replaced ruptured disk. |
| Storage tank and regenerator vessel sodium leaks. | Cracks resulting from design and fabrication defects. | Repaired by grinding or cutting out cracked areas followed by welding. Tank inlet also modified. |
| Sodium valve leaks. | Bellows failures. | Replaced. |
| Sodium pump flooded with sodium. | Internal sodium valve leak & oil leak past missing O-ring. | Repaired. |
| Frequent steam motor failures. | Various mechanical failures including rings, connecting rods and timing gear. | Repaired as failures occurred. Operating pressure reduced. Not operated since May, 1985. |
| Lightning storm damaged heliostat electronics. | Inadequate lightning protection. | Electronics repaired. Protection added to prevent reoccurrence. |
| Heliostat mirror corrosion. | Moisture in mirror support. | Mirror modules modified to reduce moisture intrusion. |
| Receiver tube distortion. | Overheating during start-up. | Straightened tubes. Start-up procedure modified. |
| Heliostat beams burned eastern part receiver & receiver door. | Grid failure caused HAC failure; beams from one group of heliostats moved across the receiver. | Repaired. |
| Major fire resulting from large sodium leak (August 1986). | Maintenance procedure error which did not provide verification of sodium plug before cutting line. | Sodium testing never resumed. Leak & fire could have been prevented if system had contained additional isolation valves. |

- Operating and performance data for comparison with distributed collector system was provided.
- High heliostat and receiver reliability was achieved ("conventional" component failures were the primary cause of plant outages).
- Areas requiring additional development were identified.
- International cooperation in the development and application of solar power systems was achieved.

### 2.3.4E Summary of Advanced System Experiments

The major system characteristics of the advanced system experiments are summarized in Table 2.18. It can be seen that there were many similarities in plant configurations. All employed a north field and a cavity receiver (although the IEA-CRS project also tested an exposed panel). All utilized 2-tank thermal storage of the receiver coolant. The steam Rankine cycle was used for power conversion: the IEA-CRS project chose a steam motor and the others employed a steam turbine.

It should be noted that the MSEE and MSS/CTE projects were both conducted at the CRTF and utilized the same heliostat field, tower and thermal storage. The major changes from MSEE to MSS/CTE were the substitutions of what was then considered to be a commercial model receiver design and new controls. Power conversion to electricity was not included in MSS/CTE.

The major difference between the four advanced system experiments was the heat transfer fluid used for the receiver coolant and thermal storage medium. Both the nitrate and Hitec salts were tested as well as liquid sodium. The lower temperature limit of Hitec can be seen in the table. Data from these experiments could be used to help select the preferred medium.

### 2.3.5 Reference System Selection

Experimental results from the system experiments are not adequate by themselves to allow selection of the preferred system configuration to be used as the reference for the central receiver concept. Accordingly, analytical system studies will be utilized to help make this selection.

### 2.3.5A System Studies

Four studies were used in the assessment of this system concept and the selection of a reference configuration. They are briefly described below.

### 2.3.5A.1 A Handbook for Solar Central Receiver Design

A handbook on central receiver technology for solar thermal power plants was published by Sandia National Laboratories, Livermore (Falcone, 1986). It contains a description and assessment of the major components in central receiver systems configured for utility-scale production of electricity using Rankine-cycle steam turbines. It also describes procedures to size and optimize a plant and discusses examples from recent system analyses. Information concerning site selection, cost estimation, construction, and operation and maintenance is also included. Procedures for calculating levelized energy costs are presented with examples and cost sensitivities.

Three central receiver configurations were evaluated and compared. The receiver heat transfer fluids were: molten nitrate salt and sodium. The salt system was analyzed with both an external receiver and a cavity receiver: the sodium system

**Table 2.18:** Advanced System Concepts

| | MSEE | MSS/CTE | Themis | IEA-CRS |
|---|---|---|---|---|
| PLANT RATING | 750 K$W_e$ | 5 M$W_t$ | 2.5 M$W_e$ | 500 K$W_e$ |
| **HELIOSTATS** | | | | |
| Number | 211 | 211 | 201 | 93 |
| Size (m$^2$) | 37.2 | 37.2 | 53.7 | 39.3 |
| Total Area (m$^2$) | 7849 | 7849 | 10,740 | 3655 |
| Field Configuration | North | North | North | North |
| **RECEIVER** | | | | |
| Configuration | Cavity | Cavity | Cavity | Cavity & Exposed |
| Tower Height m (ft) | 61 (200) | 61 (200) | 100 (328) | 43 (141) |
| Coolant | Nitrate Salt | Nitrate Salt | Hitec Salt | Sodium |
| Outlet Temp. $^oC$ ($^oF$) | 566 (1050) | 566 (1050) | 430 (806) | 530 (986) |
| **THERMAL STORAGE** | | | | |
| Type | 2-Tank | 2-Tank | 2-Tank | 2-Tank |
| Medium | Nitrate Salt | Nitrate Salt | Hitec Salt | Sodium |
| Rating (MWHr$_t$) | 7 | 7 | 40 | 5.5 |
| **POWER CONVERSION** | | | | |
| Type | Turbine | None | Turbine | Steam Motor |
| Working Fluid | Steam | | Steam | Steam |
| Inlet Temp. $^oC$ ($^oF$) | 504 (940) | | 410 (741) | 500 (932) |
| Inlet Press (bar) | 72 | | 40 | 100 |
| Heat Rejection | Air Cooler | | Dry Tower | Wet Tower |

was only evaluated with an external receiver. The predicted energy costs for all three systems were within six percent; the salt system with an external receiver produced the lowest cost energy. The differences between the energy costs for the three systems are far less than the uncertainty of the estimates.

### 2.3.5A.2 Characterization of Solar Thermal Concepts for Electricity Generation

This study was conducted by Pacific Northwest Laboratory (PNL) for the United States Department of Energy (William et al., 1987). The goal of the study was to provide a relative comparison of several alternative concepts for solar thermal electric power generation. One trough, one dish and the four major central receiver alternative system concepts were evaluated. The projections of the study were for the late 1990's time frame and were based on the capabilities of the technologies that could be expected to be achieved with further development.

Existing design reports were used to characterize system designs and provide baseline performance estimates. This data was incorporated into SOLSTEP, a PNL computer code, to provide optimum sizing of system elements, annual system performance, and levelized energy costs from the plant. Experimental results from the various central receiver system experiments were not utilized in making these predictions. This could be justified, perhaps, under the ground rule that several commercial plants of a similar design would precede the plant being evaluated. This, of course, presumes that any uncertainties will have been resolved positively.

The four central receiver concepts evaluated and the predicted energy costs were:

- water/steam with an external receiver and oil/rock storage ($0.07/KWhr)
- molten salt with a cavity receiver and salt storage ($0.05/KWhr)
- sodium with an external receiver and sodium storage ($0.05/KWhr)
- sodium with an external receiver and salt storage ($0.06/KWhr)

This study predicted that all of the advanced system concepts would have lower energy costs than water/steam. Again the differences were less than the uncertainties in the estimates, and little experimental data was utilized in making these predictions. The small differences in energy costs from the three advanced system concepts was largely due to the lower cost of thermal storage when salt was utilized.

### 2.3.5A.3 The PHOEBUS Project

A European joint venture under the name of PHOEBUS was formed in 1986 by a number of companies and organizations that have been actively engaged in research, development and construction of solar central receiver plants for electricity production (Fricker, 1987, Ministerio de Industria Y Energia, 1987, and Phoebus Executive Summary of Phase IA Work, 1988). It is the venture's objective to finance, design, construct and commission a 30 $MW_e$ central receiver power plant. They consider that this size is necessary to confirm the operation and performance of the plant and its potential for reaching commercial production costs.

Initial activities in this project included a review of the accomplishments and problems encountered in the previous full system experiments. This information was used to make a preliminary system concept evaluation with the following ranking of concepts in order of preference:

**Table 2.19:** Candidate Features Evaluated In Phase I of Utility Studies

| Feature | Candidates |
|---|---|
| Plant Ratings | 7.5 MW$_e$ and Larger |
| Heliostats | Glass/Metal and Stretched Membrane |
| Heliostat Field Configuration | Surround or North |
| Receiver Configuration | External or Cavity |
| Receiver Coolant | Liquid Sodium or Molten Nitrate Salt |
| Thermal Storage Medium | Molten Nitrate Salt Only |
| Thermal Storage Capacity | 0 to 9 Hrs at Full Plant Rated Output |
| Plant Type | Solar-Only or Solar/Fossil Hybrid |

1. Salt/Steam Turbine (based on THEMIS)
2. Water/Steam (based on Solar One)
3. Open Air Receiver/Steam Turbine
4. Sodium/Steam Turbine (based on IEA-CRS)
5. Gas Receiver/Gas Turbine/Steam Turbine (bottoming cycle)
6. Sodium Receiver/Salt Storage/Steam Turbine

This initial preference was consistent with the other system studies; however, as the Phoebus study progressed, greater interest was given to the open air receiver/steam turbine concept (Phoebus, 1988).

### 2.3.5A.4 Utility Solar Central Receiver Study

Two utility teams, one led by Pacific Gas and Electric (PG&E) and the other by Arizona Public Service (APS), conducted cooperative studies cofunded by the U.S. Department of Energy, and the utilities. The Electric Power Research Institute (EPRI) provided supplemental funding. The objective was to chart a course for commercialization of the solar central receiver. Phase I, completed in September, 1987, conducted extensive trade studies to select a preferred system configuration (Arizona Public Service Company, 1988, Arizona Public Service Company, 1988a, and Pacific Gas and Electric Company, 1988). The candidate features considered are listed in Table 2.19.

Water/Steam was not considered as a candidate because of the expected advantages of the advanced concepts. This is probably justified because of the greater potential of the advanced system concepts, even though that potential has not been demonstrated experimentally.

Sodium was excluded as the thermal storage medium on the basis of prior trade studies conducted by Pacific Gas and Electric which showed that the

**Table 2.20:** Reference Central Receiver System Concept

| | |
|---|---|
| Plant Rating | 200 MW$_e$ |
| Plant Type | Solar Only |
| Solar Multiple[a] | 1.8 |
| Heliostat Design | 150 m$^2$ stretched membrane |
| Total Reflective Area | 1.8x10$^6$ m$^2$ |
| Field Configuration | Surround |
| Receiver Type | External Cylinder |
| Receiver Coolant | Molten Nitrate Salt |
| Thermal Storage Type | Hot and Cold Tanks |
| Thermal Storage Medium | Molten Nitrate Salt |
| Storage Capacity | 5-6 Hours |
| Power Conversion | Reheat Steam Turbine |
| Land Area | 1000 hectare (3.9 mi$^2$) |

[a]Ratio of the maximum solar thermal power collection rating to that required to match the plant rating.

sodium/salt combination was superior to the all-sodium configuration when thermal storage was employed and that some storage was preferable to none.

The stretched membrane heliostat design was selected based on its lower potential cost. The high plant rating and the large solar multiple reflect the economies of scale. This is also the major reason for selection of the external receiver with the surround field; the maximum size of a cavity receiver module with a north field is much less (under 100 MW$_e$) than the preferred value (200 MW$_e$). Selection of the external receiver configuration and the large plant size was confirmed by both of the teams. Similarly, the selection of molten salt as the receiver coolant was confirmed by both teams.

A risk assessment was made for the selected design, and candidate development programs were formulated (Hillesland, Jr. and Weber, 1988, Harder et al., 1988, and Pacific Gas and Electric Company, 1988a).

## 2.3.5B Reference Central Receiver Design Concept

The preferred central receiver system design selected in the utility study will be used as the reference design representative of the central receiver concept. This selection is consistent with the other system studies cited. The major features of this design are summarized in Table 2.20.

### 2.3.6 Central Receiver Technology Status

The status of central receiver technology will be summarized here using the results of these prior developments, particularly the system experiments. Technical developments will be summarized for the overall system and for the major subsystems and components. Performance measurements for the system experiments will be reviewed to help establish a basis for estimating energy costs from a central receiver system in the future. The open technical issues will provide the basis for assessing the system's development status.

### 2.3.6A  Technical

The technical status of the central receiver is summarized below for the system and major subsystems.

### 2.3.6A.1  Central Receiver System

- System technical feasibility has been confirmed for all concepts of interest.
- Decoupling of power conversion from insolation transients by thermal storage in the advanced system concepts was demonstrated. This provides the following options for operations:
    - power generation can follow load demand or
    - power generation can be at the plant rating and hence at maximum efficiency.
- System designs are relatively complex. Based on the requirements identified in the system experiments, actual systems require:
    - many loops, valves and lines,
    - extensive and complex trace heating and
    - extensive instrumentation and controls.
- System operations are complex. This is due to:
    - overnight conditioning requirements (particularly for the advanced concepts),
    - daily startup and shutdown,
    - operation at high turndowns from design conditions (only the receiver loop for advanced concepts) and
    - transient operation through cloud passage (again only the receiver loop for advanced concepts).
- The requirements to precondition all lines and equipment to temperatures above the fluid freezing point contributes to the more complex designs and operations of the advanced concepts. This is more pronounced for molten salt than for sodium because of the higher freezing point.
- The sodium system is more complex than the salt system because of safety considerations and the requirements of fluid maintenance.
- Plants were operated effectively by utility operators who also identified many areas for system improvement. Thus, the operational complexity noted above did not seriously impact normal utility operator skill requirements.
- Operations and maintenance requirements for water/steam systems are reasonably well understood. Some of this data base is applicable to advanced system concepts.
- A preferred advanced system concept employing molten salt has been selected.
    - Conceptual designs have been developed for both a first and an Nth commercial plant.

- This allows concentration of further development along a single path.
- This selection appears justified and will be described in the following subsection; however, it is not confirmed by the results of the system experiments evaluated herein.
- More stringent quality control is required with the advanced system concepts.
- The plants have had a negligible environmental impact. (This could still be an open issue for sodium plants.)

## 2.3.6A.2  Heliostats

- Glass/metal heliostat development is largely complete.
- Heliostat availability was good during most of the system experiments.
- Proper lightning protection is required for any heliostat field.
- Periodic heliostat washing is required at most sites to maintain reflectivity, and it is cost-effective.
- The balance of maintenance requirements for glass/ metal heliostats are well understood.
- A stretched membrane heliostat concept with potential advantages over glass/metal heliostats has been identified.
  - Costs could be as low as one half those of glass/metal heliostats.
  - Early tests showed that performance could be similar to glass/metal designs.
  - Life of the polymer film is still an open issue.
- Development of the stretched membrane heliostat has just started.
  - Two 50 m$^2$ prototypes have been built and tested.
  - Development should benefit from the preceding glass/metal heliostat program.

## 2.3.6A.3  Receiver

- Receivers were generally rugged structurally, even when exposed to conditions beyond their design limits. For example:
  - Salt was frozen and thawed within several of the MSEE receiver panels.
  - Most receivers were exposed to higher than the design heat fluxes. This was a deliberate test in the IEA/CRS project.
  - Most receivers were operated through rapid temperature transients which produced larger than design stresses.
- Two receivers (Solar One and MSS/CTE) did experience significant tube leaks.
- Thermal and hydraulic performance predictions were generally confirmed by the experiments.
- Conservative design margins are required for the receiver temperatures and heat fluxes due to variations in beam quality and aim point of the heliostat field.
- The design of electric trace heating for advanced concepts is very complex with many active control loops required.
- Electric trace heating was very unreliable in the advanced system experiments.
  - MSEE/CTE had less problems than other experiments. However, the pump and valve test portion of this project did experience substantial problems.
  - Experience with sodium loops in the nuclear program has been better than the solar experience (Holmes, 1985).

- Most salt system experiments experienced problems due to the high freezing point of the heat transfer fluid.
  - Startup was often delayed due to low equipment or pipe temperatures.
  - Blockages due to frozen salt were experienced.
- Replacement of failed trace heaters is costly and impacts plant availability.
- Blanket-type insulation is required to prevent convective air flow for complex equipment shapes and bends with trace-heated advanced concepts.
- Good receiver transient response was achieved when flux gages were used with the control algorithm.
- The maximum turndown from rated flow achieved with salt receivers was four to one.
- Thermal efficiency of an exposed receiver panel was higher than for a cavity receiver of the same rating (IEA-CRS project) (Baker, 1986).
- Repainting of receiver panels has been required to maintain absorptivity.
- Most receivers tested were fabricated from alloy I 800; the current preferred material is type 316 stainless steel.
- A direct absorption receiver (DAR) concept has been identified for use with molten salt. The potential advantages over tube-type receivers are:
  - Substantially lower cost.
  - Reduced complexity.
  - Higher efficiency.
- The DAR is still only a concept.
  - Applications analyses have been completed.
  - Limited tests have been conducted.
  - A salt experiment is planned at the CRTF.

### 2.3.6A.4 Thermal Storage

- Most thermal storage systems performed as designed.
- The advantages of storage at the receiver outlet temperature, inherent in the advanced concepts, were demonstrated.
- Both the sodium and oil storage systems experienced serious fires.
- Tank leaks were experienced in several experiments.
- Molten nitrate salt appears to be the preferred thermal storage medium.
- Hot salt storage in an externally-insulated stainless steel tank is preferred. The more complex, internally-insulated, membrane-lined tank is probably not justified.
- The trace heating problems noted for the receiver also apply to molten salt thermal storage.

### 2.3.6A.5 Power Conversion and Balance of Plant

- Both salt-heated and sodium-heated steam generators were tested and performed adequately.
- Much commercial equipment is unsuitable for cyclic operation.
- Electric startup heaters were unreliable and a high parasitic power load.

### 2.3.6A.6 Control

- Distributed digital control was very effective where it was employed: Solar One, MSEE and MSS/CTE.

**Table 2.21:** Solar One Power Production

|  | Insolation[a] | Gross Elec[a] | Net Elec[a] | Net Effic.(%) |
|---|---|---|---|---|
| 1984/85 | 2411 | 168 | 103 | 4.3 |
| 1985/86 | 2544 | 216 | 147 | 5.8 |
| 1986/87 | 2446 | 216 | 139 | 5.7 |

[a]Units are KWHr/m$^2$-yr

- A high degree of automated operation was achieved.
- Much instrumentation was unreliable.

### 2.3.6B Performance

The most dramatic result from the system experiments is that the power produced was much lower than predicted. Solar One was the only plant with a net positive output of electricity, and it produced substantially less than the predicted output. Reasons for this disappointing performance include:

- Equipment outages. Most experiments encountered numerous unforeseen problems and incidents that drastically reduced plant availability. The effect varied substantially between the different experiments - from Solar One which achieved reasonably good availability to some experiments which had very little full system operation.
- Low insolation. Insolation was lower than expected for many of the experiments.
- Operational outages. Even when the full system was available, useful energy collection was much less than expected. Contributing to this effect were:
  - Lengthy and often delayed morning startups.
  - An inability to collect energy during intermittent cloudiness and high wind conditions.
  - Losses during restart operations following cloud passage.
- High parasitic power. This effect was highly exaggerated because of the very small size of most of the experiments.

Solar One provided the most extensive collection of performance data, and it is the most reliable base for performance predictions. Performance over the three-year power production phase from August, 1984 through July, 1987 is shown in Table 2.21. The tabulation is in units of KWHr/m$^2$-yr to facilitate comparisons with other systems. The target usually used for good annual performance is a net output of 350 KWHr$_e$/m$^2$-yr which corresponds to a net efficiency of 14 per cent at a site with Barstow insolation, about 2500 KWHr/m$^2$-yr.

It was seen that all of the central receiver systems required a higher than expected level of daily insolation in order to produce any net output of electricity. This daily insolation threshold for positive output was illustrated in Figure 2.13 (Solar One), Figure 2.22 (MSEE) and Figure 2.27 (THEMIS). This threshold effect illustrates a major difference between predicted and measured output from these plants. Such an effect was not predicted for any of the experimental plants nor is it reflected in predictions made for the annual net output of commercial plants.

The following conclusions can be drawn from an evaluation of the performance of these system experiments.

- The performance predicted for the commercial designs of the advanced system concepts (typically 14 per cent annual conversion efficiency from sunlight to net electricity) has not been confirmed experimentally.
- Data available from the system experiments does not rule out the predicted performance experimentally.
- For water/steam systems, the performance of a once-through-to-superheat receiver, as in Solar One, is likely to be higher than a recirculating-boiler receiver due to having lower thermal inertia.
- The performance *potential* of the advanced system concepts is likely to be higher than that of a water/steam system. This is not confirmed experimentally.
- Any analytical prediction of net annual output must be based on a detailed operations analysis.
- Parasitic power requirements remain a significant uncertainty.
- Experimental verification of annual output will be required.

## 2.3.7 Energy Costs

Upper and lower bounds on the energy costs to be expected from the reference central receiver plant will be estimated to provide an assessment of the value of this system concept.

### 2.3.7A Methodology

The bounds to the energy costs will be developed such that: the lower bound represents the lowest credible cost of energy that could be produced by the system if all of the present uncertainties in costs and performance were resolved favorably in a subsequent development program. This will represent the maximum potential of the system. The upper bound to energy costs will represent the cost of energy that could be expected from the mature system using the existing data base; the predicted performance will be based on existing experimental data, and costs will be estimated using contingencies appropriate to the maturity of the technology and the level of detail included in the design (Electric Power Research Institute, 1986). This bound will be considered to represent the high confidence, confirmed energy cost of the concept. It is hoped to estimate these bounds such that the ultimate cost of energy will fall within these bounds with a probability of 90 percent.

The levelized energy costs will be calculated in constant dollars using methods recommended by EPRI (ibid. Appendix A. 1986). Annual operation and maintenance costs will be expressed as equivalent capital costs using the constant dollar discount rate of 6.1% per year (ibid. Appendix A. 1986).

In order to facilitate comparison with alternative concepts, energy costs will also be expressed in terms of the standard solar plant (in $ per watt peak rating) with an annual capacity factor of 25 percent and operation and maintenance costs of 1 cent per kilowatt hour.

**Table 2.22:** Cost Estimate for 200 MW$_e$ Commercial Plant (1st Quarter 1987, $1000)

| Account | Conting.(%) | Cost |
|---|---|---|
| Land | 0 | 2,351 |
| Structures and Improvements | 10 | 4,652 |
| Collector ($75/m$^2$) | 0 | 142,533 |
| Receiver | 15 | 50,655 |
| Thermal Storage | 15 | 39,989 |
| Steam Generation | 15 | 23,639 |
| Electric Power Generation | 10 | 88,142 |
| Master Control | 15 | 2,221 |
| Total Direct Cost | | 354,182 |
| Indirect Cost (15% of Direct) | | 53,127 |
| AFUDC [a] (11.2% of Direct and Indirect) | | 45,619 |
| Total Capital Cost | | 452,928 |

[a]Allowance for Funds Used During Construction

### 2.3.7B Costs

The cost estimates for the commercial plant design from the utility study (Arizona Public Service Company, 1988, Arizona Public Service Company, 1988a, and Pacific Gas and Electric Company, 1988) will be used to establish the bounds on the system costs. The early stage of this system's development and the exploratory nature of most of the system experiments result in significant uncertainties in predicting system costs. The uncertainty for a water/steam system would be less than for the reference system because Solar One data would be more applicable.

The PG&E cost estimates for the 200 MW$_e$ commercial plant (APS results were similar) are given in Table 2.22 together with the cost contingency factors utilized (Pacific Gas and Electric Company, 1988). These values were selected in the utility study as the most likely costs following completion of system development and commercialization; hence, they are speculative.

The total capital cost from Table 2.22 corresponds to $2265/KW$_e$ (with a solar multiple of 1.8) or, in terms of the total reflective area of the heliostat field, $249/m$^2$. The upper and lower bounds for expected system capital cost will be developed in terms of these parameters.

### 2.3.7B.1 Lower Bound Estimate to System Cost

Most of the cost estimate in Table 2.22 appears optimistic considering the low level of detail contained in the conceptual design. The following two accounts appear to be the only credible sources of significant cost savings:

- Reduce the collector system cost from $75/m$^2$ to $50/m$^2$.
- Reduce the indirect costs from 15% to 10%.

Applying these adjustments to the cost estimate in Table 2.22 reduces the total capital cost to $375 million. Thus, the estimate for the lower limit of the central receiver system cost becomes $1876/KW$_e$ (solar multiple of 1.8) and $206/m$^2$ of reflective surface area.

### 2.3.7B.2 Upper Bound Estimate to System Cost

The upper bound to the system capital cost was developed by applying the process contingencies representative of the technology status and a project contingency appropriate to the level of project definition (Electric Power Research Institute, 1986). The following adjustments were made to the costs in Table 2.22:

- The collector subsystem contingency was adjusted from 0% to 100%. This should also cover the case of using glass/metal heliostats in the event that stretched membrane heliostat development is not successful.
- The process contingency for the receiver was adjusted from 15% to 50%. The very small scale of the advanced receivers built to date and the limited application of all design requirements in a conceptual design requires a 50% contingency on the receiver capital cost.
- The thermal storage process contingency was changed from 15% to 25%. A contingency of 15% should be adequate for the tanks, piping and the heat transfer/thermal storage medium, but a much larger contingency (perhaps 50%) would be required for the pumps and valves. Quality control requirements and trace heating will further increase cost contingencies. An overall contingency of 25% appears appropriate.
- The steam generator process contingency was changed from 15% to 40%. The small size of the steam generators built for prototype testing and the limited design definition require a contingency of about 40%.
- The 10% contingency on the electric power generation subsystem was retained. It was assumed that any additional costs required to meet the requirements of cyclic operation would be covered under the project contingency.
- The 15% contingency applied to the master control system was retained. Instrumentation costs, which are more uncertain at this time, are included within each subsystem.
- A project contingency of 25% was added to all direct and indirect costs except for the collector.
- A 50% contingency was added to the annual operations and maintenance cost. This increased it from $5.6 million to $8.4 million. At the lower level of annual efficiency (8%), the O&M cost for the standard plant is $3.6 million per year; this leaves an excess of $4.8 million per year for the high-cost, low-performance case. A discount rate of 6.1% per year (constant dollars) was used to assign an

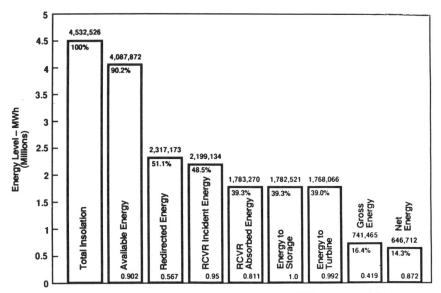

**Figure 2.30:** Annual efficiency predicted for 200 MW$_e$ commercial plant.

equivalent capital cost at plant startup. This yields an additional $65.4 million as the equivalent capital cost representative of the annual O&M costs.

Applying these adjustments to the cost estimates in Table 2.22 results in a total equivalent capital cost of $807 million. Thus, the estimate for the upper bound to system cost becomes $4350/KW$_e$ and $443/m$^2$ of reflective surface.

### 2.3.7C  Performance

The annual efficiency waterfall calculated in the utility study for the selected commercial plant is shown in Figure 2.30. The net annual efficiency of 14.3 per cent net electricity from direct normal insolation is more realistic than many previous studies which predicted values around 20 percent. An upper bound to this annual efficiency will be estimated as the highest conceivable value. A lower bound will be estimated based on results from the system experiments.

### 2.3.7C.1  Upper Bound Estimate for Annual Efficiency

The efficiency prediction in the utility study, used as the basis for this estimate, did not utilize a detailed operations analysis. So, most of the assumptions are likely to be on the optimistic side. The only factors that could conceivably be higher are the heliostat field efficiency, the receiver efficiency and the plant availability. Increasing the field efficiency from 56.7 percent to 60 percent, the receiver efficiency from 81.1 percent to 86 percent and plant availability from 90.2 percent to 95 percent would increase the annual net plant efficiency to 17 percent. This value will be used as the upper bound for the annual efficiency of the central receiver system.

### 2.3.7C.2  Lower Bound Estimate for Annual Efficiency

The best year's annual efficiency for Solar One and the thermal collection efficiency of MSEE during the 28 day power production run were used as the two experimental data sources for an estimate of the lower bound value for annual efficiency. The estimates are as follows:

- Solar One corrected to the calculated heliostat field and power conversion efficiencies of the commercial plant design: 8% annual net efficiency
  - Solar One best year efficiency: 5.8%
  - Field correction factor: 0.981
  - Power conversion correction factor: 1.40
  - 5.8% × 0.981 × 1.40 = 8.0%
- The case above using the calculated parasitics of the commercial plant design: 10.2% annual net efficiency
  - Solar One net to gross output: 0.681
  - Calculated factor for commercial plant design: 0.872
  - 8.0% × 0.872/0.681 = 10.2%
- MSEE thermal collection efficiency with the power conversion efficiency and the parasitic power factor calculated for the commercial plant: 8.0% annual net efficiency
  - MSEE thermal collection efficiency: 22%
  - Calculated conversion efficiency: 0.419
  - Calculated parasitics factor: 0.872
  - 22% × 0.419 × 0.872 = 8.0%

No one of these extrapolations provide the desired confidence level alone. The MSEE-based result utilizes the thermal collection efficiency of a molten salt experiment. However, it assumes that the thermal collection efficiency of an external receiver with a rating of 936 $MW_t$ would be as high over a full day's operation as a cavity receiver with a rating of 5 $MW_t$. This estimate also depends upon achieving the calculated efficiencies for steam generation, power conversion and particularly parasitic power. Against these factors is the expectation that learning will improve plant design and operation and result in higher efficiencies with the mature commercial system.

The Solar One-based estimates assume that operation of the salt system can be as effective as a water/steam system. (The hope is for greater effectiveness.)

The use of estimates based on both Solar One and MSEE provides the confidence level required for selecting 8 percent as the lower bound on the annual efficiency of the central receiver system. Additionally, if this value could not be achieved with the reference salt system, there is a high confidence that the estimated performance would be achievable with a water/steam system.

### 2.3.7D Results

Combining the estimates made above yields the following sets of values for the bounds on central receiver cost and performance.

- Upper Bound: Cost = $206/m², Annual Efficiency = 17%.
- Lower Bound: Cost = $443/m², Annual Efficiency = 8%.

These sets of values form the boundaries for the central receiver line plotted in Figure 2.31. With an annual fixed charge rate of 0.105, annual insolation of 2500 KWHr/m² − yr and the standard allowance of 1 cent/KWHr for operation and maintenance, the range in energy costs is 6–24 cents/KWHr. It can be seen that

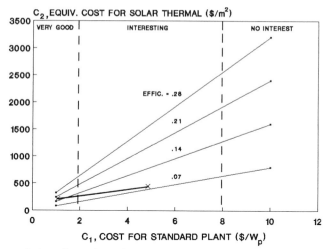

**Figure 2.31:** Central receiver compared with the standard solar plant.

the potential of the central receiver system is very good, although the experimental results obtained to date are inadequate to confirm the most favorable regime of cost and performance.

This good potential justifies consideration of further development of the central receiver concept.

### 2.3.8 Development Requirements

Despite the accomplishments of central receiver development to date, significant technical and cost uncertainties remain. They will be addressed here to identify the further development requirements of this concept.

### 2.3.8A Major Technical Issues

The major technical issues and uncertainties that must be resolved in any central receiver development program are:

- Annual net power production. This is the major uncertainty of this system concept. Its impact is so great that any development program should resolve this issue prior to commitment to full development of the commercial product. Issues which must be addressed include:
  - equipment availability,
  - operational availability and
  - parasitic power usage.
- Trace heater reliability. Either more reliable components or alternative methods of maintaining the system hot during non-operating periods are required.
- Instrumentation reliability. Improved performance and/or reliability are required for the following instrumentation for use with molten salt: flow meters, fluid level indicators and pressure transducers. Also, more reliable flux gages are required for use in receiver control. The measurement of reflected sunlight from the receiver should be considered as an alternative to flux gages.
- Equipment lifetime, particularly the number of cycles and cycle life of the receiver, is uncertain.

- Equipment, including water/steam components, must be developed and/or qualified for cyclic operation.
- Substantial scale up of equipment is required.
  - Receiver rating: 190 to 1
  - Steam generator rating: 170 to 1
  - Near-commercial-size salt pumps and valves are being developed in the ongoing MSS/CTE project.
- Development of the stretched membrane heliostat is required for the lowest cost system.
- Operation and maintenance (O&M) requirements and data are needed to reduce the uncertainty of O&M costs.
- Although not included in the preferred system design selected in the utility study, the potential of the direct absorption receiver warrants further investigation of its prospects, and, if these remain positive, full development.
- A more detailed system design, incorporating results and lessons learned from the preceding system experiments, is required to:
  - identify any additional development requirements and
  - provide the framework for a focused development program and the basis for evaluation of the results.

### 2.3.8B  Reference Development Program

The candidate development program outlined here was formulated to provide a gross measure of requirements, cost and schedule. A more detailed analyses would be required to define the most effective and efficient development program for the central receiver.

It is recommended that a phased development program be followed. The first phase would be directed toward resolving the major technical issues outlined above and reducing the uncertainty band for energy cost to within a factor of 2 from the current value of 4. The timing and funding of the second phase can then be matched better to the expected value of the commercial product and the date that it would become commercially viable. This requires reducing the uncertainties in both cost and performance to within about 40 percent.

### 2.3.8B.1  Development Phase

Activities can be grouped into the three categories of component development, system experiment and system evaluation.

*Component Development*

Carry out the component development required for the commercial, molten salt central receiver system. Utilize the system experiment outlined below to confirm component performance in a full system. Component development should include:

- Salt pumps and valves.
- Stretched membrane heliostat.
- Trace heating.
- Molten salt flow meters, fluid level indicators and pressure transducers.
- Flux gages or alternative(s).
- The direct absorption receiver.

*System Experiment*

Conduct a careful, well-instrumented system experiment to confirm component performance in a full system and to reduce the uncertainty in annual performance to within an uncertainty band of 40 percent (with a 90% confidence level).

A comparison of the performance results from Solar One with all of the other smaller-scale system experiments shows that an experiment at the scale of Solar One (10 MW$_e$) is required in order to reduce the uncertainty in annual performance to within the desired range of 40 percent. Since the testing of Solar One has been completed, it is strongly recommended that the facility be converted to a system experiment of the molten salt central receiver. This option was evaluated and costs and schedules were prepared for a flat plate receiver located to the south of the heliostat field (Sandia National Laboratories, 1985c). The configuration selected for the commercial plant, an external receiver with a surrounding heliostat field, should be used for the system experiment at Solar One.

Operation should be for a minimum of three years to obtain adequate data on component performance, annual power production and operation and maintenance requirements and costs.

*System Evaluation*

A continuing definition of the commercial plant design is required in order to:

- Identify all requirements for component development and the system experiment.
- Guide the component development program and the test plan for the system experiment.
- Interpret the results of the component development program and the system experiment to predict the energy cost from the commercial plant.

The results of this system evaluation will be used either to justify an immediate start on the next phase or to delay it until justified by the predicted economics.

The duration of this development phase is estimated to be about 7 years. A representative schedule is shown in Figure 2.32. A very preliminary cost estimate for the full phase is about $100 million. This is based on the Sandia estimate of about $60 million for the system experiment at Solar One (Sandia National Laboratories, 1985c) with the additional amount for component development, the extended 3-year test program and system evaluation.

## 2.3.8B.2 Scaleup and Demonstration Phase

It is assumed that two plants will be sufficient to provide the transition from the 10 MW$_e$ System experiment to the 200 MW$_e$ commercial plant.

*30–60 MW$_e$ Plant*

An intermediate size plant is employed to confirm equipment and system operation and performance in a scaled-up configuration. Energy costs for the commercial plant should be predictable to within a 40 percent band using the results of this plant's operation.

*Demonstration Plant*

Operation of a full commercial-scale plant will be required prior to booking genuine commercial orders. It could be called either a demonstration plant or a first commercial plant.

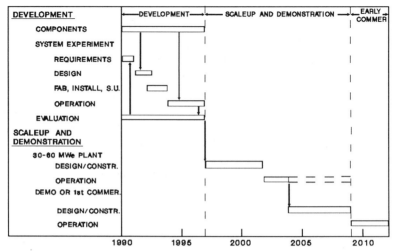

**Figure 2.32:** Recommended central receiver development program.

The duration of the scaleup and demonstration phase, shown in Figure 2.32, would be about 12 years through startup of the demonstration plant.

The costs for this phase would be dependent upon: (1) the technical and scaleup issues remaining after completion of the 10 $MW_e$ system experiment and (2) the expected economics of the plant compared to the energy costs from fossil-fueled plants at that time. In the most optimistic case, the intermediate-scale plant might be omitted, and the full cost of the first commercial plant amortized by energy sales at favorable prices and/or other incentives. This would impose no additional costs to carry the system to commercialization. In the most conservative case, the full cost of the intermediate-scale plant (approximately $200 million) would be charged as well as the excess first plant costs (also estimated at $200 million). This would add $400 million to the cost of commercializing this system. The actual cost would be expected to fall between these two values; so, the 90% confidence bounds for this phase will be taken as 0 and $400 million.

### 2.3.8B.3 Duration and Cost

Combining the foregoing estimates would yield the following approximate range of values for the limits on the development program's duration and cost.

- Development duration: 15–20 years
- Development program cost: $100–500 million

It can be seen in Figure 2.32 that commercialization does not occur until about 2010. No significant installed capacity or savings in fossil fuels could be achieved until 2020 and beyond. Thus, the time required to carry this system to full commercialization is probably comparable to the time when it could be economically viable and needed. Since more than 20 years will be required to complete development and commercialization, continued development of the system should not be dependent upon current fuel prices.

### 2.3.9 Conclusions

- The central receiver system has received substantial development support.
- Major accomplishments have been achieved in component and system development.
- A preferred system concept has been selected which uses molten nitrate salt as the thermal transport and storage medium.
- The system has the potential for energy costs as low as 6 cents/KWhr.
- There is a wide uncertainty band (about 4 to 1) in the expected energy costs.
- The system is only suitable as a large installation ($\geq 100$ MW$_e$).
- Significant development is still required.
- The potential of the system warrants continued development in a phased program.
- The time required to develop a commercial system is comparable to the time when the system may be needed because of increases in fossil fuel costs or environmental issues.

## 2.4 Dish Systems

Dish systems employ a point-focusing reflector to concentrate sunlight at the focal point. The concept shares the capability of very high concentration with the central receiver system. It differs in that the sunlight is concentrated onto receivers distributed throughout the collector field. Thus, the energy must be collected from the field before it can be dispatched to the user. Accordingly, it is often called a distributed receiver system. Potential advantages of this system type for utility-scale electric power production include:

1. Highest possible optical efficiency due to always being pointed directly at the sun.
2. High thermal and power conversion efficiencies due to the high concentration of sunlight.
3. The capability of mounting small heat engine/alternator sets at the focus of the dish. This can provide a modular system and minimum thermal losses upon system cooldown.

Development of dish components and systems is much less extensive than for the central receiver. These developments will be reviewed to assess the status of dish systems.

### 2.4.1 System Concepts

Two distinct types of system concepts have been utilized with dish concentrators. One employs centralized conversion of thermal energy to electricity; the other generates electricity right at the dish.

### 2.4.1A Central Generation

The dish system concept employing central generation of electricity is similar to the central receiver system except that energy is collected from the field in the form of a heated fluid rather than as reflected sunlight. The major elements are:

- The concentrator is a two-axis tracking reflector assembly that collects and concentrates sunlight at the focal point.

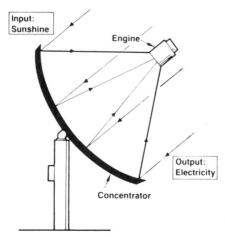

**Figure 2.33:** Principle of dish system with distributed power generation.

- The receiver absorbs the reflected sunlight and converts it into thermal energy.
- The heat transport subsystem circulates a heat transfer fluid through these distributed receivers, extracts the absorbed energy and transports it to a central location.
- The thermal storage subsystem stores the collected energy as sensible heat.
- The power conversion subsystem converts the thermal energy into electricity.

The requirement to circulate throughout the concentrator field limits the choice of heat transport fluid to those that would not freeze upon cooldown to ambient temperature. This in turn limits the maximum temperature of the working fluid, hence power conversion efficiency.

The alternatives of trace heating the thermal collection piping or of employing thermochemical transport have been proposed. Neither have been developed for dish systems and either would add substantial complexity.

### 2.4.1B Distributed Generation

The dish system concept employing distributed generation locates power conversion equipment at each dish concentrator. This allows collection of energy from the field to be in the form of electricity rather than that as a heated fluid. The principle of this concept is illustrated in Figure 2.33.

The concentrator can be the same as in the central generation concept, but it must be matched to the rating of the Power Conversion Unit (PCU).

The power conversion unit is normally a close-coupled, packaged assembly that is mounted on the concentrator. It consists of:

- A receiver, located at the focus of the concentrator, that absorbs the reflected sunlight and heats the working fluid of the heat engine.
- A heat engine which converts the thermal energy into mechanical work.
- An alternator or generator coupled to the heat engine that converts the work into electricity.
- A heat rejection system which rejects the waste heat from the engine to the atmosphere.

**Figure 2.34:** Distributed Receiver Test Facility.

- Controls to match the engine's operation to solar availability.

It should be noted that there is no thermal storage. Any energy storage, if employed, would need to be derived from the electrical output, such as battery storage. This option will not be considered in this assessment. Plants with distributed generation will be evaluated in the sun-following mode; they could be hybridized to operate with fossil fuels during periods without sunshine.

Thermal transport is minimal with this concept; instead, electricity is collected from each engine in the field.

The most striking feature of this system concept is the relatively complex power conversion unit which must be replicated thousands of times throughout the field for a utility-scale power plant. Its capital cost, operating reliability and maintenance are major issues to be evaluated in this system concept assessment.

### 2.4.2 Component Development

The Distributed Receiver Test Facility (DRTF), located adjacent to the CRTF in Albuquerque, New Mexico, is the major center for component testing of dish systems (Cameron, 1985 and Cameron, 1986). It was constructed in 1984 to assume the functions previously provided by the Jet Propulsion Laboratory at their Parabolic Dish Test Site in California's Mojave Desert. The DRTF is operated by Sandia National Laboratories for the United States Department of Energy. It is shown in Figure 2.34.

### 2.4.2A  Dish Concentrators

The dish concentrator is a two-axis tracking assembly that reflects and concentrates sunlight onto the focal point of the dish. It is functionally and physically similar to the heliostat except for the following two differences:

- Pointing Direction. The dish concentrator must always be pointed directly towards the sun during operation.
- Reflector Module Curvature. Since each dish concentrates to its own point focus, the reflector modules must approximate a truncated paraboloid.

Dish concentrators have received much less development support than has been devoted to heliostats. However, many of the lessons learned from the heliostat development program have been applied to the dish. Most notably, the evolution to larger heliostat sizes and the transition from glass/metal to stretched membrane designs are paralleled in the dish development program (Hewett, 1986). Some of the concentrators built and tested are shown in Figure 2.35.

### 2.4.2A.1  First Generation Concentrators

Four different first generation concentrators were built and tested; one was used as a test bed for further development and three were employed in full system experiments. Table 2.23 lists the major characteristics of these concentrators. Two test bed concentrators (Goldberg, 1980) were installed at the Jet Laboratory Parabolic Dish Test Site in 1980 and moved to the DRTF in 1984-1985. The Shenandoah concentrators were installed in Shenandoah, Georgia during 1981 (Kinoshita, 1985). The Sulaibyah system was installed outside Kuwait City, Kuwait in 1981 (Zewen, 1986). The White Cliffs station is located in New South Wales, Australia and was constructed in 1981 (Kaneff, 1983).

The major attention in these designs was directed toward proof-of-principle and performance rather than low cost production. The three system experiments, Shenandoah, Sulaibyah and White Cliffs, employed concentrators with relatively small reflector area. While silvered glass was used as the reflective material in three designs, the Shenandoah dish employed an aluminized plastic film. There was substantial variation in the reflector assembly, structure and mounting among the four designs.

### 2.4.2A.2  Second Generation Concentrators

The next generation of concentrators was designed toward commercial cost goals and volume production. Major features are summarized in Table 2.24. Two of the systems, MDC/USAB (Rogan, 1985) and LEC-460 (Strachan, 1987), were commercial ventures without government financial support. The Vanguard concentrator was built and tested by Advanco as part of the Vanguard project (Washom, 1984). It and the MDC/USAB programs will be described later in this section. The PKI concentrator was built as a prototype for the Small Community Solar Experiment at Osage City, Kansas (Barber, 1985). Evaluation tests of the PKI concentrator at the DRTF found a number of problems of which the most severe was low performance (Cameron and Strachan, 1987). Following these tests, it was replaced by a more advanced concept (the LaJet innovative concentrator) in that project.

Concentrators were generally larger than first generation designs. Among the unique features evaluated in this generation were: 1) a glass/steel laminate mirror

LEC-460

Test Bed Concentrator

PKI-135

LaJet Innovative

**Figure 2.35:** Dish concentrators.

in the MDC/USAB design, 2) stretched membrane reflector assemblies focused by an internal vacuum in the LEC-460 design and 3) a very large mirror array in a venetian blind arrangement in the PKI design. Truss frames were generally used for structural support.

### 2.4.2A.3 Advanced Concentrators

The latest generation of dish concentrators in the United States was developed under the Department of Energy's Innovative Concentrator Program being managed by Sandia National Laboratories, Albuquerque. They are described in Table 2.25; the advanced membrane dish produced by Schlaich and Partner of Stuttgart, Germany is also described in Table 2.25.

**Table 2.23:** First Generation Concentrators

| | Test Bed Concentrator | Shenandoah | Sulaibyah | White Cliffs |
|---|---|---|---|---|
| Manufacturer | E-Systems | Solar Kinetics | Messerschmitt Boelkow-Blohm | Australian National University |
| Type | Parabolic Dish | Parabolic Dish | Parabolic Dish | Parabolic Dish |
| Concentrator Diameter (m) | 11 | 7 | 5 | 5 |
| Reflector Area (m²) | 97.1 | 38.0 | 18.3 | 19.8 |
| Reflector Material | Silvered Glass | Aluminized Polymer (FEK-244) | Silvered Glass | Silvered Glass |
| Reflector Assembly | 224 Mirror Facets Mounted to Foam Glass Substrate | 21 Aluminum Sheet Metal Petals | Reinforced Plastic Dish | Fiberglass Shell (2300 Mirror Tiles) |
| Structure/ Mounting | Parabolic-Shaped Tubes, Alidade Structure | Aluminum Ribs, Steel Hubs, Tripod Mount Structure | Concrete Pedestal | Tubular Steel Frame, Steel Pipe Pedestal |
| Tracking Axes | Elevation-Azimuth (Circular Track) | Polar-Declination | Polar-Declination | Elevation-Azimuth |
| Concentration Ratio | 3600 | 234 | 678 | 1000 |
| Number Built | 2 | 114 | 56 | 14 |

It can be seen in the table that all concentrators have reflective areas larger than 150 m². All three U.S. designs employ a silvered polymer (ECP-300X) as the reflective material. The Acurex and SKI designs employed sheet metal reflective panels which were structurally integrated with the support structure to reduce structural weight (and cost) relative to the second generation designs. The LaJet innovative concentrator employed 95 circular, stretched membrane reflective facets and other improvements over the LEC-460 model. It had nearly four times the reflective area of the earlier unit. The Schlaich and Partner dish was the largest and employed a single stretched membrane focused by internal vacuum.

Test results from this generation of concentrators are extremely limited. One Schlaich and Partner Dish was installed in Stuttgart and two at the solar village outside of Ryadh, Saudi Arabia. Performance was less than predicted. The Acurex

**Table 2.24:** Second Generation Concentrators

| | Vanguard | MDC/USAB | LEC-460 | PKI |
|---|---|---|---|---|
| Manufacturer | Advanco | McDonnell Douglas | LaJet | Power Kinetics |
| Type | Parabolic Dish, Multiple-Facets | Parabolic Dish, Multiple-Facets | Stretched Membranes, Multiple-Facets | Square Dish, Secondary Concentrator |
| Concentrator Diameter (m) | 10.6 | 11.0 | 9.5 | 13.2 |
| Reflector Area (m$^2$) | 91.4 | 91.0 | 43.7 | 135 |
| Reflector Material | Silvered Glass | Silvered Glass | Aluminized Mylar (ECP-91) | Silvered Glass |
| Reflector Assembly | 336 Mirror Facets Bonded to Foam Glass Substrate | 82 Curved Mirrors Bonded to Stamped Steel Back Structure | 24 Reflector Membranes on Aluminum Frames, Shaped by Vacuum | 360 Mirror Facets in Venetian Blind Arrangement |
| Structure/ Mounting | Truss Frame/ Pedestal Encased in Concrete | Trusses/Beams/ Steel Pedestal | Truss Structure/ Concrete Pier | Space Frame/ Secondary Support System/ Boxbeam Track |
| Tracking Axes | Elevation-Azimuth | Elevation-Azimuth | Polar-Declination | Elevation: Wheel and Drag Links Azimuth: Rotation on Track |
| Concentration Ratio | 2700 | 8000 | 24 to 2000 (changed by vacuum) | 700 |
| Number Built | 2 | 6 | 700 | 1 |

dish was installed at the DRTF in December, 1986. It immediately suffered structural failure and was not repaired. The LaJet innovative concentrator was installed at the DRTF in July, 1987. It suffered structural damage due to winds in December, 1987 prior to actual operation. It was scheduled to be repaired, reinforced and tested during 1989. The SKI advanced concentrator was never built as a full unit; only a single gore section was fabricated.

**Table 2.25:** Advanced Concentrators

|  | Acurex Innovative | LaJet Innovative | SKI Advanced | Advanced Membrane |
|---|---|---|---|---|
| Manufacturer | Acurex | LaJet | Solar Kinetics | Schlaich & Partner |
| Type | Parabolic Dish | Stretched Membranes, Multiple Facets | Parabolic Dish | Single Stretched Membrane Dish |
| Concentrator Diameter (m) | 15.0 | 19.6 | 14.0 | 17.0 |
| Reflector Area (m$^2$) | 177 | 164 | 154 | 227 |
| Reflector Material | Silvered Polymer (ECP-300X) | Silvered Polymer (ECP-300X) | Silvered Polymer (ECP-300X) | Silvered Glass Mirror Tiles |
| Reflector Assembly | 60 Panels of Aluminized Steel Bonded to Stamped hat Section Back Sheet | 95 Membranes Mounted on Curved Space frame. Shaped by Vacuum | 30 Curved Aluminum Parabolic Gore Panels. Sandwich with Corrugated Core. | Double Steel Membranes Stretched over Steel Ring. Shaped by Vacuum. Mirror Tiles Bonded to Front Front |
| Structure/ Mounting | Rib Trusses/ Rings/Hub/ Tripod Support | Spaceframe Tripod Assembly/ Anchor Fittings | 6 Radial Arms, 2 Rings, Hub/ Tubular Steel Pedestal | Steel Support Girders on Ring Turntable base. |
| Tracking Axes | Elevation-Azimuth | Polar-Declination | Elevation-Azimuth | Elevation-Azimuth: Rotation on Wheels |
| Concentration Ratio | 2434 | 1100 | 1226 | 600 |
| Number Built | 1 | 1 | 1 Gore | 3 |

It can be seen in Table 2.25 that the ECP-300X silvered polymer was used as the reflector material for the U.S. dishes. However, tests of this polymer at the DRTF during 1987 showed excessive degradation (Cameron, 1988). Current preference for the reflective surface is a sol-gel protected silver deposited on a stainless steel substrate (Mahoney et al., 1987).

Major attention is currently being given to stretched membrane concentrators as most likely to achieve the goals of low production costs and adequate performance. Solar Kinetics, Inc. has designed a single membrane dish and produced a 1.8 meter diameter model membrane assembly. Together with the LaJet innovative concentrator, both single and multiple membrane designs are available for further development and evaluation.

### 2.4.2B Engines for Distributed Generation

The most challenging requirement of the dish system concept with distributed generation is the development of a complete Power Conversion Unit (PCU) on the scale of a single dish (25-50 $KW_e$) with the following characteristics:

- Low Capital Cost (Goal is $300/ $KW_e$ or $7500 for a 25 $KW_e$ unit)
- Low Maintenance Cost (Goal is $10/m$^2$-yr or about $900 per year for a 25 $KW_e$ unit)
- High Conversion Efficiency (Goal is 41% for the PCU and 28% for the system).
- Capability of unattended operation.

Different engine cycles have been investigated for this applicatin with most attention being given to the Stirling cycle (Linker, 1986).

### 2.4.2B.1 Stirling Cycle Engines

The Stirling cycle engine was invented in 1816 by the Reverend Robert Stirling, a Scottish minister.

The Stirling engine has many similarities to the conventional internal combustion engine; it has an engine block with cylinders and pistons. It differs, however, in that heat is supplied externally and continuously to heat a gas (normally hydrogen or helium), which is contained in a completely closed system. The heat energy is converted to mechanical energy by alternating compression and expansion of the confined gas. Heat is added to the working fluid during the expansion stage, and heat is rejected during the compression stage.

Residual heat energy is recycled through a regenerator. The regenerator stores a large portion of the heat of the working gas after expansion and returns the heat to that gas as the gas reverses direction.

The pistons in the Stirling engine have two functions: they move the gas back and forth between the hot and the cold locations (displacement), and they extract mechanical work from the engine. There are two basic classes for the displacer and power piston configuration: kinematic and free-piston.

- The kinematic engine has the pistons attached to an output drive shaft.
- The free-piston engine has displacer and power pistons that are free to move within the engine and are not physically attached to each other.

Since heat is added and removed during the isothermal expansion and compression strokes, the cycle closely approximates an ideal Carnot cycle. Because of the resulting high efficiency and low emissions due to external combustion, substantial development effort has been devoted to the kinematic engine over the last two decades for automotive and auxiliary power applications. First General Motors and then Ford had Stirling engine development programs for automotive propulsion. The Ford program was terminated in 1978 to devote their resources to nearer-term technology. The free-piston engine is under development for space power.

**Figure 2.36:** USAB 4-95 Power Conversion Unit.

*USAB 4-95 Engine*

This is the most fully-developed kinematic Stirling engine suitable for the solar application. It is rated at 40 KW$_e$ at 3000 RPM and 25 KW$_e$ at 1800 RPM (the solar version). It has four cylinders; each has a displacement of 95 cubic centimeters. (Hence the 4-95 designation.) The cylinders are arranged in a square pattern with two crankshafts coupled through gears to a single output drive shaft.

United Stirling AB of Malmo, Sweden began Stirling engine development in 1968. They designed the 4-95 engine for automotive applicatin in 1975 and tested the first model in 1976. Modifications for solar use began in 1980 (Holgerson and Percival, 1982). It employs hydrogen as the working fluid. Heat is added at 700 °C and rejection is by an air-cooled radiator using water as the heat transport fluid. Power control is accomplished by varying the pressure of the working gas within the engine. Thus, the working gas hot and cold temperatures remain constant, and efficiency remains high, at part load.

The solar version was tested by the Jet Propulsion Laboratory on their Test Bed Concentrator during 1981 and 1982 (Nelving, 1983). The 4-95 was subsequently used for the Vanguard and MDC/USAB programs which will be described later. More than fifty 4-95 engines have been built and tested with a cumulative test time of over 80,000 engine-hours (15,000 hours with solar heat) and more than 19,000 hours on a single engine. A picture of the 4-95 power conversion unit is shown in Figure 2.36.

*MTI-ASE Engine*

Beginning in 1978, the USAB 4-95 technology was transferred to the United States for automotive application under funding provided by the U.S. Department of Energy. Mechanical Technology Incorporated (MTI) in Latham, New York began development of the Automotive Stirling Engine (ASE) in 1978 under the direction of the National Aeronautics and Space Administration Lewis Research Center (NASA-LeRC). Seven first-generation automotive engines were produced with the same configuration and much of the technology of the USAB 4-95 engine (designated as R-40 at that time). These engines accumulated over 15,000 hours of test time including 1100 hours powering an automobile (Beremand and Shaltens, 1986). An improved ASE engine (the Mod II) was designed beginning in 1983. The first Mod II engine was tested in January 1986. Major improvements for the automotive application included: (1) a more compact V-configuration which fits into a small engine compartment, (2) a design for low-cost, high-volume production and (3) improved technology (Nightingale, 1986).

*USAB V4-275 Engine*

United Stirling also developed a larger kinematic Stirling engine designated as V4-275. It is similar in technology to the 4-95; it employs four cylinders, each with a swept volume of 275 cubic centimeters. The pistons are arranged in a V-configuration connected to a single drive shaft. The maximum engine rating is 100 $KW_e$; it is derated to 50 $KW_e$ output for solar applications.

This engine was developed for submarine propulsion using an air-dependent (liquid oxygen) combustion system. The solar version was tested with the Schlaich and Partner advanced membrane dish in Stuttgart, West Germany and, as a two-module installation, in Saudi Arabia. Performance results from these tests have not been published.

*STM4-120 Engine*

Another kinematic Stirling engine with potential solar application is being developed by Stirling Thermal Motors of Ann Arbor, Michigan (Meijer, 1987). This engine design was derived from that produced by the Ford Motor Company under license from N. V. Philips. The STM4–120 engine uses a variable swashplate drive to convert the reciprocating motion of the pistons to rotation of the output shaft. The swashplate is mounted on a part of the shaft which is tilted at an angle from the main shaft axis. The swashplate angle variation is accomplished with a rotary actuator. Varying the swashplate angle changes the stroke of the pistons from zero to the maximum stroke. Thus, the device provides control of the output power by variation of the swept volume of the pistons.

The STM4–120 utilizes a pressurized crankcase which eliminates the requirement for a pressure seal on the piston rods. It also employs heat pipes for the thermal input; this allows flexibility in the solar receiver design.

The first engine achieved a 100 hour endurance test and was delivered in 1989 to Sandia National Laboratories in Albuquerque, New Mexico. Development work is proceeding and it is hoped to test an engine on a test bed concentrator at the DRTF during 1990.

### V-160 Engine

A two cylinder kinematic Stirling engine is under development by Stirling Power Systems (SPS) Corporation in Ann Arbor, Michigan. SPS is a former subsidiary of USAB of Sweden. The V-160 employs much of the same technology as the 4-95.

The two cylinder of the V-160 engine are arranged in a V-configuration and have a total swept volume of 160 cubic centimeters. Helium is employed as the working fluid. The maximum engine rating is 15 KW at 3600 RPM.

SPS and Schlaich and Partner are developing a Solar Power Pack$^{TM}$ using the V-160 Stirling engine and a smaller model of the advanced membrane dish.

### MTI Free-Piston Stirling Engine

Mechanical Technology, Inc. (MTI) of Latham, New York has designed a free-piston Stirling engine for terrestrial solar application (Mechanical Technology Incorporated, 1988). The design was derived from their previously-built Space Power Demonstration Engine (SPDE). Because of their extensive experience in Stirling engine development, the NASA Lewis Research Center (LeRC) is managing this project for SNLA under a cooperative interagency agreement with the DOE.

The SPDE was a major milestone in free piston engine development in that it confirmed the feasibility of achieving a power output in the 25 $KW_e$ range from a single (2 module) engine. The largest previous free piston engine built and tested had an output of about 3 KW. The SPDE had a maximum working fluid temperature of 377 °C. Designed for an output of 25 $KW_e$, the maximum achieved was 17 $KW_e$. The mechanical output was near predictions, but the linear alternator efficiency was low due to magnetic flux leakage to the surrounding structure. These losses are understood and can be reduced in subsequent designs, primarily by use of non-ferrous materials.

The MTI design for terrestrial solar application employs a single piston engine with helium as the working fluid. Solar input is provided through a liquid sodium heat pipe solar receiver. A direct output of electricity is achieved via a linear alternator using permanent magnets. The engine/alternator is contained within an all-welded pressure vessel which provides hermetic sealing. Non-contacting seals and hydrodynamic gas bearings (spinning pistons) are employed to reduce wear. It will be capable of unattended operation with automatic startup and shutdown of the system. The engine is designed for 30-year life with a single overhaul at 20 years.

The conceptual design was completed in 1987. It is planned that either this design or the one proposed by STC (described below) will be built and tested at the DRTF during 1992.

### STC Free-Piston Stirling Engine

Stirling Technology Company (STC) of Richland, Washington, in parallel with MTI, designed a free-piston Stirling engine for terrestrial solar use (Stirling Technology Company, 1988). System requirements and project management by LeRC are the same as in the MTI contract.

The STC also employs a single piston design and helium working fluid (with a freon buffer). The working gas is hermetically sealed by using a pair of nested bellow seals. Solar input is via a liquid potassium reflux boiler. The major difference from the MTI design is that the engine output is in the form of hydraulic power which is transmitted to a ground-based, variable displacement, hydraulic pump/motor which drives a rotary alternator.

Automated operation and engine lifetime objectives are the same as for the MTI design.

This conceptual design was also completed in 1987. Either this or the MTI design will be built for testing at the DRTF in 1992.

### 2.4.2B.2 Other Engines for Distributed Generation

Other engine cycles which received significant development support for distributed generation with the dish concentrator include an organic Rankine cycle and the reciprocating steam engine. Development of these two concepts continued into 1988 for use in two small Community Solar Experiments.

*Organic Rankine Cycle Engine (ORC)*

A Rankine cycle engine employing toluene was under development by Barber-Nichols Engineering for use in the 100 $KW_e$ Small Community Solar Experiment at Osage City, Kansas (Barber, 1985, Cameron and Strachan, 1987, and Panda et al., 1985). Their design used a single stage axial flow turbine with a toluene inlet temperature of 400 °C. Four modules of 25 $KW_e$ were to be used for the Osage City experiment.

The first model of this engine was tested with solar input in 1982 for 33 hours. Bearing and inverter problems were found which were corrected in the next model. Development testing of the next ORC engine with an improved control system was conducted at the DRTF during 1986 and 1987. Additional problems of fluid leaks, startup anomalies and controls were encountered. The next engine, was tested at the DRTF in a ground-mounted, electrically-heated facility and achieved over 100 hours of successful operation. Development was terminated in 1988 due to budget limitations.

*ANU Steam Engine*

A reciprocating, uniflow steam engine was built by the Australian National University for a 25 $KW_e$ experimental solar power plant (Kaneff, 1983). This plant utilized 14 of the ANU 5 meter diameter dishes described earlier. It began operation in 1982 at White Cliffs, New South Wales, Australia. The steam engine was built by ANU using parts from two diesel engines with some parts made in their workshop.

The small Community Solar Experiment at Molokai, Hawaii planned to employ a 50 $KW_e$ version of this engine. Five modules were to be built using 306 $m^2$ PKI square dish collectors to produce a total output of 250 $KW_e$. The design included an oil-fired input for superheat and operation during low insolation periods.

An engine efficiency of 20 percent was expected. This project was also terminated in 1988.

## 2.4.3 System Experiments with Central Power Generation

Two system experiments employing central power generation were evaluated: the Shenandoah solar total energy project in Shenandoah, Georgia and the Sulaibyah solar thermal power station at Sulaibyah, Kuwait.

**Figure 2.37:** The STEP Plant at Shenandoah, Georgia.

## 2.4.3A Shenandoah Solar Total Energy Project

The Solar Total Energy Project (STEP) at Shenandoah, Georgia is an industrial solar total energy full system experiment employing dish concentrators (Kinoshita, 1985, Ney and Weidenbach, 1983, Ney, 1984, Fair, 1985, Hicks, 1985, Cummings, 1985, Ney, 1987, Stine and Heckes, 1988, and Georgia Power Company, 1988). It is configured as a hybrid system with a gas-fired heater. The electrical rating is 400 KW$_e$. The project was initiated in 1977; startup occurred in 1982. The plant is still operational at the Georgia Power Company Solar test site in Shenandoah. An aerial view of the plant is shown in Figure 2.37.

The overall U.S. Department of Energy objectives for the Solar Total Energy Project at Shenandoah, Georgia were to:

- Produce engineering and development experience on large-scale solar total energy systems as preparation for subsequent commercial size applications.
- Assess the interaction of solar energy technology with the application environment.
- Narrow the prediction uncertainty of the cost and performance of solar total energy systems.
- Expand solar engineering capability and experience with large-scale hardware systems.
- Disseminate information and results.

  Sponsors of the project were:

- United States Department of Energy

- Georgia Power Company
  The participants in the project were:

- Technical Management:        Sandia National Laboratories.

- Design Team:        General Electric (Valley Forge)
  Lockwood-Greene Architects &
  Engineers
  Scientific Atlanta
  Dow Corning Corporation
  Mechanical Technology, Inc.

- Construction Team:        Dow Corning Corporation
  L.B. Samford, Inc.
  B&W Mechanical Contractors
  Joe North, Inc.
  General Electric (Daytona Beach)
  Solar Kinetics, Inc.

- Site Team:        Georgia Power Co.
  Bleyle Corporation of America.
  Georgia Institute of Technology
  Heery & Heery, Architects &
  Engineers
  Shenandoah Development Co.
  Owens-Corning Fiberglass Corp.
  Westinghouse Electric Corporation

## 2.4.3A.1 Plant Description

The Solar Total Energy Project (STEP) uses the solar energy collected to partially meet electrical, air conditioning, and process steam needs in an industrial application. The energy collection process uses silicon heat transfer fluid (HTF) which is circulated through the receiver tubes of parabolic dish collectors. The solar energy collected is normally supplemented with energy provided by a natural gas-fired heater and then delivered to steam generator heat exchangers to produce superheated steam to drive a conventional turbine/generator set. Steam is extracted from the turbine to provide process steam for knitwear pressing and the low pressure exhaust steam is used in an absorption chiller to produce chilled water for air conditioning. A schematic of the system is shown in Figure 2.38. Major subsystems are described in Table 2.26.

The startup operations for the system normally begins with the fossil fuel heater. This heater is required in order to meet the early morning energy needs of the plant prior to the availability of solar-derived energy. The fossil-fueled heater initially provides all of the thermal energy input to the steam generator. The collector field warm-up is accomplished by circulating the heat transfer fluid through the collector field until it reaches operating temperature. Once the operating temperature is reached, the hot heat transfer fluid is combined with the output from the fossil-fueled heater for delivery to the steam generator. At this point the system is operating

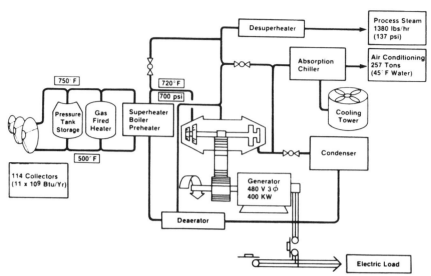

**Figure 2.38:** STEP schematic, Shenandoah, Georgia.

in the hybrid mode. As the solar input to the system increases, the fossil-fueled heater contribution may be reduced to maintain a load-following steam production rate. The plant was originally provided with a thermal storage tank which gave additional operating flexibility; however, this tank was removed from service when it became necessary to increase the fluid pressure above the tank pressure limit.

The plant can also operate in the solar-only mode; however, a size mismatch between the collector field and the thermal utilization subsystem results in continuous part-load operation and poor performance. Consequently, the solar-only operating mode is rarely used.

### 2.4.3A.2 Performance

Extensive performance test data were collected during a 30-day commercial operations test and a 14-day continuous operations test conducted in the summer of 1985. This testing showed the mismatch between the collector field and the thermal utilization system. Throughout the testing, the operations were dominated by the fossil-fueled heater operation with the collector field supplying only 25% and 18% of the total thermal input to the system for the 30-day and 14-day test periods respectively. For solar-only operation, the plant has only been able to generate 135 $KW_e$ gross and about 22 $KW_e$ net under ideal steady state conditions compared to the rated value of 400 $KW_e$.

As a dominantly fossil-fueled system, efficiencies are poor compared to conventional generating methods. The poor efficiencies for fossil-fueled operation are primarily due to low heater efficiencies and the heater location in the heat transfer fluid loop rather than in the water/steam loop. The system as installed and operated is not capable of achieving any economic advantages over conventional energy systems.

**Table 2.26:** Shenandoah Plant

Collector Subsystem

| | |
|---|---|
| Type | Two-axis tracking dishes |
| Number | 114 |
| Total Reflective Area | 4328 m$^2$ (46,562 ft$^2$) |
| Reflectivity (clean) | .82 |
| Reflective material | FEK-244 (aluminized acrylic film) laminated to aluminum |
| Drives | Polar - - 2 jackscrews in tandem (1/10 hp AC motor each) |

Absorber Subsystem

| | |
|---|---|
| Configuration | Conical cavity |
| Heat Transport Fluid | Syltherm-800$^{TM}$ |
| Operating Temperatures | 260 to 363$^o$C (500 to 685$^o$F) |
| Control | Field outlet temperature |

Thermal Storage Subsystem
(Blocked out of system since 1986)

| | |
|---|---|
| Type | Single tank, thermocline |
| Rated Storage Capacity | 1.6 MWhr$_t$ |
| Storage Medium | Syltherm-800$\underline{TM}$ |
| Operating Temperatures | 260 to 363$^o$C (500 to 685$^o$F) |

Steam Generator Subsystem

| | |
|---|---|
| Type | Syltherm$^{TM}$-heated pool-type boiler with preheater and superheater |
| Operating Temperatures (Syltherm$^{TM}$) | 363$^o$C (685$^o$F) inlet; 260$^o$C (500$^o$F) outlet |
| Main Steam Conditions | Solar 346$^o$C (655$^o$F), 48 bar (700 psi) |
| | Fossil 382$^o$C (720$^o$F), 48 bar (700 psi) |

**Table 2.26:** Shenandoah Plant (con't.)

Fossil Fuel Subsystem

| | |
|---|---|
| Function | Heats Syltherm[TM] for hybrid or fossil-only operation |
| Fuel | Natural gas |
| Rating | 2.3 $MW_t$ (8 MBTU/Hr) Output |

Power Conversion Subsystem

| | |
|---|---|
| Prime Mover | Four stage, high-speed (42,450 rpm) extraction steam turbine |
| Inlet Conditions (fossil) | 382°C (720°F), 48.3 bar (700 psi) |
| Outlet Conditions | Extraction at 177°C (350°F) (process steam & feedwater heating) Exhaust to 110°C (230°F) air-cooled condenser. |
| Generator Nameplate Rating | 400 $KW_e$ |
| Load | Local grid |

Controls

| | |
|---|---|
| Type | Initial: Centralized digital; Now: Distributed digital |
| Start-up & Shutdown Sequencing | Some steps automated |
| Console Displays | Custom graphics, touch-screen control |
| Safety Provisions | Defocus field on high temperature |

**2.4.3A.3 Significant Incidents**

Significant incidents encountered during operation of the STEP are summarized in Table 2.27.

**2.4.3A.4 Lessons Learned**

Lessons learned from the STEP project are summarized below:

- In order to minimize operator errors a formal operator training program covering startup and operating procedures is required.
- Partial load efficiencies of fossil fuel supplemental heaters must be carefully evaluated in optimizing the system design. Low fossil fuel heater efficiencies at partial loads significantly reduced overall plant efficiency. Also, the fossil fuel

**Table 2.27**: Significant Issues - STEP

| Incident | Cause | Solution/Repairs |
|---|---|---|
| Low condensate pump pressure delayed all but low level power generation for five months. | Pump rotating wrong direction (arrow on pump housing pointed in wrong direction). | Reversed pump rotation. |
| Inadequate boiler feed pump flow. | Negative NPSH. | Replaced line between deaerator and pump with larger line. |
| Steam generator tube/ tube sheet separation resulted in water-contaminated HTF. | Thermal shock due to operator error pumping cool HTF into hot steam . generator | Repaired by welding tubes to tube sheet. |
| Pneumatic controller, solenoid valve, & instrumentation failures. | Water and chemicals in instrument/air lines. | Repaired/replaced Failed components. |
| Collector field not able to operate at rated temperature. | Inability to balance flow through collectors. | Operated at reduced temperature. |
| Tracking problems damaged collectors. | Inadequate collector controls. | Redesigned control system replacing closed loop control with open loop. |
| Erratic & unreliable temperature measurements. | RTD failures. | Replaced RTD's with thermocouples. |
| Lightning damaged collector electronics. | Inadequate collector grounding (grounding straps bolted to painted surfaces). | Grounding deficiencies corrected. |
| Stalled collectors. | Low collector drive motor starting torque and/or overheated drive motors. | Manually assisted collector or waited for motor to cool if overheated. |
| Collector drive motor failures. | Moisture intrusion. | Drive motor windings sealed. |
| Collector tube leaks. | Corrosion resulting from rain water-soaked insulation. | Replaced carbon steel tube with stainless steel and modified moisture barrier. |
| Collector reflective film delamination. | Moisture. | None implemented. |
| Steam generator steam side tube corrosion. | Oxygen intrusion. | Replaced boiler tubes and maintained water level in boiler above tube bundle to minimize corrosion. |

**Table 2.27:** Significant Issues - STEP (con't.)

| Incident | Cause | Solution/Repairs |
|---|---|---|
| High heat transfer fluid loss rate. | System operated below fluid equilibrium pressure. | Increased system operating pressure. |
| Collector receiver tube blockage. | Heat transfer fluid breakdown and resultant solidification. | Increased system operating pressure, protected against over-temperature, and eliminated water contamination. |

heater should be placed in the steam loop rather than the energy collection loop to improve efficiency.

- Improved reliability is required in auxiliary systems such as the water treatment system.
- Adequate margin must be provided between the nominal operating temperature and the upper limit of the heat transfer fluid. Lack of tight temperature control capability required a significant reduction in the field outlet temperature from the design temperature.
- The centralized control and instrumentation system originally installed for the plant is not recommended for future systems. The distributed control concept is preferred.
- The closed-loop collector tracking control system originally installed was inadequate and required replacement with an open-loop tracking method.
- Exposed components must be qualified for the site environment. Rain and high humidity contributed to collector drive motor and other component failures.
- Plant design documentation, for both hardware and software, must be maintained to facilitate repairs, maintenance and upgrades.
- The economic and performance issues associated with syltherm-800$^{TM}$ heat transfer fluid requires further evaluation before using in future solar systems.
- Additional strategically-located isolation valves are required to simplify maintenance procedures.
- Plant design must more carefully consider the effects of parasitic loads, plant availability, solar availability and integrated system performance.

**Figure 2.39:** Sulaibyah Solar Power Station.

### 2.4.3A.5 Major Accomplishments

A summary of major accomplishments of the Shenandoah STEP project follows:

- Advanced the technology of parabolic dish collectors.
  - Open loop controls.
  - Reflector film test and evaluation techniques to minimize debonding of film from substrate.
- Demonstrated the solar total energy concept.
- Provided engineering and operating data which will allow future system capabilities to be predicted with better accuracy.
- Identified system deficiencies on a small scale prior to the commitment of resources for a large commercial size system.
- Developed techniques to handle Syltherm$^{TM}$ as a heat transfer fluid.
- Reconfigured system for greater efficiency/effectiveness.
- Many lessons were learned which are applicable to other solar systems.
- Showed importance of matching solar collector capacity to system load.

### 2.4.3B Sulaibyah Solar Power Station

The solar power station at Sulaibyah, Kuwait was an agricultural solar total energy full system experiment employing dish concentrators (Zewen, 1986, Zewen et al., 1983, and Moustafa et al., 1983). It was configured as a hybrid system with an oil-fired heater. The electrical rating was 100 KW$_e$. The plant was commissioned in 1981 and operated as a test facility through 1987. The plant is shown in Figure 2.39.

The objectives of the project were to:

1. Verify that the full system, including the user subsystems, works.

2. Obtain performance data for each subsystem and for total system interaction sufficient to verify design and identify uncertainties.
3. Identify areas that need additional development.
4. Document performance results and evaluations.
   Sponsors of the project were:
- Kuwait Ministry for Electricity and Water
- German Federal Ministry for Research and Technology
- Kuwait Institute for Scientific Research (KISR)
- Messerschmitt-Boelkow-Blohm (MBB)
   Major project participants and their roles were:
- Kuwait Institute for Scientific Research, Kuwait City:
  - Project management
  - Power station assembly
  - Civil and structural activities
  - Integrated system operation
  - Equipment refurbishment
- Messerschmitt-Boelkow-Blohm, Munich, Germany:
  - Project management
  - Engineering and design
  - Construction, test and support personnel
  - Collector field, control and instrumentation
- Linde, Cologne, Germany:
  - Organic Rankine power conversion system
  - Waste heat recovery system
- Cillichemie, Heilbronn, Germany:
  - Reverse osmosis desalination unit
- Krupp Industrietechnik, Bremen, Germany:
  - Multi-stage flash desalination unit

### 2.4.3B.1 Plant Description

The Sulaibyah solar power station provided both electrical and thermal energy for agricultural applications including greenhouses, desalination and water pumping for irrigation. The energy collection process used a heat transfer oil which was heated by circulation through the receivers located at the focus of parabolic dish concentrators. The power conversion subsystems used the organic Rankine cycle with toluene as the working fluid. The thermal energy user subsystems employed both the thermal energy collected from the field and that rejected by the power conversion subsystem. The major subsystems of the plant are described in Table 2.28.

### 2.4.3B.2 Performance

Performance data reported for the plant are very limited. Daily electrical output was reported to be about one half of the predicted value. The thermal inertia of the plant was high and the morning startup time was much longer than expected. A minimum threshold insolation value of 400 W/m$^2$ was required for plant operation. It is not possible to assess the annual performance of the plant because of a lack of data on fossil fuel usage and auxiliary power consumption.

**Table 2.28:** Sulaibyah Plant

### Collector Subsystem

| | |
|---|---|
| Type | Parabolic dishes |
| Number | 56 |
| Total Reflective Area | 1025 m$^2$ (10,945 ft$^2$) |
| Drives | Polar-mounted drive system |
| Power Supply | Diesel generator |

### Absorber Subsystem

| | |
|---|---|
| Rating | 739 KW thermal |
| Configuration | Spherical surface (Diameter=17 cm) |
| Heat Transport Fluid | Synthetic oil |
| Operating Temperatures | 235°C (455°F) inlet; 345°C (653°F) outlet |

### Thermal Storage Subsystem

| | |
|---|---|
| Type | 1 tank, thermocline |
| Rated Storage Capacity | 700 KWHr thermal |
| Storage Medium | Synthetic oil |
| Operating Temperatures | 235°C (455°F); 345°C (653°F) |

### Steam Generator Subsystem

| | |
|---|---|
| Type | Tube in shell; separate preheater, evaporator and superheater |
| Rating | 673 KW$_t$ |
| Operating Temperatures (Oil) | 196°C in, 345°C out |
| (Toluene) | 323°C, 15 bar |

### Fossil Fuel Subsystem

| | |
|---|---|
| Fuel | Oil |
| Rating | 750 KWt |
| Operating Temperatures (Oil) | 230°C in, 350°C out |

**Table 2.28:** Sulaibyah Plant (con't.)

Power Conversion Subsystem

| | |
|---|---|
| Prime Mover | Single stage radial in-flow turbine |
| Working Fluid | Toluene |
| Inlet Conditions | 320°C, 15 bar |
| Outlet Conditions | 242°C, 0.4 bar |
| Generator Nameplate Rating | 153 KW$_e$, 415 V, 3 Phase, 50 Hz |
| Load | Local users (no grid) |

Controls

| | |
|---|---|
| Type | Microprocessor controlled |
| Start-up & Shutdown Sequencing | Automatic/semi-automatic |
| Data Acquisition | Separate HP87 System |
| Console Displays | Standard formats |
| Safety Provisions | Independent hard-wired alarm and shutdown system |

### 2.4.3B.3 Significant Incidents

Significant incidents encountered during plant operation are summarized in Table 2.29.

### 2.4.3B.4 Summary

Major accomplishments of the project were:

- Demonstrated the technical feasibility of an agricultural solar total energy system.
- Demonstrated system operation in all modes.
- Achieved high degree of automated operation including start-up and shut-down sequences.
- Identified areas requiring additional development.
- Operated with no measurable degradation of the heat transfer and working media.

### 2.4.3C Other Systems

Two other full systems with central power generation were built and operated. These systems are noted below.

### 2.4.3C.1 Solarplant 1

This system is a privately-owned, privately-financed solar electric power plant with a rating of 4.92 MW$_e$ (net). It is located near Warner Springs, California and

**Table 2.29:** Significant Issues - Sulaibyah

| Incident | Cause | Solution/Repairs |
|---|---|---|
| Absorber coating degradation. | Improper curing. | Modify the process. |
| Absorber leaks. | Improper welding. | Repaired. |
| Failure of electronic tracking system. | Spikes on voltage. | Modified power supply. |
| Failure of main pump of energy conversion system. | Transportation damage. | Replaced. |
| Oil leaks from dish tracking unit. | Incorrect bearings. | Replaced. |
| Frozen oil in collector loop. | Low temperatures. | Provided automatic overnight thermal protection. |
| Slide ring seal failed. | Improper material selection. | Replaced. |
| Damaged absorber. defocusing | No flow. | Automatic system. |
| Failure of sensors. | Short lifetime. | Replaced. |
| Failure of encoders. | Environment. | Replaced. |
| Failure of evaporator. | Design. | Rebuilt. |
| Failure of tracking computer. | Overvoltage. | Replaced Computer. Modified power supply. |

provides power to the San Diego Gas and Electric grid. It was built by the LaJet Energy Company of Abilene, Texas. The plant started operation in January, 1985.

Solarplant 1 employs 700 of the LaJet LEC-460 solar dish concentrators described in Table 2.24. Water/steam is the thermal collection fluid in the field and a steam turbine is used for power generation.

Although sufficient data for an independent assessment was not available, significant equipment and operational problems, particularly related to daily cycling and excessive startup time, were reported (McGlaun, 1987). It was noted that startup after an extended shutdown could take half a day.

During 1987, LaJet formed a joint venture with Cummins Engine to convert Solarplant 1 to a solar/diesel combined cycle. Two 1 $MW_e$ diesel generator sets with exhaust heat recovery were installed. The recovered heat was to be used to keep the steam headers and turbine warm. This may have mitigated some of the operational problems that the system experienced.

### 2.4.3C.2 White Cliffs

This 25 KW$_e$ solar power station is installed at White Cliffs, New South Wales, Australia. It was sponsored by the government of New South Wales and designed and built by the Australian National University. The project was commissioned in 1979 and operation began in 1983.

Fourteen dish concentrators, described in Table 2.23, are installed. Water/steam is the heat collection fluid from the field. A reciprocating steam engine is used for power conversion.

While sufficient data for an independent assessment is not available, a high solar threshold for operation is reported; net electric power is normally not available when the insolation is less than 400 W/m$^2$ (Kaneff, 1984).

### 2.4.3D  Summary of System Experiments with Central Power Generation

The major characteristics of the system experiments with central power generation are summarized in Table 2.30.

Both systems were configured as total energy systems, supplying both electricity and process heat. Both employed single-tank, thermocline storage and both had a fossil-fueled heater for hybrid operation. The Shenandoah plant used a four-stage, high-speed extraction steam turbine. The Sulaibyah plant employed a single-stage, radial inflow turbine with toluene as the working fluid.

Both systems operated in a technically successful manner and produced extensive engineering and operational experience. Performance was significantly below predictions in both cases. Economic viability of the concept was not an objective of these projects and it was not demonstrated.

### 2.4.4 System Experiments with Distributed Power Generation

Two system experiments employing distributed power generation were evaluated. Both employed the USAB 4-95 kinematic Stirling engine and were conducted sequentially. A third experiment using the USAB V4-275 was conducted, but insufficient data is available for evaluation.

### 2.4.4A  Vanguard I

Vanguard I was a 25 KW$_e$ prototype solar dish Stirling engine module located at Rancho Mirage, California (Washom, 1984, Holgersson, 1984, Advanco Corporation, 1984, Washom, 1985, and Droher and Squier, 1986). Operational tests were conducted from February 1984 through July 1985. The module is shown in Figure 2.40.

The initial objectives of the Vanguard project under a cooperative agreement with the U.S. Department of Energy were to:
1. Identify an early market for dish systems and provide an implementation plan for marketing a dish/Stirling system.
2. Determine all functional, performance, cost/schedule and programmatic requirements for the dish/Stirling system.
3. Design, fabricate and test a prototype (Vanguard I) for the selected market.
4. Prepare a plan for commercial product development.

**Table 2.30:** System Experiments with Central Generation

| | Shenandoah | Sulaibyah |
|---|---|---|
| PLANT TYPE | Total Energy (Industrial) | Total Energy (Agriculture) |
| RATING, ELEC./THERMAL: | 400KW$_e$/2000KW$_t$ | 100KW$_e$/400KW$_t$ |
| **CONCENTRATOR** | | |
| Description | See Table 3-1 | See Table 3-1 |
| Number | 114 | 56 |
| Total Area (m$^2$) | 4328 | 1000 |
| **RECEIVER** | | |
| Configuration | Conical Cavity | Spherical Surface |
| Coolant | Syltherm-800 TM | Synthetic Oil |
| Outlet Temp. $^0$C($^0$F) | 399 (750) | 345 (653) |
| **THERMAL STORAGE** | | |
| Type | Single Tank Thermocline | Single Tank Thermocline |
| Media | Syltherm-800 TM | Synthetic Oil |
| Rating (MWHr$_t$) | 1.6 | 0.7 |
| **FOSSIL FUEL SUBSYSTEM** | | |
| Rating (MW$_t$) | 2 | 0.75 |
| Fuel | Gas | Oil |
| **POWER CONVERSION** | | |
| Type | 4-Stage Extraction Turbine | Single-stage Radial Turbine |
| Working Fluid | Steam | Toluene |
| Inlet ($^0$C, bar) | 382, 48.3 | 320, 15 |
| Heat Rejection | Process or Air-Cooled Condenser | Process or Air-Cooled Condenser |

**Figure 2.40:** Vanguard I.

Additional objectives of the Vanguard I extended test and evaluation plan were to:

1. Determine the ability of the Vanguard solar parabolic dish/Stirling engine module to produce electric power over a sustained period.
2. Determine the performance characteristics of system components and of the module by obtaining operating data under a variety of test conditions.
3. Collect and analyze data on installation, operation, maintenance, auxiliary load requirements, and reliability of the module and auxiliary components.
4. Develop performance correlations among module output, solar insolation, and other weather conditions to facilitate future assessment of similar modules at other geographic locations.
5. Provide an independent assessment of the performance of the Vanguard module. Sponsors of the project were:
   - United States Department of Energy
   - Advanco (under cooperative agreement with the DOE)
   - Southern California Edison Company
   - Electric Power Research Institute (extended test and evaluation) Major participants and their roles were:
   - Advanco Corporation: Team leader and concentrator subsystem.
   - United Stirling, AB (USAB): Power conversion system.
   - Rockwell, Energy Systems Group: System integrator.
   - Electrospace Systems, Inc.: Solar tracking hardware and software.
   - Onan Corporation: Generator and electrical equipment.

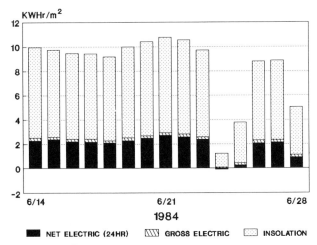

**Figure 2.41:** Vanguard I power production test run.

- Winsmith (Division of UMC Industries) and Sumitomo: Dish gear drive.
- Southern California Edison Co.: Utility design interface.
- Georgia Institute of Technology: Solar optics.
- Modern Alloys, Inc.: Concentrator foundation and installation.
- Energy Technology Engineering Center: Test plan, data evaluation and final project report.

### 2.4.4A.1 Description

The Vanguard I module used a parabolic dish to concentrate solar energy onto the receiver of a Stirling engine mounted at the focus of the dish. The Stirling engine was coupled to an electric generator which supplied power to the Southern California Edison Company's grid. The major subsystems of the module are described in Table 2.31.

The dish Stirling system is rather unique among the solar thermal power generation systems in that it has one of the shortest startup times from the time the module starts to track the sun to the time that useful power is produced. This is because the dish Stirling system is closed-coupled and has very little thermal inertia (no long runs of piping or thermal storage systems to heat up each day). The full startup sequence in the morning is completed in minutes.

Throughout the day, the dish tracks the sun automatically. The hydrogen gas pressure within the engine/receiver is varied to maintain a constant receiver temperature as the insolation varies. The grid supplies the excitation of the induction generator. When the generator speed slightly exceeds synchronous speed (1800 RPM), power is supplied to the grid.

Temporary shadowing of the dish by a cloud represents a power transient in which the input is suddenly terminated and then suddenly resumed. During cloud passage, the dish continues to track the sun. When insolation returns, the engine temperature and speed increases automatically and power production is resumed.

**Table 2.31:** Vanguard I Module

Collector Subsystem

| | |
|---|---|
| Type | Parabolic Dish Concentrator |
| Number | One |
| Total Reflective Area | 86.7 m$^2$ |
| Mirror Modules | 336 back-silvered fusion glass mirrors cold-sagged and bonded to a spherically ground foam glass substrate. |
| Drives | Exocentric Gimbal (Azimuth & 45$^{\circ}$) |
| Power Supply | Commercial Grid, back up generator |

Receiver Subsystem

| | |
|---|---|
| Rating | 73.1 KW$_t$ |
| Configuration | Cavity, circular aperture diameter: 20 cm (7.9 in) |
| Heat Transport Fluid | Hydrogen |
| Operating Temperatures | Gas: 700$^{\circ}$C (1292$^{\circ}$F); |
| | Max Metal: 750$^{\circ}$C (1382$^{\circ}$F) |
| Control | Outlet Temperature by controlling hydrogen pressure. |

Power Conversion Subsystem

| | |
|---|---|
| Prime Mover | USAB Model 4-95 MKII Stirling Engine |
| Speed | 1800 RPM |
| Inlet Conditions | 700$^{\circ}$C (1292$^{\circ}$F); Max Pressure 20 MP$_a$ (2900 psi) |
| Heat Rejection | Air-cooled water radiator |
| Generator Type | Induction |
| Nameplate Rating | 30.0 KW$_e$, 480 V, 3 Phase, 60 Hz |
| Engine Starter | Induction motor |
| Load | Utility grid |

**Table 2.31:** Vanguard I Module (con't.)

| | |
|---|---|
| Maximum Net Efficiency: | 29.4% |
| Average 24 Hr Net Efficiency: (15 day power production test) | 22.7% |
| Average On-Sun Net Efficiency: (18 mo., operating days) | 22.8% |
| Average 24 Hr Net Efficiency: (18 mo., operating days) | 18.5% |
| Average 24 Hr Net Efficiency: (18 mo., all days) | 9.7% |
| Equipment Availability (18 mo): | 72.0% |

**Table 2.32:** Vanguard Measured Performance (18 month test program)

Controls

| Type | |
|---|---|
| System: | Supervisory control system (Capability for 32 modules) |
| Power Conversion Unit: | Dedicated microprocessor |
| Start-up & Shutdown Sequencing | Automatic |
| Data Acquisition | Autodata 9 DAS, Keithley DAS |
| Console Displays | Standard video display terminal |
| Safety Provisions | Alarms, emergency slew-off, water-cooled aperture plate |

**2.4.4A.2 Performance**

Performance of the Vanguard I module was outstanding. It set numerous records for solar to net electricity conversion efficiency. These are included in Table 2.32 which summarizes performance throughout the 18-month test program. Performance during a 15-day power production test run are shown in Figure 2.41.

This high performance is mainly due to the module's short startup time and rapid response to insolation transients. High conversion efficiency, good part load performance and low parasitics all contribute to this favorable performance.

This performance is all the more impressive in that it was achieved by a first prototype unit during its first 18 months in the field.

**2.4.4A.3 Maintenance**

Engine maintenance requirements and costs are a major uncertainty of this system concept. All significant incidents and maintenance conducted during the

18-month test program are reported in Droher and Squier, 1986. While useful in identifying potential problem areas, the data are insufficient to make any quantitative evaluation of maintenance requirements. A great deal more data from a larger number of engines operating over a longer period of time will be required to make a maintenance assessment with any confidence. The USAB projections for the Mean Time Before Failure (MTBF) for the engine components are:

| Engine Component | MTBF (hrs operation) |
|---|---|
| Piston rings | 4,000 |
| O-rings | 4,000 |
| Seals | 8,000 |
| Check valve | 8,000 |
| Cavity | 16,000 |
| Heater | 16,000 |
| Regenerator | 16,000 |
| Cooler | 16,000 |
| Drive Unit | 16,000 |
| Cylinder liner | 16,000 |
| Piston/piston rod/dome | 16,000 |
| Auxiliary equipment | 16,000 |
| Generator | 16,000 |

**2.4.4A.4  Significant Incidents**

Significant incidents encountered during Vanguard I operation are summarized in Table 2.33.

**2.4.4A.5  Summary**

This was a highly successful project which produced a great deal of useful information. It established the potential viability of the dish Stirling system concept for solar thermal electric power production.

A summary of major accomplishments of the Vanguard project follows:

- The project achieved all of its specific technical objectives.
- Demonstrated outstanding performance. Set efficiency records for conversion of solar energy to electricity.
- Demonstrated unattended operation.
- Developed/demonstrated supervisory control for the engine and dish. Designed control system for 32 modules.
- Developed a new dish concentrator.
- Conducted an effective test program.
- Developed useful data on installation, operation and maintenance.

**Table 2.33:** Significant Incidents - Vanguard I

| Incident | Cause | Solution/Repairs |
|---|---|---|
| The movement of the concentrator during gravity slew was sluggish. | The weight of the gravity slew mechanism was too little. | Installed a relay timer on the skewed axis motor to isolate it from the controller. Increased weight. |
| Metal tubes enclosing the power & instrument cables broke. | Inadequate design. | An open heavy-duty cable track was installed. |
| Movement in the gimble reached its maximum travel position which resulted in early shutdown. | The power track support tube hit the pedestal tongue syncho shaft. | The pedestal tongue was notched; the power track and limit switch were relocated. |
| Mirror delamination. | Vibration from air compressor & power conversion unit. | Air compressor was placed on snubbers. No solution to the power conversion vibration. |
| High differential temps. between quads. | Broken check valve. | Check valve was replaced. |
| Several thermocouples failed. | They were not adequately protected. | The thermocouples and sleeves were protected with kaowool insulation. |
| Receiver failed. | Receiver tubes were painted with black paint which insulated the thermocouples and caused a false low-temperature reading. The controller then allowed the pressure and temperature in the tubes to rise which blew out a chunk of tube material. | Receiver was replaced. |
| Temperature in quadrant 4 was high. | Check valve failure. | Overhauled engine and replaced seals. |
| Recurring failures of the oil pump drive peg. | Tolerances between the drive peg and front cover of the oil pump were not within specs. | A new design of drive peg and a new oil pump and front cover were installed. |
| Excessive engine noise. | Wear on unhardened gears between engine and alternator. | Generator/crankshaft gears were replaced with hardened gears. |

**Figure 2.42:** MDC/USAB dish Stirling module.

- Developed preliminary cost estimates and conducted an economic evaluation.
- Identified and recommended further development requirements.

A commercial 30 MW$_e$ installation was being planned as a follow-on to Vanguard I. However, during the course of this project, USAB entered into an exclusive joint venture agreement with McDonnell Douglas Corporation for dish Stirling system development and commercialization. As a result, no follow-on activities to Vanguard I by the Advanco organization were possible.

### 2.4.4B MDC/USAB Dish Stirling System

The MDC/USAB dish Stirling system was a 25 KW$_e$ module employing the USAB 4-95 Stirling engine (Rogan, 1985, Hallet et al., 1985, Coleman and Raetz, 1986, and McDonnell Douglas Astronautics Company, 1986). A picture of a MDC/USAB dish Stirling module is shown in Figure 2.42.

The MDC/USAB project started in November 1983 and terminated June 1986. Six complete modules were built, and hardware was produced for two additional modules, extra test engines and spares. Three modules were tested at the McDonnell

Douglas solar test site in Huntington Beach, California starting in December 1984. Single modules were installed in 1985 at the Southern California Edison Solar One site and at the Georgia Power Company's Shenandoah solar test site. These companies continued operation of the modules through 1988. The system design and rights and much of the hardware was bought by the Southern California Edison (SCE) Company in 1987. SCE conducted additional testing of the system at a reduced scope until late 1988.

The MDC/USAB dish Stirling module was developed under a joint venture agreement between McDonnell Douglas Corporation (MDC) and United Stirling AB (USAB) of Malmo, Sweden. The engine had previously been converted and further developed for the solar application under United States Department of Energy support through the Jet Propulsion Laboratory's solar dish program and the Vanguard project. There was no U.S. government involvement or support of the MDC/USAB project. The objective of the joint venture was to:

- Develop and commercialize a dish Stirling solar power plant using the USAB 4-95 Stirling engine.

  The major participants and their roles were:
- McDonnell Douglas Corporation (MDC): Prime, concentrator development and manufacture, module integration, installation, marketing and commercialization.
- United Stirling, AB (USAB): Power conversion unit development and manufacture.
- Southern California Edison Co.: Purchased and operated one module, purchased system rights in 1986 and remaining hardware in 1987 from MDC.
- Georgia Power Company: Purchased and operated one module.

### 2.4.4B.1 Description

The MDC/USAB dish Stirling module used a parabolic dish concentrator to power a Stirling engine mounted at the focus of the dish. The concentrator was designed for low-cost under mass production. The engine was the same USAB 4-95 unit that was used in the Vanguard project. Major subsystems of the MDC/USAB system are described in Table 2.34.

A cutaway diagram of the USAB 4-95 Stirling engine, with the major components indicated, is shown in Figure 2.43. A picture of the power conversion unit containing the receiver, engine, heat rejection system, generator and controls was shown in Figure 2.36.

Operation of the MDC/USAB dish Stirling module was virtually the same as for the Vanguard I system described earlier except that no starter motor was used. Power from the grid turned the generator as an induction motor. As the Stirling engine warmed up it powered the motor/generator; when 1800 RPM was passed, it delivered positive power to the grid.

Startup time and transient response were similar to Vanguard I.

### 2.4.4B.2 Project Results

Because this was a private commercial venture, technical and cost data were not published. This limits the value of the project for evaluation of the technology. Excellent performance was achieved for this module in tests at the MDC solar test site at Huntington Beach, California and at the SCE Solar One Test Center at Daggett, California. A full-day efficiency waterfall chart from a Huntington Beach

**Table 2.34:** MDC/USAB Dish Stirling System

Collector Subsystem

| | |
|---|---|
| Type | Parabolic Dish Concentrator |
| Number | One per module |
| Total Projected Area | 90 m$^2$ |
| Reflector | 82 mirror facets with back-silvered thin glass bonded to stamped aluminized steel. |
| Drives | Azimuth: Harmonic drive with helicon gear |
| | Elevation: Ball screw jack with helicon hear |

Receiver Subsystem

| | |
|---|---|
| Rating | 76 KW$_t$ |
| Configuration | Cavity, circular aperture diameter of 20 cm (7.9 in) |
| Heat Transport Fluid | Hydrogen |
| Operating Temperatures | Hydrogen: 700$^o$C (1292$^o$F); |
| | Max metal: 750$^o$C (1382$^o$F) |
| Pressure | 40 to 200 bars |
| Control | Outlet temperature by controlling hydrogen pressure. |

Power Conversion Subsystem

| | |
|---|---|
| Prime Mover | USAB Model 4-95 MKII Stirling Engine |
| Speed | 1800 RPM |
| Inlet Conditions | 700$^o$C (1292$^o$F); Max Pressure 20 MP$_a$ (2900 psi) |
| Heat Rejection | Air-cooled water radiator |
| Generator Type | Induction |
| Nameplate Rating | 30.0 KW$_e$, 480 V, 3 Phase, 60 Hz |
| Load | Utility grid |

**Table 2.34:** MDC/USAB Dish Stirling System (con't.)

<u>Controls</u>

| | |
|---|---|
| Type System: | Supervisory control system |
| Power Conversion Unit: | Dedicated microprocessor |
| Start-up & Shutdown Sequencing | Automatic |
| Console Displays | CRT |
| Safety Provisions | Emergency slew-off |

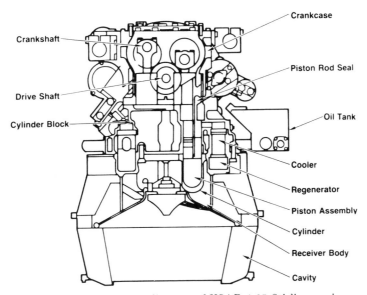

**Figure 2.43:** Cutaway diagram of USAB 4-95 Stirling engine.

test and an input-output chart for the unit at Solar One (both from unpublished data) are shown in Figure 2.44 and Figure 2.45 respectively. The low threshold of daily insolation for positive power output, seen in Figure 2.45, is most impressive. Insufficient test data exists to evaluate maintenance requirements and costs, but routine, unattended operation was maintained throughout the SCE test program.

A summary of the major accomplishments of the project follows:

- The two private companies made a major commitment to the development of solar electric power.
- They initiated a substantial development program.
- A new dish concentrator designed for low-cost mass production was developed.
- Some improvements were made to the USAB 4-95 Stirling engine PCU.

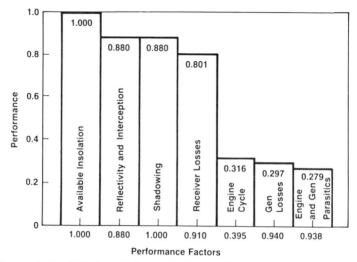

**Figure 2.44:** Full day efficiency for MDC/USAB dish Stirling module.

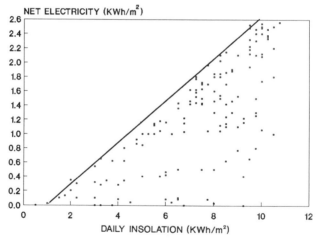

**Figure 2.45:** MDC/USAB dish Stirling daily input-output (January-June 1986 at Solar One Site).

- Significant engine test time was accrued.
- The modules operated unattended.
- A field test program to operate 8 modules for 33 months, 2 at Huntington Beach and 6 at different utilities, was started.
- Utility companies were involved in the tests and (planned) evaluation.
- Detailed production cost estimates were developed based on factory layouts, production processes, materials, man-hours, etc. '

The project was terminated in June 1986 because of the delay perceived in the timing of a commercial market caused by the decline in world oil prices.

### 2.4.5 Reference System Selection

#### 2.4.5A Central versus Distributed Power Generation
Selection of the reference dish system concept involves the successive selection between central and distributed power generation, and then, selection of the specific system concept.

On the basis of the system experiments and system studies evaluated, distributed power generation was selected over central power generation for the following reasons:

- The system experiments with distributed power generation achieved substantially higher energy production per unit concentrator area than the potential of the systems with central generation.
- High thermal inertia, with a daily cooldown and warmup of the piping system and distributed receivers, is a serious limit on the annual energy collection capability of a system with central generation.
- The potential of high temperatures (hence high conversion efficiency) of the dish system is limited by the thermal collection fluids available with central generation. Fluids that don't freeze at ambient temperatures have either of the following limitations:
  - Moderate maximum temperature capability or
  - High pressure and/or change of phase (with water).
- The distributed power generation system allows unattended operation.
- The distributed power generation system is modular with a wider range of application.

This section is consistent with that of the PNL system study (William et al., 1987).

#### 2.4.5B Reference Dish System Design Concept
The Dish Stirling system was selected as the reference dish system design concept for the following reasons:

- High performance was demonstrated in the Vanguard and MDC/USAB programs. This performance is approximately a factor of two higher than for the alternatives considered.
- Unattended operation has been demonstrated.
- Rapid warmup/startup and good response to cloud transients have been achieved.
- Part load efficiency is high.
- The engine and power conversion unit (USAB 4-95 unit) are more fully developed than the alternatives.

The only system with sufficient development and data to evaluate with any confidence is the MDC/USAB system design (or the earlier Vanguard one). Thus, this system will be selected as the reference. It is described in Table 2.35. It is expected that subsequent improvements will be made in Stirling engines and/or concentrators; the reference concept would be modified accordingly.

**Table 2.35**: Reference Dish Stirling System Concept

| | |
|---|---|
| RATING (per module) Net | 25 KW$_e$ |
| CONCENTRATOR DESCRIPTION | See Table 3-2 (MDC/USAB) |
| RECEIVER | |
| Configuration | Cavity -- Directly Heat Engine Working Fluid |
| Heat Transfer Fluid | Hydrogen |
| THERMAL STORAGE | None |
| POWER CONVERSION | |
| Type | USAB 4-95 Kinematic Stirling Engine with an Electric Generator |
| Working Fluid | Hydrogen |
| Power Level Control | Hydrogen Gas Pressure |
| Inlet Temperature | 700°C (1292°F) |
| Heat Rejection | To Air through Water-Cooled Radiator |

## 2.4.6  Energy Costs

Upper and lower bounds on the energy costs to be expected from the reference dish Stirling system will be estimated employing the same methodology described for the central receiver system in Section 2.3.7A.

### 2.4.6A  MDC/USAB Commercialization Study

The McDonnell Douglas Corporation (MDC) conducted a comprehensive evaluation of solar technology during 1982 with the objective of identifying the system(s), if any, with good commercial prospects. They selected the dish Stirling system with a preference for the USAB 4-95 engine which was most fully developed. In 1983 McDonnell Douglas and United Stirling AB (USAB) formed a joint venture to commercialize the dish Stirling system. MDC assumed responsibility for commercialiation of the system as well as development of the dish concentrator based on their large area glass/metal heliostat technology. USAB was responsible for completion of the 4-95 engine development and the provision of Power Conversion Units (PCUs) for field testing and demonstration. The program was a fully commercial undertaking between the two companies. There was no involvement or funding from the U.S. Department of Energy or their national laboratories.

The commercialization studies continued throughout the period of product development. Substantial effort was devoted to determining the actual costs to be expected during commercial production and to controlling these costs in the design

**Table 2.36:** Dish Stirling System Cost Estimates

| | DOE Long Term Goal | PNL Study Commercial | MDC Study Commercial | MDC Study Early |
|---|---|---|---|---|
| Concentrator, ($/m²) | 130 | 120 | 120 | 240 |
| Power Conversion Unit, $/m² | 152 | 66 | 110 | 220 |
| ($/KW$_e$) | (555) | (240) | (400) | (800) |
| Balance of System, ($/m²) | 27 | 48 | 45 | 90 |
| | ---- | ---- | ---- | ---- |
| Total Capital, ($/m²) | 309 | 234 | 275 | 550 |
| Annual O&M, ($/m²-yr) | 10 | 20 | NA | NA |

phase. Detailed manufacturing plans were developed in parallel with the design. Production costs were used in conjunction with engineering requirements in establishing the product design. The A. T. Kearney Company provided the factory and equipment requirements, manufacturing processes, material usage and costs and manpower requirements for producing the dish concentrator. They provided direct input to the product design through a project office at the same location as the MDC engineering, design and prototype fabrication operations. Similarly, the Volvo Company in Sweden provided detailed production cost estimates for the 4-95 Stirling engine.

The development plan of MDC/USAB included field testing of eight modules for a minimum of 33 months to obtain operation and performance data and to help determine maintenance requirements and costs.

These studies provide a basis for commercial cost projections (McDonnell Douglas Astronautics Company, 1986). The initial field testing provided useful operation and performance data but was inadequate to assess maintenance costs.

### 2.4.6B  Costs

The MDC/USAB system study described above and the PNL study (William et al., 1987) described earlier will be used as the primary source of cost data. Their results are shown in Table 2.36 together with the Department of Energy's long term goals (U.S. Department of Energy, 1984). The DOE and PNL costs are in 1984 dollars while the MDC costs are in 1985 dollars.

It is not surprising that the PNL and MDC full commercial results are close. Both used the same system and many of the same data sources. The major difference between the two estimates is the Power Conversion Unit. PNL used an estimate provided directly from United Stirling.

MDC used a more detailed estimate including input from the Volvo production cost analysis. The MDC cost also included their installation and integration of the

PCU with the concentrator and their markup. A production cost estimate by Deere and Company for the ASE Mod II engine, will be used as the basis for the lower bound on Stirling engine costs (Beremand and Slaby, 1988).

These estimates will be the basis for the upper and lower bounds on the system concept's costs.

### 2.4.6B.1 Lower Bound Estimate to System Cost

A 25 KW$_e$ module size with a 28% annual net conversion efficiency and an effective reflector area of 78.2 m$^2$ per module will be used as the model for estimating the lower bound to the system's cost. The estimates, and the bases, for the major cost elements per module are:

*Dish Concentrator cost: $7820*

A concentrator cost of $100/m$^2$ was used to represent the potential lower limit costs of stretched membrane concentrators. (This corresponds to the $50/m$^2$ assumed for the lower limit of stretched membrane heliostats in Section 2.3.)

*Power Conversion Unit Cost: $6000*

The Deere and Company Cost estimate for the ASE Mode II engine was about $3000. The balance of the power conversion unit should add no more than another $3000 to this value. This is also the same value used in the PNL study.

*Balance of System Cost: $3520*

The balance of system cost for the full commercial unit from the MDC study was $45/m$^2$ which yields $3520/module.

*Capital Cost Equivalent of the Differential Annual Operation and Maintenance Cost: $3200*

The DOE goal of $10/m$^2$ per year for O&M was used as the lowest potential value for this cost element; this results in $782/yr. The 1 cent/KW$_e$Hr allowed for the standard plant is $547/yr. The difference of $235/yr was discounted to present value at 6.1%/yr.

The total equivalent capital cost is $20,540 per module or $263/m$^2$. This was taken as the lower bound of system cost representing the potential of the concept.

### 2.4.6B.2 Upper Bound Estimate to System Cost

The same module size and rating were used for the upper bound to system cost, but an annual efficiency of 16% was employed for the standard plant O&M costs which is used as the offset for the differential O&M cost. Estimates per module for the major elements and their bases are:

*Dish Concentrator Cost: $15,640*

A concentrator cost of $200/m$^2$ was used. This is a 100% contingency on the cost used for the lower bound and is consistent with the early stage of development.

*Power Conversion Unit Cost: $12,000*

This estimate was based on applying a 20% contingency to the costs for the full commercial unit in the MDC study. It also represents a 100% contingency on the value used as the lower bound.

*Balance of System Cost: $7040*

This estimate corresponds to the early commercial system cost estimate in the
MDC study. It represents a 100% contingency on the value chosen for the lower
bound for this cost element.

*Capital Cost Equivalent of the Differential*

*Annual Operation and Maintenance Cost: $38,310*

A value of $40/m$^2$ was used for this estimate. It employs a 100% contingency on
the PNL cost estimate and is four times the value chosen as the lower bound. This
is necessary because of the limited data available to support this cost estimate. The
O&M cost at $40/m$^2$ is $3128 per year. In this case the offset for the standard plant
(1 cent/KW$_e$Hr) is $313 per year. The difference is $2815 per year. Discounting to
present value at 6.1%/yr yields $38,310.

The total equivalent capital cost is $72,990 per module or $933/m$^2$. This will
be taken as the upper (high confidence) bound to the systems' cost. It should be
noted that the annual operation and maintenance cost represents more than 50% of
the equivalent capital cost. This factor is so large because: (1) the data base is so
sparse that a large contingency is required and (2) the cost is expected to increase
at the rate of inflation, so the constant dollar discount rate was applied to represent
the present value of this annual expenditure.

### 2.4.6C  Performance

High annual performance is the major strength demonstrated by this concept in
the system experiments evaluated. Performance data from the 18 month Vanguard
test program were summarized in Table 2.32; the daily output over a 15 day power
production test run was shown in Figure 2.41.

Results for the MDC/USAB units were similar, but the sustained output was
significantly higher. The average 24 hour net efficiency for the SCE unit installed at
Daggett, California was 14.8% for 6 months of unattended operation from January
through June, 1986. The input-output plot of the daily results over the six-month
period was shown in Figure 2.45.

These results are particularly impressive when it is remembered that these were
early, first-of-their-kind units during their first phases of operation. The system
immaturity can be seen on the input-output plot for the SCE unit (Figure 2.45).
The wide dispersal of daily output under the envelope line illustrates the module's
technical immaturity. Daily output for a more mature module would cluster much
closer to the envelope line (with a substantial increase in the average efficiency from
the 14.8 percent measured here).

The very moderate parasitic power requirements can be seen by comparing the
net and gross electric output in Figure 2.41. The low daily insolation threshold for
system operation can be seen in Figure 2.45.

These results will be used to estimate upper and lower bounds on the net annual
efficiency for this system concept.

### 2.4.6C.1  Estimate of Upper Bound of Annual Efficiency

A mature system with its daily output along the envelope line of Figure 2.45
would yield about 25 percent net annual efficiency. Increase in the maximum fluid
temperature to 800 °C and other efficiency improvements could increase this value

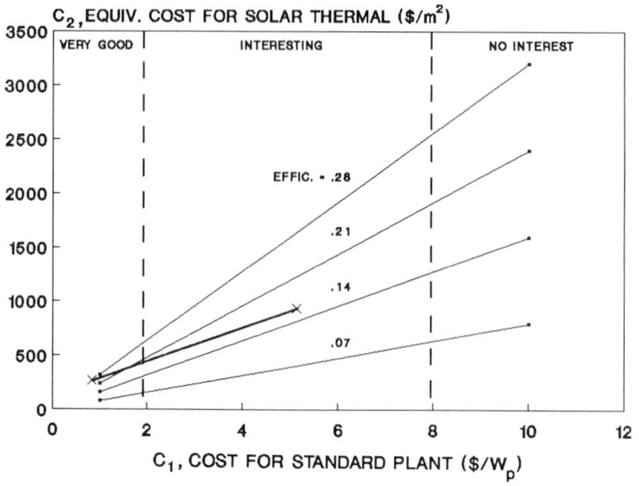

**Figure 2.46:** Dish Stirling system compared with the standard solar plant.

to about 30 percent. Application of a 93 percent availability factor would reduce this to 28 percent. This value will be used as the estimated upper bound for net annual efficiency.

### 2.4.6C.2 Estimate of Lower Bound of Annual Efficiency

There is high confidence that the 14.8 percent efficiency achieved over a six-month test period could be increased to at leat 16 percent with system development and maturity. This value will be used as the estimated lower bound for net annual efficiency.

### 2.4.6D Results

Combining the estimates made above yields the following sets of values for the bounds on the Dish Stirling system's cost and performance:

- Upper Bound: Cost = $263/m$^2$, Annual efficiency = 28%.
- Lower Bound: Cost = $933/m$^2$, Annual efficiency = 16%.

These sets of values form the limits for the Dish Stirling line plotted in Figure 2.46. With an annual fixed charge rate of 0.105, annual insolation of 2500 KWHr/m$^2$-yr and adding 1 cent/KWHr for the standard plant's O&M cost, the range of energy costs is 5–25 cents/KWHr. It can be seen that the potential of this sytem is very good. At the same time, the uncertainty (with 90% confidence) is very large (about a factor of five in energy cost). The very good potential justifies consideration of further development of the Dish Stirling system.

### 2.4.7 Development Requirements

The Dish Stirling system is at a very early stage of development. Although the integrated MDC/USAB system was developed and subjected to some field testing, substantial questions remain as to its ultimate economics. Alternative concentrator and Stirling engine designs have been identified which might have better prospects for achieving the system's potential but they are at a much earlier stage of development.

### 2.4.7A  Technology Status

The major accomplishments of the previous development programs and the current technology status will be summarized here for the MDC/USAB system, concentrators and engines.

### 2.4.7A.1  MDC/USAB System

Achievements from this program, and the precursor Vanguard project provide most of the data supporting the potential of this system concept. Major accomplishments from these programs were:
- High performance was demonstrated
    - Peak net efficiency of over 29% was measured.
    - Sustained net efficiency of over 14% for a 6 month period was achieved.
- Unattended operation was demonstrated.
- Substantial engine development was accomplished.
    - More than 50 units have been built and tested.
    - Cumulative operating time was over 80,000 engine-hours.
    - 19,000 hours of operation were logged on a single engine.
- A commercial Power Conversion Unit (PCU) was designed, built and tested.
- Engine/PCU production planning (and costing) was performed by Volvo and others.
- A dish concentrator customized to the USAB 4-95 PCU was developed.
- Integration of the Dish and PCU was started.
- Six complete modules were built.
- Some field testing was performed.
    The major technical issues remaining are:
- Maintenance requirements and costs for the engine and PCU
    - Identification
    - Reduction
- Receiver design. The tubes for each piston occupy one quadrant of the receiver. This imposes severe alignment requirements on the concentrator in order to balance the input. A design concept employing a heat pipe-type input, as in the STM4–120 and free piston engine designs, would mitigate this problem.
- Concentrator design and cost.
- Field alignment of the concentrator.
- Integration of concentrator and PCU control hardware.
- Design and integration of single modules into a multiple module plant.

This extensive listing of major issues illustrates the early stage of this system's development.

### 2.4.7A.2  Concentrators

The status of advanced concentrator development is summarized below:
- Advanced concentrator designs have been produced for the following concepts:
    - Structurally integrated reflective panels
    - Multiple-facet stressed membranes
    - Single stretched membrane

- One structurally integrated design was built but failed.
- A structurally integrated gore section for another design was built.
- The LaJet LEC-460 second generation, multiple facet stretched membrane concentrator is developed, but its area of 43.7 m$^2$ is only about one half that required for a 25 KW$_e$ module. However, this provides a technology base for the LaJet advanced innovative concentrator which has a reflector area of 164 m$^2$.
- Three large single facet stretched membrane concentrators were built in Germany and Saudi Arabia (the same system). Test results and data are not available for evaluation.
- Several scale-model mirrors were built in the United States for a single facet stretched membrane concentrator.

This tabulation shows that the concentrator is also at a very early stage of development.

### 2.4.7A.3  Advanced Stirling Engines

Stirling engines which could represent improvements over the USAB 4-95 engine are:

- STM4-120. This engine has achieved 90% of its rated output in bench testing. Development is continuing to integrate the engine into a PCU and conduct solar testing at the DRTF.
- Free-piston Stirling engines. Conceptual designs are complete for both the MTI and STC concepts. Solar testing could begin in 1992. If developed for space power, a terrestrial version could be derived from it for the solar program.

### 2.4.7B  Priorities

It is apparent from the foregoing technology review that further development is required for all major elements of this system. In fact, selection of preferred designs for the engine, the receiver and the dish concentrator remain to be accomplished. The necessary development activities were prioritized in order to help formulate an effective development plan. (This is particularly important in a period with limited budgets.) The priorities together with the rationale are:

1. Selection of the specific Stirling engine configurations.
    - The large uncertainty in engine maintenance costs must be reduced to ensure that the product being developed will be economically viable. Other expenditures are not justified until this issue is resolved.
    - The engine will require the longest development time in order to achieve a low-cost, low-maintenance product.
2. Definition of the Complete Power Conversion Unit (PCU).
    - The receiver and the balance of the PCU should be designed to meet the specific requirements of the selected engine.
    - The concentrator requirements should be identified.
3. Selection of the dish Concentrator.
    - The concentrator should be designed to meet the requirements identified for the PCU.
    - Iteration of concentrator and PCU requirements and design should be provided.
4. Development of the PCU/Concentrator as an integrated unit.

- Minimize program cost by development a single product along a single path.
- Provide trade offs as required.
- Continue product improvement of the Stirling engine.

### 2.4.7C Development Plan Outline

The following plan outline was prepared to provide a gross indication of requirements, cost and duration for development of the dish Stirling system. A more detailed study would be required to formulate the most effective and efficient development program plan.

It should be noted that the residual hardware from the MDC/USAB development program, including a number of dish concentrators and 4-95 engines, is not being used at this time. Considering the substantial testing that will be required for the development of this system, acquisition of this hardware would enhance the effectiveness and efficiency of the program. The use of this hardware will be assumed in the following plan.

The priorities described above form the bases of this phased development plan outline.

### 2.4.7C.1 Phase I, Stirling Engine Evaluation and Selection

The objectives of this phase are: (1) experimental determination of engine failure rates and repair costs to ensure that annual maintenance costs will be within the allocation (or alternatively, reassessment of the economic viability of the concept) and (2) selection of the specific Stirling engine to be used for this system. The specific development activities are:

*Program Definition and Preparation*

- Scope the full program.
- Prepare allocations for cost, performance and availability for the system and major components.
- Provide a capability to field test Stirling engine PCUs with dish concentrators. A minimum capability of 4 dishes should be provided.
- Procure the available 4-95 engines, the STM4–120 (if available) and sufficient spare parts. Upgrade auxiliary components to reduce the down time not related to basic engine failures.
- Assess whether a parallel fossil fuel burning capability would be an advantage to the test and/or development program. Provide this capability if justified.
- Prepare test and evaluation plans including the specific data requirements.
- Define and procure the Data Acquisition System.

*Field Testing and Evaluation*

- Conduct the field test program according to the test plan.
- Identify those engine components with high failure rates that require redesign or upgrading.
- Test upgraded and/or new engines as they are provided.
- Develop, test and evaluate engine maintenance.
- Update product definition and projected economics.
- Select the preferred Stirling engine.

*Stirling Engine Development Support*

- Upgrade engine reliability as indicated from the test program.
- Solarize alternative engines for testing where they become available from other development programs and/or are required to achieve the maintenance cost allocation. Solarizing the MTI ASE engine is an early possibility if the 4-95 testing indicates that a kinematic engine might achieve the maintenance cost allocation. The free piston engine is a longer-term possibility if the initial development program is successful.

The duration and cost of this phase are dependent upon the test results and which engine, if any, offers viable economics. The two bounds would be:

- If the 4-95 engine proved adequate, with minimum product improvement, the cost of this phase could be as low as $5 million and the duration as short as 4 years.
- If extensive improvements are required to the kinematic engine or the free piston engine is selected, the cost would more likely be around $100 million and the duration about 10 years. This also assumes that some of the advanced engine development costs are defrayed by other applications, such as space power.

### 2.4.7C.2 Phase II, Module Development

The objectives of this phase is to develop and demonstrate a dish Stirling module that is economically viable. The specific activities are:

*Requirements Definition*

- Review the technology status to select the design concept for the concentrator, receiver and control system.
- Utilize the results from Phase I testing to define the preferred components and configuration of the PCU.
- Prepare a preliminary design for the module.
- Prepare requirements specifications for all major components.
- Prepare test and evaluation plans.

*Development*

- Develop test configuration of the PCU.
- Develop the concentrator.
- Continue engine development.
- Upgrade components as indicated by the test program.

*Field Testing and Evaluation*

- Conduct the field test program.
- Identify components requiring upgrading.
- Test upgraded modules as they are provided.
- Update module design and projected economics.
- Provide demonstrated results to potential users and/or investors.

The duration and cost of this phase also depends upon the scope remaining after Phase I and the test results within Phase II. The cost and duration would be expected to range between $40–100 million and 4 to 10 years.

### 2.4.7C.3  Phase III, Multi-Module System

The objective of this phase is to develop and demonstrate the multi-module configuration required for a utility-scale plant. Specific activities are:

- Define the multi-module configuration.
- Add the required hardware and software.
- Incorporate any other product improvements identified in Phase II.
- Field test and demonstrate the system.

The duration and cost ranges for this phase are estimated to be 2 to 5 years and $5–20 million.

### 2.4.7C.4  Duration and Cost

Combining the foregoing estimates would yield the following approximate range of values for the limits on the development program duration and cost.

- Development duration: 10–20 years.
- Development cost: $50–200 million.

## 2.5  Trough Systems

Trough systems employ a line-focusing reflector to concentrate sunlight onto a receiver. It is a distributed receiver system with central power generation. Because the focus is along a line rather than to a point, the concentration factor of the sunlight is much lower than for either the central receiver or dish systems (typically 40–80 compared to 500–1000). Thus, this concept is expected to have lower collection and power conversion efficiencies than either the dish or central receiver. Potential advantages of this system type for utility-scale electric power production are:

1. The trough reflector and drive system has the potential for very low cost.
2. The system can be configured to be relatively simple and well within the current state-of-the-art.

Troughs are the most fully developed of the three solar thermal system concepts. Much of this development was directed toward systems producing industrial process heat. A major installation of commercial solar thermal electric power plants is now in operation in California; they utilized favorable power purchase agreements and tax credits. Accordingly, only a brief review will be made of the development background. The technology status will be derived primarily from the installed commercial plants.

### 2.5.1  System Definition

The parabolic trough collector assembly consists of:

- Reflector panels which are parabolic shaped mirrors that concentrate the sunlight onto the receiver.
- A receiver assembly that absorbs the reflected sunlight and heats a heat transfer fluid. The receiver is normally a tube with a high absorptance/low emissivity coating enclosed within a glass tube. The low emissivity coating reduces radiative thermal losses and the covering tube reduces convective losses. The absorber

tubes must have flexible end connections to accomodate their motion during the daily tracking cycle.

- A drive unit to track the sun in one axis.
- Structural supports and foundations.
- Controls.

Troughs are mounted so that the absorber is horizontal. They can be oriented either east-west or north-south. The north-south orientation normally provides slightly higher yearly energy but its winter output is very low. The east-west orientation provides a more constant output throughout the year.

The system employed for electric power production generally consists of:

- The collector subsystem including the receiver assembly.
- A heat transport subsystem which circulates a heat transfer fluid through the receivers and transports the collected thermal energy to a central location.
- A thermal storage subsystem which stores the collected energy as sensible heat.
- A power conversion subsystem which converts the thermal energy into electricity and rejects the waste heat.

## 2.5.2 Development Background

A major development program on trough systems was initiated in the mid-1970's. Sandia National Laboratory, Albuquerque (SNLA) provided technical management for the U.S. Department of Energy.

### 2.5.2A Research and Development

SNLA initiated and managed numerous design studies and prototype hardware contracts on the major components and full assemblies. A number of alternate candidates were built and tested for the: concentrating trough, reflective surface, absorber tube selective coating, drives, controls, structural supports and foundations. Parallel analytical and design studies included: field and piping layout, thermal losses and operational simulation. Numerous applications studies were conducted to guide this development.

Testing and evaluation of trough components, subsystems and components were accomplished at the Midtemperature Solar Systems Test Facility (MSSTF) operated by SNLA in Albuquerque, New Mexico (Workhoven, 1980). The MSSTF consisted of two independent installations: (1) the Subsystem Test Facility (STF), supporting research and development on components, subsystems and systems and (2) the Collector Module Test Facility (CMTF) which characterized performance of components, subsystems and systems in support of commercial programs and collector development. The CMTF provided three independent heat transport fluid loops containing (1) Therminol-66, (2) Syltherm 800$^{TM}$ and (3) water.

Results from the first generation trough development programs were collected and utilized in the Performance Prototype Trough Development Project conducted by SNLA from 1979 through 1982. The objectives of this program were to: (1) improve system performance, (2) improve durability and component life and (3) utilize designs compatible with mass production processes. Four types of reflector structures were built and tested. These were:

**Figure 2.47:** MISR Test Site at SNLA.

- Stamped sheet metal
- Sheet molding compound
- Sagged glass/steel frame
- Honeycomb sandwich

These programs carried trough system technology to an advanced state of development.

### 2.5.2B  Modular Industrial Solar Retrofit (MISR)

The purpose of the MISR program was to develop modular trough systems that provide the solar equivalent to packaged boilers for industrial process heat (Alvis, 1980 and Leonard et al., 1983). These pre-engineered modules could then be retrofit by industry to provide process steam up to 300 °C (572 °F).

Five modular test systems were designed and constructed by different contractors. Four of the systems were tested at the MISR test site (Figure 2.47) adjacent to the CRTF and DRTF in Albuquerque, New Mexico. A fifth system was tested at the Solar Energy Research Institute in Golden, Colorado. A variety of options for system components were used by the different contractors, but all systems utilized skid-mounted equipment for the heat transport and steam generation subsystems.

Each test module was evaluated separately. First, an acceptance inspection was performed to insure the contractor installed each system element as designed. Second, function and safety tests were performed to insure each system operated as designed. Third, each module was qualified by a two-week continuous operation test. During this period, the systems were operated automatically. Only routine maintenance was allowed. The systems had to operate for 14 days, seven days of which had to have direct normal insolation of greater than 4.0 $KWh/m^2/day$. Repair or unscheduled maintenance of the module caused a qualification test restart. Fourth, the collectors were life tested by rotating the collectors from stow to maximum rotation angle for 3000 cycles at operating temperature.

The results of the test evaluations showed that each of the five systems passed the qualification tests including unattended operation; all systems were deemed

**Figure 2.48:** Coolidge Solar Irrigation Facility.

equivalent. Final reports on each of the system qualification tests were published in 1986 and revised in 1987 (Cameron and Dudley, 1987, Cameron and Dudley, 1987a, Cameron and Dudley, 1987b, Cameron et al., 1987, and Cameron and Dudley, 1987c).

### 2.5.2C  Industrial Process Heat Program

A user and applications-oriented program was begun by the DOE in 1976; this program paralleled the technology development program outlined above. Technical management for the trough projects was provided by SNLA.

The objective of this program was to evaluate the technical feasibility of applying solar thermal energy for industrial plants. This "real-world" data was utilized to supplement the information developed in the test facilities.

Seven major trough installations were built under this program out of a total of 16 projects (Pappas et al., 1983). Collector areas ranged from approximately 500 m$^2$ to 5000 m$^2$.

This program produced useful data, made a major contribution to the development of trough systems and helped to reduce the cost of trough collectors by the volume purchased.

### 2.5.2D  Coolidge Solar Powered Irrigation Project

An early full system experiment providing electrical output (150 KW$_e$) from a trough system was the Coolidge Solar Irrigation Project (Duffy et al., 1980, Torkelson and Larson, 1981, Torkelson and Larson, 1982, Larson, 1983, and Larson, 1983a). It was located on a farm near Coolidge, Arizona. The plant operated from October 1979 through September 1982. An aerial view is shown in Figure 2.48.

The objectives of the Coolidge Solar Irrigation project were:
1. Quantify energy collection and conversion performance.
2. Determine equipment reliability.
3. Identify desired improvements.
4. Quantify operating and maintenance requirements.
   The project was sponsored by the United States Department of Energy.
   The major participants and their roles were:
 • Acurex Corporation: Design, prime contractor and solar collectors.
 • Sundstrand Corporation: Organic Rankine cycle power generation unit.
 • Sullivan and Masson Consulting Engineers: Design.

- University of Arizona: System operation, performance data and documentation.
- Sandia National Laboratories, Albuquerque: Test and evaluation plan, consultants for performance tests, technical support and assistance in documentation.

### 2.5.2D.1 Plant Description

The Coolidge solar irrigation plant used a field of line-focusing parabolic trough collectors to heat a heat transfer oil which was circulated through the field. The thermal energy collected was converted to electricity using the organic Rankine cycle with toluene as the working fluid. Thermal energy storage of the hot oil, sufficient for 5 hours of operation of the power conversion subsystem, was provided. A schematic of the system is shown in Figure 2.49. Major subsystems of the plant are described in Table 2.37.

During the third year of operation, the system was able to operate automatically on routine days.

### 2.5.2D.2 Performance

The available solar energy, the thermal energy collected, natural gas usage and the generated electrical energy were determined daily from January 1980 through September 1982. The performance of the plant over this period is shown in Table 2.38. The electrical energy due to solar was obtained by multiplying the plant's total output by the solar thermal energy collected divided by the total thermal energy used. A boiler efficiency of 0.7 was used for the gas-fired heater. The efficiencies tabulated are the solar-derived electrical energy output divided by the total direct normal solar energy available.

An improvement in plant output and efficiency over the three years is apparent on the table. The collector field availability was particularly high.

### 2.5.2D.3 Significant Incidents

A tabulation of significant incidents encountered during operation of the Coolidge solar irrigation system is given in Table 2.39.

### 2.5.2D.4 Summary

The Coolidge facility was operated for ee years by the University of Arizona with technical support from SNLA. During the last year, the operating staff was reduced to one full-time technician. During that year, the plant operated automatically on routine, incident-free days. However, operator attendance was required during startup of the power conversion subsystem for safety reasons. In October, 1982, the plant was deeded to the owner of the farm who continued system operation for farm irrigation.

### 2.5.2E International Energy Agency Distributed Collector System (IEA-DCS)

The International Energy Agency-Distributed Collector System (IEA-DCS) was a 500 $KW_e$ full system experiment located at Tabernas, Spain adjacent to the IEA-CRS plant and the CESA-1 site (DFVLR, 1981, Kalt et al., 1982, Kalt and Dehne, 1982, Laguia, 1982, Hansen, 1984a, Grasse, 1985, Sandgren and Andersson, 1985, Instituto de Energias Renovables/CIEMAT, 1987, and SSPS Monthly Plant Operation Reports and Daily Evaluation Summaries, 1986). It shared support facilities with the IEA-CRS plant. The plant was operated from mid-1981 until August 1986. An aerial view of the plant is shown in Figure 2.50.

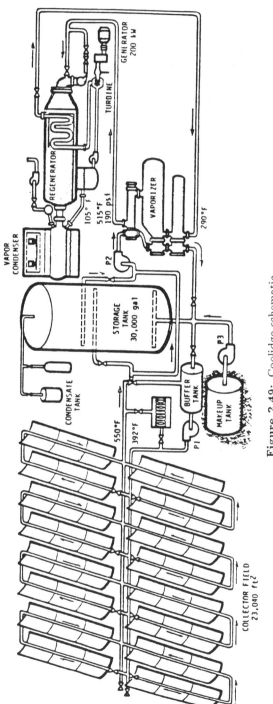

Figure 2.49: Coolidge schematic.

**Table 2.37:** Coolidge Solar Irrigation Project

| | |
|---|---|
| PLANT RATING, NET | 150 $KW_e$ |
| **COLLECTOR** | |
| Number of Troughs | 48 |
| Total Area ($m^2$) | 2140 |
| Reflective Surface | Aluminum (Original) Aluminized FEK-244 Film (Added in 1981) |
| Orientation | North-South |
| Concentration Ratio | 36 |
| **THERMAL TRANSPORT** | |
| Fluid | Caloria $^{TM}$ HT 43 |
| Inlet/Outlet Temp.$^o$C ($^o$F) | 200(392)/288(550) |
| **THERMAL STORAGE** | |
| Type | Single Tank Thermocline |
| Media | Caloria HT 43 |
| Rating ($MWHr_t$) | 5 |
| **FOSSIL FUEL SUBSYSTEM** | |
| Rating ($MW_t$) | 1 |
| Fuel | Natural Gas |
| **POWER CONVERSION** | |
| Type | Single-Stage Impulse Turbine |
| Working Fluid | Toluene |
| Inlet Conditions | 268$^o$C (514$^o$F), 10.5 bar (152 psi) |
| Heat Rejection | Water-Cooled Condenser |

The objectives of the project were:

- Determination of design viability
- Evaluation of operational behavior
- Extrapolation for advanced commercial designs

**Table 2.38:** Coolidge Performance

|  | Jan 1980–<br>Sept 1980 | Oct 1980–<br>Sept 1981 | Oct 1981–<br>Sept 1982 | Total |
|---|---|---|---|---|
| Electrical Output<br>due to Solar (MWHr$_e$) |  |  |  |  |
| Gross | 97.1 | 133.2 | 170.0 | 400.3 |
| Net | 56.0 | 75.1 | 119.1 | 250.2 |
| Average Efficiency (%) |  |  |  |  |
| Gross | 2.45 | 2.47 | 3.50 | 2.90 |
| Net | 1.41 | 1.39 | 2.45 | 1.81 |
| Collector Field<br>Availability (%) | 89 | 93 | 98 |  |

- Comparison of performance with CRS plant
- Maximization of yearly electric output
- Minimization of plant outages and losses
- Learn system aspects and gain operational experience

The project was sponsored by the following agencies of the nine participating member countries of the International Energy Agency:

- German Federal Ministry of Research and Technology
- U.S. Department of Energy
- Spanish Ministry of Industry and Energy
- Italian National Research Council
- Belgian Ministry of Scientific Research
- Swiss Federal Office of Energy
- Austrian Federal Ministry of Science and Research
- National Swedish Board of Energy Source Development
- Greek National Energy Council of the Ministry of Coordination
  The major project participants and their roles were:
- DFVLR (Germany): Operating agent (through 1984).
- CIEMAT-IER (Spain): Operating agent (from January, 1985).
- Compania Sevillana De Electricidad S.A. (Spain): Plant operation (through 1985).
- MOMPRESA (Spain): Plant operation (from January, 1986).
- Acurex Corp. (U.S.A.): Overall project management, system engineering, system integration, plant acceptance testing, collector field.
- MAN (Germany): Final detailed design, power conversion system, master control, data acquisition system, 2 collector fields.
- Technicas Reunidas (Spain): Open procurement, documentation coordination, general plant erection, storage subsystem erection and civil work.

**Table 2.39:** Significant Incidents - Coolidge

| Incident | Cause | Solution/Repairs |
|---|---|---|
| Loss of reflectivity of collector panels after 1 year. | Degradation. | Replaced coilzak reflective surfaces with aluminized acrylic film (FEK-244). |
| Reflectivity was lower in collectors that were washed than those left dirty. | Hard water. | A water softener was installed. |
| Collector tracking system failures. | Moisture, thermal stresses and electronic component problems. | Redesigned sensors and control boards were installed. Modified sensor cases and changed photodiode encapsulation. |
| Vacuum leakage caused extra pump operation. | Leaky caps on regeneration tank sensor fittings. | Recapped fittings. |
| Caloria fires (2). | Flexhoses developed leaks. Pump seal leak. | Inspected for leaks. Flexhoses replaced. |
| Generator failed to operate. | Relay failure. | Replaced relay. |
| Toluene contaminated by Caloria. | Operating error during routine replacement. | Removed from vaporizer unit. |
| Toluene boost pump seal failures (about every six months). | Wear out. | Seal replaced. |
| Receiver tube coating degraded. | Temperature and aging. | None applied. |
| Excessive collector drive motor failures (nearly 1/2 of the motors during the first year). | Rear shaft bearing and other failures. | Returned to manufacturer, repaired or replaced. |
| Collector pump and many other motors failed. | Wear out. | Replaced. |

**Figure 2.50:** IEA-DCS Plant.

### 2.5.2E.1 Plant Description

The IEA Distributed Collector System utilized one field of one-axis tracking trough collectors and two fields of two-axis tracking trough collectors to heat oil (Santotherm 55) which was in turn used to generate steam to drive a conventional turbine generator set. The second field of two-axis tracking collectors was added during the test program (April 1984). Thermal energy storage was provided by storing hot oil. No fossil-fueled heater was employed. The major subsystems of the plant are described in Table 2.40.

Daily system start-up began when the insolation reached 400–450 $W/m^2$. The ambient temperature oil in the collector field piping was circulated at minimum flow, until the field outlet temperature reached 275 °C. When this temperature was reached, the hot oil was supplied to the top of either of the thermocline storage tanks to begin storage charging. The outlet temperature of each field continued to increase until it reached the normal operating temperature of 295 °C. The oil flow was then regulated to maintain this outlet temperature.

The system was normally operated in the charging mode for several hours before the steam generator and the remaining components of the power conversion system were started. The delayed start-up of the power conversion subsystem was necessary to compensate for the low solar multiple of the plant (approximately 1). This operating strategy provided sufficient stored thermal energy to sustain full-rated power production until the approximate time that the collector fields were shut down.

**Table 2.40:** IEA-DCS System

| | |
|---|---|
| PLANT RATING, NET | 500 KW$_e$ |

COLLECTOR

| | |
|---|---|
| Acurex Field (1-Axis) | 480, 5.58 m$^2$ each |
| M.A.N. Fields (2-Axis) | 154, 32 m$^2$ each |
| Total Area | 7602 m$^2$ |
| Reflector (All) | Back-Silvered Glass |
| Orientation (Acurex) | East-West |

THERMAL TRANSPORT

| | |
|---|---|
| Fluid | Santotherm 55 Oil |
| Inlet/Outlet Temp.,$^o$C ($^o$F) | 225(437)/295(563) |

THERMAL STORAGE

| Types (2) | Single-Medium Thermocline | Dual-Medium Thermocline |
|---|---|---|
| Media | Oil | Oil/Cast Iron |
| Rating (MWHr$_t$) | 3.6 | 0.37 |

POWER CONVERSION

| | |
|---|---|
| Type | Seven-Stage Extraction Turbine |
| Working Fluid | Water/Steam |
| Inlet Conditions | 283$^o$C (541$^o$F), 25 bar (362 psi) |
| Heat Rejection | Wet Cooling Tower |

## 2.5.2E.2 Performance

Plant performance, summarized in Table 2.41, was less than originally expected. An efficiency of 9% had been predicted for the net electric output. The main reasons for the low performance were the plant's high thermal inertia, higher than expected minimum insolation required to operate the collector field (400–450 W/m$^2$ instead of 350 W/m$^2$), and the less than predicted power conversion system efficiency. The 1984 collector field expansion included modifications to reduce the thermal inertia and to improve plant operability. Rather than using buffer tanks to temper the

**Table 2.41:** IEA-DCS Operation/Performance Summary

| Year | Power Production (Hours) | Gross Electric (MWHr) | Net Electric (MWHr) |
|---|---|---|---|
| 1983 | 512 | 225 | 80 |
| 1984 | 326 | 125 | (38) |
| 1985 | 67 | 32 | Neg. |
| 1986 (to Aug) | 8 | 3 | Neg. |

oil recirculating during start-up, the new installation was designed to use the cool oil at the bottom of the dual media storge tank as a buffer during start-up. This design reduced the quantity of oil in the field which needed to be heated during start-up. The collector field expansion increased the field aperture area by 42%, increasing the solar multiple to slightly greater than 1. With the expansion, the plant demonstrated a 2.5% net clear day efficiency.

The annual power production performance of the plant did not represent the full capabilities of the plant. During 1983 the plant operating objectives were to maximize electric output and minimize plant outages; however, non-optimized operation, operating complications, and outages resulting from design infancy prevented true utility-like operation. Since 1984 the plant was only staffed for 5 day per week operation and the objectives were primarily focused on obtaining detailed subsystem and component information.

### 2.5.2E.3 Significant Incidents

Significant issues encountered during operation of the IEA-DCS plant are summarized in Table 2.42.

### 2.5.2E.4 Lessons Learned

Some of the major lessons learned from the IEA-DCS project are summarized below:

- The one-axis tracking Acurex collector was a more efficient design than the two-axis tracking MAN collector. This was primarily the result of the lower losses associated with the shorter length of interconnecting piping in the Acurex field. The two-axis tracking, pedestal-mounted trough collector requires further development in order to compete with the one-axis tracking designs.
- The thermal inertia of the oil and piping in the collector field resulted in slow plant start-up. The effect of thermal inertia must be included in plant sizing and performance predictions.
- Cool oil flow resistance through the dual medium storage tank does not allow the pumps to keep up with high flow demands. Oil viscosity changes must be considered in pump sizing and component design.
- A direct overview of the field should be provided from the control room to detect collector field problems more quickly (e.g., leaks, smoke, fire, etc.).

**Table 2.42:** Significant Incidents - IEA-DCS Plant

| Incident | Cause | Solution/Repairs |
|---|---|---|
| Flex hose oil leaks in MAN field. | Over stressed. | Replaced hose. Modified installation to reduce stress. |
| 1500 gallon oil spill. | Bellows failure at inlet to MAN field due to over stress. | Replaced all bellows with ones with increased displacement range. |
| Numerous minor oil leaks. | Gasket seal failures, weld leaks, joints loosened. | Repaired/replaced. |
| Feedwater pump failures. | Mechanical failures. | Repaired. |
| Mirror facets fell off MAN collectors. | Support failure. | Installed holders to prevent facets from falling if support failed. |
| Acurex field mirror delamination. | Glass/steel bond failure. | Mirrors replaced. |
| Water leakage into thermal oil. | Steam generator water leak. | Repaired. |
| Oil pipe rupture in MAN collector module. | Defective swivel joint which allowed oil to bypass collector. | Repaired. |
| Inadvertent discharge of fire protection system foam. | Fire detector sensor calibration. | Replaced. |

- Design performance cannot be routinely achieved if design insolation is only rarely reached. Insolation levels and availability must be carefully considered in order to avoid overly optimistic performance expectations.
- Additional instrumentation and more accurate calibration would have allowed a more complete evaluation of plant performance. Special emphasis needs to be placed on instrumentation selection, location, installation and calibration in order to produce meaningful test results.
- Collector field drives must be sized to take the collectors to stow position if high winds develop during operation.
- The black chrome coating on the Acurex field receiver tubes deteriorated without any apparent performance effect; this causes doubts on the benefit of this expensive coating.
- Mirror delamination in the Acurex field required extensive rework. Mirror bonding must be qualified for the design life and environment.

- With the exception of the problem of mirror delamination, maintenance of the one-axis tracking Acurex field proved to be easy.
- The MAN collector field was not sufficiently developed for commercial applications. Problems with the swivel joints, detached mirrors and defective electronics boxes would all need to be resolved.
- Normal maintenance of the MAN collectors was difficult and time consuming. Adequate maintenance access needs to be provided in any collector design.
- The MAN collectors soil more rapidly than the Acurex collectors due to condensation on the glass surface. Insulation to reduce condensation can reduce the frequency of mirror cleaning.
- Improvements in collector field environmental protection details such as sealing of control and electronics boxes, adding paint or other protective coatings in areas subject to corrosion, etc. are required to extend collector life.
- Oil line drains and vents would need to be relocated in order to simplify repairs and minimize air in the oil system.
- Steam generator leakage introduced water into the oil, forcing an extended shutdown. Provisions for dealing with this type of occurrence are required.
- Test systems such as the IEA-DCS should be provided with a turbine bypass to allow operation if the turbine is not available.
- The water treatment plant was undersized. Seasonal supply/demand changes must be considered in water treatment plant sizing.

**2.5.2E.5 Summary**

A summary of the major accomplishments of the IEA-DCS project were:

- Provided side-by-side operation and performance comparison of two-axis tracking and one-axis tracking parabolic trough collectors.
- Provided operation and performance data for comparison with the central receiver system located on the same site.
- Provided operation and performance comparisons between single medium and dual media thermocline thermal storage.
- Identified collector field design improvement areas.
- Provided data to aid in predicting the operation and performance of commercial-size plants.
- Identified system design deficiencies which can be corrected in future plants to improve operability and performance.
- Promoted international cooperation in the development and application of solar power systems.

**2.5.3 Solar Electric Generating System (SEGS)**

A series of solar electric power plants based on trough collectors are being placed in service on the Southern California Edison Company grid (Kearney, 1986, Kearney and Price, 1987, Jaffe et al., 1987, and Kearney, 1988). The developer made use of favorable power purchase agreements under the Public Utility Regulatory Policy Act (PURPA) and tax benefits available at the federal and California state levels. The plants were sold to third party financial ventures structured to maximize the value of the tax benefits and cash flow from electricity sales. Additionally, the developer provided performance guarantees of the plants' output and low-interest,

**Table 2.43:** SEGS Plants

| SEGS Plant Number | Rating (MW$_e$, Net) | Date In Service | SCE Power Purchase Agreement |
|---|---|---|---|
| I | 13.8 | 12/84 | Negotiated |
| II | 30 | 12/85 | " |
| III | 30 | 12/86 | Standard Offer #4* |
| IV | 30 | 12/86 | " |
| V | 30 | 9/87 | " |
| VI | 30 | 9/88 | " |
| VII | 30 | 12/88 | " |

*Standard Offer #4 provides fixed energy and capacity payments. The on-peak period rates are particularly high, over 30 cents/KWHr from noon to 6 PM during the key summer months from June through September.

non-recourse note financing to help leverage the investors' rate of return. By these means, the technology described earlier in this section has been commercialized.

The solar collector fields are designed and manufactured by Luz Industries, Israel, who also performs the bulk of the conceptual and preliminary system design engineering of the plants. Luz Engineering Corp. coordinates engineering activities in the U.S., performs operations and maintenance functions at the sites and conducts the marketing and financial activities for the projects.

The Luz parabolic trough system design utilized the extensive design and development activities carried out by the U.S. Department of Energy over the years 1975-1984. The Luz contribution has been to complete this development into practical systems and to commercialize them.

A listing of the first seven SEGS plants is given in Table 2.43. A picture of SEGS I and II, located adjacent to Solar One in Daggett, California, is shown in Figure 2.51; SEGS III, IV and V located at Kramer Junction, California are shown in Figure 2.52.

It can be seen that these first seven plants represent nearly 200 MW$_e$ of net installed capacity. The use of a parallel, fossil heat source allows these plants to obtain full capacity credit and to maximize power production during peak periods, thus achieving maximum revenue. The current low price for natural gas further enhances economics.

### 2.5.3A  Plant Descriptions

While the SEGS program made full use of the previous DOE development results, significant further development was accomplished by Luz within the SEGS program. This is illustrated in the evolutionary collector development shown in Table 2.44. The progression to larger troughs, higher concentration ratios and improved emissivity coatings is apparent in this table. Collector LS-1 was used

**Figure 2.51:** SEGS Plants I and II.

**Figure 2.52:** SEGS Plants III, IV and V.

**Table 2.44:** Characteristics of LUZ Trough Collectors

|  | LS-1 | LS-2 | LS-3 |
|---|---|---|---|
| Area (m$^2$) | 128 | 235 | 545 |
| Aperture Width (m) | 2.55 | 5.00 | 5.76 |
| Receiver Diameter (m) | .042 | .070 | .070 |
| Overall Length (m) | 50.2 | 47.1 | 95.2 |
| Concentration ratio | 61 | 71 | 82 |
| Optical Efficiency | .734 | .737 | .772 |
| Receiver Emissivity | .30 | .24 | .15 |
| Receiver Absorptivity | .94 | .94 | .97 |
| Mirror Reflectivity | .94 | .94 | .94 |
| Peak Collector Efficiency (%) | 66 | 66 | 68 |
| Annual Thermal Efficiency (%) | 51 | 50 | 53 |

for SEGS I and part of SEGS II. LS-2 was used for the balance of SEGS II and SEGS III through VI. SEGS VII utilized LS-3. A similar system evolution can be seen in the system description of SEGS I, III and VII given in Table 2.45. Schematics of SEGS I and III are shown in Figure 2.53 and Figure 2.54 respectively.

The operation of SEGS III will be described as representative of these plants. The SEGS III system normally operates for approximately six to eight continuous hours per day with the plant shut down overnight.

The principal operating modes are:
1. Low pressure turbine stage only using solar generated steam.
2. High pressure turbine stage only using steam from the gas-fired boiler-superheater.
3. Hybrid mode utilizing both stages of the steam turbine and both solar- and gas-derived steam.

During mode 1, each solar field is placed into operation and the heat transfer oil is circulated in a closed loop through the fields until it reaches the desired operating temperature of 660 °F. When this occurs, the oil is automatically directed to the two parallel solar superheaters, then to the steam generators and then through the two boiler feedwater preheaters. The feedwater is preheated from 351 °F to 487 °F prior to entering the boiler. The saturated steam leaves the boiler at 498 °F and is superheated to 622 °F. This superheated steam at 630 psia drives the low pressure stage of the steam turbine. During the low pressure cycle, when the high pressure side of the turbine is not being used, 16,000 lbs/hr of low pressure steam is circulated through the high pressure stage in order to cool the turbine blades.

The total maximum oil flow to the two solar superheaters is 3,734,000 lbs/hr at 660 °F and approximately 107 psig. The exit temperature of oil from the two boiler feedwater preheaters is 479 °F. The cooled oil is then recirculated through the four solar fields to bring the temperature back up to 660 °F.

During those hours of the day that solar energy is not available, the high pressure stage of the turbine can be used (mode 2). This mode of operation is particularly

**Table 2.45:** SEGS Plant Characteristics

| | SEGS I | SEGS III | SEGS VII |
|---|---|---|---|
| PLANT RATING, NET (MW$_e$) | 13.8 | 30 | 30 |
| **COLLECTOR** | | | |
| Type | LS-1 (LS-2 added) | LS-2 | LS-3 |
| Total Area (m$^2$) | 106,512 | 203,980 | 176,580 |
| Field Configuration | North-South | North-South | North-South |
| **THERMAL TRANSPORT** | | | |
| Fluid | ESSO 500 | Monsanto VP-1 | Monsanto VP-1 |
| Field Outlet Temp.,$^oC(^oF)$ | 307(585) | 349(660) | 393(740) |
| **THERMAL STORAGE** | | NONE | NONE |
| Type | 2-Tank | | |
| Media | ESSO 500 | | |
| Rating (MWHr$_t$) | 117 | | |
| **FOSSIL FUEL SUBSYSTEM** | | | |
| Rating (MW$_t$) | 8.5 | 88 | 84 |
| Fuel | Natural Gas | Natural Gas | Natural Gas |
| Configuration | Series Superheater | Parallel Boiler/Super. | Parallel Boiler/Super. /Reheater |
| **POWER CONVERSION** | | | |
| Type | Turbine | Dual Admission Turbine | Reheat Turbine |
| Working Fluid | Water/Steam | Water/Steam | Water/Steam |
| Inlet Conditions | | | |
| Solar: Temp, $^oC(^oF)$  Press, bar(psi) | ---- ---- | 327(620) 43.4(630) | 371(700) 100(1450) |
| Fossil: Temp, $^oC(^oF)$  Press, bar(psi) | 416(780) 35.3(512) | 510(950) 105(1515) | same as solar |

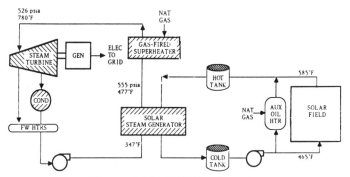

**Figure 2.53:** SEGS I schematic.

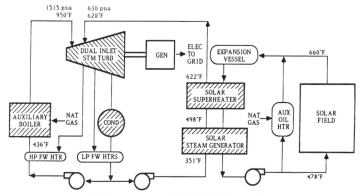

**Figure 2.54:** SEGS III schematic.

valuable during on-peak winter operation between the hours of 5 p.m. and 9 p.m., when there is no solar energy available.

The boiler is natural gas-fired, natural circulation with forced draft. It has a combination superheater, economizer and air preheater. Maximum continuous rating is 310,000 lbs/hr with a design flow rate of 289,300 lbs/hr continuous. The gas utilization efficiency of the primary boiler is 91% or greater. The maximum turndown ratio is 8 to 1.

When the unit is fired up, boiler feedwater enters the main boiler at the rate of 289,300 lbs/hr at 1690 psia and 436 °F. After circulating through the boiler and the superheater, superheated steam exits the unit at 1565 psia and a superheated steam temperature of 953.6 °F. This steam enters the high pressure stage of the turbine at 1,515 psia and 950 °F. At these conditions, the turbine generator is capable of producing 33.161 $MW_e$ of electrical power.

The hybrid mode of operation (mode 3) uses the solar field and the gas-fired boiler to operate both stages of the steam turbine. This mode is normally used during on-peak summer operation when direct normal insolation is not optimum. The

**Table 2.46:** SEGS Electricity Production (MWHr/yr)

| | SEGS I | SEGS II | SEGS III | SEGS IV | SEGS V |
|---|---|---|---|---|---|
| Net Output: | | | | | |
| 1985 | 16,402 | ----- | ----- | ----- | ----- |
| 1986 | 19,066 | 27,168 | ----- | ----- | ----- |
| 1987 | 22,165 | 24,208 | 61,589 | 63,346 | ----- |
| 1988 | 13,979 | 48,581 | 77,705 | 82,714 | 77,405 |
| Prediction: (Mature Plant) | 30,630–40,467* | 80,500 | 85,052 | 85,052 | 92,400 |

\* Field Area from 71,680 m$^2$ to 106,512 m$^2$

gas-fired boiler can be operated at a reduced load to augment the solar field input and allow the turbine generator to produce maximum power output of 33.3 MW$_e$.

## 2.5.3B Performance

The net electrical output from the SEGS plants is purchased by Southern California Edison (SCE) under the power purchase agreements cited. The values for SEGS I–V, provided by SCE (Nola, 1989), are shown in Table 2.46 together with the corresponding LUZ predictions for the mature plants (Kearney, 1986, Kearney and Price, 1987, and Jaffe et al., 1987). The range in predictions for SEGS I corresponds to changes in the plant's collector field area (71, 680 m$^2$ as built, to 95, 232 m$^2$ in July 1985, to 106, 512 m$^2$ in February 1987). The SEGS I field was reduced to 82, 960 m$^2$ in 1988 with the other 23, 552 m$^2$ added to SEGS II. The improvement in the plants' performance after SEGS I and II is apparent in the table.

The monthly net solar output of SEGS I and SEGS III are shown in Figure 2.55 and Figure 2.56 respectively. The power output and natural gas usage were supplied by Luz (Kearney, 1988). The prorating of output to solar and fossil sources was according to the heat energy supplied by each source without consideration of the supply temperature. The effect of the north-south field layout in peaking the output in the summer compared to winter is apparent in these figures. Net solar electrical output becomes negligible in mid-winter. Conveniently, annual maintenance is scheduled at this time.

The net electricity produced and the natural gas usage for the first six months of 1988 were supplied by LUZ for SEGS I through V (Kearney, 1988). These values were used to calculate the net electric output due to the solar input and the corresponding efficiencies. The calculations and results are shown in Table 2.47. The natural gas usage was divided into that used for electric power production directly and that used for other purposes such as plant morning startup, freeze protection and steam blanketing. The net electricity produced by natural gas was calculated by dividing the gas usage by the boiler heat rate and multiplying by 0.9 to account for plant parasitics. This was calculated separately for the gas used for electricity production and for the total gas used. The second case is considered more

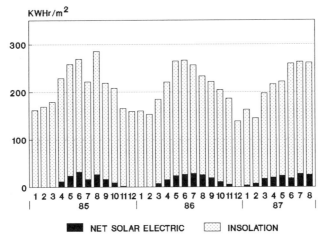

Figure 2.55: SEGS I performance (solar only).

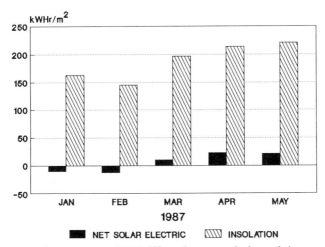

Figure 2.56: SEGS III performance (solar only).

representative of the solar output from these plants since solar heat would have been required for the other uses if gas had not been used. This would have reduced the solar energy available for electric power production. The electricity due to the solar input was calculated by subtracting the net electricity due to burning gas from the plant's total net output. The net efficiencies were calculated by dividing the "solar" electricity by the solar energy intercepted by the field using a six-month insolation value of 1250 $KWHr/m^2$.

The improvement in performance by the later plants is apparent in this table. This was the second year of operation for SEGS III and IV and the first year of operation for SEGS V.

### 2.5.3C Significant Incidents

Significant incidents encountered during operation of SEGS I and II are summarized in Table 2.48. Most of these problems were eliminated in the later plants

**Table 2.47:** SEGS Efficiencies (SOLAR) (January - June 1988)

| | SEGS I | SEGS II | SEGS III | SEGS IV | SEGS V |
|---|---|---|---|---|---|
| Net Electricity (GWHr) | 6.11 | 23.50 | 35.74 | 36.00 | 36.91 |
| Gas Used (MMBTU) Elec. Prod. | 34,416 | 93,521 | 75,501 | 79,762 | 75,251 |
| Total | 34,416 | 122,688 | 109,353 | 108,739 | 107,933 |
| Boiler Heat Rate (Btu/KWHr) | 10,840 | 9,145 | 9,139 | 9,139 | 9,139 |
| Net Electricity From Gas (GWHr) (0.9 x gas used / heat rate) | | | | | |
| Elec. Prod. only | 2.86 | 9.20 | 7.44 | 7.85 | 7.41 |
| All Gas | 2.86 | 12.07 | 10.77 | 10.71 | 10.63 |
| "Solar" Electricity (GWHr) Elec. Prod. Gas only | 3.25 | 14.30 | 28.30 | 28.15 | 29.50 |
| All Gas | 3.25 | 11.43 | 24.97 | 25.29 | 26.28 |
| Collector Field Area($m^2$) | 82,960 | 165,376 | 203,980 | 203,980 | 233,120 |
| Intercepted Solar (GWHr), ($1250 \times Area \times 10^{-6}$) | 103.7 | 206.7 | 255.0 | 255.0 | 291.4 |
| Net Efficiency (%) Elec. Prod. Gas Only | 3.1 | 6.9 | 11.1 | 11.0 | 10.1 |
| All Gas | 3.1 | 5.5 | 9.8 | 9.9 | 9.0 |

through design or other modifications. This is a major reason for the annual performance improvements seen in the later SEGS plants.

## 2.5.3D  Lessons Learned

Some of the lessons learned from the SEGS program are summarized below:

- Integrated design of the solar field and the power block is important.
- Good quality equipment is more important than lowest cost.
- Adequate quality control is required in both the manufacturing process and the field installation.

**Table 2.48:** Significant Incidents - SEGS

| Incident | Cause | Solutions/Repairs |
|---|---|---|
| High failure rate of field controllers (25% first year). | Faulty boards. | Replaced. |
| Field supervisory control system inadequate. | Deficient hardware and software. | Redesigned and retrofit. |
| Erratic field communications | Failure of cables and connectors. | Repaired. Replaced cables with more rugged units. |
| High failure rate of sun sensors. | Faulty adhesive. | Repaired in field. Corrected for new units. |
| High failure rate of inclinometers. | Units from one supplier faulty. | Replaced with acceptable units. |
| Local controllers damaged by lightning. | Frequent event in major lightning storms. | Replaced boards. Redesigned. LS-2 controllers. |
| Mirror breakage. | Separation of support pad from mirror due to faulty adhesive. | Preventive reinforcement of existing pads. |
| High failure rate of collector gear drives. | Inadequate torque and under-strength casing. | Replaced all units with redesigned drive. |
| Glass tube over receiver breakage high (6.7% in first year). | External impact. | Replaced. |
| Oil leakage in connectors between receiver tubes. | Installation, asbestos gaskets. | Retorqued flange bolts. Replaced gaskets with graphite-impregnated type. |
| A large number of major oil leaks in field and fires (24). | Weld failures due to inadequate weld penetration. | Rewelded all flanges. Improved quality control for LS-2. |
| High flexible hose leakage rate (18% leakage in first year). | Thermal expansion and tracking rotation. | Repaired. |
| Oil fire on top of cold tank. | Insulation became contaminated with oil leaking from flanges and vents; autoignited. | Extinguished; repaired. |
| Oil fire near hot tank. | Rupture of pipe expansion joint. | Brought under control in 25 minutes. Replaced valves and tank insulation. |

**Table 2.48:** Significant Incidents - SEGS (con't.)

| Incident | Cause | Solutions/Repairs |
|---|---|---|
| 2nd oil fire near cold tank. | Rupture of pipe expansion joint. | Extinguished. |
| Excessive leakage through pump and valve seals. | Improper valve packing, gaskets and seal types. | Replaced with proper materials and types. |
| Feedwater pump breakdown. | Thrust bearing failure. | Operated at half capacity with one pump; replaced. |
| Moisture carry-over to superheater. | Deficiencies in design & operation. | Modified moisture separator; lowered steam drum water level. |
| Low feedwater quality. | Quality of source well lower than specifications. | Added equipment; improved controls. |

- Even with a highly-developed technology base, the first and second plants experienced significant problems and performed below predictions.
- A dedicated organization was effective in mitigating these problems as much as possible.
- Lessons applicable to specific components were covered in the major incidents section.
- Learning within a single organization was very effective in applying the experience gained from SEGS I and II to improve SEGS III and beyond.

### 2.5.4 Energy Costs

The methodology to be employed in estimating the energy costs from future trough plants will be similar to that employed in the preceding two sections for the central receiver and the dish systems. The uncertainties should be substantially less because of the more advanced status of this technology. Consequently, the values of the upper and lower bounds to the projected energy costs should be much closer to each other than for the other two systems.

Cost and performance data from the SEGS program will be used for the future energy cost estimates. It should be noted that these estimates are made for solar-only operation at some future time when solar is cost-competitive with fossil-fueled plants. Consequently, no credit is taken for burning low-cost fossil fuel in these plants. This will make these energy costs higher than they would be if this low-cost fuel were available and utilized.

**Table 2.49:** SEGS Plant Prices

| | | |
|---|---|---|
| SEGS I: | $4493/KW$_e$; | $865/m$^2$ |
| SEGS III: | $3500/KW$_e$; | $515/m$^2$ |
| SEGS X (Projected): | $2000-2250/KW$_e$; | $355-400/m$^2$ |

### 2.5.4A  Costs

Prices quoted for representative SEGS plants by Luz are shown in Table 2.49 (Kearney, 1988). It should be noted that these are actually selling prices, dictated by the market, not costs. However, they represent the best available information and will be used for these cost estimates.

### 2.5.4A.1  Lower Bound to System Cost

The lowest values projected by LUZ for their SEGS X plant ($2000/KW$_e$, $400/m$^2$) will be used as the lower bound to system cost.

### 2.5.4A.2  Upper Bound to System Cost

The actual price for SEGS III ($3500/KW$_e$, $515/m$^2$) will be used as the upper bound to system cost.

### 2.5.4B  Performance

Performance projections will be based on the results that were shown in Table 2.47.

### 2.5.4B.1  Lower Bound for Annual Efficiency

The efficiencies calculated for SEGS III and IV in Table 2.47 were the bases for assigning the lower bound to annual efficiency for a trough plant operating in the solar-only mode. The burning of natural gas in these hybrid plants tends to increase the calculated performance relative to a solar-only plant due to the longer duty cycle achieved with hybrid operation. All daily warmup and cooldown losses would need to be supplied by solar energy in the solar-only mode. Accordingly, the high confidence, lower bound estimated for the annual net plant efficiency will be 9 percent.

### 2.5.4B.2  Upper Bound for Annual Efficiency

The relative maturity of the trough-based solar electric technology limits the further improvements in efficiency that might be expected from the fully-mature system. LUZ has identified a number of improvements that they plan to incorporate in later SEGS plants. These include:

- Increase of the plant size from 30 MW$_e$ to 80 MW$_e$.
- Use of higher operating temperatures to increase the efficiency of the power conversion cycle.
- Elimination of steam blanketing of the turbine which will reduce natural gas usage for this purpose.
- A heat transfer fluid with a lower freezing temperature is being considered to reduce gas consumption for freeze protection.
- Use of turbines with shorter startup time requirements.

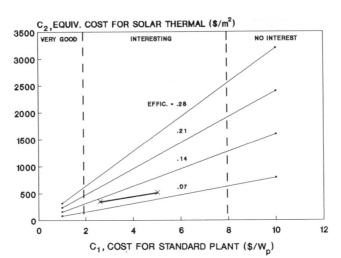

**Figure 2.57:** Trough system compared with standard plant.

The combination of product improvements to achieve higher efficiency could result in a net annual plant efficiency of 12 percent. This value will be used for the upper bound for annual efficiency.

### 2.5.4C  Results
Combining the above estimates yields the following sets of values for the bounds on trough system cost and performance.
- Upper Bound: Cost = $355/m$^2$, Annual Efficiency = 12%
- Lower Bound: Cost = $515/m$^2$, Annual Efficiency = 9%

These sets of values form the boundaries for the trough line plotted on Figure 2.57. With an annual fixed charge rate of 0.105, annual insolation of 2500 KWHr/m$^2$ yr and a standard allowance of 1 cent/KWHr for operation and maintenance, the range of energy costs is 13.4–25 cents/KWHr. This is consistent with the potential of 15 cents/KWHr arrived at in the PNL study (William et al.,1987).

### 2.5.5  Conclusions

The following conclusions can be made about trough solar thermal electric technology:
- SEGS, as a hybrid plant burning natural gas, is a commercially-viable system under the incentives provided by the standard offer number 4 power purchase agreement and federal and state tax incentives. Current incentives are less favorable.
- Energy costs projected for the mature system in the solar-only mode are 13 to 25 cents/KWHr.
- This technology is sufficiently attractive so that it should be preserved until fuel prices would make it economically competitive.
- Further development can be achieved as product improvements.

**Table 2.50:** Comparison of System Strengths and Weaknesses

|  | Strengths | Weaknesses |
|---|---|---|
| TROUGH | Commercial | Limited Performance Potential |
|  | Growth by Product Improvement |  |
|  | Relatively Simple System |  |
| CENTRAL RECEIVER | Potential for Low Cost | High Performance Not Confirmed |
|  | Potential for Good Performance | Complex System–High Freezing Temperature Fluid |
|  | Utility-Scale System | Development/Demonstration Size is Large/ Costly |
| DISH | Very High Performance | Maintenance Costs Uncertain |
|  | Modular System | Preferred Engine Design Uncertain |
|  | Development Can be in Small Steps | Limited Development Base |

# 2.6 System Comparison

A brief comparison of the reference configurations of each of the system concepts will be made.

## 2.6.1 Qualitative Comparison

The major strengths and weaknesses of the three selected system concepts are given in Table 2.50.

Evaluation of the trough system is quite straightforward. It is an operating commercial system with a growth potential through identified product improvements. It is well within the state-of-the-art; no technology breakthroughs are required. Its major weakness is that its energy cost potential is about a factor of two higher than the potential of either of the other two systems. However, the potential energy cost (lower bound) of the trough is substantially lower than the upper limits esimated for the energy costs for the other two systems, and trough technology is developed.

Evaluation of the central receiver is much more complex. It has very good potential for low energy costs and is inherently a utility-scale system. In fact, the referenced utility study predicted energy costs very close to the lower bound chosen here. However, the uncertainties in performance and cost are high. Solar One was the only central receiver system to produce a positive net output of electrical energy, and that output was substantially less than the original prediction. While there are good reasons for the underperformance of the advanced system experiments, they cannot be used as confirmation of the more optimistic predictions made for future commercial plants.

This uncertainty in annual performance was expressed by the solar thermal industry in a 1985 survey conducted for Sandia National Laboratories by the Solar Energy Industries Association. After expressing "great confidence in their technology," central receiver companies stated: "Central receiver total annual energy production and the reliability and availability of the system are large unknowns which must be cleared up before investors will be willing to accept the risks of supporting commercial-scale installations" (Erickson and La Porta, 1986).

Costs of the central receiver are also highly uncertain. The utility study cost estimates were based on less than a conceptual level of design detail and utilized very low contingency factors. A budgetary cost estimate would approximate the upper limit selected for the system's costs in this evaluation.

Another weakness of the reference central receiver system concept is that the system is very complex and many of the problems posed by the high freezing temperature of the molten salt heat transfer fluid have not been resolved to a commercial plant's requirements.

An impediment to commercialization of the central receiver system is that its inherent size is large. This makes any development and demonstration program lengthy and costly. This contrasts to the short "generation" time found in the SEGS program which was so successful in implementing design modifications in successive trough plants.

The selected dish system concept is even more difficult to evaluate, partially because it has received the least development support. It has the very highest potential of any of the systems, but it also has the greatest uncertainty. Even the first modules deployed had outstanding performance and reasonably good availability. But, maintenance costs are highly uncertain because of a lack of data. Engine concepts that could offer low maintenance costs have been identified but are at a very early stage of development. The national solar effort, by itself, could probably not justify the cost of a complete engine development in the foreseeable future. Fortunately, there are a number of other applications, such as space power systems, automotive propulsion and auxiliary power, that could share the development costs. The concept's high performance potential justifies continued attention.

In contrast to the central receiver, the module size of the dish system is small. This allows development and product improvement to be accomplished in small, less costly, steps once an engine is available.

### 2.6.2 Performance

The Solar One central receiver system and the SCE dish Stirling module were operated side-by-side on the same site in Daggett, California. A comparison of their performance, normalized to the collector area, is given in Table 2.51 for the first six months of 1986. It should be kept in mind that the Solar One system was more mature than the dish at the time of this comparison. Solar One began its fourth

**Table 2.51:** Performance Comparison Solar One Test Site (Jan. - June 1986)

| | Insolation (KWHr/m$^2$) | Net Electricity (KWHr/m$^2$) SOLAR ONE | DISH STIRLING |
|---|---|---|---|
| January | 160.1 | 8.0 | 29.1 |
| February | 153.0 | 6.3 | 7.1 |
| March | 184.3 | 10.0 | 32.6 |
| April | 220.7 | 14.7 | 35.0 |
| May | 265.4 | 19.8 | 40.1 |
| June | 267.2 | 12.2 | 41.0 |
| Total (6 months) | 1250.7 | 71.0 | 184.9 |
| Net Efficiency | | 5.7% | 14.8% |

year of operation (second year of its power production phase) during this period while this was part of the first year in the field for the Dish Stirling module.

It can be noted that the Dish Stirling system outperformed Solar One by a factor of 2.6. This is despite having comparable thermal to electric conversion efficiencies. Some understanding of the reasons for this difference can be gained by comparing the daily input-output charts shown in Figure 2.58. The envelope of the dish Stirling input-output chart is superposed on the Solar One chart to illustrate the large loss due to this threshold effect. The major difference that can be observed is the daily insolation threshold value required for a net positive output. This can be attributed to differences in:

- Thermal inertia
- Startup times
- Response to insolation transients
- Operation on intermittent cloudy days
- Parasitics

All of the central receiver system experiments analyzed displayed this threshold effect, even in their thermal energy collection (since there was little net electricity produced). Minimizing the impact of this effect must be a major goal of central receiver system designers. Until it is demonstrated experimentally, the high performance potential of the central receiver remains a speculative projection.

### 2.6.3 Summary

A summary comparison of the three systems is given in Table 2.52. The progressive improvement in the potential for producing low cost energy can be seen in moving from the trough to the dish system. A corresponding increase in the uncertainty band is also apparent.

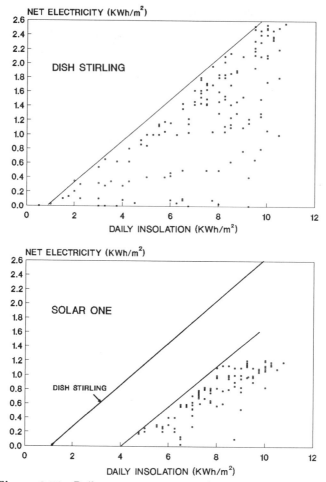

**Figure 2.58:** Daily output comparison (January - June, 1986).

## 2.7 Summary and Conclusions

Solar Thermal electric technology consists of the three technologies associated with the major system categories reviewed: the central receiver, the dish system and the trough system. Each will be summarized separately.

### 2.7.1 The Central Receiver

The central receiver has received the most substantial development support of any of the solar thermal electric system concepts. The current status of this concept and its future prospects are summarized below.

Several "generations" of heliostat hardware were built, tested and improved upon in successive iterations. A number of full system experiments were built and operated, the most notable being the 10 MW$_e$ Solar One installation at Daggett, California on the Southern California Edison Company grid. System experiments

**Table 2.52:** System Comparison

|  | Trough | Central Receiver | Dish Stirling |
|---|---|---|---|
| Cost ($/m$^2$) | 355-515 | 206-443 | 263-933 |
| Net Annual Eff. (%) | 9-12 | 8-17 | 16-28 |
| Energy Cost Equiv. to standard plant at ($/Wp) | 2.6-5.0 | 1.1-4.8 | 0.8-5.1 |
| Levelized Energy Cost (Cents/KWHr) | 13-25 | 6-24 | 5-25 |
| Development Cost ($ Million) | 0 | 100-500 | 50-200 |
| Development Time (Yr) | 0 | 15-20 | 10-20 |

included first generation systems, which employed water/steam as the thermal transport fluid (like Solar One), and advanced concepts, which used liquid sodium or molten salt for thermal transport. These experiments confirmed the technical feasibility of the concepts tested.

A preferred central receiver concept, which employs molten nitrate salt for thermal transport and storage, has been selected as representative of the central receiver system. The predicted performance and cost for this plant are superior to the other central receiver concepts evaluated.

Future electrical energy costs from this system are expected to be between 6 and 24 cents per kilowatt-hour; this corresponds to electricity from the standard solar plant costing $1.1 to 4.8 per peak watt. Major factors causing this large uncertainty ·in the energy cost predictions are:

- Net annual electrical energy production is highly uncertain. Data from system experiments employing molten salt for thermal transport are inadequate to either confirm or contradict the performance predictions. In fact, daily (24 hour) net positive output of electrical energy was never achieved by any of the advanced system experiments (employing either molten salt or liquid sodium for heat transport) despite very positive advance predictions. This result can be attributed to inadequacies of the experiments rather than to a fundamental deficiency of the technology, but it leaves the performance predictions unconfirmed. Consequently, current predictions of high performance are based on analytical simulations without experimental verification.

- The high freezing temperature of the molten salt heat transport and storage fluid, 221 °C (430 °F), creates a number of technology issues which have not been resolved fully.

- The cyclic operation of the receiver and "convetional" equipment exceeds the capability of many commercial-grade components.

- System costs are uncertain due to the early stage of equipment development and the conceptual level of the system design.

Significant development remains to bring this concept to commercial readiness. Highest priority should be given to reduction in the uncertainty of the annual net electricity output and resolution of technical issues associated with the molten salt fluid and the resulting impact on system costs.

The potential of this system concept warrants continued development in a phased program.

### 2.7.2 Parabolic Dish Systems

Parabolic dish technology has received much less development support than the central receiver. First and second generation concentrators were built and evaluated, but there has been very little hardware testing of the more advanced concepts. However, dish concentrator development did benefit from the more extensive heliostat program. System experiments employing central power generation were tested, but only single modules were tested with power generation at the dish.

The preferred system concept employs a Stirling cycle engine mounted on each dish for power generation. Tests on single modules have shown very high performance. A major factor contributing to this high performance is the direct coupling between the solar receiver and the engine. This limits thermal losses during the startup and shutdown cycles that are required with a variable solar input.

Future electrical energy costs from this system are expected to be between 5 and 25 cents per kilowatt-hour; this corresponds to electricity from the standard solar plant costing $0.8 to 5.1 per peak watt. Major factors contributing to this high level of uncertainty are:

- Operation and maintenance costs are highly uncertain. In fact, more than 50% of the equivalent capital cost in the upper bound to the system's cost represents the present value of the annual expenditures for operation and maintenance.
- System capital costs are uncertain because of the early stage of development. The specific engine and concentrator designs remain uncertain.

Significant development remains to bring this concept to commercial readiness. The reference kinematic Stirling engine has had substantial development, but maintenance requirements and costs are uncertain due to limited field testing. Resolution of this issue by extended testing is required in order to assess the adequacy of this design for the solar application. If required, potentially superior engine concepts have been designed, but they are at a much earlier stage of development. If sufficiently low maintenance costs can be confirmed for one of the candidate engine designs, development should proceed with (1) completion of engine development, (2) development of the concentrator, (3) integration of the engine and concentrator into a system module, (4) development of a multi-module system and (5) demonstration.

The potential of this system concept warrants support of the first stage of engine development and testing.

### 2.7.3 Parabolic Trough Systems

Parabolic trough systems are in an entirely different category than either of the preceding system concepts. Several large hybrid trough plants are operating in California as a commercial venture and more plants are being installed. They made use of: (1) a relatively advanced technology base, (2) the ability to burn natural gas to supply 25% of the plants' thermal energy, (3) favorable power purchase agreements and (4) federal and state tax incentives.

Following startup of the first plant in late 1984, subsequent plant designs incorporated improvements based on the lessons learned from the first plant's operation. This has been a most effective product improvement program and has brought the trough system to an advanced state of development.

Future electrical energy costs from this system in the solar-only mode are expected to be between 13 and 25 cents per kilowatt-hour; this corresponds to electricity from the standard solar plant costing $2.6 to 5.0 per peak watt. There is a much lower range of uncertainty for this concept than for either the central receiver or dish systems. This is, of course, due to the greater maturity of this technology. However, the potential of this concept is substantially lower than that of either the central receiver or the dish. This is due to the lower concentration of sunlight achievable with the trough which results in lower collection temperatures and lower power conversion efficiency. The confirmed energy cost is in the same range as the confirmed energy costs for the other two systems.

Because of its advanced state of development, further development is not required.

### 2.7.4 Overall Conclusion

Solar thermal electric technology has the potential for providing a low-cost renewable energy source for utility-scale electric power production. The concepts with the greatest potential are the central receiver and the dish Stirling systems. Neither is technically ready now for commercial deployment. However, the time required for the development of this technology is likely to be comparable to the time remaining before fossil fuel prices increase to a level that would make these alternatives economically attractive. Some level of support from public funds will be required to continue development of this technology. Neither the market nor the product is sufficiently near-term to attract significant corporate or investment capital (at least in the United States). This technology could provide a much-needed renewable energy supply in the future; however, the substantial investment already made in its development could be lost if continuity of development is not maintained.

# References

Advanco Corporation (1984), *Solar Parabolic Dish Stirling Engine System Module FY 1985 Test Plan*, DOE-AL-16333-3, El Segundo, California.

Alpert, D.J. et al. (1988), *Optical Performance of the First Prototype Stretched-Membrane Mirror Modules*, SAND 88-2620, Sandia National Laboratories, Albuquerque, New Mexico.

Alpert, D.J. and R.M. Houser (1988), *Performance Evaluation of Large Area Glass-Mirror Heliostats. Proceedings of the Fourth International Symposium on Research, Development and Applications of Solar Thermal Technology*, Santa Fe, New Mexico.

Alvis, R.L. (1980), *Modular Industrial Solar Retrofit Project (MISR).*, *Proceedings of the Line-Focus Solar Thermal Energy Technology Development, A Seminar For Industry*, SAND80-1666, Albuquerque, New Mexico, pp. 87–92.

Anderson, J.V. et al. (1986), *Direct Absorption Receiver System Assessment, Solar Energy Research Institute*, Final Draft, Golden, CO.

Arizona Public Service Company (1988), *Alternate Utility Team Utility Solar Central Receiver Study* Volumes 1 and 2, DOE/AL/38741-2, Phoenix, AZ.

Arizona Public Service Company (1988a), *Arizona Public Service Utility Solar Central Receiver Study* Volumes 1 and 2, DOE/AL/38741-1, Phoenix, AZ.

Avellaner, J. et al. (1985), *CESA-1 Project Status Report.*, *Proceedings of the Second International Workshop on the Design, Construction and Operation of Solar Central Receiver Projects*, Varese, Italy, D. Reidel Publishing Company, Dordrecht, pp. 73-82.

Avellaner, J., F. Sanchez, and C. Ortiz (1985), *Informe de la Primera Fase de Operacion y Ensayos* (in Spanish), Ministerio de Industria y Energia, Madrid, Spain.

Babcock and Wilcox (1984), *Molten Salt Steam Generator Subsystem Research Experiment*, SAND 84-8177, Barberton, Ohio.

Babcock and Wilcox (1986), *Molten Salt Electric Experiment Steam Generator Subsystem* Final Report, Sandia National Laboratories, SAND 85-8181, Livermore, California.

Baker, A.F. (1986), *International Energy Agency (IEA) Small Solar Power Systems (SSPS) Sodium Cavity and External Receiver Comparison.*, *Proceedings of the Third International Workshop June 23-27, 1986, Konstanz, Federal Republic of Germany*, Vol. 1, Berlin, Springer-Verlag, pp. 235-245.

Baker, A.F. (1987), *Daily Performance Summaries*, unpublished data, Sandia National Laboratories, Livermore, California.

Balanza, R. and J. Roman (1985), *CESA-1 Control, System and Cycle Operation and Status Report.*, CESA-1 Project Status Report., *Proceedings of the Second International Workshop on the Design, Construction and Operation of Solar Central Receiver Projects*, Varese, Italy, D. Reidel Publishing Company, Dordrecht, pp. 221-225.

Barber, R.E. (1985), *Small Community Experiment #1 Osage City, Kansas.*, *Proceedings of the Distributed Receiver Solar Thermal Technology Conference*, SAND 84-2454, Albuquerque, NM, pp. 13-20.

Bator, P.A. (1987), *Molten Salt Subsystem/Component Test Experiment Pump and Valve Test Results to Date.*, *Proceedings of the Solar Thermal Technology Conference*, Albuquerque, NM, p. 166.

Becker, M., (ed.) (1982), *IEA SSPS CRS Measurement Campaign in Almeria*, DFVLR, Cologne, Germany.

Becker, M., (ed.) (1983), *SSPS Central Receiver System Midterm Workshop*, Tabernas, Spain.

Becker, M. (ed.) (1986), *Solar Thermal Central Receiver Systems*, *Proceedings of the Third International Workshop*, June 23-27, 1986, Konstanz, F.R.G., Vols. 1 and 2, Springer-Verlag, Berlin.

Becker, M., H. Ellgering and D. Stahl (1983), *Construction Experience Report for the Central Receiver System of the International Energy Agency Small Solar Power Systems Project*, SR-2, DFVLR, Cologne, Germany.

Beninga, K. and B.L. Butler (1987), *Stretched-Membrane Heliostat Performance & Cost*, presented to Solar Central Receiver Utility Studies Technical Review Core Group, SAIC.

Beninga, K.J., B.L. Butler and J.A. Sandubrae (1988), *Improved Stretched Membrane Heliostat Mirror Module Development. Proceedings of the Fourth International Symposium on Research, Development and Applications of Solar Thermal Technology*, Santa Fe, NM.

Beremand, D.G. and R.K. Shaltens (1986), *NASA/DOE Automotive Stirling Engine Project Overview 1986*, NASA Lewis Research Center, DOE/NASA/50112-66; NASA TM-87345, Cleveland, Ohio.

Beremand, D.G. and J. Slaby (1988), *A Stirling Briefing.*, Power Technology Industry Review Volume I, NASA Lewis Research Center, Cleveland, Ohio.

Bezian, J.J. and B. Bonduelle (1985), *THEMIS Heliostat Field.*, Proceedings of the Second International Workshop on the Design, Construction and Operation of Solar Central Receiver Projects, Varese, Italy, D. Reidel Publishing Company, Dordrecht, 106-111.

Boeing Aerospace Company (1985), *Technical Assessment of Solar Thermal Facilities*, Electric Power Research Institute Contract RP2003-3.

Bohn, M.S. et al. (1986), *Direct Absorption Receiver Experiments and Concept Feasibility*, Solar Energy Research Institute, SERI/TR-252-2884, Golden, CO.

Bonduelle, B. (1986), *THEMIS Project: Description, Results and Lessons Learned.*, Briefing at SNLA CRTF.

Bonduelle, B. and A.M. Cazin-Bourguignon (1986), *THEMIS Receiver: Thermal Losses and Performance.*, Proceedings of the Third International Workshop June 23–27, 1986 Konstanz, FRG, Vol. 1, Berlin, Springer-Verlag, pp. 273–282.

Bonduelle, B. and B. Rivoire (1987), *Centrale Experimentale THEMIS, Results et Projection*, (THEMIS final report in French), CNRS, AFME, Font-Romen, France.

Borgese et al. (1983), *EURELIOS, the 1 MW$_e$ Helioelectric Power Plant of the European Community Program; Proceedings of the International Workshop on the Design, Construction and Operation of Solar Central Receiver Projects*, SAND82-8048, pp. 101–138.

Boutin, V. and B. Rivoire (1986), *THEMIS Annual Energy Estimate.*, Proceedings of the Third International Workshop June 23–27, 1986 Konstanz, FRG, Vol. 1, Berlin, Springer-Verlag, pp. 383–391.

Bucher, W. (1984), *SSPS-CRS First Period of Operation*, SR-4, DFVLR, Cologne, Germany.

Butler, B.L. et al. (1986), *Development of Stressed Membrane Heliostat Mirror Module* Final Report, Science Applications International Corporation, Draft SAIC-86/1872, San Diego, Calfornia.

Burns & McDonnell Engineering Company (1983), *10 MW$_e$ Solar Thermal Central Receiver Pilot Plant: Report on Lessons Learned & Project Documentation*, prepared for Electric Power Research Institute, RP-2003-2.

Cameron, C.P. (1988), Sandia National Laboratory, Personal communication, Albuquerque, NM.

Cameron, C.P. (1985), *The Distributed Receiver Test Facility Status and Plans. Proceedings of the Distributed Receiver Solar Thermal Technology Conference*, SAND84-2454, Albuquerque, NM, pp. 103–110.

Cameron, C.P. (1986), *Distributed Receiver Field Tests. Proceedings of the Solar Thermal Technology Conference*, SAND86-0536, Albuquerque, NM, pp. 183–192.

Cameron, C.P. and V.E. Dudley (1987), *Acurex Solar Corporation Modular Solar Retrofit Qualification Test Results*, Sandia National Laboratories, SAND85-2616, Revised, Albuquerque, NM.

Cameron, C.P. and V.E. Dudley (1987a), *The BMD Corporation Modular Solar Retrofit Qualification Test Results*, Sandia National Laboratories, SAND85-2617, Revised, Albuquerque, NM.

Cameron, C.P. and V.E. Dudley (1987b), *Custom Engineering, Incorporated Modular Solar Retrofit Qualification Test Results*, Sandia National Laboratories, SAND85-2618, Revised, Albuquerque, NM.

Cameron, C.P. and V.E. Dudley (1987c), *Solar Kinetics, Incorporated Modular Solar Retrofit Qualification Test Results*, Sandia National Laboratories, SAND85-2620, Revised, Albuquerque, NM.

Cameron, C.P., V.E. Dudley, and A.A. Lewandowski (1987), *Foster Wheeler Solar Development Corporation Modular Solar Retrofit Qualification Test Results*, Sandia National Laboratories, SAND85-2619, Revised, Albuquerque, NM.

Cameron, C.P. and J.W. Strachan (1987), *Distributed Receiver Test Facility Activities., Proceedings of the Solar Thermal Technology Conference*, SAND87-1258, Albuquerque, NM, pp. 115–121.

Champion, R.L. (1984), *Heliostat Development Program. Proceedings of the Department of Energy Solar Central Receiver Annual Meeting*, SAND85-8202, San Diego, California, pp. 145–148.

Chavez, J.M., C.E. Tyner, and W.A. Coliche (1988), *Direct Absorption Receiver Flow Testing and Evaluation. Proceedings of the Fourth International Symposium on Research, Development and Applications of Solar Thermal Technology*, Santa Fe, NM.

Coleman, G.C. and J.M. Friefeld (1980), *Pilot Plant Receiver Panel Testing at the Central Receiver Test Facility.* Department of Energy Large Solar Central Power Systems Semiannual Review, SAND80-8505, Albuquerque, NM, pp. 205–212.

Coleman, G.C. and J.E. Raetz (1986), *Field Performance of Dish/Stirling Solar Electric Systems., Proceedings of the Solar Thermal Technology Conference*, SAND86-0536, pp. 316–325.

Concentrating Solar Collectors: Key Technical Issues Workshop (1986), Albuquerque, New Mexico.

Corvi, C. and G. Dinelli (1985), *Technical Aspects of the EURELIOS Plant Operations., Proceedings of the Second International Workshop on the Design, Construction and Operation of Solar Central Receiver Projects, Varese, Italy*, Dordrecht: D. Reidel Publishing Company, pp. 10–20.

Cummings, R.D. (1985), *STEP Heat Loss Test Results*, Cummings Engineering, Wilmington, MA.

Darsey, D.M. et al. (1977), *Solar Thermal Test Facility Experiment Manual*, Sandia National Laboratories, SAND77-1173, Albuquerque, NM.

Delameter, W.R. and N.E. Bergen (1986), *Review of Molten Salt Electric Experiment: A Solar Central Receiver Project*, Sandia National Laboratories, SAND86-8449, Livermore, California.

DeLaquil, III, P. and J.V. Anderson (1984), *The Performance of High-Temperature Central Receiver Systems*, SAND84-8233, Sandia National Laboratories, Livermore, California.

Department of Energy Solar Central Receiver Semiannual Meeting (1980), SAND80-8049, San Francisco, California.

Department of Energy Solar Central Annual Meeting (1983), SAND83-8018, Albuquerque, New Mexico.

DFVLR (1981), *Small Solar Power System 500 KW$_e$ Distributed Collector System Tabernas (Almeria) Plant Design* Book No. 1, Cologne, Germany.

Doyle, J.F., P.B. Bos, and J.M. Weingart (1986), *Solar Thermal Central Receiver Integrated Commercialization Analysis*, Final Report, 3 Volumes, SAND86-8176, Polydyne, Inc. and Associates, San Mateo and Berkeley, California.

Droher, J.J. and S.E. Squier (1986), *Performance of the Vanguard Solar Dish-Stirling Engine Module*, Electric Power Research Institute, EPRI AP-4608, Palo Alto, California.

Drouot, L.P. (1983), *The THEMIS Program. Proceedings of the International Workshop on the Design, Construction and Operation of Solar Central Receiver Projects*, SAND82-8048, pp. 21–27.

Duffy, D., M. Matteo, and D. Rafinejad (1980), *Design, Construction and Operation of a 150 KW Solar-Powered Irrigation Facility*, Sandia National Laboratories, ALO-4159-1, Albuquerque, NM.

Electric Power Research Institute (1985), *Water-Steam Rankine-Cycle Solar Central Receiver Power Plant: Cost Estimates*, EPRI AP-3801, Palo Alto, California.

Electric Power Research Institute (1985), *Studies on Water-Steam Rankine-Cycle Solar Central Receiver Power Plants*, EPRI AP-3982, Palo Alto, California.

Electric Power Research Institute (1986), *TAG$^{TM}$ - Technical Assessment Guide* Volume 1: Electricity Supply 1986, EPRI P-4463-SR, Palo Alto, California, pp. 3–1 to 3–14.

Electric Power Research Institute (1987), *TAG$^{TM}$ – Technical Assessment Guide*, Volume 3: Fundamentals and Methods, Supply - 1986, EPRI P-4463SR, Volume 3, Palo Alto, California.

EPRI (1985), *Solar Thermal Workshop on Technical Problems and Issues*, Overland Park, Kansas.

Erickson, J.J. and C. LaPorta (1986), *Review of the Solar Thermal Power Industry: Future Outlook*, Sandia National Laboratories, SAND 86-8183, Albuquerque, NM, p. 51.

Etievant, C., A. Amri, M. Izygon, and B. Tedjiza (1987), *Central Receiver Plant Evaluation I) Insolation Data From Odeillo and Targasonne*, Sandia National Laboratories, SAND86-8185, Livermore, California.

Etievant, C., A. Amri, M. Izygon, and B. Tedjiza (1988), *Central Receiver Plant Evaluation II) Themis Collector Subsystem Evaluation*, Sandia National Laboratories, SAND87-8182, Livermore, California.

Etievant, C., A. Amri, M. Izygon, and B. Tedjiza (1988a), *Central Receiver Plant Evaluation III) Themis Receiver Subsystem Evaluation*, Sandia National Laboratories, SAND88-8101, Livermore, California.

Etievant, C., A. Amri, M. Izygon, and B. Tedjiza (1988b), *Central Receiver Plant Evaluation IV) Themis Thermal Storage Subsystem Evaluation*, Sandia National Laboratories, SAND88-8100, Livermore, California.

Etievant, C., A. Amri, M. Izygon, and B. Tedjiza (1988c), *Central Receiver Plant Evaluation V) Themis System Performance Evaluation*, Sandia National Laboratories, SAND88-8102, Livermore, California.

Fair, D. (1985), *Georgia Power Company Solar Total Energy Project Test Report for Thirty Consecutive Day Test*, Georgia Power Company, Shenandoah, GA.

Falcone, P.K. (1986), *A Handbook for Solar Central Receiver Design*, Sandia National Laboratories, SAND86-8009, Livermore, California.

Faas, S.E. (1986), SNLA; Personal Communication and trip report from visit to THEMIS.

Faas, S.E. and W.S. Winters (1986), *An Evaluation of Heliostat Field/Receiver Configurations*, SAND86-8007, Sandia National Laboratories, Livermore, California.

Fricker, H. (1987), *The Phoebus Project*, 30 MW$_e$ *Solar Demonstration Plant*, Albuquerque, NM.

Georgia Power Company (1988), *Solar Total Energy Project Summary Report*, SAND87-7108, Shenandoah, Georgia.

Goldberg, V.R. (1980), *Test Bed Concentrator. Proceedings of the First Semi-Annual Distributed Receiver System Program Review January 22-24, 1980*, Jet Propulsion Laboratory, DOE/JPL-1060-33, Pasadena, California.

Grasse, W. (1985), *SSPS Results of Test & Operation 1981 - 1984*, SR-7, DFVLR, Cologne, Germany.

Grasse, W. (1986), *The Central Receiver System (CRS) of the SSPS Project: Subsystem and System Performance.* In Solar Thermal Central Receiver Systems, *Proceedings of the Third International Workshop June 23-27, 1986 Konstanz, Federal Republic of Germany, Vol. 1*, Berlin, Springer-Verlag, pp. 23–45.

Grasse, W. and M. Becker (1983), *Central Receiver System (CRS) in the Small Solar Power Systems Project of the International Energy Agency., Proceedings of the International Workshop on the Design, Construction and Operation of Solar Central Receiver Projects*, SAND82-8048, pp. 75–99.

Grasse, W. and F. Ruiz (1985), *Central Receiver Solar Power Plant of the IEA/SSPS Project - Summary of Results and Experiences After Three Years of Testing. Proceedings of the Second International Workshop on the Design, Construction and Operation of Solar Central Receiver Projects, Varese, Italy*, Dordrecht: D. Reidel Publishing Company, pp. 21–34.

Gretz, J. (1981), *EURELIOS, the 1 MW$_e$ Helioelectric Power plant of the EC in Adrano, Sicily, Italy. Proceedings of Department of Energy Solar Central Receiver Semiannual Meeting*, Sandia National Laboratories, SAND82-8048, Livermore, California, pp. 51–63.

Gretz, J. (1986), *Concept and Operating Experiences with EURELIOS.*, Solar Thermal Central Receiver Systems, *Proceedings of the Third International Work-*

*shop June 23–27, 1986 Konstanz, Federal Republic of Germany, Vol. 1*, Berlin, Springer-Verlag, pp. 65–79.

Gretz, J., A. Strub, and W. Palz (1984), *Thermo-Mechanical Solar Power Plants - EURELIOS, The* 1 MW$_e$ *Experimental Solar Thermal Electric Power Plant in the European Community*, Dordrecht: D. Reidel Publishing Company.

Gretz, J., A.S. Strub, and A. Skinrood (eds.) (1985), *Thermo-Mechanical Solar Power Plants, Proceedings of the Second International Workshop on the Design, Construction and Operation of Solar Central Receiver Projects*, Varese, Italy, 4–8 June 1984, Vols. 1 and 2, D. Reidel Publishing Company, Dordrecht, Holland.

Hallet, R.W., R.J. Holl, and C. Bratt (1985), *The Dish Stirling System for Solar Electric Power Production.*, McDonnell Douglas Astronautics Company, Huntington Beach, California.

Hansen, J. (1984), *SSPS-CRS Advanced Sodium Receiver Construction Experience Report*, SR-5, DFVLR, Cologne, Germany.

Hansen, J. (1984a), *SSPS-DCS Supplement Construction Experience Report*, DFVLR, IEA-SSPS Operating Agent SR 6, Cologne, Germany.

Harder, J.E., Pascal De Laquil III, et al. (1988), *Risk Identification, Assessment, and Mitigation for the Utility Solar Central Receiver Studies. Proceedings of the Fourth International Symposium on Research, Development and Applications of Solar Thermal Technology*, Santa Fe, NM.

Hewett, R. (1986), *Dish Solar System State-of-the-Art the Concentrator Subsystem*, Solar Energy Research Institute, Preliminary Draft, Golden, CO.

Hicks, T.H. (1985), *Georgia Power Company Solar Total Energy Project Test Report for Continuous Fourteen Day Commercial Operations*, Georgia Power Company, Shenandoah, GA.

Hillairet, M.J. (1983), *2500 KW THEMIS Solar Power station at Targasonne. Proceedings of the International Workshop on the Design, Construction and Operation of Solar Central Receiver Projects*, SAND82-8048, pp. 29–44.

Hillesland, Jr., T. and E.R. Weber (1988), *Utilities' Study of Solar Central Receivers. Proceedings of the Fourth International Symposium on Research, Development and Applications of Solar Thermal Technology*, Santa Fe, NM.

Holgerson, S. and W.H. Percival (1982), *The 4-95 Solar Stirling Engine - A Progress Report*. Presented at U.S. Department of Energy Automotive Technology Development Contractor Coordination Meeting, Dearborn, MI.

Holgersson, S. (1984), *United Stirling's Solar Engine Development - The Background for the Vanguard Engine. Proceedings Fifth Parabolic Dish Solar Thermal Power Program Annual Review December 6–8, 1983*, Indian Wells, California, Jet Propulsion Laboratory, DOE/JPL 1060-69, Pasadena, California, pp. 95–101.

Holl, R.J., D.R. Barron, and S.A. Saloff (1989), *Molten Salt Solar Electric Experiment Volume 1: Testing, Operation and Evaluation*, Final Report, Contractor report by McDonnell Douglas, Palo Alto, California: Electric Power Research Institute, EPRI GS-6577.

Holmes, J.T. (1985), *Electric Heating for High-Temperature Heat Transport Fluids*, Sandia National Laboratories, SAND85-0379, Albuquerque, NM.

Holmes, J.T. et al. (1980), *Operating Experience at the Central Receiver Test Facility (CRTF)*, Sandia National Laboratories, SAND80-2504C, Albuquerque, NM.

Instituto de Energias Renovables/CIEMAT (1987), *IEA Small Solar Power Systems Project Task 1* Final Report, Madrid, Spain.

Interatom (1982), *Central Receiver System Solar Power Plant Almeria/Spain Plant Description*, Germany.

International Workshop on the Design, Construction, and Operation of Solar Central Receiver Projects (1982), SAND82-8048, Claremont, California.

Jaffe, D., S. Friedlander, and D. Kearney (1987), *The LUZ Solar Electric Generating Systems in California, LUZ Engineering Corporation*, Los Angeles, California.

Kalt, A. and H. Dehne (1982), *Operational Strategies for the SSPS DCS Plant. Proceedings of the DCS Midterm Workshop*, Almeria, Spain.

Kalt, A., M. Loosme, and H. Dehne (1982), *Distributed Collector System Plant Construction Report*, DFVLR, IEA-SSPS Operating Agent, SSPS SR 1, Cologne, Germany.

Kaneff, S. (1983), *The White Cliffs Solar Power Station. Proceedings Fourth Parabolic Dish Solar Thermal Program Review Nov. 30-Dec. 2, 1982*, Jet Propulsion Laboratory, DOE/JPL-1060-58, Pasadena, California, pp. 299–317.

Kaneff, S. (1984), *White Cliffs - Operating Experience. Proceedings Fifth Parabolic Dish Solar Thermal Power Program Annual Review December 6-8, 1983, Indian Wells, California*, Jet Propulsion Laboratory, DOE/JPL 1060-69, Pasadena, California, pp. 146–158.

Kearney, D.W. (1986), *SEGS I White Paper 1985 Operation and Performance*, LUZ Engineering Corporation, Westwood, California.

Kearney, D. (1988), LUZ Engineering Corporation, Personal Communications, Westwood, California.

Kearney, D.W. and H.W. Price (1987), *Overview of the SEGS Plants.*, American Solar Energy Society Solar '87 Conference, Portland, Oregon.

Kesselring, P. and C.J. Winter (1987), Forward to Frederico G. Casal. *Solar Thermal Power Plants*, Springer-Verlag, Berlin, p. vii.

Kiceniuk, T. (1985), *Development of an Organic Rankine-Cycle Power Module for a Small Community Solar Thermal Power Experiment*, DOE/JPL-1060-80, Jet Propulsion Laboratory, Pasadena, California.

Kinoshita, G.S. (1985), *The Shenandoah Parabolic Dish Solar Collector*, Sandia National Laboratories, SAND83-0583, Albuquerque, NM.

Kolb, G.J. and U. Nikolai (1988), *Performance Evaluation of Molten Salt Thermal Storage Systems*, Sandia National Laboratories, SAND87-3002, Albuquerque, NM.

Laguia, P. (1982), *Operational Behavior of the DCS Plant.*, *Proceedings of the DCS Midterm Workshop*, Almeria, Spain. pp. 439–467.

Larson, D.L. (1983), *1982 Annual Report of the Coolidge Solar Irrigation Project*, Sandia National Laboratories, SAND83-7124, Albuquerque, NM.

Larson, D.L. (1983a), *Final Report of the Coolidge Solar Irrigation Project*, Sandia National Laboratories, SAND83-7125, Albuquerque, NM.

Leonard, J.A. et al. (1983), *Session II - MISR System and Plant Interface Designs and Qualification Results. Proceedings of the Distributed Solar Collector Summary Conference – Technology and Applications*, SAND83-0137C, Albuquerque, NM, pp. 135–246.

Lewandowski, A., R. Gee, and K. May (1984), *Industrial Process Heat Data Analysis and Evaluation*, Vols. 1 and 2, SERI/TR-253-2161, Solar Energy Research Institute, Golden, Colorado.

Linker, K.L. (1986), *Heat Engine Development for Solar Thermal Dish-Electric Power Plants*, Sandia National Laboratories, SAND86-0289, Albuquerque, NM.

Lopez, C. (1986), SCE; Personal communications and trip report from visit to THEMIS.

Lopez, C. (1987 and 1988), Private communications, Solar One test Center Manager.

Luke, A.G. and C.D. Miserlis (1977), *How Steam and Electric Tracing Compare in Plant Operations.*, Oil and Gas Journal.

Mahoney, A.R. et al. (1987), *Sol-Gel Planarized Flexible Solar Mirrors. Proceedings of the Solar Thermal Technology Conference*, SAND87-1258, Albuqueque, NM, pp. 91–100.

Martin Marietta Denver Aerospace (1981), *Alternate Central Receiver Power System, Phase II* Final Report, MCR-81-1707, Denver, Colorado.

Martin Marietta Aerospace (1982), *Molten Salt Thermal Energy Storage Subsystem Research Experiment*, Volumes I and II, SAND80-8192, Denver, Colorado.

Martin Marietta Aerospace (1985), *Molten Salt Electric Experiment (MSEE) - Phase I Report* (3 Volumes), Sandia National Laboratories, SAND85-8175, Livermore, California.

Martinez, F. (1985), *CESA-1 Staffing, Operation and Maintenance Status Report*, CESA-1 Project Status Report. *Proceedings of the Second International Workshop on the Design, Construction and Operation of Solar Central Receiver Projects*, Varese, Italy, D. Reidel Publishing Company, Dordrecht, pp. 238–246.

Masele, V., F. Pharabod, and B. Rivoire (1985), *THEMIS Plant Operation Progress Report, Proceedings of the Second International Workshop on the Design, Construction and Operation of Solar Central Receiver Projects, Varese Italy*, Dordrecht: D. Reidel Publishing Company, pp. 62–72.

Mavis, C.L. (1986), *Solar Thermal Central Receiver Heliostat Technology, Proceedings of the Solar Thermal Technology Conference*, SAND86-0536, Albuquerque, NM, pp. 61–72.

Maxwell, C. and J. Holmes (1987), *Central Receiver Test Facility Experiment Manual*, Sandia National Laboratories, SAND86-1492, Albuquerque, NM.

McDonnell Douglas Company under Department of Energy contract DE-AC03-79SF10499 (1982), Master Equipment List (RADL item 2-19).

McDonnell Douglas Astronautics Company (1983), 10 MW$_e$ *Solar Thermal Central Receiver Pilot Plant, Mode 1 (1110) Test Report*, Huntington Beach, California.

McDonnell Douglas Astronautics Company under Department of Energy contract DE-AC03-79SF10499 (1985), *Pilot Plant Station Manual* (RADL Item 2-1), Volume 1, System Description, Revised.

McDonnell Douglas Astronautics Company (1986), unpublished material from MDC/USAB dish Stirling program.

McGlaun, M.A. (1987), *LaJet Energy Company Update of Solar Plant 1. Proceedings of the Solar Thermal Technology Conference*, SAND87-1258, Albuquerque, NM, pp. 142–150.

MDC/USAB dish Stirling program, unpublished material

Mechanical Technology Incorporated (1988), *Conceptual Design of an Advanced Stirling Conversion System for Terrestrial Power Generation*, DOE/NASA/0372-1; NASA CR-180890, Latham, NY.

Meijer, R.J. (1987), *The Evolution of the Stirling Engine*, Stirling Thermal Motors, Inc., Ann Arbor, MI.

Ministerio de Industria y Energia (1983), *Descripcion General del Proyecto CESA-1.* (in Spanish), CESA-84-1-2, Madrid, Spain.

Ministerio de Industria Y Energia (1987), *International 30 MW$_e$ Solar Tower Plant, Feasibility Study - Phase I Presentation of Results*, Madrid, Spain.

Moustafa, S.M., H. El-Mansy, and H. Zewen (1983), *Operational Strategies for Kuwait's 100 KW$_e$/0.7 MW$_{th}$ Solar Power Plant*. Presented at Solar World Congress, Perth, Western Australia.

Munjal, P.K. (1985), *Solar Central Receiver Preliminary Design Studies Summary and Review of Contract Results*, ATR-85(5836)-IND, The Aerospace Corporation, El Segundo, California.

Murphy, L.M. (1983), *Technical and Cost Benefits of Lightweight, Stretched-Membrane Heliostats*, Solar Energy Research Institute, SERI/TR-253-1818, Golden, Colorado.

Murphy, L.M. (1986), *Stretched-Membrane Heliostat Technology.*, J. Solar Energy Engineering **108**, pp. 230–238.

Murphy, L.M. et al. (1985), *System Performance and Cost Sensitivity Comparisons of Stretched Membrane Heliostat Reflectors with Current Generation Glass/Metal Concepts*, Solar Energy Research Institute, SERI/TR-253-2694, Golden, Colorado.

Murphy, L.M. et al. (1986), *Structural Design Considerations for Stretched-Membrane Heliostat Reflector with Stability and Initial Imperfection Considerations*, Solar Energy Research Institute, SERI/TR-253-2338, Golden, Colorado.

Nelving, H.G. (1983), *Testing of the United Stirling 4-95 Solar Stirling Engine on the Test Bed Concentrator. Proceedings Fifth Parabolic Dish Solar Thermal Power Program Annual Review*, DOE/JPL-1060-69, Indian Wells, California.

Ney, E.J. (1984), *Solar Total Energy Project Shenandoah, Georgia Site Annual Technical Progress Report for the Period July 1, 1983 through June 30, 1984*, ALO/3994-84/1, Georgia Power Company, Atlanta, GA.

Ney, E.J. (1987), *Solar Total Energy Project at Shenandoah, Georgia.*, presented at Solar Thermal Technology Conference, Albuquerque, NM.

Ney, E.J. and W.H. Weidenbach (1983), *Development of the Solar Total Energy Project (STEP) at Shenandoah, Georgia (U.S.A.)*, Solar 83 International Solar Energy Symposium, Palma de Mallorca.

Nightingale, N.P. (1986), *Automotive Stirling Engine Mod II Design Report*, Mechanical Technology Incorporated, DOE/NASA/0032-28; NASA CR-175106, Latham, NY.

Nola, S. (1989), Southern California Edison Company, Rosemead, California, Personal Communications.

Norris, Jr., H.F. (1986), *Total Capital Cost Data Base - 10 MW$_e$ Solar Thermal Central Receiver Pilot Plant*, SAND86-8002, Sandia National Laboratories, Livermore, California.

Ohlson, R.N. (1979), *A User's Experience with Current Self-Limiting Heat Tracing Cable*, 1979 IEEE Conference, Paper Number PCI-80-25.

Pacific Gas and Electric Company (1988), *Solar Central Receiver Technology Advancement for Electric Utility Applications Phase I Topical Report* Volumes 1 and 2, San Ramon, California.

Pacific Gas and Electric Company (1988a), *Utility Solar Central Receive Study Phase IIB Review Meeting*, San Ramon, California.

Panda, P.L., T. Fujita, and J.W. Lucas (1985), *Summary Assessment of Solar Thermal Parabolic Dish Technology for Electrical Power Generation*, Jet Propulsion Laboratory, DOE/JPL-1060-89, Pasadena, California, pp. 2–53 to 2–58 and 3–1 to 3–4.

Pappas, G.N. et al. (1983), *Session III - Solar Field Projects. Proceedings of the Distributed Solar Collector Summary Conference – Technology and Applications*, SAND83-0137C, Albuquerque, NM, pp. 247–357.

Pharabod, F., J.J. Bezian, B. Bonduelle, B. Rivoire, and J. Guillard (1986), *THEMIS Evaluation Report.*, Solar Thermal Central Receiver Systems, *Proceedings of the Third International Workshop June 23–27, 1986 Konstanz, Federal Republic of Germany, Vol. 1*, Berlin, Springer-Verlag, pp. 91–103.

Phoebus Executive Summary of Phase IA Work (1988).

Phoebus (1988), - *Results of the System Comparison of the 30 MW$_e$ European Feasibility Study, Proceedings of the Fourth International Symposium on Research, Development and Applications of Solar Thermal Technology*, Santa Fe, NM.

Radosevich, L.G. and C.E. Wyman (1982), *Thermal Energy Storage Development for Solar Thermal Applications*, Sandia National Laboratories, SAND81-8897, Livermore, California.

Risser, V.V. and K.W. Stokes (1987), *Herperia Photovoltaic Power Plant: 1985 Performance Assessment*, New Mexico Solar Energy Institute, EPRI AP-5229, Las Cruces, New Mexico.

Risser, V.V. and K.W. Stokes (1988), *Photovoltaic Field Test Performance Assessment: 1986*, New Mexico Solar Energy Institute, EPRI AP-5762, Las Cruces, New Mexico.

Risser, V.V. and K.W. Stokes (1989), *Photovoltaic Field Test Performance Assessment*, Southwest Technology Development Institute, EPRI GS-6251, Las Cruces, New Mexico.

Rivoire, B. and B. Bonduelle (1986), *THEMIS Solar Plant Technology Heliostat Field and Molten Salt Primary Loop*, GEST.

Rivoire, B. and B. Bonduelle (1986a), GEST; Personal Communications.

Rockwell International (1983), *Final Report, Sodium Solar Receiver Experiment*, SAND82-8192, Canoga Park, California.

Rogan, J.E. (1985), *The McDonnell Douglas/USAB Dish/Stirling System.*, *Proceedings of the Distributed Receiver Solar Thermal Technology Conference*, Sandia National Laboratories, SAND84-2454, Albuquerque, NM, pp. 31–39.

Sanchez, F. (1985), *CESA-1 Heliostat Field Evaluation Status Report.*, CESA-1 Project Status Report., *Proceedings of the Second International Workshop on the Design, Construction and Operation of Solar Central Receiver Projects*, Varese, Italy, D. Reidel Publishing Company, Dordrecht, pp. 120–123.

Sanchez, F. (1986), *Results of CESA-1 Plant.*, Solar Thermal Central Receiver Systems *Proceedings of the Third International Workshop June 23–27, 1986 Konstanz, Federal Republic of Germany, Vol. 1*, Berlin, Springer-Verlag, pp. 46–63. Sandgren, J. and M. Andersson (1985), SSPS Evaluation and Comparison on a Thermal Basis, Energy Research Commission, Sweden.

Sandia National Laboratories (1981), *Testing of the Prototype Heliostats for the Solar Thermal Central Receiver Pilot Plant*, SAND81-8008, Livermore, California.

Sandia National Laboratories (1982), *Second Generation Heliostat Evaluation Summary Report*, SAND81-8034, Livermore, California.

Sandia National Laboratories (1985), *Final Report on the Experimental Test & Evaluation Phase of the 10 MW$_e$ Solar Thermal Central Receiver Pilot Plant*, SAND85-8015, Livermore, California.

Sandia National Laboratories (1985a), *10 MW$_e$ Solar Thermal Central Receiver Pilot Plant: 1984 Summer Solstice Power Production Test*, SAND85-8216, Livermore, California.

Sandia National Laboratories (1985b), *Proceedings of the Department of Energy Solar Central Receiver Annual Meeting*, SAND85-8202, Livermore, California.

Sandia National Laboratories (1985c), *A Study of Alternative System Conversions for the Solar One Pilot Plant*, SAND85-8212, Albuquerque, NM.

Sandia National Laboratories (1987), *MSS/CTE Quarterly Review Presentations*, Albuquerque, NM.

Sandia National Laboratories (1988), *MSS/CTE Pump and Valve Test Conference*, Albuquerque, NM.

Sandia National Laboratories (1988a), *Final Report on the Power Production Phase of the 10 MW$_e$ Solar Thermal Central Receiver Pilot Plant*, SAND87-8002, Livermore, California.

Sandia National Laboratories (1988b), *10 MW$_e$ Solar Thermal Central Receiver Pilot Plant Receiver Performance Final Report*, SAND88-8000, Livermore, California.

Selvage, C. (1984), *500 KW Central Receiver System (CRS) of the Small Solar Power Systems (SSPS) Project - Almeria, Spain.*, *Proceedings of the DOE Solar Central Receiver Annual Meeting*, SAND85-8202, pp. 66–93.

Shaltens, R.K. (1987), *Comparison of Stirling Engines for Use with a 25 KW Dish-Electric Conversion System*, NASA-Lewis Research Center, Cleveland, Ohio, DOE/NASA/33408-2, NASA TM-100111, AIAA-87-9069.

Small Solar Power System 500 KW$_e$ Distributed Collector System Tabernas (Almeria) Plant Data Book No. 1, DFVLR, Cologne, Germany.

Smith, D.C. and J.M. Chavez (1988), *A Final Report on the Phase I Testing of a Molten-Salt Cavity Receiver*, Vols. I–III, Sandia National Laboratories, SAND87-2290, Albuquerque, NM.

Solar Central Receiver Technology Workshop (1986), Albuquerque, New Mexico.

Solar Energy Research Institute (1987), *Power From the Sun: Principles of High Temperature Solar Thermal Technology*, SERI/SP-273-3054, Golden, Colorado.

Solar Kinetics, Inc. (1986), *Development of the Stressed Membrane Heliostat*, Draft, Dallas, TX.

Solar Kinetics, Inc. (1987), *Point-Focus Concentrator Reflector Assembly, Phase I*, SAND87-7014, Dallas, Texas.

Solar One monthly operation and maintenance reports.

SSPS Monthly Plant Operation Reports and Daily Evaluation Summaries (1986).

SSPS Operating Agent IER/CIEMAT (1987), *IEA Small Solar Power Systems Project Task 1* Final Report, Madrid, Spain.

Stine, W.B. and A.A. Heckes (1988), *Performance of the Solar Total Energy Project at Shenandoah, Georgia*, Sandia National Laboratories, SAND86-1910, Albuquerque, NM.

Stirling Technology Company (1988), 25 $KW_e$ *Solar Thermal Stirling Hydraulic Engine System Final Conceptual Design Report*, DOE/NASA/0371-1; NASA CR-180889, Richland, Washington.

Strachan, J.W. (1987), *An Evaluation of the LEC-460 Solar Collector*, Sandia National Laboratories, SAND87-0852, Albuquerque, NM.

Thomas, R.J. et al. (1987), *Large Area Heliostat Development. Proceedings of the Solar Thermal Technology Conference*, SAND87-1258, Albuquerque, NM, pp. 203–211.

Thornton, J.P. et al. (1980), *Final Report: Comparative Ranking of 0.1–10 $MW_e$ Solar Thermal Electric Power Systems*, Vols. 1 and 2, SERI/TR-351-461, Solar Energy Research Institute, Golden, Colorado.

Torkelson, L.E. and D.L. Larson (1981), *1980 Annual Report of the Coolidge Solar Irrigation Project*, Sandia National Laboratories, SAND80-2378, Albuquerque, NM.

Torkelson, L.E. and D.L. Larson (1982), *1981 Annual Report of the Coolidge Solar Irrigation Project*, Sandia National Laboratories, SAND82-0521, Albuquerque, NM.

Torralbo, A. Munoz et al. (1983), *A Spanish Power Tower Solar System, The Project CESA-1. Proceedings of the International Workshop on the Design, Construction and Operation of Solar Central Receiver Projects*, SAND82-8048, pp. 139–152.

Tyner, C.E. and S.F. Wu (1988), *Commercial Direct Absorption Receiver Design Studies. Proceedings of the Fourth International Symposium on Research, Development and Applications of Solar Thermal Technology*, Santa Fe, NM.

U.S. Department of Energy (1984), *National Solar Thermal Technology Program, Five Year Research and Development Plan, 1986–1990*, Washington, D.C.

U.S. Department of Energy (1985), *Solar Thermal Technical Information Guide*, SERI/SP-271-2511, Washington, D.C.: Government Printing Office.

U.S. Department of Energy (1986), *Solar Thermal Technology, Annual Evaluation Report, 5 Reports for Fiscal years 1982 through 1986*. National Technical Information Service, DOE/JPL-1060-61, 2 Vols. (July 1983), SERI/PR-253-2188, August 1984, DOE/CE-T13 (July 1985), DOE/CH10093-1 (August 1986), DOE/CH10093-12 (July 1987).

Washom, B.J. (1984), *Vanguard I Solar Parabolic Dish-Stirling Engine Module* Final Report, Advanco Corporation, DOE-AL-16333-2, El Segundo, California.

Washom, B.J. (1985), *Vanguard's 1200 Hour on Grid-Connected, On Sun Operations. Proceedings of the Distributed Receiver Thermal Technology Conference*, Sandia National Laboratories, SAND84-2454, Albuquerque, NM, pp. 271–276.

Weingart, J.M. (1986), *The U.S. Solar Thermal Energy Program: Commercializing the Solar Vision* (Revised Version). Jerome Weingart and Associates, Berkeley, California, p. 15.

White, D.L. (1988), *Stretched Membrane Heliostat Mirror-Module Design Improvement., Proceedings of the Fourth International Symposium on Research, Development and Applications of Solar Thermal Technology*, Santa Fe, NM.

William, T.A. et al. (1987), *Characterization of Solar Thermal Concepts for Electricity Generation, Volume 1 - Analyses & Evaluation*, Battelle Pacific Northwest Laboratories, PNL-6128, Richland, Washington.

Wolf, S. and E.A. Hernandez (1981), *Thermal Performance and Dynamic Stability Evaluation of Solar Pilot Plant Receiver Panel Test at CRTF*, Sandia National Laboratories, SAND81-8181, Albuqueque, NM.

Workhoven, R.M. (1980), *Mid-Temperature Solar Systems Test Facility (MSSTF). Proceedings of the Line-Focus Solar Thermal Energy Technology Development, A Seminar for Industry*, SAND80-1666, Albuquerque, NM, pp. 93–113.

Zewen, H. (1986), *Messerschmitt-Boelkow-Blohm GmbH*, Munich Federal Republic of Germany, Personal communication.

Zewen, H., G. Schmidt, and S. Moustafa (1983), *The Kuwait Solar Thermal Power Station: Operational Experiences with the Station and its Agricultural Application.* Presented at Solar World Congress, Perth, Western Australia.

CHAPTER 3

# High-Efficiency III-V Solar Cells

John C.C. Fan, Mark B. Spitzer and Ronald P. Gale

## 3.1 Abstract

High efficiency solar cells have many practical applications. Concepts for obtaining high-efficiency solar cell modules are well understood. Single junction III-V cells composed of either gallium arsenide or indium phosphide have achieved high efficiencies. In particular, gallium arsenide cells have now reached 25.7% one-sun efficiency. To achieve over 30% one-sun efficiency, multi-bandgap structures may be needed with III-V cells playing a major role. A detailed description of multi-bandgap solar cells is given.

## 3.2 Introduction

Owing to the area-related balance of photovoltaic systems costs, there are significant advantages for high-efficiency photovoltaic modules versus low-efficiency modules. The balance of system costs include land costs, array structure costs and others, and these costs cannot be reduced easily. Figure 3.1 shows the relationship between flat-plate module cost (in 1987 dollars per square meter) and module efficiency as a function of levelized electricity cost. For a levelized electricity cost of $0.06 per kilowatt, the tradeoff between module cost and module efficiency is well illustrated. For example, for a 10% module, the allowable module cost is about $10 per square meter. For a 20% module, the allowable module cost is around $80 per square meter, eight times higher than that for 10% module. The cost advantage will be even greater if the area-related costs incurred during module fabrication, for example, the cost of antireflection coatings and contact fingers, the cost of cover-glass, and the cost of adhesive etc. are included in the calculation. It should be noted that Figure 3.1 is calculated using only $50 per square meter for the balance of system costs. If that cost remains high, high-efficiency cells may well be the type that would be most attractive for large-scale terrestrial applications.

It is well known that the computer chip prices have been coming down steadily every year. Chip price is one of the few things in life that inflation has not affected. The major cost improvement is due to the fact that one can get computer chips to provide the same functions with an ever decreasing chip area. Therefore, with successive years, each wafer produces more and more chips and the unit cost per chip has come down. In solar cells, it is somewhat similar. The collection efficiency is area related and sunlight is a diffuse and not a powerful energy source. Up to

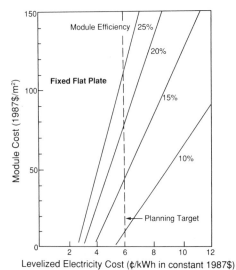

**Figure 3.1:** Module costs and efficiencies vs. 30 year levelized electricity cost for flat-plate photovoltaic structures.

a certain limit, by reducing the solar cell array size while maintaining the same amount of power in the cell by improving efficiency, it is possible to attain reduced cost through a higher level of performance.

With regard to space applications, the advantages of high efficiency and lighter modules are even more obvious. Figure 3.2 shows that the specific power requirement of NASA for future missions is ever increasing (Fan et al., 1984). In other words, it is necessary to provide higher and higher power with lighter-weight photovoltaic modules (i.e. higher power-to-weight ratios). To get higher power-to-weight ratios, one can increase the module efficiency and/or reduce the module weight. A third factor in space application is radiation sensitivity, which we shall discuss later.

Another important application for high efficiency cells is in concentrating sunlight configurations. In these configurations, sunlight is focused down by a lens or other means onto solar cells. In this case, the actual cell cost can be much higher because the major cost is in the concentrating optics. It is important that in concentrating configurations the cell efficiency should be as high as possible.

In view of the complex tradeoffs between module efficiency and a host of other parameters, including weight, radiation sensitivity, reliability and concentration ratio as well as module cost and balance-of-system cost, it is obvious that no one type of module will be ideal for all photovoltaic applications. It follows also that no one material system will be ideal for all applications. Our assessment of the most promising material systems for different applications will be given later in this chapter. We will now discuss the most likely configurations and the most probable major applications anticipated (Fan, 1984) for modules with efficiencies of 10, 15, 20, 25, 30%.

At this early stage in the development of photovoltaic technology and particularly of high-efficiency cells, only general conclusions can be drawn concerning the configurations and applications of modules at the various efficiency levels we are

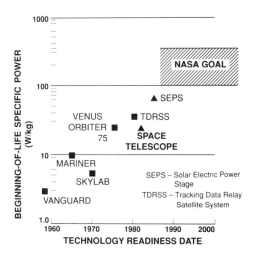

**Figure 3.2:** Specific power of space solar arrays vs. technology readiness date.

considering. For each level, Table 3.1 presents our views on whether the modules may be suitable for use in the flat plate or concentrator configurations and whether they may be applicable in terrestrial power generation, space power generation, or consumer electronics. Note that the table itself only indicates possibilities; it does not predict that the modules will ultimately be used in the configurations and applications listed for them. Two of the principal conclusions embodied in Table 3.1 deal with the limitations anticipated for modules with efficiencies of less than 20%. First, these low and medium-efficiency modules will not be suitable for use in concentrator configurations because the efficiency of their component cells will be too low for a favorable tradeoff with the capital and operating costs of concentrator systems. Second, because of their low power-to-weight ratio and low efficiency, these modules will not be suitable for most of the space applications leaving high-efficiency modules the only possible choice for such applications. With respect to terrestrial applications, although low - and medium - efficiency modules will initially play an important role, we expect that in the long term high-efficiency modules will become dominant as these modules become inexpensive. Perhaps, high-efficiency modules may become sufficiently inexpensive to justify their use in retrofitting terrestrial systems originally constructed with low-efficiency modules. Even in the area of consumer electronics, high-efficiency modules should be preferred for applications where available area is limited and substantial power is required.

In this chapter, we shall discuss research efforts directed toward developing high-efficiency cells for producing modules with efficiencies of 20–30% at AM 1. We conclude that III-V compounds should play very important roles in achieving such modules. Emphasis will be placed on cells for inexpensive flat-plate modules, although much of the research is also applicable to concentrator cells. The performance of single-junction cells, including those incorporating materials with different bandgaps, will be considered first. Next we will consider multi-junction, multibandgap cells (but not multi-junction cells fabricated from a single material).

**Table 3.1:** Configuration and utilization of photovoltaic modules of various efficiencies.

| MODULE EFFICIENCY | CONFIGURATION | | UTILIZATION | | |
|---|---|---|---|---|---|
| | FLAT-PLATE | CONCENTRATOR | TERRESTRIAL CELLS | SPACE CELLS | CONSUMER ELECTRONICS |
| LOW (10%) | X | | X | | X |
| MEDIUM (15%) | X | | X | | X |
| HIGH (20%) | X | X | X | X | X |
| VERY HIGH (25%) | X | X | X | X | X |
| ULTRAHIGH (30%) | X | X | X | X | X |

Finally, advanced concepts for achieving module efficiencies significantly above 30% will also be briefly described.

## 3.3 Single-Junction Crystalline Cells

Figure 3.3 shows a schematic diagram of a solar cell with its top surface (emitter) composed of a metal grid through which the light goes into a *p-n* junction device which produces a current and voltage, thus creating electrical power. Therefore, solar cells are solid-state devices which directly convert sunlight to electrical power. To get high conversion efficiency a solar cell must have high current, high voltage and high electrical coupling factor (usually called fill factor). The solar cell material absorbs the solar photons, generating electrons and holes. If the electrons and holes recombine before being separated by the junction, heat is generated. However, if the electrons and holes are separated by the junction before recombination, there will be current collection as well as voltage generation. The quality of the junction and the bandgap of the semiconductor material will determine the voltage values. If the material is of high quality, then the photogenerated carriers will diffuse to the junction. If the material is of very poor quality, then the photogenerated carriers actually will recombine creating heat before they reach the junction. In addition, after the carriers are separated at the junction, they still have to reach the contact grids for collection and delivery to the load. If the contact grids and contact resistance are not optimal, the fill factor can suffer.

The material structures that have been investigated for single-junction cells range from amorphous to polycrystalline to monocrystalline. However, for high-efficiency single-junction cells, only single-crystal or very-large-grained materials can be utilized. Figure 3.4 shows the theoretical maximum one-sun conversion efficiencies at AM 1 calculated as a function of energy gap $E_g$ for various operating temperatures (Fan et al., 1984). (In the rest of this chapter, unless otherwise stated, all efficiency values will be given for one-sun, AM 1 conditions.) The formulation of the calculations is given in Fan et al. (1983).

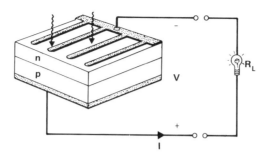

**Figure 3.3:** Schematic diagram of a solar cell.

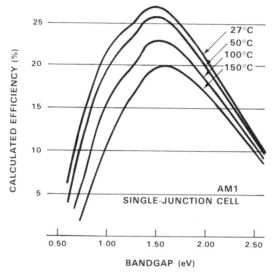

**Figure 3.4:** Calculated maximum AM 1 one-sun conversion efficiencies at 27, 50, 100, and 150°C as a function of bandgap for single-junction solar cell made of crystalline materials.

For single-junction cells, as shown in Figure 3.4, the highest calculated efficiencies are found for energy gaps of about 1.50 eV, which give values of about 27.5% at an operating temperature of 27°C. The efficiencies decrease with increasing temperature, especially for lower bandgap materials such as Si, with $E_g = 1.12$ eV. (In flat-plate configurations, normal module operating temperatures are about 40 − 50°C.) For Si the calculated maximum efficiency at 27°C is about 24%, and Si cells have recently achieved an efficiency of 23.0% (Green et al., 1988, Green, 1989). With advances in material quality and refinements in cell design, Si cells should improve a litte more. However, it is difficult to see how flat-plate modules with efficiencies much higher than 20% can be produced from Si cells, because the module efficiency is always somewhat lower than cell efficiency. Module efficiencies suffer mainly because of the variation in performance between the individual cells in each module and the shading losses from electrical interconnections and the area required by the structural framework of the module, as well as packing factor losses.

To achieve 20% module efficiency with single-junction cells will require a semi-conductor material with $E_g$ between 1.3 and 1.6 eV. Of the binary compounds meeting this requirement, GaAs ($E_g = 1.43$ eV) is currently the leading candidate. The other candidates are indium phosphide (InP) with a bandgap of 1.35 eV and cadmium telluride (CdTe) with a bandgap of 1.5 eV. GaAs cells have already attained efficiencies of 21-22% using a heteroface structure (Werthen et al., 1988) or a shallow-homojunction structure (Bozler and Fan, 1988). Recently, cells with efficiencies over 24% (Gale et al., 1988) have been obtained by using a GaAs/GaAlAs back-surface-field heterostructure (see later section in this chapter). With further refinements such as a GaInP$_2$ layer instead of a GaAlAs layer, one-sun efficiencies 25.7% have been obtained (Olson et al., 1989). The utilization of GaAs/GaAlAs or GaAs/GaInP$_2$ heterostructures is possible because these ternary alloys, whose energy gaps exceed that of GaAs, are lattice-matched to GaAs. Such heterostructures, by accomplishing either charge separation or charge confinement, can lead to a significant improvement in cell conversion efficiency. Figure 3.4 was calculated for conventional homojunction structures, and does not reflect the improvement that can result from advanced heterostructures.

Although GaAs cells are more efficient than Si cells, the higher bulk material cost and limited availability of GaAs compared to Si could be major obstacles for GaAs cells. Therefore, the amount of GaAs used in each cell must be kept low in order to reduce the material cost per cell and also to avoid limitations on material availability. This can be accomplished because GaAs absorbs sunlight very effectively, allowing GaAs cells to be much thinner than Si cells. Figure 3.5 plots the normalized photocurrent at AM 1 as a function of thickness for GaAs and Si. For GaAs a 2 $\mu$m thick layer can generate over 90% of the photocurrent produced by an infinite thickness layer. Silicon, however, requires a layer over 200 $\mu$m thick to achieve the same ratio. Obviously, light trapping concepts, in which sunlight is encouraged to pass through the cell several times, will allow thinner layers to be used. However, this affects the different materials the same way, and the relative ratio will be about the same.

For single-junction GaAs cells to be widely used in flat-plate modules, it will be necessary to make thin-film devices at reasonably low cost. To retain high efficiency, the cells must utilize high-quality, preferably single-crystal, GaAs films. If 20% modules can be achieved with 5 $\mu$m thick GaAs films, the cost of GaAs material per cell will be very low, and there should be enough Ga available for large-scale deployment of GaAs cells.

Table 3.2 lists the basic costs of the semiconductor material contained in Si and GaAs cells, obtained by multiplying the weight of the material in each cell by the current costs per gram of electronic grade Si and GaAs in polycrystalline form. For 20% efficient GaAs modules made from 5 $\mu$m thick cells, the GaAs material cost is about \$10 m$^{-2}$ or \$0.05/Wp. For 15% efficient Si modules made from 250 $\mu$m thick cells, the Si material cost is \$35 m$^{-2}$ or \$0.23/Wp. The price of raw polysilicon may come down from the current price, especially if material of lower quality than electronic grade can be used without significantly degrading module efficiencies. The price of GaAs raw material may also be reduced by larger volume production and by using material of lower quality.

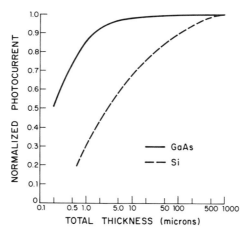

**Figure 3.5:** Normalized photocurrent computed for Si and GaAs as a function of solar cell thickness for AM 1.

**Table 3.2:** Basic material cost of Si and GaAs solar cells.

| MATERIAL | 1988 PRICE* ($/KG) | THICKNESS (μ m) | WEIGHT (g/m²) | MODULE (%) (AM1) | $/m² | $/W$_p$ |
|---|---|---|---|---|---|---|
| Si | 60 | 250 (10 mil) | 580 | 15 | 35 | 0.23 |
| GaAs | 400 | 5 | 26 | 20 | 10 | 0.05 |

*ELECTRONIC GRADE, POLYCRYSTAL

The actual material cost for solar cell modules, of course, depends on the yields of the material and device processing techniques used. Presently, the material yield for Si cells is estimated to be about 30–40%, that is, 1 kg raw Si polycrystal is required to produce cells containing 300–400 g of Si. Therefore, the net Si material cost amounts to about $100 m$^{-2}$. Referring to Figure 3.1, the allowable cost for a 15% module is only $45 m$^{-2}$, smaller than the material cost. For thin-film GaAs cells, it is difficult to estimate chemical yields, since such cells are generally grown by chemical vapor deposition (CVD). Although some CVD reactors have achieved 50% yields, current conventional reactors have yields of about 25%. On the basis of the latter value, the net material cost for GaAs modules is about $40 m$^{-2}$. Referring to Figure 3.1, the allowable cost for a 20% module is 80 m$^{-2}$, which is more than the estimated GaAs material cost. This evaluation strongly suggests that GaAs cells have a substantial advantage over Si cells with respect to material cost. However, much of this advantage will be lost if the expected reduction in the cost of gases such as AsCl$_3$, AsH$_3$ and TMG (trimethylgallium) does not materialize. Certainly, the present Si solar cells cost less than GaAs cells. This is primarily due to the

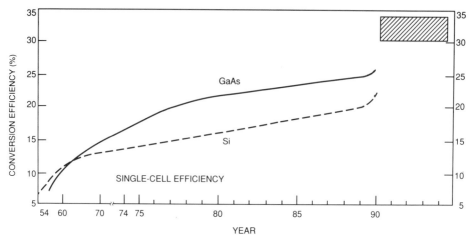

**Figure 3.6:** Measured one-sun conversion efficiencies of GaAs and Si in successive years (courtesy of John P. Benner of Solar Energy Research Institute, Golden, Colorado).

early stages of GaAs manufacturing development when compared with that of Si cells. The above calculations demonstrate that there is no intrinsic major barrier to low-cost GaAs solar cells; however, these cells must be of thin-film forms, if they are to be used in flat-plate configurations. Various techniques of producing thin-film GaAs solar cells will be described in later sections of this chapter.

It should be noted that GaAs material is very close to being the ideal photovoltaic material both for single-junction cells, and for multi-junction cells. The multi-junction cells will be described later. GaAs is the second most studied electronic material in the world, next to silicon. The material is very useful for fast electronic devices and circuits. The excellent light emitting properties of GaAs allow the commercialization of light emitting diodes and lasers, resulting in an annual commercial business in GaAs devices of over two billion dollars at the present time. Therefore, GaAs is a technologically proven material, with many important applications other than photovoltaics, and these applications will assist in the development and realization of an inexpensive photovoltaic material.

The other important III-V material is InP. Currently, there is strong interest in this material because of some exciting radiation-hardened properties of InP space solar cells. We will discuss InP cells later in this chapter, after the section on GaAs solar cells.

### 3.3.1 Bulk GaAs Solar Cells

Figure 3.6 shows the measured one-sun conversion efficiencies of GaAs and Si in successive years. Silicon started in the mid-fifties at only about 5% and kept improving. In 1988 silicon cells have reached about 20% at one sun, and in late 1989 they reached 23.0% at AM 1.5.

GaAs performance started below the efficiency of silicon until the early sixties when it crossed over. Since then, GaAs has always been more efficient than silicon. The efficiency of the GaAs cell has been keeping about 3 to 4 absolute percentage

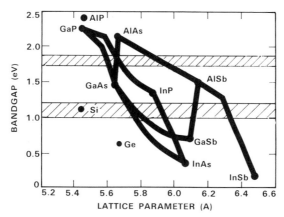

**Figure 3.7:** Different bandgap energies vs. lattice constants of different materials.

points higher than that of silicon. (This difference is predicted by calculated results shown in Figure 3.4). In 1988 GaAs was pushing around 24% at one sun. The efficiency has just reached 25.7%. Further increases in record efficiencies are expected. On the same curve we have shown that over 30% can possibly be achieved with a two-junction cell.

There are several key GaAs solar cell design considerations. One previously mentioned is the thickness of the GaAs cell. As a direct-bandgap semiconductor, GaAs need only be 2 to 4 $\mu$m thick to absorb almost all of the available solar spectrum above its bandgap of 1.4 eV. The active part of a GaAs cell is, therefore, thin which allows a variety of substrates to be used. These substrates include GaAs, Si, Ge and others discussed in later sections.

Another design consideration is the high surface recombination in GaAs (often over $10^7$ cm/sec). The high surface recombination velocity will cause the photogenerated carriers to recombine at the surface before they are separated by a $p$-$n$ junction. Two approaches have been used to minimize its effects on GaAs solar cells. The first uses a layer of high-bandgap AlGaAs over the GaAs emitter. This heterostructure has been found to greatly reduce the surface recombination velocity to below $10^4$ cm/sec. Figure 3.7 shows different bandgap energies versus lattice constants of different materials. An interesting point of GaAs and GaAlAs is that they have almost the same lattice constant, indicating that they are atomically and crystallographically very well matched. This means they will make an excellent interface for surface passivation. In addition, AlGaAs has a higher bandgap than GaAs, forming a barrier, or mirror for photogenerated carriers, thus increasing the effectiveness of collecting the photogenerated carriers. This approach is used in heteroface cells and some double heterostructure cells. These structures will be discussed in more detail in later sections. The second approach uses a very thin emitter layer to allow most of the light to be absorbed in the base of the cell and away from the surface. This structure was named the shallow homojunction cell and is discussed in more detail later.

Fabrication of these cell structures relies upon epitaxy for the deposition of the necessary layers. Initially, cells were grown by liquid-phase epitaxy (LPE)

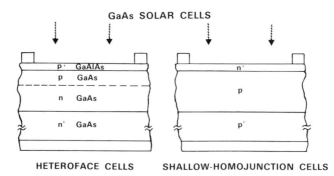

**Figure 3.8:** Schematic diagram of heteroface and shallow-homojunction GaAs solar cells.

consisting of a $p$-doped AlGaAs layer, deposited on an $n$-type GaAs substrate. The shallow-homojunction cell does not require AlGaAs, and is able to be deposited by chloride vapor-phase epitaxy (VPE). The development of organometallic chemical vapor deposition (OMCVD) provides more control over the doping, composition, and thickness of the individual layers, and allows more complex structures to be grown. These structures include AlGaAs/GaAs back-surface fields and double heterostructures. Molecular-beam epitaxy (MBE) was also used for growth of cells, but the large-area requirement for solar cells limits the potential of using MBE growth. For more details on these four different III-V growth techniques see the review by Hollan et al. (1980).

The control and flexibility of OMCVD also allows utilization of the various GaAs solar cell structures which have been developed. The best cells incorporate all of the advanced design structures. Both $p$-on-$n$ and $n$-on-$p$ structures have produced excellent results.

### 3.3.1A  Heteroface Cells

The first high-efficiency GaAs solar cell was reported by Woodall and Hovel, 1977. They used LPE to deposit a $p$-doped AlGaAs layer on an $n$-doped GaAs base, producing the heteroface cell structure. Because of outdiffusion of $p$ dopant into the base, a homojunction was formed in the GaAs. The AlGaAs layer acted as a passivating window for the front of the cell, yielding cells with reasonable quantum efficiencies in the short-wavelength region of the spectrum. With this structure, they were able to fabricate cells with efficiencies of about 17–18% at AM 1.

The heteroface cell structure (see Figure 3.8) was well suited for the existing GaAs material technology at the time, and has been further developed by many workers so that efficiencies up to 20% at AM 1 have been achieved. There were several problems with cell fabrication, however, relating to the AlGaAs window layer. First, reliable ohmic contacts to the AlGaAs layer are not easy to fabricate. Either the window layer was locally removed for grid contact, or a heavily-doped GaAs layer was grown above the window for contact purposes and later removed from the window. Second, the AlGaAs layer was prone to oxidation if the layers have large concentrations of Al. Finally, the LPE growth is not amenable to accurate control of layer thickness and uniformity.

### 3.3.1B Shallow Homojunction Cells

Another basic type of high-efficiency GaAs cell structure was named the shallow homojunction cell (Fan et al., 1978). In this structure, a thin, heavily doped n-type GaAs emitter was grown on a p-type base by VPE (see Figure 3.8). The emitter is thin in order to minimize the absorption of incident light in the emitter, thus minimizing the number of photogenerated carriers between the front surface and the junction. Most of the cell's contribution to the current comes from the base. Carrier diffusion to the junction in the base is maximized by the long diffusion length of electrons in the p-doped base. The base itself grown on a heavily doped p-type buffer layer or substrate, forms a back-surface field. This enhances the diffusion of photogenerated carriers that are generated deep in the GaAs by the longer wavelength light penetrating toward the junction.

The shallow homojunction structure solved many of the problems of the heteroface cell. With its heavily-doped GaAs emitter and substrate, the shallow homojunction was easy to contact. The cell was also very stable, with no AlGaAs needed for its fabrication. However, for maximum cell efficiency, the emitter thickness needed to be optimized. This optimization process was usually carried out by thinning the emitter during cell fabrication. A practical emitter thickness of about 500 Å is usually used, with one-sun efficiency of about 20% reported (Fan et al., 1978).

### 3.3.1C Heterostructure Cells

More advanced cell structures have been made possible by the use of the OMCVD technique to deposit GaAs and AlGaAs layers with control over doping, composition, and thickness. In fact, the development of this growth technique in the early 1980's was driven by solar cell research in general. A variety of GaAs/AlGaAs cell structures have been developed including tandem cells using alloys of indium and/or phosphorus.

One of the first improvements made to GaAs cells using OMCVD was the growth of an emitter layer in the heteroface cell instead of using diffusion to form the junction. Thus, the emitter thickness and doping could be optimized, and heteroface cells could be grown with both n-on-p and p-on-n structures. A GaAs front contact layer (cap) was also added to allow for easier contact formation and to protect the AlGaAs window during processing. After processing, the cap is removed. Efficiencies of these modified heteroface cells improved to about 22%.

Another cell improvement was the addition of a heterostructure back-surface field (BSF). This was achieved with a buried p-doped AlGaAs layer below the base in an n-on-p shallow-homojunction cell. The cells showed an excellent long-wavelength response, and exhibited improved voltage and fill factor. The next step in the GaAs cell evolution was to employ a double heterostructure. This structure was first demonstrated (Gale et al., 1984) using a heterostructure BSF with a thin AlGaAs direct-bandgap emitter (See Figure 3.9). Open-circuit voltages of 1.05V were achieved along with fill factors of 87%, for an AM 1 efficiency of 23%. Variations of this structure have already allowed GaAs solar cells to reach efficiencies above 24% AM 1.5 global. Finally, Olson et al. (1989) recently replaced the AlGaAs emitter with a $GaInP_2$ emitter (lattice matched to GaAs), and one-sun efficiency of 25.7% at AM 1.5 has been obtained.

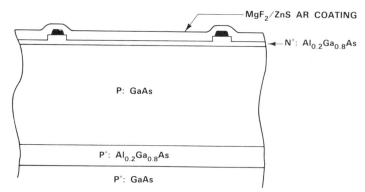

**Figure 3.9:** Schematic diagram of a double-heterostructure GaAs solar cell.

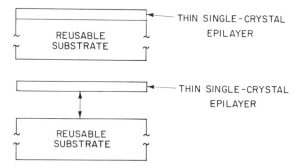

**Figure 3.10:** Schematic diagram showing the peeled-film CLEFT technology.

### 3.3.2 Thin-Film GaAs Solar Cells

As discussed earlier, there are significant advantages if GaAs solar cells can be composed of thin layers of GaAs. Two approaches are being investigated for such cells; growth on reusable GaAs substrates and growth on inexpensive substrates such as silicon. Important advances have been achieved in both areas.

### 3.3.2A CLEFT Technique

The CLEFT process permits the growth of thin single-crystal GaAs films by CVD on reusable GaAs substrates. Since many films can be obtained from one substrate, this process should permit a marked reduction in material usage and cost; since single-crystal films are used, cell efficiencies should remain high and, since the films are only a few microns thick, such cells can have very high power-to-weight ratios.

The CLEFT process is a peeled-film technique. The basic idea of peeled-film technology is to grow a thin single-crystal epilayer on a single-crystal mold, to separate the epilayer from the mold, and then to use the mold again (see Figure 3.10).

The key element of the CLEFT process (McClelland et al., 1980) is the use of lateral epitaxial growth. If a mask with appropriately spaced stripe openings is deposited on a GaAs substrate, the epitaxial growth initiated on the GaAs surface exposed through the openings will be followed by lateral growth over the mask, eventually producing a continuous single-crystal film that can be grown to any desired thickness, and with any desired cell structure. The upper surface of the

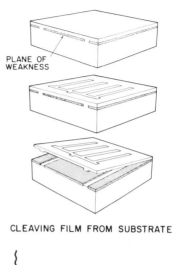

**Figure 3.11:** Schematic diagram showing a CLEFT cell peeling from a single-crystalline mold.

CLEAVING FILM FROM SUBSTRATE

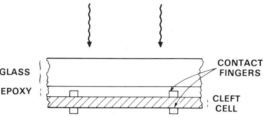

**Figure 3.12:** Schematic diagram showing a cross-section of a CLEFT cell.

film is then bonded to a secondary substrate of some other material. If there is poor adhesion between the mask material and the GaAs, the film will be strongly attached to the GaAs substrate only at the stripe openings. The film can be cleaved from the GaAs substrate without significant degradation of either (see Figure 3.11).

Figure 3.12 is a schematic diagram of the cross section of a CLEFT solar cell. The structure of this cell is unique in several respects. The entire thickness of GaAs is less than 10 $\mu$m, compared with $\sim$ 300 $\mu$m for conventional cells. The glass substrate has a dual role, since it both supports the film and serves as the cover glass for the cell. Another advantage of the structure is that there are metal contacts on both sides of the GaAs film (with the contact patterns on both sides aligned with an infrared technique). The two-sided metallization is especially useful for devices where the current flow is perpendicular to the surface, e.g., in a solar cell.

The CLEFT procedure has been developed so that conversion efficiencies have improved to over 23% at one sun AM 1.5 for 4 cm$^2$ total area cells. This represents the highest truly thin-film solar cell efficiency ever reported. In addition, the successful reuse of GaAs substrates has also been reported (Gale et al., 1988).

Another technique of peeling off thin-film GaAs layers has been successfully developed (Konagai et al., 1978). This thin-film process was reported in early 1980, where thin-film GaAs solar cell layers grown by MBE with intermediate AlGaAs layers as etching layers were separated from GaAs substrates. Solar cells were fabricated using this technique with conversion efficiencies up to 15% reported. However, this technique was not developed further, because of the difficulty of removing large-area layers and of controlling the etching process. Very recently, a major improvement in this chemical etching technique was reported (Yablonovitch et al., 1987) and using a wax-bending technique, the chemical etchant is encouraged to enter the etching channel and layers as large as 1 cm × 2 cm have been successfully separated. This technique must be further developed before it has any potential of producing large-area cells.

### 3.3.2B  GaAs Single-Junction Cells on Ge and Si Substrate

As we have described earlier, GaAs solar cells can be made lighter with Si substrates (Si is about a factor of two less dense than GaAs and thin Si wafers are much stronger than GaAs wafers of the same thickness), a very important advantage for space applications. In addition, currently Si bulk wafers are much cheaper than GaAs bulk wafers. Since all advanced GaAs solar cell structures require heterostructure epitaxial growth, it is logical to use Si substrates for GaAs cells, if efficient cells can be made. Finally, the material system of GaAs and Si is important for multijunction tandem cells (we will discuss this later in this chapter). The material challenge of this heteroepitaxy system is very severe, however. The lattice constants of GaAs and Si are different by 4%, and the thermal coefficients of expansion of GaAs and Si are different by a factor of two. These two physical differences cause many misfit dislocations, and stress-related problems.

The first GaAs-on-Si solar cells ever fabricated were reported in 1981 (Gale et al., 1981), and they were grown with a thin Ge interface layer between GaAs and Si. Small area cells with one-sun efficiency over 10% were obtained, and by 1984, much larger-area cells (up to 0.5 cm$^2$), and more efficient cells (up to about 14%) were obtained (Tsaur et al., 1982). The Ge interface was introduced because the lattice constants of GaAs and Ge are almost the same, and Ge surfaces are easier to keep clean for GaAs growth. However, with further development in Si surface cleaning, efficient GaAs solar cells have been successfully grown directly on Si substrates (Tsaur et al., 1982, Okamoto et al., 1988). In fact, the GaAs-on-Si solar cells have been improved so that one-sun efficiency of 18% AM 0 have been recently obtained (Okamoto et al., 1988). These cells were grown with elaborate defect-reduction techniques, such as thermal cycling (Tsaur et al., 1982) and strained superlattices (Okamoto et al., 1988). With further improvements in material quality, even higher efficiency values will be obtained. The GaAs-on-Si solar cells potentially can be useful for space applications. However, the cost of Si substrates (unless low-cost Si wafers can be used) must be very low for terrestrial applications.

## 3.4 Multijunction Crystalline Cells

Since the maximum practical efficiency calculated for single-junction solar cells is 27.5%, the highest module efficiency achievable with such cells will not be much above 20%. The limitation on cell efficiency arises principally because the range of photon energies covered by the solar spectrum is too broad for efficient conversion by a single material. Thus, for a semiconductor with energy gap $E_g$, solar photons with energies less than $E_g$ cannot generate electron-hole pairs. Furthermore, each photon with an energy equal to or higher than $E_g$ can generate only one electron-hole pair, so that the difference between the photon energy and $E_g$ is lost as heat.

An increase in the overall conversion efficiency can be obtained by applying the principle of spectral splitting, in which the solar spectrum is divided into two or more ranges of photon energy and in which each spectral range incident on a cell tailored to be optimally efficient over that range. The most popular approach is to have solar cells with different energy gaps stacked in tandem so that the cell facing the sun has the largest energy gap. The top cell absorbs the photons with energies equal to or greater than its energy gap and transmits less energetic photons to the cells below. The next cell in the stack absorbs the transmitted photons that have energies equal to or greater than its energy gap and transmits the rest downward in the stack etc. In principle, any number of cells can be used in tandem.

Although the tandem cell concept is well known, the development of tandem technology is still in its early stages. We have carried out a detailed analysis of the conversion efficiency of tandem structures for single-crystalline (Fan, 1984, Fan et al., 1983), polycrystalline (Fan and Palm, 1984), and amorphous cells (Fan and Palm, 1983). We shall outline the results of this analysis, primarily focussing on devices composed entirely of high quality crystalline cells, which will give the highest overall efficiencies. We shall also comment on some promising materials systems for tandem applications.

According to our computer analysis, even tandem structures with only two cells can achieve a substantial increase in conversion efficiency over single-junction cells. Further increases in efficiency are calculated for three-cell tandem structures but these are probably insufficient to justify the added complexity at this time. In the following discussion, we shall therefore consider only two-cell structures.

Two basic designs can be used for two-cell tandem devices: either the top cell can be grown on the bottom cell to form a monolithic structure or the two cells can be fabricated separately and then stacked vertically and bonded together in a hybrid structure (Fan, 1984). For monolithic devices, the materials requirements for the various semiconductor layers are very stringent. Since the two cells are connected in series (two-terminal connection), for maximum efficiency the same photocurrent must be generated in each cell. This current-matching requirement greatly limits the number of possible combinations of semiconductors and requires accurate control of growth parameters. In addition, since the largest contribution to the overall conversion efficiency comes from the top cell, the quality of the layer forming this cell must be very high. In general, however, there will be significant lattice parameter mismatch and differential thermal expansion between the materials of the two cells (e.g., in the case of GaAs and Si system). For devices using single-crystal cells, with such lattice differences the top layers obtained by standard epitaxial growth technique would be of poorer quality. Novel growth techniques will therefore be required to allow effective utilization of materials with different lattice properties in monolithic structures.

For hybrid two-cell devices, differences in lattice properties between the two semiconductors no longer present a problem. The top cell must be fabricated so that the solar photons with energies less than the bandgap of the top cell will be transmitted to the bottom cell. Usually, this requires the top cell to be very thin (e.g., by the CLEFT process). The top cell is frequently bonded to the bottom cell. In this structure the cells can be connected either in series (two terminals) or separately (four terminals) by using a bonding material that is either conducting

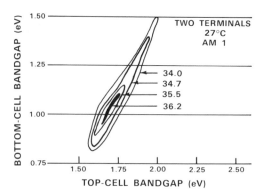

**Figure 3.13:** AM 1 isoefficiency plots for the two-cell, two-terminal tandem structure at 27°C and one sun.

or insulating. For a four-terminal structure, current matching is not necessary and therefore less-accurate process control is needed than for two-terminal structures. There are several problems, however, peculiar to the hybrid approach. The design of antireflection coatings is more complex, since optical coatings are required not only on the upper surface of the top cell but also on the bottom surface of this cell and on the upper surface of the bottom cell. In addition, the bonding materials must be optically matched to the cells, and they must be thermally conductive in order to minimize cell heating.

For any tandem device consisting of two single-crystal cells whether its structure is monolithic or hybrid and whether it has two- or four-terminal connections, for maximum overall efficiency our calculations (Fan, 1984) show that the energy gaps should be 1.75–1.80 eV for the top-cell material and 1.0–1.1 eV for the bottom-cell material. (See Figure 3.13 and Figure 3.14) For these values the calculations give a maximum overall efficiency of 36%–37%, with the top cell contributing about two-thirds of the power output. The highest practical module efficiency that can be expected is therefore about 32–33%. Because of its well-developed solar cell technology and low cost, single-crystal silicon is the material of choice for the bottom cell in a maximum efficiency tandem structure. No material with a comparable demonstrated solar cell capability is currently available for the top cell. The most promising materials at this time are two ternary III-V alloys based on GaAs: $Ga_{.7}Al_{.3}As$ and $GaAs_{.7}P_{.3}$.

For tandem devices consisting of two single-crystal cells the maximum calculated overall efficiency is higher by almost 10 absolute percentage points than the maximum of 27.5% calculated for single-junction single-crystal cells. This large difference suggests that tandem devices incorporating polycrystalline or amorphous cells could still have overall efficiencies significantly higher than the best single-junction cells, possibly permitting advantageous cost-efficiency trade-offs. For an initial exploration of this possibility, we shall consider (Fan, 1984) the four tandem structures shown schematically in Figure 3.15, which utilize "high quality" (single-crystal or very-large-grained) cells and "low quality" (polycrystalline or amorphous) cells: case 1, two high quality cells; case 2, a low quality top cell on a high quality bottom cell;

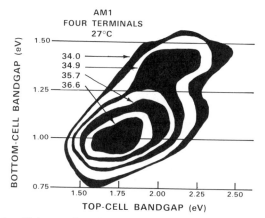

**Figure 3.14:** AM 1 isoefficiency plots for the two-cell, four-terminal tandem structure at 27°C and one sun.

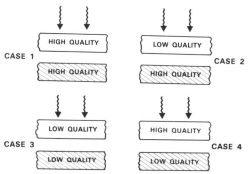

**Figure 3.15:** Schematic diagram showing the four combinations of high-and-low-quality materials in two-cell tandem structures.

case 3, two low quality cells; case 4, a high quality top cell on a low quality bottom cell.

Case 1 is the case that we have already discussed which gives the maximum overall efficiency. Case 2, with a low quality top cell, is unsuitable for high efficiency devices because the main contribution to the output of a tandem structure is made by the top cell. If the efficiency of the top cell is low enough, the sunlight absorbed by this cell will produce less power than if this radiation were incident on the high quality bottom cell. Therefore, this configuration can easily have a lower conversion efficiency than the bottom cell alone. Case 3 is also of little interest, since two inefficient cells in combination still yield a low overall efficiency.

In contrast, case 4 is quite interesting since a high quality top cell can make a good contribution to the overall efficiency, which can therefore be significantly higher than the efficiency of the bottom cell alone. However, in a two-terminal structure, because of photocurrent matching, the contribution of the top cell decreases as the quality of the bottom cell is reduced. For a relatively small reduction in bottom-cell photocurrent the overall efficiency for case 4 will actually fall below the efficiency of the top cell alone, placing a serious limitation on the usefulness of the two-terminal structure. This limitation does not apply to the four-terminal structure, in which the contribution of the top cell is independent of the efficiency of the bottom cell. However, there is still a lower limit on the quality of the material that can be used for the bottom cell in the four-terminal structure. This arises because the contribution of the bottom cell is particularly sensitive to the cell's relative red response, since the photocurrent in this cell is generated by the longer wavelength photons transmitted through the top cell. Nevertheless, for some applications not requiring the highest possible efficiency, case 4 may be an advantageous alternative to case 1, since the reduction in cost achieved by using a lower quality material for the bottom cell might more than compensate for the resulting decrease in overall efficiency. Two promising candidates for such a material are polycrystalline silicon with enhanced red response and $CuInSe_2$ ($E_g = 1.02$ eV).

### 3.4.1 Prospects for High-Efficiency Modules

In view of the complex trade-offs between module efficiency and a host of other parameters including weight, reliability and concentration ratio as well as module cost and balance-of-system cost, it is obvious that no one type of module will be ideal for all photovoltaic applications. It follows that no one materials system will be ideal for all applications. To give an overview of the present prospects for high efficiency modules, we shall present our assessment of the most promising materials systems, the most likely configurations and the most probable major appications to be anticipated for modules with efficiencies of 20%, 25% and 30%.

Table 3.3 summarizes our conclusions concerning the materials systems for these three module efficiency levels, as well as for modules with efficiencies of 10% and 15%. In each case our evaluation is based both on the properties of the present materials and on cost considerations. For 10% modules, single-junction cells using either a-Si:H or polycrystalline $CuInSe_2$ are the strongest candidates. The most promising systems for 15% modules are single-junction cells using low cost silicon sheets or ribbons, and tandem structures using a-Si:H for the top cell and either a-Si:H or polycrystalline $CuInSe_2$ for the bottom cell.

For 20% modules, single-junction cells with the highest possible efficiencies are needed; the most promising materials for such cells are single-crystal GaAs, $Ga_{.95}Al_{.05}As$ or $GaAs_{.95}P_{.05}$. Tandem cells will be required for 25% and 30% modules. In both cases the best choices for the top cells are single-crystal $Ga_{.7}Al_{.3}As$ and $GaAs_{.7}P_{.3}$. For the bottom cells the most promising materials are polycrystalline silicon and $CuInSe_2$ (Figure 3.15, case 4) for 25% modules and single-crystal silicon (Figure 3.15, case 1) for 30% modules. The III-V alloys should be used in the form of thin films, which could be prepared by the CLEFT process. A two-cell hybrid structure utilizing these materials is shown schematically in Figure 3.16. Monolithic devices with silicon as the bottom-cell material are also attractive alternatives.

### 3.4.2 Advanced Concepts for Module Efficiencies Over 30%

As discussed above, the highest practical one-sun efficiency module that can be achieved by using two-cell tandem structures for spectral splitting is expected to have an efficiency of about 32–33%. In principle, still higher efficiencies could be obtained by increasing the number of cells in tandem. Thus, efficiencies of over 50% have been estimated (Loferski, 1982) for a tandem structure composed of over 20

**Table 3.3:** Most promising materials for photovoltaic modules of various efficiencies[a] (Fan et al., 1984).

| MODULE EFFICIENCY | SINGLE-JUNCTION ( | TANDEM STRUCTURES | |
| --- | --- | --- | --- |
| | | TOP CELL | BOTTOM CELL |
| LOW (10%) | a-Si:H POLY-CuInSe$_2$ | | |
| MEDIUM (15%) | Si (sheets or ribbons) | a-Si:H | a-Si:H OR POLY-CuInSe$_2$ |
| HIGH (20%) | GaAs Ga$_{0.95}$Al$_{0.05}$As GaAs$_{0.95}$P$_{0.05}$ | POLY-CdTe | POLY-CuInSe$_2$ |
| VERY HIGH (25%) | | Ga$_{0.7}$Al$_{0.3}$As OR GaAs$_{0.7}$P$_{0.3}$ | POLY-Si OR POLY-CuInSe$_2$ |
| ULTRAHIGH (30%) | | Ga$_{0.7}$Al$_{0.3}$As OR GaAs$_{0.7}$P$_{0.3}$ | Si |

[a]MATERIALS IN SINGLE-CRYSTAL OR VERY-LARGE GRAINED FORM UNLESS OTHERWISE NOTED

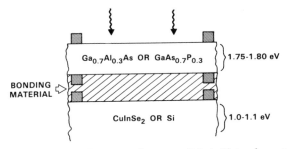

**Figure 3.16:** Schematic diagram of a two-cell hybrid tandem structure.

cells. In practice, however, system complexities will limit the number of cells to two or at most three. Therefore some other approach will be necessary if module efficiencies are to be increased much above 32% (Even with high concentration ratios which often increase the module efficiencies somewhat, efficiencies much higher than 30% are difficult to obtain).

Conceptually, two other approaches (Fan, 1985) have the potential for significantly increasing the solar cell conversion efficiency. In the first approach the solar spectrum is efficiently compressed into a narrower band (or bands) of photon energies that is then directed to an optimized cell (or cells). The principle of spectral compression is the basis for the thermovoltaic technique, which employs an IR detector to convert the black body radiation from an enclosure heated by solar radiation. Compression is achieved here by a lower temperature nature of the black body emitter than that of the sun (6,000 K). However, this technique is subject to very severe engineering problems because of the high temperatures involved to obtain useful emission.

Spectral compression is also accomplished by fluorescence concentrators employing optical conversion to lower photon energies but these devices are inefficient, unstable or both. Fluorescent concentrators utilize a broad band detector material in which the conversion efficiency is increased because photons with energies exceeding $E_g$ excite more than one electron-hole pair and/or photons with energies less than $E_g$ generate electron-hole pairs by means of multiphonon processes. Although multipair production and multiphoton excitation have been observed (Fan, 1985), so far both are extremely inefficient.

At present, both spectral compression and broad band detection represent intriguing approaches, but major breakthroughs would be required for them to make practical contributions to solar energy conversion.

Therefore, to obtain very high-efficiency cells, multijunction cells are the most practical approaches, and two-cell structures are the most viable ones at this time.

Table 3.3 summarizes some of the top and bottom cell combinations that are being investigated at present. In the sections to follow, we will review the reported results as well as other work on similar approaches (Spitzer and Fan, 1990).

### 3.4.3 Mechanically-Stacked Multijunction Cells

#### 3.4.3A GaAs Mechanically Stacked on Si

Mechanically stacked GaAs on Si cells have been experimentally pursued by a number of investigators. The respective bandgaps are 1.43 eV and 1.11 eV. The first report of a GaAs/Si tandem was made in 1984, and consisted of the stacking of a thin GaAs cell formed by the cleavage of lateral epitaxially-overgrown films for transfer (CLEFT) on a conventional Si cell (Fan et al., 1984, Fan et al., 1982).

Since then, more optimized improvements were made to this combination, so that in 1988, Gee and Virshup (1988) reported a combined four-terminal efficiency of GaAs top cell and Si bottom cell of 31% at AM 1.5 under a concentration of 347 suns. No one-sun efficiencies were reported. The thin GaAs top cell was prepared with a chemical etching technique.

One approach to attainment of a top cell band gap of 1.7 eV is to employ $Al_2Ga_8As$ for the top cell. In such an approach, the AlGaAs is grown on top of a GaAs or Ge substrate by chemical vapor deposition. The principal difficulty in this approach is that since the substrate band gap is less than 1.7 eV, the substrate absorbs light intended for absorption in the Si bottom cell. Consequently, to make this approach effective, the substrate must be removed.

Using a chemical etching technique Boettcher et al. (1987) has reported thin AlGaAs cells as high as 20% at AM 2, although the band gap of the cell thinned in this way is not stated in the reference.

A similar approach has been reported by Tobin (1987) who, in an effort to overcome various problems with removing the GaAs substrate, grew the AlGaAs cells on Ge. The Ge is both expansion-matched and lattice matched to GaAs and AlGaAs, thus good epitaxy can be expected and has been achieved. In fact, formation of GaAs cells on Ge has been reported by a number of teams (Partain et al., 1988, Patel et al., 1988, Timmons et al., 1988, Fan et al., 1979). In Tobin's work, the Ge was removed in an etch that stops at GaAs and hence does not damage the active cell. Cell efficiency of up to 13.7% was obtained (AM 1.5 spectrum, 242 suns) on an AlGaAs cell with a band gap of 1.69 eV.

There is a problem with the approach described above where the substrate is removed by etching and consequently wasted. Both Ge and GaAs substrates are expensive. To overcome this problem, two approaches are being studied. The CLEFT (Gale et al., 1988) technique is one in which the epitaxial GaAs or AlGaAs cell is mechanically separated from the substrate. The chemical liftoff technique is one (Yablonovitch et al., 1987) in which the solar cell is released from the substrate in an etch which removes only a thin (100 Å) sacrificial layer. In either case, the film is released intact, and the substrate may be reused. The CLEFT technique has been successfully applied to GaAs/Si (Fan et al., 1982) and more recently to GaAs/CuInSe$_2$ tandem fabrication, as we will describe below, and the extension to AlGaAs/Si tandem is straightforward and holds considerable promise. The CLEFT technique is developed sufficiently that 3" diameter GaAs films or AlGaAs films can be routinely separated from GaAs substrates. The chemical liftoff techniques can currently liftoff about 1 cm by 2 cm films.

### 3.4.3B  GaAs Mechanically Stacked on CuInSe$_2$

Highly efficient mechanically-stacked cells intended for space power have been formed from the combination of CuInSe$_2$ and GaAs. (Kim et al., 1988, Kim et al., 1990) An attractive aspect of the use of CuInSe$_2$ is the high radiation resistance of such cells (Burgess et al., 1988). Briefly, a GaAs heteroface cell formed by the CLEFT (McClelland et al., 1980, Gale et al., 1988), having a thickness of between 5 $\mu$m and 10 $\mu$m, is stacked on top of a polycrystalline CuInSe$_2$/CdZnS cell that is formed on a glass substrate.

The stack is made by first forming the CuInSe$_2$/CdZnS bottom cell on glass, followed by mounting a CLEFT GaAs cell on top of the bottom cell. A space qualified adhesive is used to obtain adhesion between the two cells. A second layer of space qualified adhesive is used between the top cell and the coverglass. Note that the use of the CLEFT process makes possible the removal and reuse of the GaAs substrate without generation of a large amount of waste. The authors report efficiency of over 20% with a four wire interconnect (Kim et al., 1988). Recently, a GaAs CLEFT/CuInSe$_2$ tandem cell having an area of 4 cm$^2$ demonstrated an AM 0 efficiency of 21.6% at one sun (Kim et al., 1990). Perhaps the most important result of this work is that a specific power of 440 W/kg has been obtained for this type of tandem cell with a glass cover. The extension of this work to AlGaAs/CuInSe$_2$ would yield yet higher efficiency.

### 3.4.3C  InGaAs Bottom Cells

The growth of high quality InGaAs solar cells on lattice mismatched GaAs substrates has been reported (Dietze et al., 1982, Lewis et al., 1988a). The growth is carried out using OMCVD. The band gap of the InGaAs cell formed in this way is 1.1 eV, and the resultant efficiency is 18% (one sun, AM 2). Although this work was intended for use ultimately in a monolithic tandem cell, the cell would form a good bottom cell in a mechanical stack, and recently Gee and Virshup (1988) reported an efficiency of 30.2% (AM 1.5, 350 suns) obtained by stacking a GaAs cell on top of an InGaAs cell in the same manner as had been done for the GaAs/Si stack.

This result is comparable to the 31% AM 1.5 efficiency obtained with the GaAs/Si tandem.

### 3.4.3D  GaAs Mechanically Stacked on Ge

The mechanical stacking of GaAs on Ge has been described by Partain et al. (1987). The multijunction cell was formed from a GaAs concentrator cell with an efficiency of 22.2% and a Ge concentrator cell with an efficiency (when covered by a GaAs cell) of 1.25% (159 suns, AM 0). The cells were operated in a four terminal configuration. The total efficiency of the stack was 23.4%. Although this efficiency is quite good, it will be necessary to obtain a greater contribution from the Ge cell in order to justify the added complexity inherent in a four terminal design.

### 3.4.3E  GaAsP on GaAsSb

Cape et al. (1987) have investigated GaAsP top cells grown on GaP, and GaAsSb grown on GaAs for bottom cells. In this approach, the growth of the top cell on GaP makes removal of the substrate unnecessary, since it is transparent to the radiation transmitted through the GaAsP cell. The best reported top cell efficiency was 17% (AM 1.5, 83 suns, $E_g = 1.65$ eV). In the work by Lewis et al. (1988a), the cells were considered as intermediate steps toward an eventual monolithic approach for terrestrial application. This approach does not appear to have a good near term application; however, the GaAsSb cell may be useful as a bottom cell in combination with GaAs or AlGaAs, provided an efficient cell with a smaller band gap can be achieved. Fraas et al. (1990) have recently reported a stack comprising a GaAs cell stacked on a GaSb. The four-terminal AM 1.5 (direct) efficiency of about 37% was reported with a prismatic cover at a concentration of $100\times$.

### 3.4.4  Monolithic Multijunction Cells

Two-terminal monolithic cells not only require current matching, as discussed earlier, but also require that the interface between the cells have negligible resistance. In an $n^+p/n^+p$ tandem, one forms a $p/n$ junction at the interface between cells. To overcome the series resistance that would ordinarily be introduced by this junction, research has focused on metal (Ludowise et al., 1982) tunneling (Timmons and Bedair, 1981, Miller et al., 1982) or compositional (Timmons and Chiang, 1989) interconnects to obtain high conduction in forward and reverse bias.

Most of the work on heteroepitaxy for multijunction cells has been carried out in III-V materials systems. For this reason, we will limit the discussion to III-V and column IV materials. we have organized reported results by bottom cell and substrate groups.

### 3.4.4A  GaAs Bottom Cells

The best results to date have been obtained by using GaAs as one of the cells in the multijunction stack. This is probably a result of the maturity of the GaAs crystal growth and cell formation processes. In this section we will review designs that use GaAs as both the substrate and the bottom cell.

### 3.4.4B  AlGaAs Grown on GaAs

Significant progress has been made by using an AlGaAs/GaAs multijunction (Ludowise et al., 1982, Virshup et al., 1988, Macmillan et al., 1988). The top $Al_{.37}Ga_{.63}As$ cell has a band gap of 1.93 eV; the bottom cell is GaAs (1.43 eV). The cell is grown using OMCVD on top of GaAs substrates. An AM 0 efficiency of 22.3% has been obtained in this way at one sun, when the cell is operated as a two-terminal device. For AM 1.5, the two-terminal tandem structures have efficiencies as high as 27.6% at one sun. An interesting aspect of this work is that the intercell reverse $p/n$ junction resistance has been overcome by using a metal interconnect that effectively shorts this junction.

Flores (1983) has described a cell similar to the above described cell. Flores' cell is formed by liquid phase epitaxy; the top cell comprises $Ga_{.8}Al_{.2}As$ $E_g = 1.7$ eV and the bottom cell is made of GaAs. The cell design requires operation as a three-terminal device. An AM 1.5 efficiency of 20.5% has been achieved at a concentration of 80 suns. Liquid phase epitaxy has also been investigated by Gavand for formation of $Al_{.35}Ga_{.65}As$ top cells. A top cell efficiency of 14.6% (AM 1, 25°C) was obtained (Gavand et al., 1987).

Amano et al. (1987) have reported the formation of an $Al_{.4}Ga_{.6}As$/GaAs tandem grown by molecular beam epitaxy. The top and bottom subcells are interconnected by using a $n^{++}/p^{++}$ tunnel junction formed in GaAs. The efficiency obtained in this cell is reported to be 20.2% (one sun, AM 1.5).

### 3.4.4C  GaInP Grown on GaAs

Olson et al. (1988) have recently reported the use of $Ga_xIn_{1-x}P$ for top cell formation. The band gaps available in this system are in the range of 1.34 eV to 2.25 eV. One advantage of this ternary is that since no Al is present in the active layers, the minority carrier properties are less susceptible to water and oxygen contamination. (AlGaAs layers have been found to be very sensitive to water and oxygen contamination.) A heteroface structure can be formed by using $Al_{.5}In_{.5}P$ that is lattice constant matched to GaAs. The devices made thus far comprise two-terminal $Ga_{.5}In_{.5}P$/GaAs tandems (Olson et al., 1988). A GaAs tunnel diode is used to connect the two subcells. The resultant one-sun AM 1.5 efficiency is 27.3% (Olson et al., 1989). Olson's model indicates that 25% efficiency may be attainable in the near future.

### 3.4.4D  Ge Bottom Cells

The use of Ge as both a bottom cell and as a substrate for top cell growth has recently received renewed attention. Ge is attractive because its lattice constant and thermal expansion coefficient are nearly perfectly matched to GaAs and AlGaAs, as was described in an earlier section. Ge also has the advantage that it may be available in larger areas than can be obtained presently with GaAs. Also Ge is mechanically much stronger than GaAs, and thus thin Ge wafers are possible to be used for lighter-weight space power modules.

### 3.4.4E  GaAs Grown on Ge

Investigations of the growth of GaAs on Ge for solar cell applications have been underway for many years. In early work, Fan et al. (1979) demonstrated that up to 21% AM 1 efficiency could be obtained from a shallow-homojunction GaAs cell grown on Ge.

More recently, Tobin et al. (1988) have reported the fabrication of GaAs/Ge monolithic cells intended for space operation. The GaAs heteroface top cell (p-type emitter on n-type base) is grown by OMCVD on n-type Ge substrates. The GaAs growth process allows Ga and As to diffuse into the Ge. Presumably, since the Ga solubility in Ge is higher than As, the result is a p-type diffused junction near the GaAs/Ge interface. The measurements of the performance of the tandem cell have been difficult to calibrate, owing to the spectral content and intensity of simulated AM 0 light. An approximate AM 0 measurement has been made at 35,000 feet in an airplane (AM 0.22); the efficiency was determined to be 18% (Tobin et al., 1988). Recent improvements have led to an efficiency of 28.7% at 200 suns and AM 1.5 direct (Wojtczuk et al., 1990).

### 3.4.4F  Si Bottom Cells

Silicon bottom cells are of interest for several reasons. First, Si is abundant and high quality Si wafers are relatively low in cost. Second, as noted earlier, Si has the proper band gap for an efficient current match to various practical top cells, and owing to the maturity of Si cell technology, excellent cells have been formed that would qualify as good bottom cells. However, the lattice constant of Si is not well-matched to most III-V cells. Consequently, the challenge is to obtain a high quality epitaxy for the top cell. To obtain this quality, various types of graded or buffer layers have been employed. Most of the work has been with single-junction cells as described before. However, there are some monolithic results.

### 3.4.4G  GaAs Grown on Si

Approaches to the lattice and current-matching problems described above have been addressed in 1984 by Tsaur et al., who have suggested that the lattice constant mismatch can be accommodated by compositional grading, and that the current can be matched by forming openings that are etched in the GaAs top cell to allow more light to penetrate to the Si bottom cell. This method will increase the Si current and decrease the GaAs current, and by adjusting the amount of GaAs removed, can permit current matching, but with a slightly lower theoretical efficiency. The authors actually reported a tandem GaAs-on-Si cell, with an open-circuit voltage of 1.2V. This is the first monolithic GaAs-on-Si tandem results reported. However, because of the current-matching problems, no efficiency values were given. The current matching was much improved with strip-openings. With only 25% of the GaAs area removed, the short circuit current was reported to increase by 70%.

### 3.4.4H  GaAsP Grown on Si

The growth of $GaAs_xP_{1-x}$ on Si presents interesting possibilities for monolithic tandem formation. The band gaps available span a wide range (1.42 eV to 2.25 eV). In addition, GaP is both nearly lattice matched to Si and, owing to its wide band gap, transparent in the range of interest. For this reason, GaP is attractive for the buffer layer between the top and bottom cells. So far, no tandem results have been reported; however, we will review some results of simple-junction cells grown on Si.

Several investigators are pursuing a GaAsP/Si approach. Beck et al. (1988) have investigated the growth of monolithic GaAsP cells on Si by using a GaP transition layer, followed by a strained layer superlattice, on top of which is grown a GaAsP solar cell. Vernon et al. (1987) have reported the development of a GaAs.8P.2 cell that has a band gap of 1.7 eV and is grown by OMCVD.

### 3.4.4I InGaAs Bottom Cells

Researchers have made considerable progress in the development of monolithic III-V multijunction cells. These cells consist of a Ga.8In.2As bottom cell (1.15 eV) (Dietze et al., 1982) on top of which is grown an Al.8Ga.2As top cell (1.72 eV) (Lewis et al., 1988a, Lewis et al., 1988b). The cells are grown monolithically on GaAs substrates with grading layers employed between the cells to accommodate lattice mismatch. A metal interconnect is used between the top and bottom cells; this interconnect can also function as a terminal if the cell operates in a three terminal mode. This work has led to efficiency of 16.5% (AM 1.5, one sun).

The development of InGaAs in a three-junction monolithic cascade is presently underway (MacMillan et al., 1988). The three junction structure comprises the two junction metal interconnected AlGaAs/GaAs cell described in section 3.4.4B with an InGaAs cell. The cells are all grown on a common GaAs substrate. One problem to overcome in the InGaAs cell development is the formation of graded layers that are necessary to overcome lattice mismatch between InGaAs and GaAs; since the InGaAs cell is on the back of the substrate, these graded layers must also be transparent. If these difficulties can be overcome, the cell can yield a practical AM 0 efficiency of 30%.

## 3.5 InP Solar Cells

We have reviewed quite extensively single-junction and multijunction cells primarily using Ga-based compounds. Recently, InP cells are becoming very interesting, primarily for space applications. We will now devote a section to reviewing the InP cells.

The room temperature band gap of InP is 1.3 eV and so it is in the useful range for efficient solar energy conversion. An attribute of InP that is particularly interesting for space solar cells is its radiation damage and annealing properties. In the sections to follow, we will describe both homojunction and heterojunction solar cells, and we will follow this with a review of recent measurements of InP radiation resistance. A comprehensive review was recently prepared by Coutts and Yamaguchi (1988).

A characteristic of importance in InP cell design is the high optical absorption coefficient (greater than $1~\mu m^{-1}$), similar to GaAs. To obtain high performance, the junction must therefore be quite shallow, or alternatively, the surface must have a low recombination velocity. Spitzer et al. (1987) have reported on highly efficient cells with a junction depth of only 400 Å with AM 0 efficiency of 18%. Yet higher efficiency was obtained with an ion-implanted emitter (Keavney and Spitzer, 1988). Since these emitters are thin, they are characterized by high sheet resistance and therefore require a fine-line front surface grid.

The InP surface recombination velocity has been reported to be in the range of $10^4$ cm/sec (Hoffman et al., 1980). However, InP solar cell modeling (Keavney, 1988) indicates that the recombination velocity may be somewhat higher in fabricated solar cells. Thus, whether highly efficient InP cells will require surface passivation or a wide band gap window layers remains an open question.

### 3.5.1 Homojunction Solar Cells

To date there have been three principal methods of forming homojunctions in InP: diffusion, ion implantation, and epitaxy. The best InP cells to date have been fabricated by using epitaxy to form a $p$-type region on a heavily doped substrate, with ion implantation used to form a shallow emitter (Keavney and Spitzer, 1988). Each homojunction formation method will be reviewed below.

Diffusion has been studied by a number of teams investigating solar cells as well as other structures. Yamamoto et al. (1986) have utilized S and Se diffusion in sealed quartz ampoules to obtain relatively deep junctions. In this technique, the substrate is placed in a sealed ampoule with both the diffusion source and a small amount of P; the P is placed in the tube to prevent decomposition of the InP surface during the high temperature diffusion treatment (600 to 700°C). An active area AM 1.5 efficiency of 16.5% has been achieved in this way (Yamaguchi and Uemura, 1984). Shallow junction formation by diffusion in an open tube has been reported by Ghandi and Parat (1987). The open tube utilizes an evaporated $Ga_2S_3$ source film on the InP surface, with an $SiO_2$ cap to prevent surface decomposition at the diffusion temperature. The resulting junctions had depths in the range of 400 to 700 Å.

Zn diffusion is also of some importance, particularly in $n^+pp^+$ InP solar cells in which the back $p^+$ is used to reduce back contact resistance. Zn diffusion into $n$-type InP has been investigated by van Gurp et al. (1987) who used sealed ampoule techniques with a $Zn_3P_2$ source. Dhouib et al. (1985) have also described Zn diffusion using elemental Ga + Zn + P.

Ion implantation has been investigated by a number of groups, and a review of implantation into InP was published by Davies et al. (1978). Ion implantation requires annealing to restore the crystal quality of the near-surface region. The use of various caps or a flowing $PH_3$ diffusion ambient (Davies et al., 1978, Davies, 1981) are two ways to minimize surface decomposition during a tube anneal. This process is particularly easy to utilize if $PH_3$ diffusion tubes or VPE reactors with $PH_3$ lines are available. Alternatively, a sealed ampoule with a small amount of P can be used. Spitzer et al. (1987) and Keavney and Spitzer (1988) have reported on ion implanted cells that utilized low energy $^{28}Si^+$ implantation and a $PH_3$ anneal. Under these conditions, the Si implant forms an $n$-type emitter. The resultant total area AM 0 efficiency is over 18%. A schematic diagram of such a solar cell is shown in Figure 3.17.

Junctions grown by both liquid and vapor phase epitaxy (LPE and VPE) have been used for InP solar cell formation. Turner et al. (1980) reported the use of LPE for fabrication of cells with AM 1 efficiency of 15%. More recently, Choi et al. (1987) have investigated metal-organic chemical vapor deposition (MOCVD) and LPE, and have obtained AM 0 efficiency of 15.9% with MOCVD. Typical source gases for the MOCVD process are trimethylindium [$(CH_3)_3In$] and $PH_3$. Spitzer et al. (1987) have reported on InP cells formed by MOCVD with total-area AM 0 efficiency of 17.9% (measured at NASA-Lewis Research Center). Yamaguchi et al. (1986) have

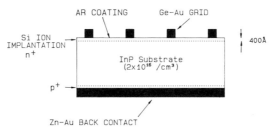

**Figure 3.17:** Schematic diagram of an ion-implanted InP solar cell (dimensions not to scale).

also reported obtaining cells with an active-area AM 0 efficiency of 18%. Keavney and Spitzer (1988) have reported cells in which the base is formed by MOCVD, and the emitter is formed by ion implantation. The total-area AM 0 efficiency of the best cells is 18.8%.

Note that epitaxy allows the formation of $n^+pp^+$ or $p^+nn^+$ structures, in which the $p$-$p^+$ and $n$-$n^+$ interfaces act as true back surface fields, because the field can be grown within the structure with a base region thickness of approximately a diffusion length. Indeed, this was the approach of Keavney (1988a) that led to AM 0 efficiency of 18.8%. Owing to the ability to form an effective back surface field, epitaxy may ultimately be the preferred method of fabrication.

### 3.5.2 Heterojunction Solar Cells

A significant amount of work on InP heterojunction cells has been carried out. In 1976, Shay et al. reported on epitaxial CdS grown on (111) InP using a double source high vacuum technique. A conversion efficiency of 12.5% was obtained. Similar results with an efficiency of 14.4% were reported by Yoshikawa and Sakai (1977).

The use of indium-tin-oxide (ITO) on InP has received attention. Harsha et al. have reported 14.4% efficiency for $n$-ITO/$p$-InP cells in which the ITO was deposited by ion beam sputtering (Harsha et al., 1977). RF sputtering has been used to deposit ITO in a similar structure, and has yielded AM 1.5 active-area efficiency of 19.1% (Coutts and Naseem, 1985). In addition, this technique has yielded excellent short circuit currents (Wu et al., 1987). The high value of short circuit currents also indicates the potential of using the substrates, without epitaxy, to form the collecting volume. Recently, Gessert et al. (1990) have reported AM 1.5 efficiency of 18.8%.

### 3.5.3 Radiation Effects

The most interesting feature of InP cells is its radiation resistance. Most of the radiation work was reported by Yamaguchi et al. (1986), (1984a), (1984). In essence, InP solar cells, under electron and X-ray radiation, have less radiation damage than GaAs cells, which in turn have smaller radiation damage than Si. This phenomenon is presumed to be a self-annealing effect in InP that anneals out radiation damage close to room temperatures. This advantage can have important implication for certain space systems.

InP solar cells have application mainly in space power systems, owing to the radiation annealing properties described above. Efficiencies competitive with GaAs cells have been demonstrated; however, it is still necessary to develop reliable contact systems, interconnects, and further improve the related cell technology. In addition, the cost of InP cells must be reduced to levels comparable to Si and GaAs. At the present time, InP substrates are approximately three times more expensive than GaAs, and about 40 times more expensive than Si. Work is therefore in progress to form InP heteroepitaxially on alternative substrates, and in particular, on GaAs (Horikawa et al., 1988) and Si (Yamamoto et al., 1986, Lee et al., 1987). Another interesting approach to this technology is the development of InP/InGaAs tandem structures (Wanlass et al., 1990). If these approaches can yield material of suitable quality, then InP cells will be able to compete with GaAs cells in terms of efficiency and cost while being superior in radiation resistance.

## 3.6 Conclusion

In this chapter we have examined the status of high efficiency solar cells. In order to obtain high conversion efficiency, advanced solar cell structures must be used, and the photovoltaic material must be carefully chosen. The ideal material should be highly absorbing in the active solar spectrum range so that it can be used in thin-film cells. It should also have large majority-carrier mobility and long minority-carrier diffusion length. The material should form excellent homojunctions, or heterojunctions with another latticed-matched material. The material should be stable, inexpensive in thin-film configuration, and should have a broad technology base (in other words, the material should also be very useful for products other than photovoltaic cells). In addition, the cells made from this material should be useful for flat-plate, concentrating, space, and tandem applications. Finally, for the maximum efficiency, the single-junction cells should be composed of a material with a band gap $Eg \sim 1.3 - 1.6$ eV and for a two-cell tandem structure, the top-cell should be composed of a material with $Eg \sim 1.5 - 1.8$ eV and the bottom cell should be composed of a material with $Eg \sim 1.0 - 1.2$ eV. The III-V materials, in particular GaAs-based materials, satisfy almost all the requirements of an ideal photovoltaic material.

Much effort has been devoted to exploring the potential of III-V material systems for photovoltaic applications. Important advances have been accomplished. New GaAs-based solar cells using heterostructures have been developed with one-sun conversion efficiency exceeding 25% at AM 1.5. Efficient thin-film GaAs solar cells have been obtained on GaAs reusable substrates (over 23% at AM 1.5) and on Si (up to 18% at AM 0). Exciting results have been achieved with tandem structures, either by mechanically stacking thin-film GaAs CLEFT cells on Si, or CuInSe$_2$ cells, or by monolithically growing GaAlAs cells on GaAs cells (combination of bandgaps of these cells are not optimal), and even better results are expected when the proper combination is used. For space applications, not only thin-film GaAs solar cells are becoming available, but also the discovery of superior radiation resistance of InP cells may allow new important applications. In conclusion, major improvements in material quality and device structures have been attained in III-V solar cells. We believe that the III-V material system is now very important for photovoltaic cells. With continuing development, this material system can satisfy most of the near-term, and long-term terrestrial and space application requirements.

# References

Amano, C., H. Sigiura, A. Yamamoto, and M. Yamaguchi (1987), Phys. Lett. **51**, 1988.

Beck, E.E., A.E. Blakeslee, and T.A. Gessert (1988), Solar Cells **24**, 205.

Boettcher, L., C. Flores, F. Paletta, and M. Martella (1987), Record of the 16th IEEE Photovoltaic Specialists Conf., IEEE, New York, p. 1512.

Bozler, C.O. and J.C.C. Fan (1977), Appl. Phys. Lett. **31**, 629.

Burgess, R.M., W.S. Chen, W.E. Delaney, D.H. Doyle, N.P. Kim, and B.J. Stanbery (1988), Record of the 20th IEEE Photovoltaic Specialists Conf., IEEE, New York, p. 909.

Cape, J.A., L.M. Fraas, P.S. McLeod, L.D. Partain (1987), *Research on Multibandgap Solar Cells*, Final Report for SERI Contract ZL-4-03123-1.

Choi, K.Y., C.C. Shen, and B.I. Miller (1987), Record of the 19th IEEE Photovoltaic Specialists Conf., IEEE, New York, p. 255.

Coutts, T.J. and S. Naseem (1985), Appl. Phys. Lett. **46**, 164.

Coutts, T.J. and M. Yamaguchi (1988), *Current Topics in Photovoltaics*, Academic Press, London.

Davies, D.E. (1981), J. Crystal Growth **54**, 150.

Davies, D.E., W.D. Potter, and J.P. Lorenzo (1978), J. Electrochemical Society **125**, 1845.

Dhouib, A., A.L. Conjeaud, B. Maloumbi, A. Gouskov, and L. Gouskov (1985), Record of the Sixth E.C. Photovoltaics Solar Energy Conf., IEEE, New York, p. 54.

Dietze, W.T., M.J. Ludowise, and P.E. Gregory (1982), Appl. Phys. Lett. **41**, 984.

Fan, J.C.C. (1984), Solar Cells **12**, 51.

Fan, J.C.C. (1985), Record of the 18th IEEE Photovoltaic Specialists Conf., IEEE, New York, p. 30.

Fan, J.C.C., C.O. Bozler, and R.L. Chapman (1978), Appl. Phys. Lett. **32**, 390.

Fan, J.C.C., C.O. Bozler, and B.J. Palm (1979), Appl. Phys. Lett. **35**, 875.

Fan, J.C.C., R.W. McClelland, and B.D. King (1984), Record of the 17th IEEE Photovoltaic Specialists Conf., IEEE, New York, p. 31.

Fan, J.C.C. and B.J. Palm (1983), Solar Cells **10**, 81.

Fan, J.C.C. and B.J. Palm (1984), Solar Cells **12**, 401.

Fan, J.C.C., B.-Y. Tsaur, and B.J. Palm (1982), Record of the 16th IEEE Photovoltaic Specialists Conf., IEEE, New York, p. 692.

Fan, J.C.C., B.-Y. Tsaur, and B.J. Palm (1983), SPIE Photovoltaics for Solar Energy Applications II, **407**, 73.

Flores (1983), IEEE Electron Device Letters **EDL-4**, 96.

Fraas, L.M., J.E. Avery, J. Martin, V.S. Sundaram, G. Givard, V.T. Dinh, T.M. Davenport, J.W. Yerkes, and M.J. O'Neill (1990), IEEE Transport Electron Devices, **ED-37**, 443.

Gale, R.P., J.C.C. Fan, B.-Y. Tsaur, G.W. Turner, and F.M. Davis (1981), IEEE Electron Dev. Lett. **EDL-2**, 169.

Gale, R.P., J.C.C. Fan, G.W. Turner, and R.L. Chapman (1984), Record of the 16th IEEE Photovoltaic Specialists Conf., IEEE, New York, p. 1422.

Gale, R.P., J.C.C. Fan, G.W. Turner, and R.L. Chapman (1988), Record of the 20th IEEE Photovoltaic Specialists Conf., IEEE, New York, p. 296.

Gale, R.P., R.W. McClelland, B.D. King, and J.V. Gormley (1988), Record of the 20th IEEE Photovoltaic Specialists Conf., IEEE, New York, p. 446.

Gavand, M., L. Mayett, B. Montegu, J.P. Boyeaux, and L. Laughier (1987), Inst. Phys. Conf. Ser. No. 91: Chapter 3, presented at the Int. Symp. GaAs and Related Compounds, Heralion, Greece.

Gee, J.M. and G.F. Virshup (1988), Record of the 20th IEEE Photovoltaic Specialists Conf., IEEE, New York, p. 754.

Gessert, T.A., X. Li, M.W. Wanless, and T.J. Coutts (in press 1990), Record of the Space Photovoltaic Research and Technology Conference, NASA-Lewis Research Center.

Ghandi, S.K. and K.K. Parat (1987), Appl. Phys. Lett. **50**, 209.

Green, M.A. (1989), Si solar call measured in SERI, personal communication by L.L. Kazmerski.

Green, M.A., A.W. Blakers, S.R. Wenham, S. Narayanan, M.R. Willison, M. Taouk, and T. Szpitalik (1988), Record of the 20th IEEE Photovoltaic Specialists Conf., IEEE, New York, p. 38.

Harsha, K.S., K.J. Bachmann, P.H. Schmidt, E.G. Spencer, and F.A. Thiel (1977), Appl. Phys. Lett. **30**, 645.

Hoffman, C.A., H.J. Gerritsen, and A.V. Nurmiko (1980), J. Appl. Phys. **51**, 1603.

Hollan, L. et al. (1980), *"The Preparation of Gallium Arsenide" in Current Topics in Material Sciences*, E. Kalis, (ed.), North Holland, Chapter 5.

Horikawa, H., Y. Ogawa, Y. Kawai, and M. Sakuta (1988), Appl. Phys. Lett. **53**, 397.

Keavney, C.J. (1988), Record of the NASA Lewis Research Center Space Photovoltaic Research and Technology Conference.

Keavney, C.J. and M.B. Spitzer (1988), Appl. Phys. Lett. **52**, 1439.

Kim, N.P., R.M. Burgess, B.J. Stanbery, R.A. Mickelsen, J.E. Avery, R.W. McClelland, B.D. King, M.J. Boden, and R.P. Gale (1988), Record of the 20th IEEE Photovoltaic Specialists Conf., IEEE, New York, p. 457.

Kim, N.P., B.J. Stanbery, R.P. Gale, and R.W. McClelland (1988), Record of the NASA Space Photovoltaic Research and Technology Conference, NASA-Lewis Research Center (in press 1990).

Konagai, M. Sagimoto, and K. Takahashi (1978), J. Crystal Growth **45**, 277.

Lee, M.K., D.S. Wuu, and H.H. Tung (1987), Appl. Phys. Lett. **50**, 1725.

Lewis, C.R., C.W. Ford, G.F. Virshup, B.A. Arau, R.T. Green, and J.G. Werthen (1988a), Record of the 18th IEEE Photovoltaic Specialists Conf., IEEE, New York, p. 556.

Lewis, C.R., H.F. Macmillan, B.C. Chung, G.F. Virshup, D.D. Liu, L.D. Partain, and J.G. Werthen (1988b), Solar Cells **24**, 177.

Loferski, J.L. (1982), Record of the 16th IEEE Photovoltaic Specialists Conf., IEEE, New York, p. 648.

Ludowise, M.J., R.A. Larue, P.G. Borden, P.E. Gregory, and W.T. Dietze (1982), Appl. Phys. Lett. **41**, 550.

Macmillan, H.F., H.C. Hamaker, G.F. Virshup, and J. G. Werthen (1988), Rec. of the 20th IEEE Photovoltaic Specialists Conf., IEEE, New York, p. 48.

McClelland, R.W., C.O. Bozler, and J.C.C. Fan (1980), Appl. Phys. Lett. **37**, 560.

Miller, D.L., S.W. Zehr, and J.S. Harris (1982), J. Appl. Phys. **53**, 744.

National Department of Energy 5-Year Plan (1987).

Okamoto, H., Y. Kadota, Y. Watanabe, Y. Fukuda, T. Oh'hara, and Y. Ohmachi (1988), Record of the 20th IEEE Photovoltaic Specialists Conf., IEEE, New York, p. 475.

Olson, J.M., S.R. Kurtz, and A.E. Kibber (1988), Record of the 20th IEEE Photovoltaic Specialists Conference, IEEE, New York, p. 777.

Olson, J.M., S.R. Kurtz, and A.E. Kibber (1989), Cells measured at SERI, private communication by L.L. Kazmeiski.

Partain, L.D., M.S. Kuryla, R.E. Weiss, J.G. Werthen, G.F. Virshup, H.F. MacMillan, H.C. Hamaker, and D.L. King (1987), Record of the 19th IEEE Photovoltaics Specialists Conf., IEEE, New York, p. 1504.

Partain, L.D., G.F. Virshup, and N.R. Kaminar (1988), Record of the 20th IEEE Photovoltaic Conf., IEEE, New York, p. 759.

Patel, R.M., S.W. Gersten, D.R. Perrachione, Y.C.M. Yeh, D.K. Wagner, and R.K. Morris (1988), Record of the 20th IEEE Photovoltaic Specialists Conf., IEEE, New York, p. 607.

Shay, J.L., S. Wagner, K.J. Bachmann, and E. Buehler (1976), J. Appl. Phys. **47**, 614.

Spitzer, M. and J.C.C. Fan (to be published in Solar Cells, 1990).

Spitzer, M.B., C.J. Keavney, S.M. Vernon, and V.E. Haven (1987), Appl. Phys. Lett. **51**, 364.

Spitzer, M.B., C.J. Keavney, S.M. Vernon, and V.E. Haven (1987), Record of the 19th IEEE Photovoltaic Specialists Conf., IEEE, New York, p. 146.

Timmons, M.L. and S.M. Bedair (1981), J. Appl. Phys. **52**, 1134.

Timmons, M.L. and P.K. Chiang (1989), Record of the 19th IEEE Photovoltaic Specialists Conference, IEEE, New York, p. 102.

Timmons, M.L., J.A. Hutchby, D.K. Wagner, and J.M. Tracy (1988), Record of the 20th IEEE Photovoltaic Specialists Conf., IEEE, New York, p. 602.

Tobin, S.P. (1987), *Development of a Thin AlGaAs Solar Cell*, Sandia Contractor Report SAND87-7098.

Tobin, S.P., S.M. Vernon, C. Bajgar, V.E. Haven, L.M. Geoffroy, and D.R. Lillington (1988), IEEE Electron Device Letters **EDL-9**, 256.

Tobin, S.P., S.M. Vernon, C. Bajgar, V.E. Haven, L.M. Geoffroy, M.M. Sanfacon, D.R. Lillington, R.E. Hart, Jr., K.A. Emery, and R.J. Matson (1988), Record of the 20th IEEE Photovoltaic Specialists Conf., IEEE, New York, p. 405.

Tsaur, B.-Y., J.C.C. Fan, G.W. Turner, F.M. Davis, and R.P. Gale (1982), Record of the 16th IEEE Photovoltaic Conf., IEEE, New York, p. 1143.

Tsaur, B.-Y., J.C.C. Fan, G.W. Turner, B.D. King, and R. McClelland and G.M. Metze (1984), Record of the 17th IEEE Photovoltaic Specialist Conf., IEEE, New York, p. 440.

Turner, G.W., J.C.C. Fan, and J.J. Hsieh (1980), Appl. Phys. Lett. **37**, 400.

Van Gurp, G.J., P.R. Bondewijn, M.N.C. Kempeners, and D.L.A. Tjaden (1987), J. Appl. Phys. **61**, 1846.

Vernon, S.M., S.P. Tobin, V.E. Haven, R.G. Wolfson, M.W. Wanlass, and R.J. Matson (1987), Record of the 19th IEEE Photovoltaic Specialists Conf., IEEE, New York, p. 108.

Virshup, G.F., B.C. Chung, and J.G. Werthen (1988), Record of the 20th IEEE Photovoltaic Specialists Conf., IEEE, New York, p. 441.

Wanlass, M.W., T.A. Gessert, G.S. Horner, K.A. Emery, and T.J. Coutts (in press 1990), Record of the Space Photovoltaics Research and Technology Conference, NASA-Lewis Research Center.

Werthen, J.G., G.F. Virshup, C.W. Ford, C.R. Lewis and H.C. Hamaker (1988), Record of the 20th IEEE Photovoltaic Specialists Conf., IEEE, New York, p. 300.

Wojtczuk, C.J., S.P. Tobin, C.J. Keavney, C. Bajgar, M.M. Sanfacon, L.M. Geoffroy, T.M. Dixon, S.M. Vernon, J.D. Scofield, and D.S. Ruby (1990), IEEE Transport Electron Devices, **ED-37**, 455.

Woodall, J.W. and H.J. Hovel (1977), Appl. Phys. Lett. **30**, 492.

Wu, X., T.J. Coutts, R.G. Dhere, T.A. Gessert, and N.G. Dhere (1987), Record of the 19th IEEE Photovoltaic Specialists Conf., IEEE, New York, p. 140.

Yablonovitch, E., T. Gmitter, J.P. Harbison, and R. Bhat (1987), Appl. Phys. Lett. **51**, 2222.

Yamaguchi, M., K. Ando, A. Yamamoto, and C. Uemura (1984), J. Appl. Phys. **44**, 432.

Yamaguchi, M. and C. Uemura (1984), Appl. Phys. Lett. **44**, 611.

Yamaguchi, M., C. Uemura, and A. Yamamoto (1984a), J. Appl. Phys. **55**, 1429.

Yamaguchi, M., Y. Yamamoto, Y. Itoh, and C. Uemura (1986), *Proceedings of the 2nd International Photovoltaic Science and Engineering Conference*, Beijing, China, p. 573.

Yamamoto, A., N. Uchida, and M. Yamaguchi (1986), Optoelectronics Devices and Technologies **1**, 41.

Yoshikawa, A. and Y. Sakai (1977), Solid State Electronics **20**, 133.

CHAPTER 4

# High-Efficiency Silicon Solar Cells

Richard M. Swanson and Ronald A. Sinton

## 4.1 Introduction

In the last five years silicon solar cells have undergone significant evolution resulting in greatly improved efficiencies. As an illustration, Figure 4.1 plots the highest reported silicon concentrator cell efficiency versus year. Also shown for comparison are gallium-arsenide concentrator cell results. Silicon one-sun cells have undergone a similar, but less dramatic, improvement.

Besides the normal progress obtainable through continued design refinement, the major factors enabling this improvement are:

- incorporation of light trapping,
- reduction of recombination through improved material quality, cleaner processing, surface passivation and reduced contact coverage fraction, and
- improvements in understanding of the fundamental physical processes important in cell operation such as recombination and carrier transport in heavily doped regions.

Early workers tended to view solar cells as $p$-$n$ junctions which happen to have light-induced generation. This led to simple models of cell operation emphasizing diffusion of carriers in low-level injected doped regions (Hovel, 1975). Cell designs used thick, rather highly doped wafers. The importance of reducing recombination, such as by having high minority carrier lifetime, was recognized early; however, the physics of recombination was not sufficiently studied to know what was limiting the lifetime, and how high a lifetime could reasonably be attained in various parts of the cell.

An apt analogy to the importance of controlling and reducing recombination is provided by a leaky bucket being continually filled with a faucet (Schwartz, 1983). The level of water represents the density of electron-hole pairs and the incoming stream represents the incident photons. One wishes to attain the highest water level (highest carrier density and, hence, voltage) but it is found that every time a leak (source of recombination) is plugged the water rises only a little until another leak takes over. So it is in solar cells; in order to achieve significant increases in output voltage it is necessary to examine every part of the cell for potential recombination paths and take steps to block those paths. This requires that the internal flow of carriers must be well modeled and understood. Our ability to do this continues to improve. At this point it seems only fair to reveal that one can now state with

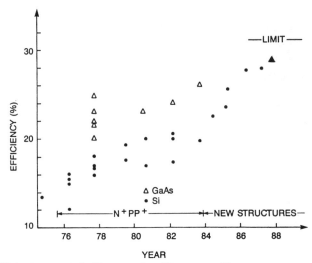

**Figure 4.1:** Highest reported silicon and gallium arsenide concentrator cell efficiencies versus date.

some certainty that the practical limit efficiency for one-sun silicon solar cells at room temperature is between 24 and 25 percent, and that for concentrator cells lies between 30 and 31 percent. An implementation of a concentrator system with a restricted acceptance angle for the incident light may extend these efficiencies somewhat towards the limit of 36–37% (Campbell, 1986). Thus, barring some unforeseen breakthrough, laboratory cell efficiencies are rapidly approaching their limits. Nevertheless, much work remains to consolidate these recent improvements into practical, production-worthy designs.

This article discusses these results. Particular emphasis is directed toward concentrator cells which operate in high-injection conditions, as this is where the bulk of the authors' experience lies. Most of the material presented in this chapter, however, is equally applicable to flat-plate cells. Concentrator cells have historically led the drive to higher efficiency because concentrator economics can tolerate the complex processing and high-quality materials that are necessary to achieve those high efficiencies. Many of the high-efficiency concepts which first appeared on concentrator cells, such as passivated emitters, have subsequently found their way into the entire range of solar cell applications including one-sun cells.

Section 2.3.4D presents a simplified view of basic cell operation while Section 4.3 contains the details of the relevant semiconductor device physics. Finally, Section 4.4 discusses the design of high-efficiency cells.

## 4.2 Basic Cell Operation

We assume that the reader is familiar with the fundamentals of both semiconductor device operation and solar cells. Excellent texts on the basics of solar cell operation can be found in references (Green, 1982), (Fahrenbruch and Bube, 1983), and more particularly Green, 1987). Rather than trying to re-do what has been done in these texts, we present here a simplified viewpoint of device operation which has been central to our thinking.

The analysis of silicon solar cells is often couched in terms of the current density equations, drift plus diffusion, and the continuity equation. Such an approach obscures the physics of operation, however, because the operation of a well-designed solar cell is controlled by generation and recombination. The need for current transport is responsible only for certain second-order losses in the cell. This Section presents an analysis of cell operation in terms of an integral formulation of the continuity equation, which brings forth the balance between generation and recombination as determining the output of current, and quasi-Fermi potentials as determining the output of voltage.

### 4.2.1 The Semiconductor Device Equations

The standard equations used to model steady-state carrier transport in silicon devices, when heavy doping effects can be neglected, are:
1. current transport equations

$$\vec{J}_n = -q\mu_n n\vec{\nabla}\psi_i + qD_n\vec{\nabla}n = -q\mu_n n\vec{\nabla}\phi_n \tag{4.1}$$

$$\vec{J}_p = -q\mu_p p\vec{\nabla}\psi_i - qD_p\vec{\nabla}p = -q\mu_p p\vec{\nabla}\phi_p \tag{4.2}$$

2. continuity equations

$$\vec{\nabla} \cdot \vec{J}_n = q(r - g_{ph}) \tag{4.3}$$

$$\vec{\nabla} \cdot \vec{J}_p = -q(r - g_{ph}) \tag{4.4}$$

3. Poisson's equation

$$\nabla^2\psi_i = -\frac{q}{\epsilon}(p + N_D^+ - n - N_A^-) \tag{4.5}$$

4. carrier density equations

$$n = n_i e^{q(\psi_i - \phi_n)/kT} \tag{4.6}$$

$$p = n_i e^{q(\phi_p - \psi_i)/kT}. \tag{4.7}$$

The symbols used have their standard meanings as given below:

| | |
|---|---|
| $r = r_{th} - g_{th}$ | net thermal recombination rate per unit volume (Here, $r_{th}$ is the recombination rate and $g_{th}$ is thermal generation rate.) |
| $g_{ph}$ | photogeneration rate per unit volume |
| $\vec{J}_n$ | electron current density |
| $\vec{J}_p$ | hole current density |
| $n$ | electron concentration |
| $p$ | hole concentration |
| $\mu_n, \mu_p$ | electron and hole mobilities |

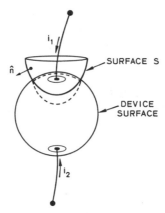

**Figure 4.2:** A general two terminal device.

| | |
|---|---|
| $D_n$, $D_p$ | electron and hole diffusivities |
| $\phi_n$, $\phi_p$ | electron and hole quasi-Fermi potentials |
| $\psi_i$ | potential referenced to the intrinsic level |
| $\epsilon$ | static dielectric permitivity |
| $q$ | magnitude of electron charge |
| $N_D^+$, $N_A^+$ | ionized doping density |

Using Equations (4.1) through (4.4), the electron and hole current densities can be eliminated. The resulting equations, along with Poisson's Equation (4.5), define three coupled non-linear differential equations in three unknowns, $n$, $p$, and $\psi_i$. Equations (4.6) and (4.7) are superfluous but will prove useful in relating terminal voltage to carrier densities. Given appropriate boundary conditions these equations can be solved — in principle. The search for reasonable analytic solutions to these equations is hopeless, however; either some simplifications must be made or numerical techniques used. Several approximate analytic approaches will be presented below and in Section 4.4.

### 4.2.2 An Integral Approach to the Terminal Current

Once a solution for $n$, $p$, and $\psi_i$ is found, either numerically or through judicious analytic approximation, the terminal current can be found in either one of two ways; by a differential or an integral method.

### 4.2.2A Differential Method

Consider the general two terminal device, shown in Figure 4.2, with an imaginary surface $S$ that surrounds terminal 1. The terminal currents are $i_1$ and $i_2 = -i_1$. In the steady-state, $i_1$ may be found by

$$i_1 = \int_S (\vec{J}_n + \vec{J}_p) \cdot \hat{n}\, dS \qquad (4.8)$$

where $\hat{n}$ is an outward unit vector normal to $S$ and $J_n$ and $J_p$ are given by Equations (4.1) and (4.2). Any surface can be used to calculate $i_1$ that surrounds contact 1, but if one doesn't use *exact* solutions for $\vec{J}_n$ and $\vec{J}_p$ (or equivalently, for $n$, $p$ and $\psi_i$) then different surfaces may give different answers.

### 4.2.2B Integral Method

Using this method, one integrates the continuity Equations (4.3) and (4.4), over the device volume,

$$\int_V \vec{\nabla} \cdot \vec{J}_n \, dv = q \int_V (r - g_{ph}) \, dv \tag{4.9}$$

$$\int_V \vec{\nabla} \cdot \vec{J}_p \, dv = -q \int_V (r - g_{ph}) \, dv. \tag{4.10}$$

Next, Gauss' divergence theorem is used to convert the left-hand side to surface integrals over the device's surface, $S_{tot}$.

$$\int_{S_{tot}} \vec{J}_n \cdot \hat{n} \, dS = q \int_V (r - g_{ph}) \, dv \tag{4.11}$$

$$\int_{S_{tot}} \vec{J}_p \cdot \hat{n} \, dS = -q \int_V (r - g_{ph}) \, dv \tag{4.12}$$

The surface integration is divided into three regions; $S_1$ (contact 1), $S_2$ (contact 2), and $S$ (remainder of device) giving

$$\int_{S_{tot}} \vec{J}_n \cdot \hat{n} \, dS = \int_{S_1} \vec{J}_n \cdot \hat{n} \, dS + \int_{S_2} \vec{J}_n \cdot \hat{n} \, dS + \int_S \vec{J}_n \cdot \hat{n} \, dS \tag{4.13}$$

$$\int_{S_{tot}} \vec{J}_p \cdot \hat{n} \, dS = \int_{S_1} \vec{J}_p \cdot \hat{n} \, dS + \int_{S_2} \vec{J}_p \cdot \hat{n} \, dS + \int_S \vec{J}_p \cdot \hat{n} \, dS. \tag{4.14}$$

At contact 1 the current is

$$i_1 = -\int_{S_1} \vec{J}_n \cdot \hat{n} \, dS - \int_{S_1} \vec{J}_p \cdot \hat{n} \, dS. \tag{4.15}$$

The minus sign results because the normal vector $\hat{n}$ is outward while the current is referenced inward. Now, solving for $-\int_{S_1} \vec{J}_p \cdot \hat{n} \, dS$ in Equation (4.14), inserting this into Equation (4.15), and using Equation (4.12) gives,

$$i_1 = \int_{S_2} \vec{J}_p \cdot \hat{n} \, dS - \int_{S_1} \vec{J}_n \cdot \hat{n} \, dS + \int_S \vec{J}_p \cdot \hat{n} + q \int_V r \, dv - q \int_V g_{ph} \, dV \tag{4.16}$$

Let us suppose that contact 1 is $p$-type and contact 2 is $n$-type so that $J_p$ and $J_n$ are minority carrier current densities in the integrations in Equation (4.16). Since, in a solar cell, positive current out of the $p$-type contact corresponds to positive output power it is convenient to define the terminal current $I = -i_1 = i_2$. Then Equation (4.16) can be written in the suggestive form

$$I = I_{ph} - I_{b,rec} - I_{s,rec} - I_{cont,rec} = I_{ph} - I_{rec} \tag{4.17}$$

where

$$I_{rec} = I_{b,rec} + I_{s,rec} + I_{cont,rec} \tag{4.18}$$

and

$$I_{ph} = q \int_V g_{ph} \, dv \qquad \text{photogeneration current}$$
$$I_{b,rec} = q \int_V r \, dv \qquad \text{bulk recombination}$$
$$I_{s,rec} = \int_S \vec{J}_p \cdot \hat{n} \, dS \qquad \text{surface recombination} \qquad (4.19)$$
$$I_{cont,rec} = \int_{S_2} \vec{J}_p \cdot \hat{n} \, dS - \int_{S_1} \vec{J}_n \cdot \hat{n} \, dS \qquad \text{contact recombination.}$$

In other words, if one properly accounts for recombination, one can say that the output current equals the photogeneration current minus the total recombination current.* It should be stressed that Equation (4.17) is an exact result of the continuity equation.

The position of the surface over which one integrates to get the contact recombination is completely arbitrary, as long as it surrounds the contact. If the integration surface is moved away from the metal-semiconductor contact, then the volume in the region between the integration surface and the metal must not be included in the recombination integral. This a convenient maneuver when the region around the contact is heavily doped and low-level injected. Consider, for example, a metal contact to a heavily doped $n$-type region. If the integration surface is moved in from the surface to eliminate this region, the hole current at the new surface is then comprised of hole recombination in the doped layer plus the hole surface recombination. (This is of course what would have been calculated using by including the region in the original volume.) This current is then easily calculated, using conventional low-level-injection theory, as a function of the hole density at the edge of the integration region.

The integral approach has the advantage that any errors in $n$ and $p$ tend to be averaged out when performing the integrations. It also does not depend on knowing $\vec{\nabla} n$ and $\vec{\nabla} p$. This is an advantage because most analytical approximations, or numerical calculation techniques, yield better results for $n$ and $p$ than for their gradients. For solar cells, using Equation (4.17) approaches the essence of what is desired more closely than the differential method which requires knowledge of the carrier fluxes throughout the device.

### 4.2.3 The Terminal Voltage

A typical advanced solar cell has highly doped regions near the contacts (for ohmic contact and reduction of contact recombination), and a more lightly doped base. This is illustrated in the band diagram in Figure 4.3. The horizontal coordinate on this diagram, position, is not necessarily a spatially straight line from

---

* Sometimes one sees the terminal current written as the short-circuit current minus the "dark current." The short circuit current will be the photogeneration current minus whatever recombination occurs at short circuit. In this case the dark current is then any additional recombination obtaining when the device is not at short-circuit. In the simplest case of low-level injection, the equations defining the carrier density are linear and the dark current becomes independent of light intensity and equal to the recombination of the cell in the dark at the same junction potential. Such a cell is said to obey the "superposition principle." We will see below that narrow base cells also obey the superposition principle. In all other cases such a separation cannot be rigorously made. Equation (4.17) remains true, however.

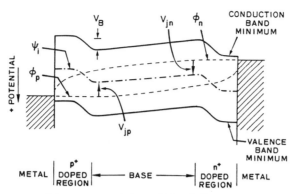

**Figure 4.3:** Solar cell band diagram.

front to back, but just represents a continuous physical path from the $p^+$ to $n^+$ contact areas. The terminal voltage (less any drop in the contacts and metal bus) is $\phi_p$ at the $p^+$ contact minus $\phi_n$ at the $n^+$ contact.

To a very good approximation $\phi_n$ can be assumed constant through the $n^+$ region, its space charge region, and into the edge of the quasi-neutral base near the $n^+$ contact at position 2. The same applies to $\phi_p$ in the $p^+$ region up to position 1. This is because these regions are heavily doped and hence have an abundance of majority carriers. Referring to Figure 4.3 it is easily seen that the output voltage is

$$V = V_{jn} + V_{jp} + V_B + V_c + V_m \tag{4.20}$$

where

$$V_{jn} = \psi_i - \phi_n = kT/q \ln(n/n_i) \text{ [evaluated at point 2]} \tag{4.21}$$

$$V_{jp} = \phi_p - \psi_i = kT/q \ln(p/n_i) \text{ [evaluated at point 1]} \tag{4.22}$$

$$V_B = \psi_i(1) - \psi_i(2) = -\int_1^2 \vec{\nabla}\psi_i \cdot d\vec{l} = \int_1^2 \vec{E} \cdot d\vec{l} \tag{4.23}$$

$$V_c \text{ [contact voltage]} \tag{4.24}$$

$$V_m \text{ [metal grid voltage]}. \tag{4.25}$$

In normal operation, $V_B$, $V_c$, and $V_m$ are all negative and represent the base, contact and metal resistive loss respectively. Detailed methods of calculating $V_c$ and $V_m$ are available in the literature (Basore, 1984).

The importance of Equations (4.20) through (4.25) lies in the need to have solutions for $n$, $p$, and $\psi_i$ only in the base semiconductor region. The details of what happens in the $n^+$ and $p^+$ semiconductor regions are unimportant for determining voltage. This is fortunate for, as discussed below, what happens in these regions is difficult to model because Equations (4.1) through (4.7) do not apply in very highly doped regions. It should be kept in mind that the above voltage and current equations are valid regardless of whether the base is high-level injected (i.e., has a minority carrier density greater than the doping density) or not.

Notice that if one neglects the base, contact and metal voltage drops, and assumes that $n$ and $p$ are constant throughout the base then Equations (4.21) through (4.25) imply that

$$V \approx \frac{kT}{q} \ln(pn/n_i^2). \qquad (4.26)$$

In this case, $\phi_n$ and $\phi_p$ are constant throughout the base and

$$V = \phi_p - \phi_n. \qquad (4.27)$$

These equations describe the essence of device voltage — the terminal voltage is just the separation of quasi-Fermi potentials.* Equivalently, the $pn$ product appears as the major determinant of voltage.

Transport losses, which are hopefully small in a high-efficiency cell, will decrease the voltage below this ideal as shown by Equation (4.20). If the cell can be well modeled as having a series resistance, $R_s$, so that $V_B + V_c + V_m \approx -IR_s$, then the above equations imply

$$V \approx \frac{kT}{q} \ln(pn/n_i^2) - IR_s. \qquad (4.28)$$

In many cases this is a reasonable approximation; however, in cells operating in high-level injection, a more rigorous calculation of the base voltage is required. This will be covered in subsequent sections.

### 4.2.4   Carrier Photogeneration

A solar cell's first function is to absorb light to produce electron-hole pairs. From Equation (4.17) it is immediately apparent that to maximize efficiency one must design the cell to produce as many electron-hole pairs as possible. The number of electron-hole pairs produced in the silicon by a given amount and spectrum of incident light is simply a function of:

- how many of the photons are transmitted through the cell front surface into the bulk, and
- the path length that these photons can travel once in the bulk silicon before leaving or being absorbed in a way which does not produce electron-hole pairs.

Silicon is weakly absorbing since it has an indirect bandgap. Thus, for practical solar cells, a substantial volume of silicon is usually devoted simply to absorbing light. In virtually all cells, this volume of silicon is the most lightly doped region. The amount of light absorbed depends upon the cell thickness and path that the light takes through the cell. The details of the optimization to maximize this photogeneration are discussed in Section 4.4, Cell Design Engineering. Current, state-of-the-art silicon cells incorporating light-trapping produce about 42 mA of current for each 100 mW of incident sunlight (AM 1.5 direct spectrum).

---

* Thermodynamically, this is equivalent to saying that the terminal voltage is the difference of the electron and hole electromotive force.

## 4.2.5 Sources of Recombination

In this section we explore the impact of various types of recombination on the terminal characteristics. The phenomenological equations describing the common recombination processes are presented. Detailed discussion of the experimental determination of the various coefficients involved is delayed until Section 4.3, The Physical Details of Recombination and Transport.

To calculate terminal current the various recombination terms in Equation (4.17) need to be evaluated. Various regions of the cell require different approaches.

### 4.2.5A Diffusion Region Recombination

Part of the integration in Equation (4.17) for the bulk recombination involves the $n^+$ and $p^+$ diffused regions where no solution, as yet, exists. The same applies to contact recombination. This problem can be side-stepped by defining the surface of the device to skirt the highly doped regions.* The new surface of integration is then just inside the neutral base under the contact regions. It is easy to see in a physical manner why this is possible because the net hole current entering the $n$-type diffusion area, for example, either recombines within the diffusion region or at the contact. Thus the same result is obtained by either evaluating the hole current at the edge of the space charge region or calculating the net recombination in the diffused area plus contact. (The photogeneration in the diffused regions under the contacts is being neglected.)

It will be shown in Section 4.3 that the minority carrier hole injection current density into a highly doped $n$-type region can always be written, provided the doped region is everywhere low-level injected, as

$$J_p = J_{0n} \left( \frac{pn}{n_i^2} - 1 \right) \tag{4.29}$$

where $J_{0n}$ is a temperature-dependent constant, to be called the diffusion saturation current, and $p$ and $n$ are evaluated at the internal edge of the neutral highly doped region (del Alamo, 1984). The $pn$ product will be constant to within a very small error across the space charge region, so the $pn$ product at the edge of the space charge region in the neutral base may be used instead. Equation (4.29) may still be used to calculate the injection current; however, any recombination in the space charge region will be thereby neglected if this region is also excluded from the volume recombination calculation. In high efficiency cells the space charge region recombination is normally very small.

$J_{0n}$ can be considered in two ways; it can be calculated using some particular model of transport in heavily doped regions or it can simply be measured for the diffusion employed and taken as an experimental parameter.

Similarly

$$J_n = J_{0p} \left( \frac{pn}{n_i^2} - 1 \right) \tag{4.30}$$

for the electron current density entering a $p$-type diffused region.

---

* As mentioned above, the location of this surface is arbitrary.

### 4.2.5B  Recombination in the Base

Recombination in the base of silicon solar cells can occur through a variety of mechanisms. Radiative recombination is the inverse of the photogeneration process responsible for the operation of solar cells and hence must occur to some degree. Other, non-radiative mechanisms usually dominate in silicon. A wide variety of material defects, both intrinsic and extrinsic (or impurity related) can catalyze the recombination process. The three-particle process known as Auger recombination, which is discussed more fully in Section 4.3, becomes important at high carrier densities.

**Radiative recombination.** The simplest to model, if least important, form of recombination is radiative recombination. The process is proportional to the excess $pn$ product:

$$r = B(pn - n_i^2) \tag{4.31}$$

where $B$ is called the radiative rate coefficient. B has an experimental value of $9.5 \times 10^{-15}$ cm$^3$/s (Schlangenotto, 1974). Direct-gap semiconductors such as GaAs typically have radiative coefficients that are several thousand times larger. The weak, phonon-mediated radiative process in silicon means that other recombination paths usually dominate.

**Defect mediated recombination.** Any type of crystalline defect can potentially act as a catalyst for electron-hole recombination. The physical processes involved in this recombination are not well understood. The carriers must lose their energy through some mechanism. This could be radiative or non-radiative; the non-radiative process is usually thought to involve energy exchange with lattice distortions around the defect. Due to the lack of physical understanding, defect-mediated recombination is usually modeled using the phenomenological approach of Shockley, Read and Hall (Sze, 1981). Under low-level injection conditions, this gives a recombination rate proportional to the excess minority carrier density;

$$r = \frac{n - n_0}{\tau} \tag{4.32}$$

where $n_0$ is the electron density in equilibrium and $\tau$ is the low-level lifetime. More complicated expressions are obtained if one includes the effects of majority carrier capture.

Much of the early work on single crystal silicon cells was concerned with increasing the base lifetime, $\tau$, through reduction of crystalline defects. The use of high-quality, float-zone silicon combined with the incorporation of state-of-the-art semiconductor processing has reduced defect levels in laboratory cells to the point that bulk, defect-mediated recombination is no longer the dominant loss. Recombination in low-cost, commercial cells, particularly poly-crystalline cells, is still dominated by defect-mediated recombination, although progress continues. Recent results in hydrogen passivation (Schmidt, 1982) and phosphorus gettering (Narayanan, 1986) have allowed improved cell performance.

This work will concentrate on results achievable in high-quality silicon and, hence, defect mediated recombination will play a small role. One type of recombination that is intrinsic to semiconductors, and hence cannot be reduced through improved technology, is Auger recombination.

**Auger recombination.** In an Auger recombination event the excess energy of an electron-hole pair is given to a third free particle, either an electron or a hole. The recombination rate is then

$$r = C_n(n^2 p - n_o^2 p_o) + C_p(p^2 n - p_o^2 n_o). \tag{4.33}$$

$C_n$ is the $n$-type Auger coefficient and refers to the process where the excess energy is given to an electron. This process will dominate in $n$-type silicon. $C_p$ is the corresponding $p$-type Auger coefficient. In material doped to over $10^{18}$ cm$^{-3}$ experiments show that $C_n = 2.8 \times 10^{-31}$ cm$^6$/s and $C_p = 0.99 \times 10^{-31}$ cm$^6$/s (Dziewior and Schmid, 1977). When the doping, or carrier density, is less than $10^{17}$ cm$^{-3}$, the coefficients appear to be about four times larger (Sinton and Swanson, 1987a). These coefficients will be discussed more fully in Section 4.3.

### 4.2.5C Surface Recombination

The only remaining portion of the cell where recombination can occur is at the free surfaces between contacted areas. Like bulk defect-mediated recombination, surface recombination is catalyzed through defects. These defects are usually thought to be mostly the dangling bonds of surface silicon atoms not bonded to four nearest neighbors. The net recombination rate per unit area is usually assumed to be proportional to the excess carrier density so that

$$J_{s,\mathrm{rec}} = I_{s,\mathrm{rec}}/A = qs(n - n_0) \tag{4.34}$$

on $p$-type surfaces and

$$J_{s,\mathrm{rec}} = I_{s,\mathrm{rec}}/A = qs(p - p_0) \tag{4.35}$$

on $n$-type surfaces. $s$ is the phenomenological surface recombination velocity. It is very dependent on the surface preparation method. Clean, bare silicon surfaces have $s$ in the range of $10^3$ to $10^5$ cm/s. Surfaces which are "passivated" through the growth of silicon dioxide can have $s$ in the range of 1 to $10^3$ cm/s. Improvements in surface passivation have been responsible for considerable improvement in cell performance. The details of surface recombination and passivation are discussed more fully in Section 4.3.

### 4.2.6 Simplified Device Operation — The Narrow-Base Cell

In this section we use the previous concepts to analyze cell operation in a simplified manner. The main simplification is to assume constant electron and hole quasi-Fermi potentials, or equivalently, constant $pn$ product. The voltage is then given by Equation (4.26).

This might be called the "narrow-base" approximation because, if the cell is sufficiently thick, gradients in $\phi_n$ and $\phi_p$ will necessarily become significant. We are thus ignoring any voltage drops in the base region, which will be small in the narrow-base approximation (i.e., $V_B = 0$ in Equation (4.20)). It is possible to make a first order improvement in this approximation by including resistance losses through a series resistance via Equation (4.28).

### 4.2.6A  Current-Voltage Characteristics

To calculate the current we use Equation (4.17) and evaluate all the sources of recombination.

**Diffused region recombination.** In this case the $n$-contact diffused region recombination current density is given by Equation (4.29)

$$J_{n-\text{cont,rec}} = J_{on} \left( e^{qV/kT} - 1 \right) \tag{4.36}$$

and the total current is just the area of the $n$-contact regions, $A_{n-\text{cont}}$, so that

$$I_{n-\text{cont,rec}} = A_{n-\text{cont}} J_{on} \left( e^{qV/kT} - 1 \right). \tag{4.37}$$

Note how these equations have the familiar form for the diffusion current in a diode. The current is actually composed of drift and diffusion components if the diffusions are non-uniformly doped.

Similarly

$$I_{p-\text{cont,rec}} = A_{p-\text{cont}} J_{op} \left( e^{qV/kT} - 1 \right) \tag{4.38}$$

for the recombination current in the $p$-type contact diffusion.

**Recombination in the base.** To perform the remainder of the bulk recombination integration the recombination in the base needs to be evaluated. If it is assumed that the base is $p$-type and low-level injected then

$$R_n = R_p = \frac{n - n_o}{\tau_n} \tag{4.39}$$

and

$$I_{\text{base,rec}} = q \int_V \frac{n - n_o}{\tau_n} \, dV. \tag{4.40}$$

Here, $\tau_n$ includes the aggregate of all bulk recombination mechanisms: defect, Auger and radiative. Noting that, under low-level injection,

$$pn = n_i^2 e^{qV/kT} \approx N_A n \tag{4.41}$$

gives

$$I_{\text{base,rec}} = \frac{qAW n_i^2}{N_A \tau_n} \left( e^{qV/kT} - 1 \right) \tag{4.42}$$

so that base recombination has the same voltage dependence as diffused region recombination. Here $A$ is the area of the device and $W$ is the thickness of the base. We will find that anytime a region is in low-level injection and the recombination is linear in excess carrier density then the current will have this dependence on terminal voltage. That solar cells often have high-level injection and non-linear recombination effects is the main reason for introducing the above analysis.

If the base is high-level injected, then the electron and hole densities will be approximately equal by charge neutrality. This implies that

$$pn = n^2 = n_i^2 e^{qV/kT} \tag{4.43}$$

or

$$n = n_i e^{qV/2kT}. \tag{4.44}$$

Under high level injection, the defect-mediated recombination will be shown to be

$$R_p = \frac{n}{\tau_{hl}} \tag{4.45}$$

where $\tau_{hl}$ is the high level lifetime (usually larger than the low level lifetime). Assuming constant quasi-Fermi potentials in this case gives

$$I_{base,rec} = \frac{qAW n_i}{\tau_{hl}} e^{qV/2kT}. \tag{4.46}$$

Note that this results in an "ideality factor" of two for the case of defect mediated bulk recombination in a high-level injected base. This should not be confused with the case of space-charge region recombination which often exhibits an ideality factor of near two.

Radiative recombination (the recombination of an electron-hole pair to produce a photon) is not particularly prevalent in silicon but is proportional, nevertheless, to the $pn$ product and hence exhibits an "ideality factor" of one in both high-level and low-level injection. In Section 4.3.4 it will be shown that Auger recombination often dominates in highly injected bases. In high-level injection Equation (4.33) shows that the recombination will go as $R_p = C_A n^3$ where $C_A = C_n + C_p$. This gives a recombination current of

$$I_{base,rec} = qAWC_A n_i^3 e^{3qV/2kT} \tag{4.47}$$

for an "ideality factor" of $2/3$.*

**Surface recombination.** The remaining portion of the device is the insulated surface between contact regions. Surface recombination, as we will see, proves to be a complicated function of excess carrier densities and is thus harder to model. If one assumes, however, that the surface region between contacts is low-level injected, and that the recombination is proportional to the excess minority carrier density, then

$$I_{surf,rec} = qs \int_S (n - n_o) \, dA \tag{4.48}$$

or, assuming constant quasi-Fermi potentials again,

$$I_{surf,rec} = \frac{qA_{surf} s n_i^2}{N_A} \left( e^{qV/kT} - 1 \right) \tag{4.49}$$

where $A_{surf}$ is the surface area of the device.

---

* This gives an increase in voltage of 40 mV per decade of current, rather than the usual 60 mV per decade at room temperature.

If the surface is high-level injected, this becomes

$$I_{\text{surf,rec}} = qsA_{\text{surf}}n_i e^{qV/2kT}. \tag{4.50}$$

To obtain the overall recombination, the effects of all the above are added as appropriate. For example, if the cell is in low-level injection throughout, then, from Equation (4.17),

$$J = I/A = J_{ph} - \left(J_{on} + J_{op} + \frac{qWn_i^2}{N_A\tau_n} + \frac{qA_{\text{surf}}sn_i^2}{N_A}\right)\left(e^{qV/kT} - 1\right). \tag{4.51}$$

Noting that the recombination current goes as $e^{qV/\gamma kT}$ with $\gamma = 1$, we say that the cell has an "ideality-factor" of 1.

Alternatively, if the base is in high-level injection so that $n = p$, then

$$J = J_{ph} - \left(\frac{qsA_{\text{surf}}}{A} + \frac{qWn_i}{\tau_{hl}}\right)e^{qV/2kT} + (J_{on} + J_{op})e^{qV/kT} + (qWC_An_i^3)e^{3qV/2kT}. \tag{4.52}$$

Note that, under high-level injection, the cell has an ideality factor which varies depending upon which recombination mechanism dominates.

Several observations are in order. Within the narrow-base approximation, the short circuit current ($I$ when $V = 0$) is just $J_{ph}$. This means that all photogenerated carriers are collected. One says that the internal quantum efficiency is then unity. If there is recombination of photogenerated carriers at short-circuit, then this must be due to gradients in $\phi_n$ and $\phi_p$ and is thus a type of transport loss. Also, within the narrow-base approximation, all cells obey the superposition principle; at any voltage the terminal current is the generation current minus the recombination current in the dark at the same voltage.

### 4.2.6B  Efficiency and the Maximum Power Point

The efficiency, $\eta$, is defined as the maximum output power divided by the solar power input, $P_{sun}$,

$$\eta = \frac{I_{mp}V_{mp}}{P_{sun}} = \frac{J_{mp}V_{mp}}{P_{sun}/A}, \tag{4.53}$$

where $V_{mp}$ and $J_{mp}$ are the voltage and current evaluated at the point which gives maximum output power. This is often rearranged to give

$$\eta = \frac{J_{sc}}{P_{sun}/A}\frac{J_{mp}V_{mp}}{J_{sc}V_{oc}}V_{oc} = \mathcal{R}\cdot FF\cdot V_{oc} \tag{4.54}$$

where the subscript $sc$ refers to short-circuit and $oc$ refers to open-circuit. $\mathcal{R} = J_{sc}/(P_{sun}/A)$ is called the responsivity, and $FF = J_{mp}V_{mp}/J_{sc}V_{oc}$ is called the fill-factor.

In the narrow base approximation, $J_{sc} = J_{ph}$, so that the responsivity is 0.42 A/W when the one-sun photogeneration current is 42 mA per 100 mW. The responsivity in this approximation is independent of incident intensity because the photogeneration current is linear in intensity. As seen in Section 4.4, actual cells may have a responsivity that is either sublinear or superlinear in intensity.

Closed-form solutions for the maximum power point in terms of the current-voltage characteristic do not exist; however, it is straight-forward to iteratively find the maximum power voltage and current for any given set of parameters.

For cases in which the ideality factor of the curve is well defined, an approximate equation for the fill factor in terms of the open circuit voltage has been proposed by Green (1983a). This is

$$FF = \frac{v_{oc} - \ln(v_{oc} + 0.72)}{v_{oc} + 1} \tag{4.55}$$

where

$$v_{oc} = \frac{V_{oc}}{\gamma kT/q} \tag{4.56}$$

for an ideality factor $\gamma$.

Another approach that gives surprising accuracy is to estimate that the cell will be at the maximum power point when the terminal current is 95% of the short-circuit current. For a cell operating such that recombination in low-level injected regions dominates, so that the recombination is linear in excess carrier density, this indicates a reduction in $n$ to 5% of its value at open circuit (cf. Equation (4.39)). Reducing the minority carrier density by 95% (a factor of $1/20$) reduces the $p$-$n$ product by a factor of $1/20$ as well (in low-level injection). The voltage corresponding to this new $p$-$n$ product is given by:

$$V_{mp} = \frac{kT}{q} \ln(\frac{1}{20}(pn/n_i^2)) = V_{oc} - \frac{kT}{q} \ln(20). \tag{4.57}$$

This amounts to a voltage reduction of 78 mV at the maximum power point for a temperature of 300K (27°C).

The effect of a series resistance, $R_s$, can be easily included in this approximation by noting that it will cause an additional voltage drop of $I_{mp}R_s = J_{mp}\mathcal{R}_f$ where $\mathcal{R}_f \equiv \mathcal{R}_f A$. The maximum power voltage is now

$$V_{mp} = V_{oc} - \frac{kT}{q} \ln(20) - 0.95 J_{ph}\mathcal{R}_f \tag{4.58}$$

As an example, suppose that $J_{sc} = 0.037$ A at 100 mW insolation and $V_{oc} = 0.612$ V. Then the approximate efficiency is (neglecting series resistance)

$$\eta = \frac{(0.612 - 0.078) \times (0.037 \times 0.95)}{0.100} = 0.188. \tag{4.59}$$

If a specific series resistance of 0.7 $\Omega$cm$^2$ is present, then the efficiency is reduced to

$$\frac{(0.612 - 0.078 - 0.95 \times 0.7 \times 0.037) \times (0.037 \times 0.95)}{0.100} = 0.179. \tag{4.60}$$

Such a series resistance is seen to cause a 1% drop in absolute efficiency.

The preceding method of calculation efficiency is quite robust against the assumed terminal current. If a terminal current of 97% of the short-circuit current had been assumed, the calculated efficiency would be 18.7% for the case with no series resistance. Similarly, 93% yields 18.7%. The more exact Equation (4.55) gives 18.8% for this case.

All of the recombination mechanisms listed for the low-injection cell are linear in $n$. With this rule of thumb, the maximum-power-current is at 95% of the short-circuit-current, always gives a maximum-power-voltage 78 mV lower than the open circuit voltage. The corresponding results for the high-injection cells are that if base or surface recombination dominates, the maximum power point is 156 mV lower than the open-circuit voltage. If the diffused regions dominate it is 78 mV lower. If Auger recombination dominates, then the maximum power voltage is 52 mV lower. In each case, the assumption of 95% quantum efficiency introduces less than 0.3% error in absolute efficiency in comparison to the more exact methods over a wide range of the devices of interest (those with open circuit voltages greater than 600 mV).

For cases of well defined ideality factor the connection between this method and that of Equation (4.55) is seen if one realizes that then

$$J = J_{ph} - J_o e^{qV/\gamma kT} \tag{4.61}$$

so that

$$V_{mp} = \frac{\gamma kT}{q} \ln(0.05 J_{ph}/J_o) = V_{oc} - \frac{\gamma kT}{q} \ln(20) \tag{4.62}$$

The fill factor is then

$$FF = (1 - 1/20) \frac{v_{oc} - \ln(20)}{v_{oc}} \approx \frac{v_{oc} - \ln(20)}{v_{oc} + v_{oc}/20} \tag{4.63}$$

in the above approximation. This almost exactly equals Equation (4.55) when $v_{oc} = 20$, a typical value. The equation is very insensitive to the exact value for the current at maximum power because that affects only the argument of the logarithmic term and the small term in the denominator. An advantage of analyzing cells in this manner, however, is that the assumption of constant ideality factor is not needed — one simply evaluates the efficiency at 95% of $J_{ph}$.

It is interesting to analyze what this means physically. In low-injection silicon solar cells, the excess minority carrier density in the bulk at the maximum power point is about 5% of what it is under open circuit voltage conditions. In a high-injection cell dominated by bulk or surface recombination, it is also 5%. But in a high-level injected cell dominated by emitter recombination, it operates with a base carrier density 22% of its open circuit value. If it is high-level injected and Auger recombination dominates, it operates at 37% of its open circuit carrier density.[*] These figures provide some interesting insight into solar cell operation under these

---

[*] Emitter recombination is quadratic in base carrier density in high-level injected cells so that a reduction of $n$ to 22% reduces recombination to $0.22^2 = 0.05$. When Auger recombination dominates, recombination is cubic in $n$ resulting in $0.37^3 = 0.05$.

various conditions. This type of solar cell modeling, in which the voltage and current are calculated explicitly in terms of carrier density, is more intuitive than some other modeling methods. It is easily generalized to more complex solar cells, such as the backside point-contact design, and helps remind the designer that it is the carrier density at the maximum power point that is important in determining the dominant sources of recombination and efficiency.

Of course all of these approximate calculations assume that the device designer has implemented a design with negligible series resistance, both laterally through the diffused regions and in the metal grid.

This analysis fails if the condition of uniform carrier density in the bulk is not met. For uniform carrier density, several conditions must be satisfied. These are given below.

$$L \geq W \tag{4.64}$$

$$D/W \geq (s \text{ or } \frac{J_o N_A}{q n_i^2}) \tag{4.65}$$

The first condition is that the diffusion length be much greater than the wafer thickness. The second condition states that the diffusion velocity must exceed either the surface recombination velocity, or the effective surface recombination velocity in the case of a diffused surface.

Cases where the carrier density is not uniform include conventional thick solar cells with low base diffusion length, or devices requiring transport of carriers over large distances, as in backside point-contact cells. The existence of very large local current densities also causes deviations from uniformity, again as in backside point-contact cells. In these cases, this analysis is still useful as an initial approximation, indicating the mechanisms which dominate the behavior. In any case, once the design field is narrowed, a more precise model should be used to evaluate and "fine tune" the design. Often, the baseline gained with the approximate model helps guide the user through a more complex, complete numerical analysis by providing a case from which to judge the numerical result.

### 4.2.7 Limit Efficiencies

It is interesting to calculate the efficiency limits for silicon solar cells as a predictive guide to possible improvements. We first present approximate limit efficiencies calculated in the context of the narrow-base solar cells discussed here. Consider a 100 $\mu$m thick one-sun cell. The limit imposed by the amount of light absorbed is about 42 mA/cm$^2$ of current density. The other limit generally considered fundamental is the bulk Auger recombination. Most of the other recombination parameters are often considered to be technologically surmountable. For a case under high-level-injection conditions, the I-V curve will be defined by

$$J = J_{ph} - C_A n^3 W q. \tag{4.66}$$

This gives an open-circuit voltage of 744 mV, and an efficiency of 27.9% at 25 °C. This value is about one absolute percentage point lower than the efficiency calculated in the same way by Green (1984) because an ambipolar Auger coefficient

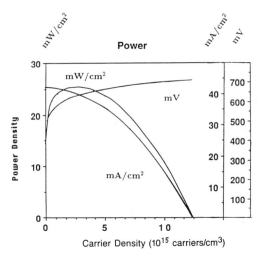

**Figure 4.4:** An example of the modeling of a cell operating near the Auger limit to the efficiency, but with n- and p-type emitter saturation current densities of $2.5 \times 10^{-14}$ A/cm$^2$. The voltage and current density as a function of the parametric variable n, the carrier density. The power density, the product of J and V, is also shown.

of $1.66 \times 10^{-30}$ was used here that is 4 times higher than Green assumed. This will be discussed more fully in Section 4.3.

Another relevant limit would be to abandon the rather optimistic assumption that all of the technologically-related parameters are surmountable and insert best estimates for these instead. When this is done, the most significant additional recombination is from the diffused areas. For example, assume that both $J_{on}$ and $J_{op}$ can be reduced to $2.5 \times 10^{-14}$ A/cm$^2$, a relatively optimistic scenario. Then

$$ J = J_{ph} - C_A n^3 W q - (J_{on} + J_{op})(\frac{n}{n_i})^2. \tag{4.67} $$

The terminal current density, voltage, and power density determined by this equation are shown in Figure 4.4 as an example of the modeling techniques of the last section. The parametric variable, the carrier density, is given by the horizontal axis. Since there are two recombination terms with differing dependencies on the carrier density, the ideality factor for this cell is ill defined. So the approximate methods are inaccurate and the maximum-power-point is best determined directly from Equation (4.67) and the associated voltage for each carrier density. Evaluation of this I-V curve yields a significantly lower open-circuit voltage of 713mV, and an efficiency of 25.5%.

The similar calculations for a 50 $\mu$m-thick concentrator cell operating at 300 suns with a responsivity of 43.2 mA/100mW indicate that in the Auger limit, an efficiency of 33% is to be expected. When the effects of emitter saturation current are considered, this is reduced to 32.2%. The effects of this emitter recombination are smaller in this case because the carrier density is much higher under concentration and the base Auger recombination truly dominates under these conditions. The assumption of a photogenerated current density of 43.2 mA/cm$^2$ assumes that the

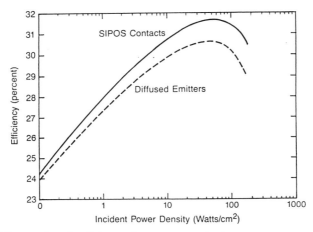

**Figure 4.5:** The predicted efficiency for a 50 $\mu$m-thick point-contact solar cell. The emitter size and shape has been independently optimized for the two cases, with SIPOS contacts and with diffused emitters.

directionality of the light is taken advantage of in the ways available for concentrating systems as described in a later section on light-trapping. In principle, current densities up to 46 mA/cm$^2$ are possible for 100 mW/cm$^2$ of AM 1.5D (air-mass 1.5 direct-normal sunlight).

For concentrator cells, other factors also reduce the expected efficiencies. The high current densities cause voltage drops within the base region, making many of the assumptions in the simplified analysis less valid. A modeled result for a point-contact cell, in which the effects of all of the recombination mechanisms are included as well as the base voltage drops and three-dimensional effects is shown in Figure 4.5. This indicates the sensitivity to the emitter saturation current by showing the attainable efficiencies for two choices of emitter technology within the constraints of the backside point-contact solar cell design (Sinton and Swanson, 1987a). In this case an efficiency of 30.5% is predicted with diffused emitters and 31.5% with SIPOS emitters.

More detailed accounts of the limit efficiencies for silicon solar cells are available. Reference (Tiedje, Yablonovitch, Cody, and Brooks, 1984) carefully details the limit efficiency for a one-sun cell with only bulk Auger and radiative losses, finding a value of 29.8%. In Green (1987), an examination of the photogeneration enhancements available from the use of concentrated light are discussed. When the maximum theoretical concentration of sunlight is assumed, efficiencies as high as 36% are possible.

# 4.3 The Physical Details of Recombination and Transport

### 4.3.1 Introduction

In order to model cell operation properly one must not only have useful methods of solving the semiconductor equations but, perhaps more importantly, accurate physical parameters such as recombination coefficients and mobilities are needed as well, all as a function of doping density. Solar cell performance depends heavily on recombination, as opposed to most other semiconductor devices where recombination is rather secondary to their operation. Therefore much of the device literature is of insufficient detail on recombination parameters which has necessitated new measurements. Band-gap shrinkage, and other heavy doping effects, are also of determining performance for diffused regions and so the solar cell community has been active in researching these effects also. This section presents some of the recent results in this area.

### 4.3.2 Transport and Recombination in Doped Regions

We have seen that it is only necessary to calculate the motion of minority carriers to determine the recombination in doped regions. In order to do this, one must know certain physical parameters such as mobility and lifetime which are often not well known. The problem is complicated by the fact that at higher doping densities the density of states is modified by so-called "heavy doping effects."

As mentioned in Section 4.1, the semiconductor equations presented there do not incorporate effects occurring at high doping densities. As discussed more fully in del Alamo (1984), this can be easily corrected for the case of minority carrier transport in low-level injected regions. Thermodynamically, the driving force for carrier transport is the gradient in electromotive force, or quasi-Fermi level. To first order, the current will be proportional to that gradient. In addition, since the minority carriers are dilute and don't interact with each other, the current should be proportional to the minority carrier density. For the case of $n$-type material, the second version of Equation (4.2) expresses this fact, where the factor $q\mu_p$ can be thought of as the proportionality constant. We thus have

$$\vec{J}_p = -q\mu_p p \vec{\nabla} \phi_p \qquad (4.68)$$

being valid in heavily doped semiconductors. Of course, one must know $\mu_p$ in order to perform calculations. Often the mobility of holes when they are majority carriers is used here, but we shall see that this is not exactly correct.

In addition, the minority carrier density in heavily doped material can be written

$$p = p_o e^{q(\phi_p - \phi_n)/kT}. \qquad (4.69)$$

This is valid regardless of the density of states of the valence band because the hole quasi-Fermi level is sufficiently far from the valence band that the non-degenerate form for the Fermi factor is appropriate. Of course, heavy doping effects mean that the equilibrium minority carrier density is no longer given by the low-density mass-action law; specifically

$$p_o \neq n_i^2/N_D. \qquad (4.70)$$

Often we will use an "effective bandgap shrinkage" defined through

$$p_o = \frac{n_i^2}{N_D} e^{\Delta E_g/kT}. \qquad (4.71)$$

This is not directly a true shrinkage in bandgap, but incorporates effects from changes in shape of the bands as well (del Alamo and Swanson, 1984).

Finally, we have the continuity equation

$$\frac{\partial J_p}{\partial x} = -qR_p = -q\frac{p - p_o}{\tau_p}. \tag{4.72}$$

A normalized excess carrier density is defined through

$$u = \frac{p}{p_o} - 1 = e^{q(\phi_p - \phi_n)/kT} - 1 \approx e^{qV/kT}. \tag{4.73}$$

Note that $u$ is a measured external parameter through the terminal voltage. The above equations can be written

$$J_p = -qp_o D_p \frac{du}{dx} \tag{4.74}$$

$$\frac{dJ_p}{dx} = -\frac{qp_o D_p}{L_p^2}u \tag{4.75}$$

where

$$D_p \equiv \frac{kT}{q}\mu_p \tag{4.76}$$

and

$$L_p \equiv \sqrt{D_p\tau_p}. \tag{4.77}$$

Examination of the above equations shows that there are actually only two independent parameters which control minority carrier motion in the steady-state, $p_o D_p$ and $L_p$. In particular, the lifetime cannot be determined using steady-state measurements without assuming a value for $D_p$ or $\mu_p$. This has caused considerable discrepancy to appear in quoted literature values for these variables (Swanson, del Alamo, and Swirhun, 1984).

We have made measurements of $L_p$ using transistor structures and $\tau_p$ using photoluminescence decay. This has allowed a determination of $\mu_p$. Similar measurements have also been performed on $p$-type material to determine $L_n$ and $\tau_n$. The results are shown in Figures 4.6, 4.7, 4.8, 4.9, 4.10 and 4.11 (del Alamo and Swanson, 1987 and Swirhun, Kwark, and Swanson, 1986).

At higher doping densities the lifetime decreases, probably due to Auger recombination. We have found that the lifetime is rather independent of sample preparation and doping technique, suggesting that traps are not playing a significant role at high doping. The measurements tend to be in rather good agreement with those of Dziewior and Schmid (1977). An Auger coefficient of $C_n = 2.8 \times 10^{-31}$ cm$^6$/s for $n$-type material, Figure 4.6, and $C_p = 1 \times 10^{-31}$ cm$^6$/s for $p$-type material, Figure 4.9, represents the results well. In this case the lifetimes can be modeled as

$$\tau_p = \frac{1}{C_n N_D^2} \tag{4.78}$$

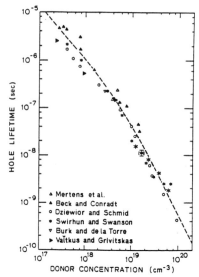

**Figure 4.6:** Minority carrier lifetime versus doping in $n$-type material (del Alamo, 1985).

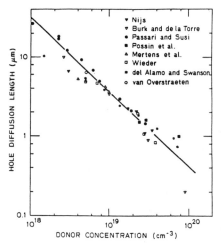

**Figure 4.7:** Minority carrier diffusion length versus doping in $n$-type material (del Alamo, 1985).

and

$$\tau_n = \frac{1}{C_p N_A^2}. \tag{4.79}$$

The measured minority carrier mobilities, shown in Figures 4.8 and 4.11, deviate substantially from values of the same carriers when they are majority carriers, the majority carrier values, being more than a factor of two higher at the highest doping densities. This effect has been predicted theoretically by Bennett (1983) and has

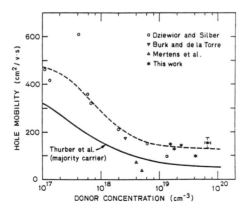

**Figure 4.8:** Minority carrier mobility versus doping in $n$-type material (del Alamo, 1985).

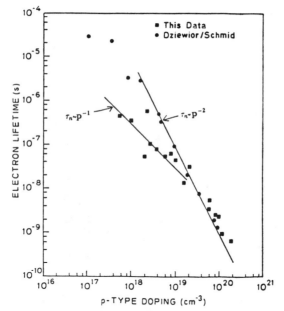

**Figure 4.9:** Minority carrier lifetime versus doping in $p$-type material (Swirhun, 1987).

also been experimentally reported by Dziewior and Silber (1979). Note that there is no evidence for a "mobility edge" as one sees in amorphous silicon, because this would result in lower minority carrier mobility.

Using transport measurements across the base region of transistors the apparent bandgap shrinkage was measured. These devices have epitaxial bases so that the doping profiles are uniform, simplifying the data interpretation. The results are shown in Figures 4.12 and 4.13.

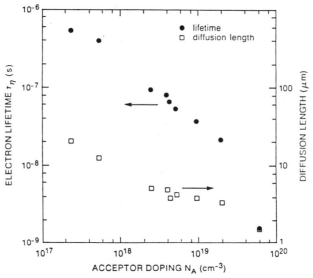

**Figure 4.10:** Minority carrier diffusion length versus doping in $p$-type material (Swirhun, 1987).

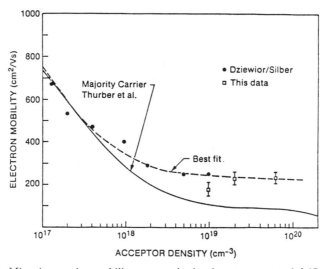

**Figure 4.11:** Minority carrier mobility versus doping in $p$-type material (Swirhun, 1987).

The apparent bandgap shrinkage for $p$-type material agrees well with the measurements of Slotboom and deGraaf (1977); however, the $n$-type data falls considerably below most of the other reported literature values. When the original data is used to correct literature values to that which would be computed using the new, higher minority carrier mobility, the agreement is nearly perfect (Figure 4.12) in this case as well.

Using the above physical parameters, it is possible to calculate the saturation current for diffusions of arbitrary profile (del Alamo, 1985). For example, Figure 4.14

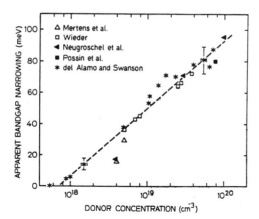

**Figure 4.12:** Apparent band gap shrinkage versus doping for $n$-type material (del Alamo, 1985).

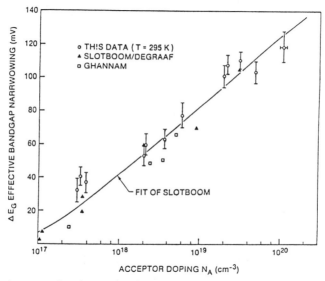

**Figure 4.13:** Apparent band gap shrinkage versus doping for $p$-type material (Swirhun, 1987).

shows the calculated saturation current for $n$-type emitters with Gaussian profiles when the surface recombination velocity is large, as it would be under metal contacts. Note that the most important factor in obtaining low saturation current is to have a large junction depth. If, however, one has passivated surfaces with a surface recombination velocity of $10^3$ cm/s then the best emitters have a low surface recombination velocity and are shallow to limit bulk recombination in the heavily doped portion. Such a calculation is shown in Figure 4.15.

### 4.3.3 Surface Recombination

Surfaces contribute significantly to recombination because they tend to have a relatively large density of recombination centers. Both the density and capture

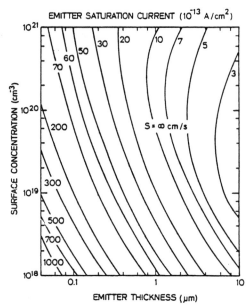

**Figure 4.14:** Calculated emitter saturation current for *n*-type emitters versus junction depth and surface concentration assuming a Gaussian profile. The surface recombination velocity is infinite (del Alamo, 1985).

cross-sections of these centers are very process dependent. In addition, the surface can be charged, affecting the surface potential and carrier densities. This makes modeling surface recombination problematic. Some measurements and calculations are presented below which indicate that surface recombination velocities can be significantly less than previously thought.

Assuming simple Shockley-Read-Hall type recombination the total recombination through the continuum of possible states at the interface between silicon and silicon-dioxide, the total surface recombination velocity for an oxide passivated surface can be calculated (Eades and Swanson, 1985). This requires knowledge of both the density and cross-section dependence on energy as well as the surface potential. Capture cross section measurements were performed using DLTS on oxidized and annealed silicon surfaces. The results are shown in Figure 4.16, Figure 4.17, and Figure 4.18. Very little band bending would be expected for these samples which have fixed charge densities less than $10^9$ cm$^{-2}$.

Using the above densities and cross sections the surface recombination velocity can be calculated. Figures 4.19 and 4.20 present the results for low-level injection.

Note that *n*-type surfaces have significantly lower recombination velocities than *p*-type. This is a direct result of the much smaller capture cross section for holes than electrons.

Because of the large capture cross section energy dependence the recombination velocity is very injection level dependent. This is illustrated in Figures 4.21 and 4.22.

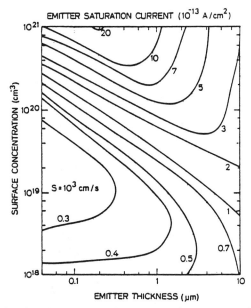

**Figure 4.15:** Calculated emitter saturation current for $n$-type emitters versus junction depth and surface concentration assuming a Gaussian profile. The surface recombination velocity is $10^3$ cm/s (del Alamo, 1985).

Since solar cells usually run at significant injection levels the relevant recombination velocity is usually much less than that encountered at very low injection levels. Measurements based on low injection levels should thus be used with caution. In addition, it is interesting to note that recombination velocities less than 10 cm/s are attainable on high quality surfaces.

### 4.3.4 Auger Recombination in High Level Injection

At high levels of injection the electron and hole densities are equal and so the Auger recombination rate becomes

$$R_n = R_p = C_n n^3 + C_p n^3 = (C_n + C_p)n^3 = C_A n^3 \qquad (4.80)$$

where $C_A = C_n + C_p$ is called the ambipolar Auger coefficient. In attempting to model the performance of point-contact silicon solar cells running in high-level injection we have been forced to adjust the typical ambipolar Auger coefficient used in device modeling (Sinton and Swanson, 1987a). Accordingly, we adapted the point-contact cell design into a test structure ideally suited for measuring this parameter.

Figure 4.23 shows the structure of a point-contact silicon solar cell. Electron-hole pairs are generated at the bottom and must diffuse the to the n and p diffused regions at the top to be collected. The cell depends on obtaining very high carrier recombination lifetime to achieve high current collection and voltage. Accordingly, procedures were developed that yield trap assisted lifetimes over one millisecond in finished devices. Surface recombination was reduced by passivating the surface

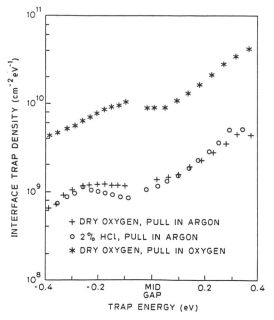

**Figure 4.16:** Interface trap density for the oxides studied. All samples were post-metallization annealed at 450°C in forming gas. The values in the lower half of the gap come from measurement of a *p*-type substrate and the upper gap values from an *n*-type sample (Eades and Swanson, 1985).

with silicon dioxide except in the metal contact areas. When this is done one finds that recombination in the diffused regions, or at the metal-semiconductor contact, dominates. To reduce this component the size of the diffused regions is reduced to a polka dot array of small regions — hence the name "point-contact cell."

Cells with varying sizes and spacings of contact diffusions were fabricated. The resulting diffused region recombination components were extracted with the results shown in Figure 4.24. As expected, this recombination is proportional to the contact coverage fraction. Measurements indicated that the lifetime in the bulk was decreasing with carrier density, and at a rate faster that that predicted using the "standard" ambipolar Auger coefficient of $4 \times 10^{-31}$ cm$^6$/s. Thus a device with very small contact coverage was fabricated — the "Auger test structure." This was to force bulk Auger to be as dominant as possible.

Figure 4.25 shows a secondary effect of having small, widely spaced contacts for this purpose — the base region carrier density, as calculated from our device model, is nearly uniform from front to back. This is a result of the reduced contact recombination and therefore the reduced diffusion of carriers to the contacts. This causes the modeled results to be quite independent of assumed transport parameters.

In order to assure that the test device is dominated by Auger recombination, the modeling code was exercised to identify the sources of recombination as a function of incident power. As seen in Figure 4.26, at all power densities over 1 watt/cm$^2$, Auger recombination dominates. This lends support to the accuracy with which

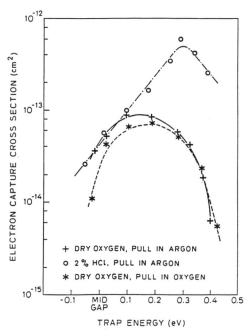

**Figure 4.17:** Electron capture cross section in the upper half of the gap (Eades and Swanson, 1985).

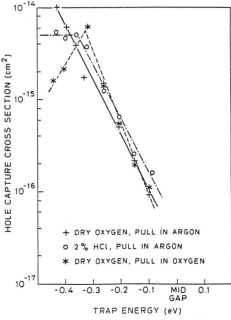

**Figure 4.18:** Hole capture cross section in the lower half of the gap (Eades and Swanson, 1985).

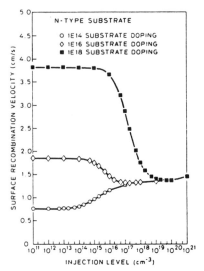

**Figure 4.19:** Surface recombination velocity in low-level injection for an *n*-type substrate at flat-band (Eades and Swanson, 1985).

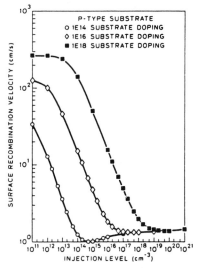

**Figure 4.20:** Surface recombination velocity in low-level injection for a *p*-type substrate at flat-band (Eades and Swanson, 1985).

the Auger coefficient can be extracted by fitting the measured characteristics to the model.

In Figure 4.27 the measured open circuit voltage of the Auger test structure is shown as a function of incident power. In order to fit this result to the model it is necessary to use an Auger coefficient of $1.66 \times 10^{-30}$ cm$^6$/s, a factor of over four times the standard value. Also shown is the result of using the standard value.

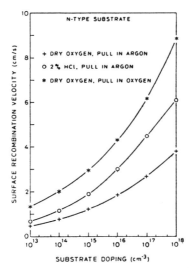

**Figure 4.21:** Surface recombination velocity versus injection for an $n$-type substrate at flat-band (Eades and Swanson, 1985).

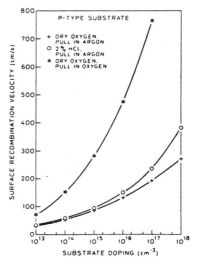

**Figure 4.22:** Surface recombination velocity versus injection for a $p$-type substrate at flat-band (Eades and Swanson, 1985).

Altering the Auger coefficient alters more than the open-circuit voltage. For example the open-circuit voltage decay after interrupting a forward current is controlled by recombination of carriers in the base. Figure 4.28 shows the decay rate of the voltage as a function of carrier density in the open-circuit configuration. The carrier density is determined from the terminal voltage. A constant base lifetime would result in a constant $dV/dt$ whereas a lifetime that is inversely proportional to

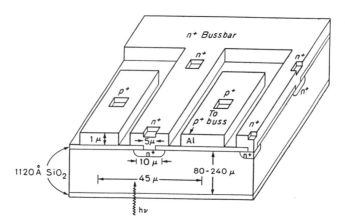

**Figure 4.23:** Structure of the point-contact solar cell.

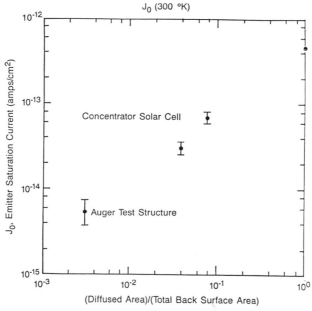

**Figure 4.24:** Measured saturation current of point-contact cells versus contact coverage fraction. The concentrator solar cell shown is the optimized 27.5% cell reported in Sinton et al. (1986).

carrier density squared, as in Auger recombination, yields a parabolic dependence. To best fit the data requires an Auger coefficient of $1.66 \times 10^{-30}$ cm$^6$/s once again.

Finally, as the incident intensity increases, the carrier density at the front of the cell increases causing a loss of carriers by Auger recombination, even at the short-circuit condition. This causes the short circuit current to be sublinear in incident

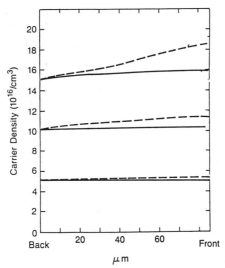

**Figure 4.25:** Calculated carrier density from front to back for the standard device, labeled "concentrator solar cell" in the previous figure, (dashed line) and Auger test structure (solid line) under open-circuit-voltage conditions. The profiles shown are those that would result in three particular backside carrier densities (hence voltages). The Auger test structure can attain high voltages at lower concentrations of incident light, and with less of a gradient in carrier density across the wafer.

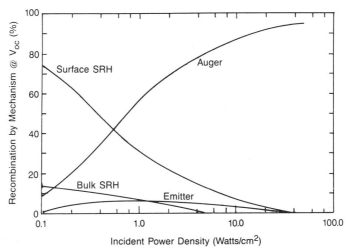

**Figure 4.26:** Calculated recombination by source versus incident solar power level for the Auger test structure. SRH refers to Shockley-Read-Hall, defect mediated recombination.

power. Although this is a less accurately modeled second order effect, as shown in Figure 4.29, it is necessary to use an increased Auger coefficient to predict the measured result using the modeling code.

In Figure 4.30 the lifetime versus carrier density, as measured with the Auger test structure, is presented as the dashed line. The solid line is from measurements

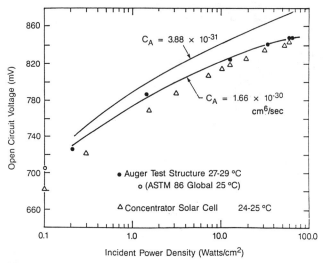

**Figure 4.27:** Measured open-circuit voltage of the Auger test structure versus incident power density.

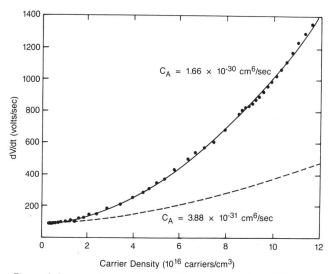

**Figure 4.28:** Rate of change of open circuit voltage versus time following a current pulse.

on high lifetime, high resistivity, highly injected material using a photoconductivity decay technique (Yablonovitch and Gmitter, 1986). It is in agreement with our results. The measured ambipolar Auger coefficient is once again $1.66 \times 10^{-30}\,\mathrm{cm^6/s}$. This is four times that which one would obtain by summing the $n$- and $p$-type Auger coefficients determined from material doped over $1 \times 10^{19}\,\mathrm{cm^{-3}}$. Also shown are various results on low level injection in material where the carrier density is achieved by doping rather than injection. The lifetime on doped material is in

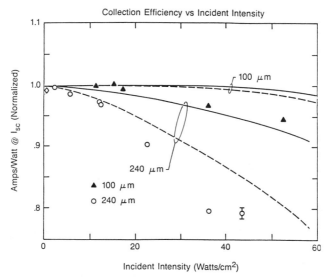

**Figure 4.29:** Measured and calculated responsivity normalized to its value at low intensity versus incident intensity. The solid lines are calculated for an auger coefficient of $3.88 \times 10^{-31} \mathrm{cm}^6/\mathrm{s}$ and the solid lines for the value developed here, $1.66 \times 10^{-31} \mathrm{cm}^6/\mathrm{s}$.

substantial agreement with that on highly-injected material. This implies that the Auger coefficient must be higher for doping, as well as injected, densities in the range of $1 \times 10^{17} \mathrm{cm}^{-3}$. In the past this has been explained by postulating that an additional recombination mechanism is operative in doped material, probably due to a doping related defect. This explanation seems highly suspect because of the agreement of results on doped material with those on injected material — the presence of the free carriers is sufficient to yield lifetimes in the observed range so that no additional recombination mechanism need be postulated. The only way of reconciling these results is to postulate that the Auger coefficient is slightly carrier density dependent. In fact, other authors have predicted very similar results in a much ignored 1981 paper (Vaitkus and Grivitkas, 1981). They see a sort of jog in the Auger coefficient in the $10^{18}$ $\mathrm{cm}^3$ range which they attribute to degeneracy effects. Whatever the cause, it is necessary to use the higher value of ambipolar Auger coefficient to correctly model electron-hole plasmas in this carrier density range.

# 4.4 Cell Design Engineering

### 4.4.1 Maximizing Photogeneration

The major improvement which has pushed the efficiency of concentrator cells from about 20% to 28% over the last 8 years has been light-trapping. A major problem with silicon is that it doesn't absorb light very well. Even a wafer 1 mm thick only absorbs about 95% of the light available for producing electron-hole pairs. One way to absorb more of the light in a thin layer is to utilize light-trapping. Because of the difference in index of refraction between silicon and air, light which

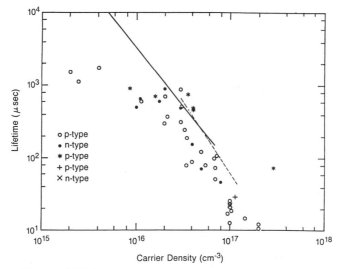

**Figure 4.30:** Measured lifetime on high-level injected electron-hole plasmas versus injected carrier density. A comparison is shown between the results from measurements on specialized solar cells (dashed line)(Sinton and Swanson, 1987a), and photoconductivity decay measurements (solid line)(Yablonovitch and Gmitter, 1986). Also included are data where the carrier density is from doping. This data is (from top to bottom in the legend): (Huber, Bachmeier, Wahlich, and Herzer, 1986), both *p-* and *n-*type; (Ciszek, 1987) for *p-*type, and (Dziewior and Schmid, 1977) for *p-* and *n-*type.

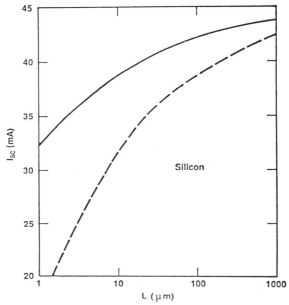

**Figure 4.31:** The short-circuit current as a function of cell thickness with textured Lambertian surfaces (solid) and polished surfaces (dashed)(Tiedje, Yablonovitch, Cody, and Brooks, 1984).

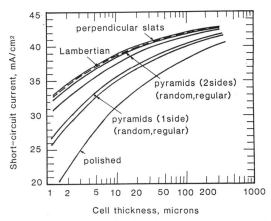

**Figure 4.32:** The calculated short circuit density as a function of the silicon cell thickness for various light-trapping schemes (Green and Campbell, 1987). AM 1.5 Global, 97mW/cm$^2$.

is inside the silicon must be within 16° of normal to the surface in order to escape. Otherwise, it will reflect off of the interface and traverse the wafer again. Essentially, it is trapped. This enhancement is shown in Figure 4.31 indicating the short-circuit currents possible for silicon cells of different thicknesses with and without light-trapping (Tiedje, Yablonovitch, Cody, and Brooks, 1984). This silicon is assumed to have unity front surface transmission and a perfect backside mirror.

One way to achieve light trapping is to texturize the front surface of the wafer. This leads to a somewhat randomized scatter of the light as it enters the wafer. If this light is assumed to scatter around until it is isotropic then the formulation of the enhancement caused by the light trapping is straightforward. If an ideal backside mirror is in place then this enhancement is found to be equivalent to increasing the wafer thickness by a factor $4n^2$, about 50 (Yablonovitch and Cody, 1982). Alternative methods of calculation utilize ray tracing to evaluate specific structures.

The most common engineering solution to incorporate light-trapping is to use anisotropic etching to form pyramids or V-grooves on the surface of a (100) oriented wafer. The slopes on these features are well defined as the (111) planes, which have an angle of 54.7° to the wafer surface. This makes the problem very amenable to ray-tracing, especially for the case of normally-incident light.

The results from one such study are shown in Figure 4.32 (Green and Campbell, 1987). This figure indicates that the limit of Lambertian scattering, also shown in Figure 4.31, is closely approached by the case of pyramidal texturization on both sides of a wafer. A slightly better scheme, superior to both Lambertian and double-sided texturization, is the use of V-grooves on both sides of the wafer. Those on the rear are oriented perpendicularly to the front-side grooves. Pyramids on the front-side only, a popular implementation, are significantly less effective. For example, they would produce almost 3 mA less current on a 50 $\mu$m-thick cell.

Much of the difference between the six texturization schemes shown in the figure would be diminished in the case of non-normal incidence for the light, a non-specular backside reflector, or a distribution of angles on the texturized surfaces.

An implementation of texturization onto solar cells must take account of many technological details of the texturized surface. Four texturized surfaces are shown in Figure 4.33. These were formed by different amounts of isotropic etching after creating the features with an anisotropic etch. Notice that these examples span the range from regular pyramidal texturization towards a more Lambertian surface.

Each of these wafers was measured to have a different light absorption characteristic and surface recombination velocity. Additionally, the surface recombination velocities were found to be a sensitive function not only of the texturization details, but the anneal schedule as well. These results are summarized in Figure 4.34.

The absorption of incident light in these wafers is controlled by two effects:

1. The degree of light-trapping. The light-trapping of weakly absorbed light will diminish as the wafer is isotropically etched and becomes more planar.
2. The double-bounce incidence that the pyramidal surface provides will also cease to function as the topography flattens.

Both of these effects are clearly seen in the reflectance data shown in Figure 4.35. A loss in the double-bounce incidence shows up as an increase in the reflectance for short wavelengths. The loss of light-trapping is indicated by increased reflection at long wavelengths near the bandgap for silicon (1,110 nm).

An even more direct demonstration of the effects of light-trapping is shown in Figure 4.36. This comparison of the spectral responsivity of a texturized and an untexturized solar cell clearly indicates the enhancements of both the double-bounce incidence and the light-trapping.

The photogeneration increases arising from light-trapping occur rather uniformly within the cell. In order that this increased current can appear as terminal current, the solar cell must have a very high internal quantum efficiency. The line shown in Figure 4.36 is the theoretical limit if there were no reflection or recombination losses. An actual measurement of the cell reflection indicated that this cell does have internal collection efficiency greater than 98% between 350 nm and 950 nm wavelengths.

A new and important wrinkle was introduced in a paper by Campbell and Green where it was pointed out that more effective light-trapping than that discussed above was possible (Campbell and Green, 1986). An enhancement of $(4n^2)/(\sin^2\theta)$ is available where $\theta$ is the acceptance angle of the light. The essence of this idea is that if you know the angle of incidence of the light, then you can engineer where it goes in order to utilize it most effectively. Above, it was argued that light-trapping from a texturized surface works because only light within 16–17° of normal will escape. Consider the case shown in Figure 4.37. This shows (for illustration purposes) an extreme case of engineering where the light will go. In the two cells, the same number of photons is incident per cm² of cell area. However, in the second case, (b), the path length presented to the photons is $20,000\ \mu$m, to be compared with top case, (a), in which the photons traverse only 100 $\mu$m of silicon. This difference amounts to an increase in short-circuit current density of about 15%. The penalty paid is in achieving the 100X concentration necessary for case (b). Considerations related to obtaining the necessary light concentration introduce factors of $\sin\theta$ into the analysis. For solar energy, the half angle of the sun defines a maximum concentration ratio. This in turn limits the enhancement factor to $4n^2\sin^{-2}(0.27°)$. Of course, absorption leading to electron-hole production is not the only absorption which is enhanced. Parasitic, free-carrier absorption and non-ideal mirrors limit the practicality of this enhancement technique to values much less than the theoretical limit.

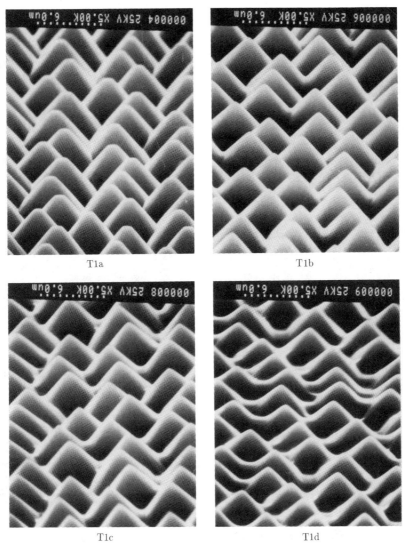

T1a

T1b

T1c

T1d

**Figure 4.33:** Texturized surfaces formed by anisotropic etching of (100) silicon followed by various times of isotropic etching (King and Swanson, 1988).

| Dependence of Surface Recombination Velocity (SRV) of Texturized Surfaces on Pyramid Rounding in HNO$_3$ : NH$_4$F Solution | | | | | |
|---|---|---|---|---|---|
| Sample ID | Pyramid Rounding Process | Forming Gas Anneal Only | | After Aluminum Anneal | |
| | | Untexturized SRV $(\frac{cm}{s})$ | Texturized SRV $(\frac{cm}{s})$ | **Untexturized SRV $(\frac{cm}{s})$** | Texturized SRV $(\frac{cm}{s})$ |
| T1a (w = 280$\mu$m), V2a (w = 81$\mu$m) | 20 sec in 50:1 HNO$_3$:NH$_4$F | 12.9 | 65.4 | 9.31 | 15.3 |
| T1b (w = 277$\mu$m), V2b (w = 81$\mu$m) | 2 min in 20:1 HNO$_3$:NH$_4$F | 19.2 | 51.7 | 10.65 | 38.9 |
| T1c (w = 271$\mu$m) V2c (w = 72$\mu$m) | 4 min in same 20:1 solution | 12.7 | 36.0 | 25.9 | 29.1 |
| T1d (w = 265$\mu$m) V2d (w = 70$\mu$m) | 8 min in same 20:1 solution | 13.6 | 38.9 | 2.91 | 1.90 |

**Figure 4.34:** The measured surface recombination velocity for the various texturization profiles indicated on the last figure. Texturized surfaces have higher surface recombination velocities than untexturized prior to aluminum anneals (King and Swanson, 1988).

Two practical implementations of this idea are shown in Figure 4.38 and Figure 4.39. The first is the prismatic cell cover, which refracts the light from over the metal front-surface grid lines into the cell (O'Neill, 1985), shown in Figure 4.38. A similar structure using reflective optics was suggested in the original paper (Campbell and Green, 1986). Another idea with the same result is to put the solar cell in a reflective box with an input aperture for light which is smaller than the cell area, Figure 4.39. This increases the probability that the light will be absorbed rather than escape (Sinton and Swanson, 1987a). The enhancement possible is shown in Figure 4.40. In fact, these ideas for decreasing the acceptance angle for light could be used instead of texturizing in cases when texturizing is difficult to achieve. A method which reduced this acceptance angle to 17° can give a planar cell absorption equivalent to Lambertian texturization.

## 4.4.2 Effects on Cell Performance

There are several advantages to using light-trapping on relatively thin solar cells rather than making 1 mm thick cells in order to absorb the available light.

The first, a practical concern, is that you can make more thin solar cells than thick ones from a given amount of silicon.

Second, many of the parasitic recombination mechanisms are related to the cell volume. These can be virtually eliminated by thinning the cells. This causes gains in efficiency right down to the thickness where the cell is too thin to absorb the majority of the light, even with light-trapping techniques. This principle is essentially that

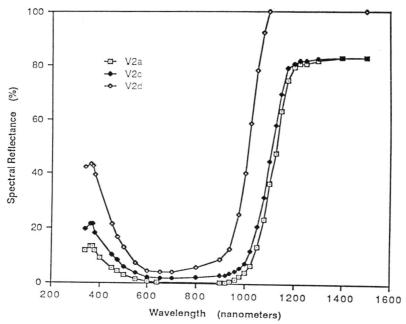

**Figure 4.35:** Reflectance data from several wafers between 70 and 80 $\mu$m thick. V2d is untexturized. V2c had a longer isotropic etch after texturization than V2a (King and Swanson, 1988).

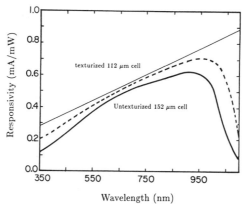

**Figure 4.36:** The spectral responsivity for two cells processed at the same time. The texturized cell shows a lower reflectance loss, since the light gets a double bounce. Also the texturized cell has a vastly improved response to light near the bandgap energy (Sinton et al., 1986).

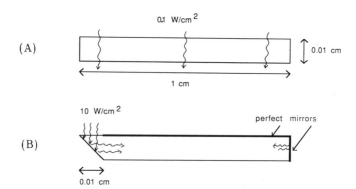

**Figure 4.37:** An extreme case of engineering a solar cell to maximize the photogeneration per unit input power. The same number of photons are incident on both of these cells. In case (B), the path length traversed by these photons is greater by a factor of 200 than in case (A); this increases the short-circuit current density by about 15%. Unfortunately, case (B) requires that the light be concentrated by a factor of 100.

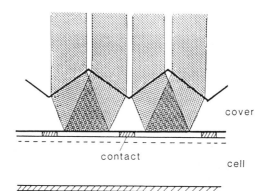

**Figure 4.38:** A prismatic secondary concentrator, in which the light is refracted onto the non-metallized areas. Since the metallized areas can act as nearly perfect mirrors to photons already within the cell, this scheme enhances the near-bandgap response (Green, 1987).

if you have starting material for which the diffusion length is less than the cell thickness, and you can't change the diffusion length, then use a thinner cell.

The third reason for using thin cells when possible is that in order to collect the photogenerated minority carriers and produce power, a gradient in carrier density must exist to drive the carrier diffusion. Over long distances, this gradient represents a loss in terminal voltage. The closer the collecting junction to the photogeneration, the better.

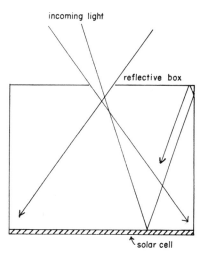

**Figure 4.39:** A reflective box which reduces the effects of cell reflectance and improves the near bandgap response by not letting photons escape until they are incident upon the cell several times on average (Sinton and Swanson, 1987b).

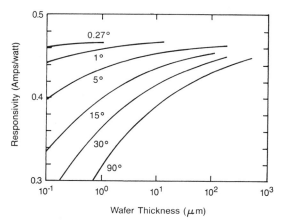

**Figure 4.40:** The responsivity vs. cell thickness for texturized cells as a function of the cell thickness (adapted from Campbell and Green, 1986). The curves refer to cells with varying acceptance angles.

These effects are clearly seen in a modeling study on backside point-contact solar cells. For lower lifetime substrates, a 200˙ μm thick texturized cell which is 18% efficient would compare to an equivalent 28% efficient cell at 50 μm. In a system which imposed an acceptance angle of 15°C, the efficiency would increase to nearly 30%. These results are given in Figure 4.41.

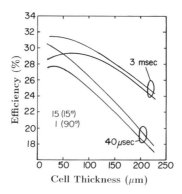

**Figure 4.41:** The efficiency vs. thickness for cells with two different values of bulk lifetime. Cells fabricated on low-lifetime material show improvement as the base is thinned. This improvement is even more dramatic if the light-trapping is enhanced by reduced acceptance angles (Sinton and Swanson, 1987b). The upper curve for each pair is for a 15° acceptance angle and the lower curve for a 90° acceptance angle. These acceptance angles correspond to 15:1 and 1:1 cell:aperture area ratios in the scheme of Fig 4.39.

Another effect of a common light-trapping implementation, front-side silicon patterning, is the fore-shortening of the photogeneration profile. In cells operating in the "thin-base" limit, this is a second order effect in that if the cell already has a unity quantum efficiency, the exact position of photogeneration only changes the terminal voltage slightly. However, in the case of a cell with a low internal base quantum efficiency, generating the electron-hole pairs very close to the front-side junctions can have a significant beneficial effect.

### 4.4.3 Emitters: Optimizing the Diffused Regions in the Cell

Most high-efficiency silicon solar cells are dominated by the recombination in the diffused regions, or at the metal contacts to these diffused regions. This fact can be demonstrated by modeling of the type described earlier.

The front emitter in a conventional solar cell presents the most difficult optimization problem in the design. A proper treatment requires, at the very minimum, a detailed knowledge of the doping dependence of two quantities, the product of the equilibrium minority carrier density with the minority carrier mobility, and the diffusion length (del Alamo, 1985). Once these parameters are known, then an emitter model which solves the transport equations developed in Section 4.3 can predict the minority carrier characteristics of a particular diffusion. The important quantities for solar cell design are internal quantum efficiency for illuminated emitters and emitter saturation current, $J_o$, for all emitters. Additional parameters from the point of view of designing practical solar cells are the metal-silicon contact resistance and the sheet resistance.

Once all of these facts are in, it is clear that there is no single superior solar cell emitter design. A different design is necessary for each application. A good illustration of this is the diversity of emitters in use today. Clearly, this is a very significant design detail for which no clear consensus exists.

### 4.4.4 Minimizing Emitter Recombination

It is widely recognized that recombination in the diffused regions, especially at the metal-silicon interface, is the dominant parasitic loss mechanism in most silicon solar cells. A review of the literature indicates that there are three methods for specifically addressing this issue.

- In the point-contact solar cell, this recombination is reduced by minimizing the emitter area, both of dopants and metal-silicon contacts.
- A more common method is illustrated in the PESC solar cell by Green et al. Here, a passivated planar diffused region is used, with an absolute minimum of metal-silicon contact area. An example is shown in Figure 4.49.
- The last method is to use a fundamentally different type of emitters such as polysilicon, SIPOS (Semi-Insulating Polysilicon, or charge-induced emitters.

A discussion of the optimization of diffused $n$-type emitters can be based upon the modeling of del Alamo (del Alamo, 1985). In this work, the fundamental transport parameters were determined using test devices fabricated on uniformly doped epitaxial layers. These parameters were then compared to previous work, and a consensus was found based upon a careful comparison of the assumptions made in interpreting and reporting the previous data. The minimum number of directly measured parameters was then applied via an emitter model to predict the performance of several common emitter profiles. This section draws heavily upon the experimental and modeling results of del Alamo.

Figure 4.42 indicates the internal quantum efficiency for an emitter illuminated with sunlight with an AM 1 spectrum. A silicon-silicon dioxide surface recombination velocity of 1000 cm/sec was assumed. The internal quantum efficiency plotted here is the fraction of the electron-hole pairs generated in the emitter that are collected rather than recombining. A more directly useful quantity would be the absolute current that is absorbed and recombined in the emitter. This could have been calculated directly, but can also be found by multiplying the internal quantum efficiency of the emitter by the amount of current generated there. This is shown in Figure 4.43. This plot shows contours of potential short-circuit-current that is lost to absorption and recombination in the front emitter. The contour of 0.5 mA/cm$^2$ indicates a loss of about 1.2% of the potential cell current. It is interesting to note that this contour is along a line of nearly constant sheet resistance, 71 $\pm$ 5 $\Omega/\square$ in the $10^{19}$cm$^{-3}$ decade of surface concentration.

Figure 4.14 shows the emitter saturation current as a function of the surface dopant concentration and the junction depth. This is for a surface recombination velocity of infinity, and is appropriate for metal-contacted diffusions. Figure 4.15 gives these same contours of emitter saturation current but for an assumed surface recombination velocity of 1000 cm/sec, a good assumption for oxide passivated diffusions in many cases. Note, however, that the values in the lower left hand corner are very sensitive to this assumption.

Consider the design of an optimized front-surface emitter. The first criteria is that the emitter does not absorb and recombine many carriers. This is insured by staying to the left of the 0.5 mA/cm$^2$ contour in Figure 4.43. This implies sheet resistances greater than 70 $\Omega/\square$.

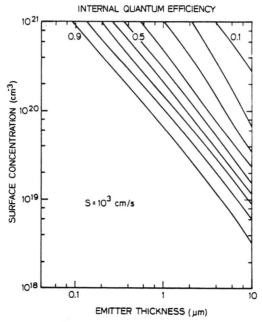

**Figure 4.42:** The quantum efficiency for gaussian, phosphorus doped emitters under an AM 1 spectrum. The passivated surface is assumed to have an S=1000 cm/sec (del Alamo, 1985). The base doping is $10^{16}\,\mathrm{cm}^{-3}$.

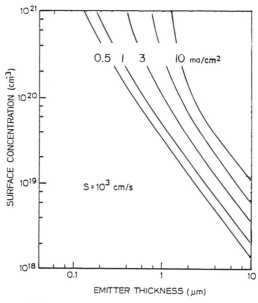

**Figure 4.43:** Contours of absolute current density lost in the emitter to photon absorption and subsequent recombination.

Low sheet resistance is another asset to a good emitter. For one-sun cells, a low sheet resistance allows a coarse grid pattern. In concentrator cells, a low sheet resistance and a fine grid pattern may both be necessary. Assume for example that you would like a sheet resistance less than 200 $\Omega/\square$. This leaves a rather narrow, well defined portion of the curve from which to choose an emitter; those that lie between 70 and 200 $\Omega/\square$. Three more criteria enter. The emitter saturation current should be as low as possible for the passivated emitter regions. This drives the optimum towards the lower portion of the plot. Additionally, the top grid needs to contact this emitter. This actual contact area may vary greatly, but must minimize the spreading resistance effects in the diffused region and be adequately large that the contact resistance is negligible. The lowest values of $J_o$ for $s = \infty$ (metal contacted emitters) are in the mid to lower-right-hand region.

All of these criteria favor rather lightly doped, very deep emitters. This result is in contrast to some of the conventional wisdom of solar cell front emitters, but is coming into a wider acceptance rather recently (Cuevas and Balbuena, 1988). The one point which may favor keeping a surface concentration above the $10^{19}$ range is the contact resistance. In one-sun cells, the metal-silicon area can be quite small, with high current densities. In concentrator cells, the area must be larger, and will certainly have high current densities. A plot of the contact resistance of Ti and PtSi contacts as a function of the surface concentration of phosphorus is shown in Figure 4.44. The contact resistance is a very strong function of this surface concentration. For perspective, consider a concentrator cell with 1% metal-silicon contact area, designed for operation at 36 W/cm$^2$ of incident power (360 suns). The current density in this case would be 14 A/cm$^2$, or 1400 A/cm$^2$ at the contacts. For Ti contacts, this would give a 0.7 mV drop for a surface concentration of $10^{20}$/cm$^3$, 3 mV at $4\times10^{19}$, but 84 mV at $2\times10^{19}$! This simple example illustrates that contact resistance can be a real problem if ignored during the design phase.

In short, the criteria outlined above for an optimized single-diffusion front-side phosphorus emitter are:

- $\Omega/\square \geq 70$ to avoid current loss.
- $\Omega/\square \leq 200$ to minimize series resistance.
- Surface dopant concentration $\leq 10^{20}$cm$^{-3}$ to obtain low emitter saturation currents for the passivated emitter regions.
- Surface dopant concentration $\geq 10^{19}$ cm$^{-3}$ to insure good contact resistance.
- Junction depth $\geq 0.1$ $\mu$m for process control of the metal-silicon interface.

The region of emitters that meet these requirements are in the shaded region of Figure 4.45. Within this shaded area, minimization of the emitter saturation current for the metal-contacted areas, as shown in Figure 4.14, drives the choice towards the lower right-hand corner. This emitter, 2 $\mu$m deep and doped around $10^{19}$cm$^{-3}$ is quite deep and lightly doped.

Several recent papers highlight this trend towards thicker front-side emitters. In work by Tobin et al., deep front emitters of varying thickness were investigated as a way to produce high-temperature stable cells (Tobin, Spitzer, Bajgar, Geoffroy, and Keavney, 1987). Recognizing that metalization spiking through the junction during contact sintering was a major constraint, emitters with surface concentrations between 1 and $2\times10^{19}$/cm$^3$ were compared on a standard solar cell. Cells with

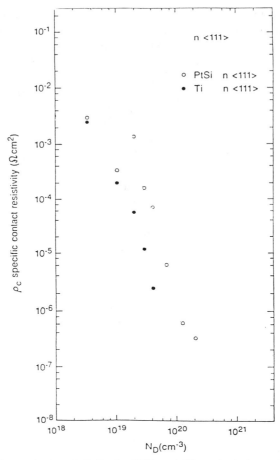

**Figure 4.44:** The contact resistivity for PtSi and Ti contacts to $< 111 >$ Si (Swirhun, 1987).

0.94 $\mu$m thick emitters were shown to be stable with respect to 15 minute anneals in $N_2$ at 600 °C. This is a significant result, indicating the process insensitivity and robust qualities of deep emitters. No significant change in short circuit current was noted in increasing the junction depth from 0.26–0.94 $\mu$m. This is in good agreement with the modeled results of del Alamo. However, since the total solar cell saturation current for these cells was about $6.6 \times 10^{-13}$ A/cm$^2$ and independent of the junction depth, it seems that the cells were base dominated so that no conclusions can be made about trends in the emitter $J_o$.

Another study in which special care was taken to determine the emitter component has been reported by Cuevas and Balbuena (1988). For example, a 1 $\mu$m thick emitter with a surface concentration of $10^{19}$ cm$^{-3}$ was found to have a $J_o$ less than $1.6 \times 10^{-13}$ A/cm$^2$. The same emitter, unpassivated, had $20 \times 10^{-13}$. These values are in good agreement with the modeling shown above. These relatively deep emitters were put to use in demonstrating a simple process yielding 19%, 4 cm$^2$ cells.

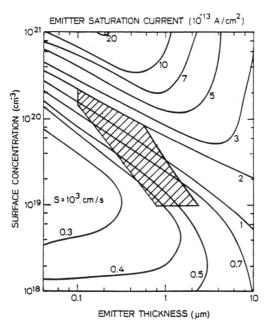

**Figure 4.45:** The region of interest for a single-diffusion optimized $n^+$ emitter on the front of a solar cell. The criteria taken into account which bound this area are the passivated region emitter saturation current (top), contact resistance (bottom), emitter quantum efficiency (right), and metal-contact process control (left). The optimization of the metal-contacted areas makes the right lower corner of the shaded region preferred.

This discussion above was based almost entirely upon the modeling of del Alamo. A prudent alternative would be to measure the emitters of interest experimentally. A collection of measured values is shown in Table 4.1, for several phosphorus and boron diffusions (Kane and Swanson, 1985). In the case of boron emitters, it was found necessary to deposit a thin aluminum layer on the surface in order to attain a large enough surface recombination velocity that the emitter is transport limited. This result is called $J_{o,Al}$ in Table 4.1. A comparison of these experimental values with those shown in the modeling indicates good agreement.

### 4.4.5 Specific Design Examples

First, consider the backside point-contact solar cell design shown in Figure 4.23. The effective emitter saturation current in this design is reduced by minimizing the total diffused area to that covered by metal contacts. This is shown with experimental data in Figure 4.24. Additionally, notice that these emitters are on the back of the cell, where little photogeneration occurs. Therefore the emitter quantum efficiency constraint does not exist. From the modeling on metal contacted emitters, the choice appears to be for very deep emitters 4–10 microns deep, with surface concentrations between $10^{19}$ and $10^{20}$ cm$^{-3}$. The same emitters can be used to advantage on the front of the solar cell, if they are in the shadow of gridlines (Sinton,

**Table 4.1:** Measured emitter saturation currents for several diffusions. (Kane and Swanson, 1985). $J_{o,pass}$ is the emitter saturation current for diffusion which have received high-quality thermal oxidations to passivate the surface. $J_{o,bare}$ is the result after etching off the passivation oxide in HF. This simulates a case with large surface recombination velocity.

Boron emitters

Substrate: $N \approx 10^{13}$ cm$^{-3}$

| $R_{sh}(\frac{\Omega}{\square})$ | $x_j(\mu m)$ | $N_{surf}(cm^{-3})$ | $J_{0,pass}(\frac{pA}{cm^2})$ | $J_{0,Al}(\frac{pA}{cm^2})$ |
|---|---|---|---|---|
| 5.3 | 4.8 | $7.0 \times 10^{19}$ | 0.35 | 0.42 |
| 37 | 2.1 | $3.6 \times 10^{19}$ | 0.07 | 0.57 |
| 63 | 3.0 | $1.3 \times 10^{19}$ | 0.09 | 1.10 |
| 240 | 1.1 | $5.4 \times 10^{18}$ | 0.13 | 1.00 |
| 463 | 1.2 | $1.3 \times 10^{18}$ | 0.06 | 1.10 |

Phosphorus emitters

Substrate: $N \approx 10^{13}$ cm$^{-3}$

| $R_{sh}(\frac{\Omega}{\square})$ | $x_j(\mu m)$ | $N_{surf}(cm^{-3})$ | $J_{0,pass}(\frac{pA}{cm^2})$ | $J_{0,bare}(\frac{pA}{cm^2})$ |
|---|---|---|---|---|
| 9.0 | 5.8 | $5.0 \times 10^{19}$ | 0.45 | 0.50 |
| 57.2 | 3.6 | $1.2 \times 10^{19}$ | 0.10 | 1.10 |
| 75.3 | 1.8 | $2.0 \times 10^{19}$ | 0.13 | 1.70 |
| 370 | 1.2 | $2.5 \times 10^{18}$ | 0.08 | 1.70 |
| 890 | 1.3 | $1.0 \times 10^{18}$ | 0.01 | 7.00 |

Kane, and Midkiff, 1988). From the modeling on passivated emitters, it is clear that these deep, heavily-doped emitters would not be appropriate in an illuminated region. Additional advantages to the dark performance for these emitters are that being deep and heavily doped, contact resistance can be very low, and the effects of the various problems which can occur at the metal-silicon interface are minimized in deep emitters (such as junction spiking). Criteria associated with three-dimensional behavior in the base need to be addressed, and puts limits on the size and spacing of the diffused regions. An optimized design for a point-contact cell has about 10% of the backside area covered by $n$-type emitter and 10% covered by $p$-type emitter. Such a cell has an effective emitter saturation current of about $0.9 \times 10^{-13}$ A/cm$^2$, relative to the total cell area. Measurements indicate that the p and $n$-type emitters have comparable effective contributions of $0.45 \times 10^{-13}$ A/cm$^2$ each.

This design is a good candidate for a demonstration of the limits of the approximate modeling described in Section 4.2. This solar cell design is one of the few cases in the literature where all of the parameters have been well characterized independently of the solar cell itself.

For the design described above and shown in Figure 4.23, the simple modeling is easily applied. For the low concentration range, the device is dominated by surface emitter recombination. In this range, operating under high-level-injection, one has

$$J_{ph} - J = \frac{J_o}{n_i^2} n^2 + 1.8 S q n. \tag{4.81}$$

The value for $J_o$ to be plugged in here is the area-weighted value discussed above, $9 \times 10^{-14}$ A/cm$^2$. The front surface of a point contact solar cell and 80% of the back is base-oxide interface, hence the 1.8 cm$^2$ per cm$^2$ of cell area giving rise to the "1.8" in the surface recombination. Assume that the surface recombination velocity has

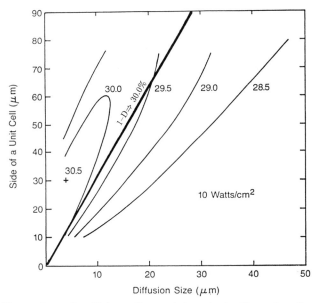

**Figure 4.46:** The contours for efficiency for a point-contact cell as a function of the emitter size and spacing. The results for an approximate, one-dimensional model are also shown, as 30% efficiency for a line of constant emitter coverage fraction (Sinton and Swanson, 1987a).

is 7.5 cm/sec (a value typically measured for oxidized and texturized surfaces), the photogeneration, $J_{ph}$, is 0.432 A/W times the incident power, $P_{sun}$, and the maximum power point is at a quantum efficiency of 95%. Then the left side of the above equation is equal to $0.05 \times J_{ph}$. Luckily, the equation is a quadratic equation, easily solved for $n$, hence $V_{mp}$. Then

$$\eta = \frac{0.95 \times J_{ph} \times V_{mp}}{P_{sun}}. \tag{4.82}$$

At 100 suns (10 Watts/cm$^2$), these assumptions lead to an efficiency of 30%. This is shown superimposed upon the results from a complete 3-dimensional simulation for a 50 $\mu$m thick cell in Figure 4.46. This turns out to be a very good approximation. But it is also clear what the limitations are. From this one-dimensional analysis, it might seem wise to reduce the emitter coverage fraction further, for example to the 0.3% of the total area indicated for the "Auger test structure" in Figure 4.24. This would be predicted to yield higher voltages and efficiencies due to the reduction in diffused region recombination. The experimental result is shown for this structure in Figures 4.47 and 4.48. This emitter area is too small to efficiently draw current from the cell. The structure becomes completely dominated by the ignored, three-dimensional effects causing a loss of current at high power densities. This effect would be completely missed by the one-dimensional model. However, a complete three-dimensional model is able to explain these effects fairly accurately as indicated in Figure 4.48.

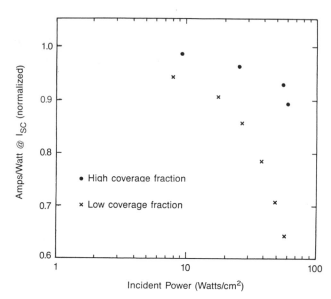

**Figure 4.47:** For a solar cell with too small contacts which are too widely spaced, the quantum efficiency under high illumination quickly drops off. A one-dimensional analysis which simply reduced the emitter saturation current by the coverage fraction of the diffusion would miss this important point.

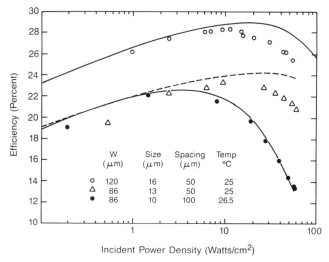

**Figure 4.48:** Here, three dimensional modeling is shown which does a pretty good job of predicting the effects of having less than optimum emitter area (Sinton and Swanson, 1987a).

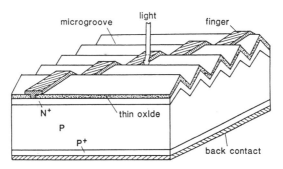

**Figure 4.49:** A low-resistivity concentrator solar cell with 25% efficiencies at 10 W/cm$^2$ (Green et al., 1987).

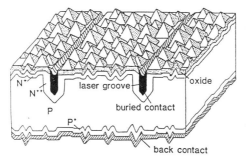

**Figure 4.50:** A schematic diagram of a laser grooved cell (Green, 1987).

The point-contact cell, being highly three-dimensional, is a worst-case challenge for a simple modeling scheme. However, even in this case, it is clear that the range of efficient operation is the same range where the simplifying assumption, namely uniform carrier density, works best. It is interesting to note that a one-dimensional model works well on optimized three-dimensional structures. Essentially, the three-dimensional effects work against higher efficiency, hence in an optimized design are minimized. A simple model can be a good guide but must be used judiciously.

In the 25% efficient PESC (for Passivated Emitter Solar Cell) concentrator design by Green et al., a planar junction is used in the front of the solar cell, Figure 4.49 (Green, Zhao, Blakers, Taouk, and Narayanan 1987). This junction is characterized by a sheet resistance of 50 $\Omega/\square$. From the curves on quantum efficiency, it can be seen that this is a good choice for a concentrator cell, since it has a low sheet resistance, but not so low as to suffer diminishing quantum efficiencies. An emitter with this sheet resistance may cause a loss of 1–2 mA/cm$^2$. The exact process schedule and resulting geometry for the emitter is not given, but for the sake of discussion presume that it is 4000 A thick, gaussian, and has a surface dopant density of 10$^{20}$ cm$^{-3}$. Then we would predict an emitter saturation current of 2$\times$10$^{-12}$ A/cm$^2$ for the 2% of metal-contacted area and 2$\times$10$^{-13}$ A/cm$^2$ for the 98% of the area which is oxide-passivated. This yields an effective emitter saturation current of about 2.4$\times$10$^{-13}$ A/cm$^2$ for the cell front-side. Presuming that the Al back surface field has a performance similar to the optimum for deep $p^+$ diffusions,

$4 \times 10^{-13}$ A/cm$^2$, then this cell design would have an effective emitter saturation current density of $6.4 \times 10^{-13}$ A/cm$^2$, if fabricated on a long lifetime substrate (perhaps a weak assumption here). This is consistent with the approximate value calculated from the one-sun open-circuit voltage, namely $4.5 \times 10^{-13}$ A/cm$^2$. Note that a one-sun solar cell could have less metal-silicon area and lower doping in the front-side $n^+$ emitter since the series resistance effects are less important. Hence, much better values for the emitter $J_o$ can be (and have been) demonstrated under those conditions (Blakers and Green, 1985).

A truly optimized design might use deep, heavily-doped diffusions under the metal lines, minimizing the $J_o$ and contact resistance there, but have deep lightly doped emitters elsewhere, in the region of photogeneration. This design has been implemented by several groups, perhaps most cleverly in the laser-grooved cell (Green, 1987). This cell is fabricated without lithography, and has very thick metal gridlines defined by the laser grooves (Figure 4.50).

These examples discussed above all utilized rather conventional integrated circuits emitter diffusion technology. One very encouraging recent development has been the fabrication of very efficient cells without subjecting the bulk of the wafer to high temperatures and, hence, potential lifetime-reducing contamination. In work reported by Wood et al., a wafer was implanted with a boron junction using a non-analyzed glow-discharge apparatus utilizing BF$_3$ (Wood, Westbrook, and Jellison, Jr., 1987a and Wood, Westbrook, and Jellison, Jr., 1987b). This diffusion was then annealed by an eximer laser, which melts a thin surface layer. Subsequent solar cell open-circuit voltages indicate that the total solar cell $J_o$ is about $6 \times 10^{-13}$ A/cm$^2$. This is for a $p^+$ emitter with a sheet resistance of 34 $\Omega/\square$. Significantly, these solar cells have 97% of the short circuit current of cells fabricated with a passivated emitter having a factor of 6 higher in sheet resistance. Most likely, this indicates a quantum efficiency for the heavily doped emitters approaching unity. Some of these cells have measured efficiencies of 18.9%. These do not have any of the light-trapping features described earlier. There is every indication that the technology demonstrated is capable of efficiencies of about 21–22% if applied to a texturized, high lifetime wafer. This would be very interesting for a fabrication sequence in which the highest temperature that the wafer is subjected to is 500 C and no patterning of the passivating oxide prior to metalization is necessary. The same authors have demonstrated 19.6% efficient cells by following the above procedure with a thermal oxidation at 850 C.

Another process by which solar cells can be fabricated at low temperature has been demonstrated by Jaeger and Hezel (1987). In this cell, no diffused emitter or back-surface-field exist. Instead, a front $n^+$ region is formed by an inversion layer which results from a high fixed charge density introduced as cesium contamination prior to a deposition of an overlaying silicon nitride layer. Contact to the emitter is made by an MIS structure, Al-thin SiO$_2$-Si. The back surface is directly contacted by a metal grid, but is passivated between gridlines by another silicon-nitride layer. This solar cell has achieved voltages at one-sun of 590 mV. It is extremely difficult to separate the recombination components in this cell because it is not clearly dominated by the boundary condition at any single surface. The authors concluded that the front and back of the solar cell had comparable amounts of recombination. This analysis assigns an emitter saturation current of $3 \times 10^{-12}$ to the front emitter.

The efficiency obtained for these cells, 15% with front illumination and 13% with rear illumination is especially interesting considering that the substrates were 2 $\Omega$-cm $p$-type, 100 $\mu$m thick and untexturized. With such a good base quantum efficiency, they could benefit from light-trapping, perhaps improving into the 18% range. Additionally, these cells are intended for use as bifacial cells. This is an interesting idea in which the scattered light incident onto the cell backside is utilized as well as the direct light. Although this doesn't fit well into the laboratory definitions for "efficiency," it can significantly boost the power out of a flat-plate collector, perhaps making these comparable to higher efficiency cells in the final analysis (Cuevas, Luque, Eguren, and del Alamo, 1981 and Silard and Nani, 1988).

## References

Basore, P.A. (1985), *Optimum Grid-Line Patterns for Concentrator Solar Cells Under Nonuniform Illumination*, Solar Cells **14**, 249-260.

Bennett, H.S. (1983) *Hole and Electron Mobilities in Heavily Doped Silicon: Comparison of Theory and Experiments*, Solid State Electronics **26**, 1157.

Blakers, A.W. and M. A. Green (1985) *Oxidation Condition Dependence of Surface Passivation in High Efficiency Silicon Solar Cells*, Appl. Phys. Lett. **47**, 818-820.

Blakers, A.W. and M. A. Green (1986) *20% Efficient Silicon Solar Cells*, Appl. Phys. Lett. **48**, 215-217.

Campbell, Patrick and Martin A. Green (1986), *The Limiting Efficiency of Silicon Solar Cells under Concentrated Sunlight IEEE Trans Electron Devices* **ED-33**, no. 2, pp. 234-239.

Caughey, D.M. and R. E. Thomas (1967) *Proc. IEEE* **55**, 2192.

Ciszek, T.F. (1987) *Material Considerations for High-Efficiency Silicon Solar Cells*, Solar Cells.

Cuevas, A., A. Luque, J. Eguren, and J. del Alamo (1981), *High-Efficiency Bifacial Back Surface Field Solar Cells*, Solar Cells **3**, 337-340.

Cuevas, A. and M. Balbuena (1988), *19% Efficient Thick Emitter Silicon Solar Cells*, *Proceedings of the 8th European Photovoltaic Solar Energy Conference*, Florence, Italy.

Dannhauser, L. (1972), *Die Abhängigkeit der Trägerbeweglichkeit in Silizium von der Konzentration der Freien Ladungsträger*, Solid State Elec. **15**, 1371-1375.

del Alamo, J.A. and R.M. Swanson (1984), *The Physics and Modeling of Heavily Doped Emitters*, IEEE Trans. Elec. Dev. **ED-31**, No. 12, pp. 1878-1895.

del Alamo, J.A. (1985), *Minority Carrier Transport in Heavily Doped n-Type Silicon*, Ph.D Dissertation, Stanford Universty, Stanford CA.

del Alamo, J.A. and R.M. Swanson (1987), *Measurement of Steady-State Minority-Carrier Transport Parameters in Heavily Doped n-Type Silicon*, IEEE Transactions on Electron Devices **ED-34**, No. 7, pp. 1580-1589.

Dziewior, J. and W. Schmid (1977), *Auger Coefficients for Highly Doped and Highly Excited Silicon*, Appl. Phys. Lett. **31**, no. 5, p. 346.

Dziewior, J. and D. Silber (1979), *Minority-Carrier Diffusion Coefficients in Highly Doped Silicon*, Appl. Phys. Lett. **35**, 170.

Eades, W.D. and R.M. Swanson (1985) *Calculation of the Surface Generation and Recombination Velocities at the Si-SiO$_2$ Interface*, J. Appl. Phys. **58**, No. 11, pp. 4267-4276.

Fahrenbruch, A.L. and R.H. Bube (1983), *Fundamentals of Solar Cells*, Academic Press, New York.

M.A. Green (1982), *Solar Cells*, Prentice-Hall, Inc., Englewood, NJ.

Green, M.A. (1983), *Accuracy of Analytical Expressions for Solar Cell Fill Factors*, Solar Cells **7**, 337-340.

Green, M.A. (1983), *Minority carrier lifetimes using compensated differential open-circuit voltage decay*, Solid State Electronics **26**, no. 11, pp 1117.

Green, M.A. (1984), *Limits on the Open Circuit Voltage and Efficiency of Silicon Solar Cells Imposed by Intrinsic Auger Processes*, IEEE Trans. Elec. Dev. **ED-31**, No. 5.

Green, M.A. and P. Campbell (1987), *Light Trapping Properties of Pyramidally Textured and Grooved Surfaces*, Proceedings of the 19th IEEE Photovoltaic Specialist Conference, pp. 912-917.

Green, M.A. (1987), *High Efficiency Silicon Solar Cells*, Trans Tech Publications, Switzerland.

Green, M.A., J. Zhao, A.W. Blakers, M. Taouk, S. Narayanan (1986), *25% Efficient Low Resistivity Silicon Concentrator Solar Cells*, Elec. Dev. Lett. **EDL-7**.

Hovel, H.J. (1975), *Solar Cells*, in Semiconductors and Semimetals **11**, Academic Press, New York.

Huber, D., A. Bachmeier, T. Wahlich, and H. Herzer (1986), *Minority Carrier Diffusion Length and Doping Density in Nondegenerate Silicon*, in *Semiconductor Silicon 1986*, H. R. Huff, T. Abe, and B Kobessen Eds., p. 1022.

Irwin, J.C. (1962), *Resistivity of Bulk Silicon and Diffused Layers in Silicon*, Bell Sys. Tech. J. **41**, 387.

Jaeger, K. and R. Hezel (1987), *Bifacial MIS Inversion Layer Solar Cells Based on Low Temperature Silicon Surface Passivation*, Proceedings of the 19th IEEE Photovoltaic Specialist Conference, pp. 388-391.

Kane, D.E. and R.M. Swanson (1985), *Measurement of the Emitter Saturation Current by a Contactless Photoconductivity Decay Method*, Proceedings of the IEEE 18th Photovoltaic Specialist Conference, pg. 578.

King, R.R. personal communication, also, R.M. Swanson (1988), *Studies in Advanced Silicon Solar Cells*, Sandia National Laboratories Contractor Report SAND88-7026.

Kleinmann, O.A. (1956), *The Forward Characteristics of the PIN Diode*, Bell System Technical Journal, pp. 685-706.

S. Mathews and R.L. Walker (1964), *Mathematical Methods of Physics*, W.A. Benjamin Inc., p. 322.

Moll, J.L. (1964), *Physics of Semiconductors*, McGraw Hill Inc., p. 150.

Narayanan, S., S.R. Wenham, and M.A. Green (1986), *High Efficiency Polycrystalline Silicon Solar Cells using Phosphorus Pretreatment*, Applied Physics Letters **48**, 873-875.

O'Neill, M.J. (1985) *A Low-Cost 22.5X Linear Fresnel Lens Photovoltaic Concentrator Module which Uses Modified, Mass-Produced, One-Sun Silicon Cells*, Proceedings of the 18$^{th}$ IEEE Photovoltaic Specialists Conference, Las Vegas, pp. 1234-9.

Pankove, J.I. (1971), *Optical Processes in Semiconductors*, Prentice-Hall Inc., p. 111.

Schlangenotto, H., H. Maeder, and W. Gerlach (1974), *Temperature Dependence of the Radiative Recombination Coefficient in Silicon*, Physica Status Solidi **21a**, 357-367.

Schmidt, W., K.D. Rasch, and K. Roy (1982), *Improved Efficiencies of Semiconductor and Metallurgical Grade Cast Silicon Solar Cells by Hydrogen Plasma Treatments*, *Proceedings of the 16^{th} IEEE Photovoltaic Specialists Conference*, San Diego, pp. 537-542.

Lammert, M.D. and R.J. Schwartz (1977), *The Interdigitated Back Contact Solar Cell: A Silicon Solar Cell for Concentrated Sun-light*, IEEE Trans. Elec. Dev. **ED-24**, No. 4, pp. 337-342.

Schwartz, R.J., M.S. Lundstrom and R.D. Nasby (1981), *The Degradation of High-Intensity BSF Solar-Cell Fill Factors Due to a Loss of Base Conductivity Modulation*, IEEE Trans. Elec. Dev. **ED-28**, No. 3, pp. 264-269.

Schwartz, R.J. (1983), private communication.

Silard, A. and G. Nani (1988), *Bifacial Solar Cells on Large-Area Low-Cost Silicon Wafers*, IEEE Elec. Dev. Lett. **9**, No 1.

Sinton, R.A., Y. Kwark, S. Swirhun, R.M. Swanson (1985), *Silicon Point Contact Concentrator Solar Cells*, Elec. Dev. Letters **EDL-6**, no. 8, pp. 405-407.

Sinton, R.A., Y. Kwark, P. Gruenbaum, R.M. Swanson (1985), *Silicon Point Contact Concentrator Solar Cells*, *Proceedings 18th IEEE Photovoltaic Specialist Conference*, pp. 61-65.

Sinton, R.A., Y. Kwark, J.Y. Gan, and R.M. Swanson (1986), *27.5% Silicon Concentrator Solar Cells*, Electron Device Letters **EDL-7**, no. 10.

Sinton, R.A. and R.M. Swanson (1987), *Recombination in Highly-Injected Silicon* Trans. Elec. Dev. **ED-34**, no. 6.

Sinton, R.A. and R.M. Swanson (1987a), *Design Criteria for Si Point-Contact Concentrator Solar Cells*, IEEE Trans. Elec. Dev. **ED-34**, No. 10.

Sinton, R.A. and R.M. Swanson (1987b), *Increased Photogeneration in Thin Silicon Concentrator Solar Cells*, IEEE Elec. Dev. Lett. **EDL-8**, No. 11.

Sinton, R.A., D.E. Kane, N. Midkiff, unpublished. *Cells of 1.5625 cm^2 area with efficiencies of 23% at 14 Watt/cm^2 of incident sunlight have been demonstrated. These have the deep diffusions under the frontside grid.* (measured at Sandia National Laboratories, July 1988.

Slotboom, J.W. and H.C. de Graaff (1977), *Bandgap Narrowing in Silicon Bipolar Transistors*, IEEE Trans. Elec. Dev. **ED-24**, 1123.

Svantesson, K.G. and N.G. Nilsson (1979), *The Temperature Dependence of the Auger Recombination Coefficient in Silicon*, J. Phys. C: Solid State Physics **12**, 5111-5120.

Swanson, R.M. (1979), *Silicon Photovoltaic Cells in TPV Conversion*, EPRI Report ER-1272.

Swanson, R.M. (1983), *Point-Contact Silicon Solar Cells*, EPRI Report AP-2859.

Swanson, R.M., J. del Alamo, and S. Swirhun (1984), *Measuring and Modeling Minority Carrier Transport in Heavily Doped Silicon*, Int. Conf. on Heavy Doping and the Metal-Insulator Transition, (Satellite Conf. of the 1984 International Conference on the Physics of Semiconductors, IUPAP).

Swanson, R.M. (1986), *Point-Contact Solar Cells: Modeling and Experiment* Solar Cells **7**, no. 1.

Swirhun, S.E., Y.H. Kwark, and R.M. Swanson (1986), *Measurement of Electron Lifetime, Electron Mobility and Band-Gap Narrowing in Heavily Doped p-Type Silicon*, IEEE International Electron Devices Meeting, Los Angeles, December 7-10, 1986, Technical Digest, pp. 24-27.

Swirhun, S.E. (1987), *Characterization of Majority and Minority Carrier Transport in Heavily Doped Silicon*, Ph.D dissertation, Stanford University, Stanford, CA.

Sze, S.M. (1981), *Physics of Semiconductor Devices, Second Edition*, John Wiley & Sons, p. 35.

Tiedje, T., E. Yablonovitch, G.D. Cody, B.G. Brooks (1984), *Limiting Efficiency of Silicon Solar Cells*, IEEE Trans. Electron Devices **ED-31**, 711-761.

Tobin, S.P., M.B. Spitzer, C. Bajgar, L. Geoffroy, and C.J. Keavney (1987), *Advanced Metalization for Highly Efficient Solar Cells*, *Proceeding of the 19th IEEE Photovoltaic Specialist Conference*, pg. 70, New Orleans, LA.

Vaitkus, Yu. and V. Grivitkas (1981), *Dependence of the Rate of Interband Auger Recombination on the Carrier Density in Silicon*, Sov. Phys. Semicond. **15**, No. 10, pp. 1102.

Weaver, H.T. (1982), *Ineffectiveness of Low High Junction in Optimized Solar Cell Designs*, Solar Cells **5**, 275.

Wood, R.F., R.D. Westbrook, and G.E. Jellison, Jr. (1987a), *18% Efficient Intrinsically Passivated Laser Processed Silicon Solar Cells*, Appl. Phys. Lett. **50**, No. 2.

Wood, R.F., R.D. Westbrook, and G.E. Jellison, Jr. (1987b), *High-Efficiency Intrinsically and Extrinsically Passivated Laser Processed Silicon Solar Cells*, *Proceedings of the 19th IEEE Photovoltaic Specialist Conference*, pp. 519-524.

Yablonovitch, E. and G.D. Cody (1982), *Intensity Enhancement in Textured Optical Sheets for Solar Cells*, IEEE Trans. Elec. Dev. **ED-29**, No. 2.

Yablonovitch, E. and T. Gmitter (1986), *Auger Recombination at Low Carrier Densities*, Applied Physics Letters **49**, No. 10, p. 587.

CHAPTER 5

# CuInSe$_2$ and CdTe
## Scale-up for Manufacturing

Kenneth Zweibel and Richard Mitchell

## 5.1 Introduction

Two candidate polycrystalline thin film materials, CuInSe$_2$, and CdTe, have emerged as potential leaders in the effort to develop low-cost photovoltaics. Until recently, most progress in these materials was on a laboratory scale. For instance, small CuInSe$_2$ cells have reached 14.1% (ARCO Solar), Photon Energy and Ametek have achieved 12.3% and 11% efficiencies for CdTe. However, both technologies are now moving toward commercial products and are being developed for power modules. Issues relating to their scale-up to manufacturing are increasingly germane.

Figure 5.1 shows the rather dramatic recent progress of CuInSe$_2$ and CdTe square-foot modules. At 11.1% aperture-area efficiency, the CuInSe$_2$ module made by ARCO Solar (Mitchell et al., 1989) has become the most efficient module of that size for any potentially low-cost thin film. Mitchell and coworkers have also reported a 9.7% efficient 4-square-foot CuInSe$_2$ module producing 37.8 W–something like the power output we have come to expect of crystalline silicon modules!

In particular, CuInSe$_2$ has become a leading thin film material. In addition to its high performance, initial outdoor tests on encapsulated ARCO Solar modules at SERI (Figure 5.2) have resulted in no degradation–confirming the reputation for stability CuInSe$_2$ has had because of prior tests on cells. High performance and apparent stability have won CuInSe$_2$ many new adherents. One lingering question has been its cost in manufacturing. This review is a start at answering questions about that issue.

CdTe is also progressing toward respectable module efficiencies (now over 7%). Its apparent potential for very low cost production is perhaps its greatest strength. This review describes some of the very inexpensive processes being developed to make CdTe.

Processing issues related to devices are made more numerous by the multiplicity of approaches that exist (1) for making the absorber materials themselves (i.e., the CuInSe$_2$ or CdTe) and (2) for making the other layers (windows, heterojunctions, contacts) needed to complete a device. Each of these layers may be made by several fabrications methods, each varying in complexity, cost, materials usage, yield, safety, etc. Part of the purpose of this review is to enumerate the different methods of

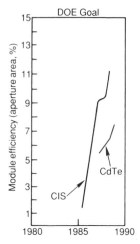

**Figure 5.1:**  Progress in module efficiency for CdTe and CuInSe₂

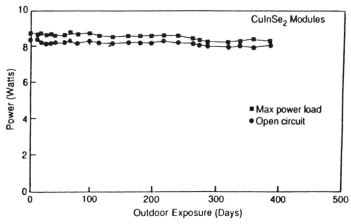

**Figure 5.2:**  Initial outdoor tests of ARCO Solar CuInSe₂ modules at SERI show no charge in efficiency.

making the various layers while pointing out their salient characteristics and needs for further development.

This article builds upon an excellent review published in a previous volume (IV) of this publication (Bloss et al., 1988). Bloss et al. focused on cell fabrication and analysis, also giving much relevant historical background that will not be repeated here. The reader is referred to it as a complement to this publication.

We have attempted to select and examine the key processes for making CuInSe₂ and CdTe as well as for fabricating window materials (e.g., dip-coated CdS) and oxide and metal contacts. For example, we have sections on selenization of Cu and In to make CuInSe₂ and on electrodeposition and spraying to fabricate CdTe (among others). We also have sections on depositing Mo contacts and fabricating ZnO and SnO₂. We have attempted to survey the entire process spectrum from substrate to finished cells. We have also touched upon materials issues such as availability and toxicity. As part of each section, we have summarized the status and issues of the various approaches as well as given a set of conclusions concerning their advantages and disadvantages in relation to the alternatives. These polycrystalline thin film technologies are on the verge of extensive commercialization. If trends continue, they may come to dominate photovoltaics because of their high efficiencies, stability, and low production costs.

## 5.2 CuInSe$_2$

### 5.2.1 Selenization to Form CuInSe$_2$ Films

Selenization processes are defined as those which use a Se containing fluid, usually $H_2Se$, to react with Cu and/or In to form $Cu_2Se, In_2Se_3$, and/or $CuInSe_2$ after the deposition of the Cu and In layers. The basic process for selenization to form $CuInSe_2$ is a chemical reaction following metallic deposition. Distinct Cu and In metal layers are deposited, with both the Cu and In layers being of precise thicknesses. This stacked film is then subjected to a Se-rich environment. Selenization has been investigated using a liquid Se source in the form of $H_2SeO_3$ by SERI (Chen et al., 1985). However, the most common method for selenization is to anneal the films in a $H_2Se$ atmosphere.

The selenization processes known to have achieved near 10% cells are those reacting $H_2Se$ with distinct Cu and In films deposited by sputtering, evaporation, or electroplating. Other procedures that have not yet produced cells of reasonable efficiency include: the addition of a Se layer or layers (Cu-In-Se stacked layers), the deposition of $Cu_2Se$ and $In_2Se_3$ layers, Cu/In alloy deposition, and the multiple layering of films (e.g. Cu-In-Cu-In-Cu-In....). Each of these processes includes an anneal in $H_2Se$ to optimize the $CuInSe_2$ film properties. In some cases, they still show promise, and some are covered in another section (Section 5.2.2, CuInSe$_2$: Other Deposition Processes).

### 5.2.1A History and Present Status

Research into selenization methods followed earlier work in the sulfidization of RF-sputtered Cu and In films by Grindle (1979) and MBE-grown Cu-In films by Binsma and Van Der Linden (1982a,b). These efforts led to high quality $CuInS_2$ films. The possibility of using selenization of these same Cu-In films to form $CuInSe_2$ was suggested by their success.

The sequential deposition of Cu and In layers for selenization was investigated by Chu et al. (1984 a,b) with SERI support and independently by Kapur et al. (Patent: *Forming a Compound Semiconductive Material*, 1986). Their initial reported results in the fabrication of CuInSe$_2$ films by electroplating and evaporation stimulated other researchers to pursue selenization for the fabrication of CuInSe$_2$ devices.

There are many methods of depositing Cu and In for subsequent $H_2Se$ selenization. Some examples are: the sequential sputtering of Cu and In layers, the electrodeposition of a Cu-In layer from one electrolyte bath to form a Cu-In alloy, the sequential electrodeposition of Cu and In layers, and the sequential evaporation of Cu and In layers. In all cases, the sequential layering of Cu and In films (rather than alloying) has produced the best solar cells. This discussion will focus on three

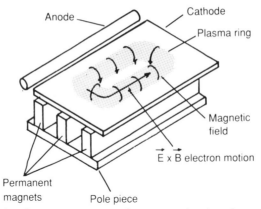

Anode
Cathode
Plasma ring
Magnetic field
$\vec{E} \times \vec{B}$ electron motion
Permanent magnets
Pole piece

**Figure 5.3:** Schematic of magnetron sputtering target showing the magnets for confinement of the ions.

methods of depositing Cu and In films: sputtering, electrodeposition, and evaporation.

### 5.2.1A.1 Sputtering

Sputtering involves the creation of vapor species through their kinetic ejection from a Cu or In target surface by the bombardment of that surface with energetic, nonreactive ions. Ejection is caused by momentum transfer between the impinging ions and the Cu or In atoms of the target surface. The sputtered atoms, which are mostly neutral, condense on a substrate to form a metal film (Bunshah et al., 1982).

Sputtering for the coverage of large areas is most often done with a magnetron sputtering source. In magnetron sputtering, the bombarding species are ions produced in a plasma discharge that is magnetically confined to a region adjacent to the target. This increases the efficiency with which ions sputter atoms from the target and minimizes undesirable bombardment of the deposited films. The substrates are arranged to intercept the sputtered Cu or In flux as indicated in the schematic (Figure 5.3).

There are numerous efforts in the area of sputtering of CuInSe$_2$. However, only ARCO Solar has reported results on devices that exceed 10% for sputter-deposited CuInSe$_2$. The ARCO Solar process is proprietary, and details are not available. (See patents: Choudary et al., 1986, Ermer et al., 1987, Gay and Kapur, 1984, Kapur et al., 1986, Love and Choudary, 1984, and Ermer and Love, 1989). ARCO Solar has achieved an active-area efficiency of 14.1% on a 3.5 cm$^2$ device (FF=0.67%, J$_{sc}$ = 41.0 mA/cm$^2$, and V$_{oc}$ = 0.508 volt). For large area (938 cm$^2$ aperture-area, 55-cell module) ARCO Solar attained efficiencies of 11.1% with FF=0.64, I$_{sc}$ = 637 mA, and V$_{oc}$ = 25.38 volts (SERI-measured, see Figure 5.4). Their continued research (Mitchell and Liu, 1988, Mitchell et al., 1988) is supported by SERI funding.

### 5.2.1A.2 Electroplating

At least four organizations (ARCO Solar, ISET, SMU, and Weizmann Institute) have been involved in the fabrication of CuInSe$_2$ through the electrodeposition of Cu and In layers and subsequent heat treatment in H$_2$Se.

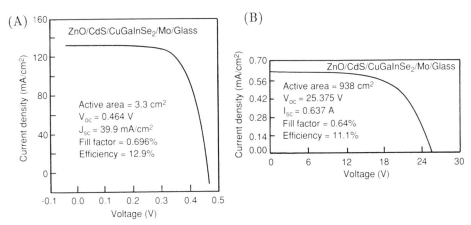

**Figure 5.4:** ARCO Solar CuInSe₂ cell (A) and module (B) performance curves.

ARCO Solar began their first efforts in the electrodeposited formation of Cu and In layers to better understand the process for the selenization of Cu/In films to form CuInSe₂. This effort resulted in a patent being issued (Kapur et al. *Forming a Compound Semiconductive Material*, U.S. 45811089 A) in April of 1986. Though ARCO Solar is a pioneer in this field, they have not published results specific to electrodeposition. Information concerning their work is only available from their patent.

Researchers at Weizmann Institute first reported work in this field in November 1984 (Hodes and Cahen, 1984). They continued efforts in the electrodeposition of Cu-In alloy films from a single bath until 1986. Theirs is the only published work on the attempt to use one electrolyte to form a Cu-In for subsequent heat treatment by H₂Se. Hodes reported on efforts to electroplate Cu-In alloys on Ti in 1985 (Hodes et al., 1985 a,b; Hodes and Cahen, 1986). They focused on electrodepositing a Cu-In alloy from aqueous solutions of $Cu^+$ and $In^{3+}$ ions and thiourea. Their films were then heat treated in a H₂Se atmosphere. Cell efficiencies of 1.5% were measured using a liquid polysulfide electrolyte/semiconductor junction, with 750 nm quantum efficiencies of 0.67 measured at short circuit. Problems with the heterogeniety of the films were assumed to be responsible for poor efficiencies. Later reports (Lokhande and Hodes 1987, Hodes and Cahen 1987, Cahen and Hodes 1987) described their conclusions that CuInSe₂ films formed from the selenization of Cu-In alloy layers suffered compositional changes due to vapor transport of In₂Se away from the film. These films were annealed at temperatures that ranged from 300°C up to 640°C. It was concluded that for annealing temperatures below 550°C the films were *n*-type, and for temperatures above 600°C they were *p*-type. Although larger grains are formed at elevated annealing temperatures, the loss of In reduced cell performance. The resulting films had an excess of $Cu_xSe$ on the surface, and the bulk material was 5%–10% Cu-rich.

During 1984, electrodeposition work was also reported by the Chus at Southern Methodist University. Their SERI-funded efforts were directed toward the deposition of Cu and In films from separate electrolytes. They reported selenizing the

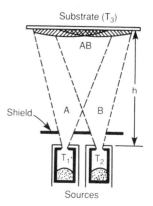

**Figure 5.5:** Schematic of an evaporation system. Each source is held at a fixed distance from the substrate (H) and the sources are held at temperatures $T_1$ and $T_2$ respectively with the substrate held at temperature $T_3$.

electrodeposited Cu-In layers individually between deposition steps and together after sequential deposition by heat treatment in $H_2Se$ at 630°C (Chu et al., 1984 a,b). All of the films were large grain $CuInSe_2$ with a preferred (112) orientation. They did not report results on any devices.

Since 1984, SERI-funded selenization research has been carried out by International Solar Electric Technology (Kapur et al., 1985). With their electrodeposition method for sequentially preparing Cu and In films, they have achieved a reported efficiency of over 10%, but their typical efficiencies are lower due to ongoing adhesion problems (see below).

ISET made $CuInSe_2$/CdS cells (1 cm²) with an efficiency of 8% (no AR) in 1985. These single-phase $CuInSe_2$ films were typically 2 $\mu$m thick with a Cu/In/Se ratio of 24.5/25.5/50.0. All of the films were $p$-type with carrier concentrations in the range of $10^{15}$ to $10^{16}$. Films had a sheet resistance of between 50–500 killiohms/square; and a bulk resistivity of around 50-$10^3$ ohm-cm. The surface had a 0.5-0.8 $\mu$m roughness when deposited on electron-beam evaporated Mo. More recent ISET devices fabricated by the electrodeposition process typically have around 400 mV $V_{oc}$, a $J_{sc}$ of 30 mA/cm², a FF of 60%, and 8% efficiencies (Kapur et al., 1986). Devices of over 10% have now been reported only rarely, with one small area device recently reported at 11.0% efficiency (Kapur et al., 1987 a,b; Kapur et al., 1988).

### 5.2.1A.3 Evaporation

Another approach to the formation of Cu and In for subsequent selenization is evaporation (shown in Figure 5.5).

There has been reported work by SMU, ISET, ARCO Solar, the University of Stuttgart (F.R.G), and IEC in the evaporation of Cu-In films for selenization to $CuInSe_2$. In the 1984 papers discussed earlier, Chu et al. reported evaporating Cu and In layers which were selenized by heat treatment in $H_2Se$ at 630°C. Devices of reasonable performance were not produced, and Chu has not reported further in this area. In an effort to investigate film adhesion properties during selenization, ISET has fabricated evaporated Cu and In layers for selenization. Early devices have shown reported efficiencies of 7%. ARCO Solar mentions Cu-In evaporation in their 1986 patent. They have not published subsequent results. Stuttgart has reported in 1988 on the evaporation of Cu and In and selenization with elemental Se sources. The Se was introduced in both a closed ampule and transported through

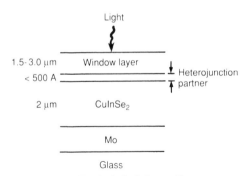

**Figure 5.6:** Typical CuInSe₂ cell structure.

an open system within an Ar atmosphere (Dimmler et al., 1988). The first device efficiencies reported from this process were around 4.1%. The Institute of Energy Conversion (IEC) is currently investigating this approach in cooperative efforts with ISET and Chronar.

### 5.2.1B Cell Structure

All groups fabricating CuInSe₂ solar cells via selenization processes use a similar cell structure (Fig. 5.6). The most common substrate is glass. This offers both low cost and an available industrial capacity for supply. A 2 $\mu$m thick Mo film is then deposited. The CuInSe₂ layer is then deposited with one of the several selenization processes, followed by an $n$-type heterojunction partner, conductive window material, grids, and an antireflection coating.

### 5.2.1C Process Description

The selenization process for the fabrication of CuInSe₂ can be divided into two steps: the deposition of Cu and In films and selenization-annealing. Sputtering, electrodeposition, and evaporation of Cu and In are discussed here. Selenization is discussed separately in the final section.

### 5.2.1C.1 Sputtering of Cu-In

Magnetron sputtering has been successfully used for many large-area, thin metallic film applications (Bunshah 1982, Chopra and Das, 1983). Magnetron sputtering targets of materials such as Cu and In can be fabricated with great reliability for large-area applications. For the fabrication of CuInSe₂, Cu and In layers for subsequent selenization can be sputtered sequentially with DC planar magnetrons at $10^{-6}$ to $10^{-7}$ torr. In a typical process, a thin Cu layer (2,000 Å) is sputtered onto a Mo-coated glass substrate, followed by a thin In layer (4500 Å) as shown in Figure 5.7. The resulting glass/Mo/Cu/In structure is then reacted in an atmosphere of H₂Se at around 400°C to form single-phase CuInSe₂. The composition of the CuInSe₂ film is determined by the Cu and In layer thicknesses and the selenization time and temperatures. The thicknesses of the Cu and In films are a result of the deposition time and the power supplied to the planar magnetron target.

The Cu layer is sputtered from a 99.99% purity Cu target. If the target is supplied with 1 amp at 1 kV, the deposition rate is around 1 $\mu$m/min. The substrate is held at the ambient temperature; and at the deposition rate, the target temperature is held around 200°C with standard water cooling.

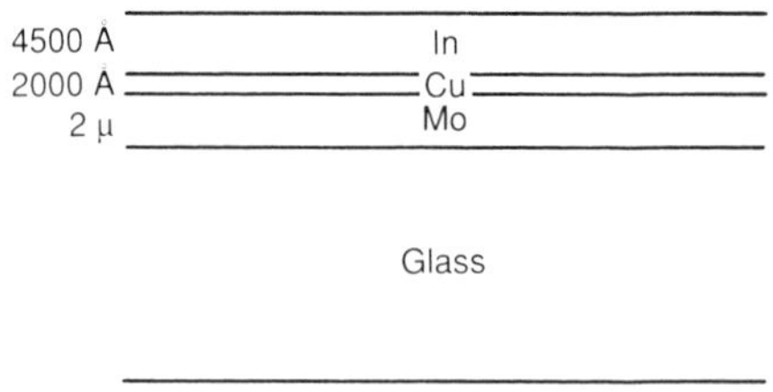

**Figure 5.7:** Schematic of Cu and In layering on Mo/glass by sputtering and electrodeposition.

The In layer is also sputtered from a target of 99.99% purity. If the target is supplied with an equivalent 1 amp at 1 kV, the In deposition rate is also around 1 $\mu$m/min. The substrate temperature is kept at 150°C. If the In-target is suspended over the substrate, care must be taken to insure that the target temperature is kept well below this temperature so that the In does not melt and splatter.

Within normal sputtering regimes, rates of Cu and In are proportional to the power delivered (to the source). Control of the deposition rate and uniformity can be quite good, because control of power can be quite precise (near 0.1%)

### 5.2.1C.2  Electroplating of Cu-In

ISET electroplates Cu and In layers from two electrolyte baths. A thin Cu layer is electrodeposited on a Mo substrate and then covered with a thin In layer (see Figure 5.8). This stacked glass/Mo/Cu/In film is then reacted in an atmosphere of $H_2Se$ at 400°C to form single-phase $CuInSe_2$. Deposition potential and time are the determining factors in the thickness of each of the electrodeposited layers.

Because of its low cost, the substrate most commonly used for this process is Mo-coated glass (although high quality films have been produced on Mo-coated alumina and various thickness of Mo foil). However, the adhesion of $CuInSe_2$ on Mo-coated glass has been a serious problem. Further investigation into the microstructure of Mo and the advantages of different deposition methods may be important in addressing this problem. (See also Section 5.2.1C.4 on selenization and Section 5.2.3A.1 on contact layers.) Another possible approach to improve $CuInSe_2$ sticking is to bombard the substrate with ions during Mo deposition. This technique may allow control of the structural stresses that the Mo/glass and Mo/$CuInSe_2$ interfaces are subjected to during the film expansion during selenization. Other factors, such as film stoichiometry, are also important in this complex problem.

In the ISET approach, a Cu layer 2000 Å thick is electrodeposited on the Mo (Kapur et al., 1985). An initial micro-burst of 80 mA/cm² is applied to the Mo substrate, which is immersed in a 0.5M solution of $CuSo_4$ in 2.0M $H_2SO_4$. A constant 40 mA/cm² current is then applied for approximately 12 seconds. The schematic equation governing deposition from this electrolyte is:

$$\text{Cathode}: \quad Cu^{++} + 2e^- \rightarrow Cu$$
$$\text{Anode}: \quad Cu \rightarrow Cu^{++} + 2e^-$$

with the cathode efficiency of $CuSO_4$ electrolyte being around 97%. (The cathode efficiency is the ratio of flux of the plated ions to the flux of electrons supplied as current to the electrodes.)

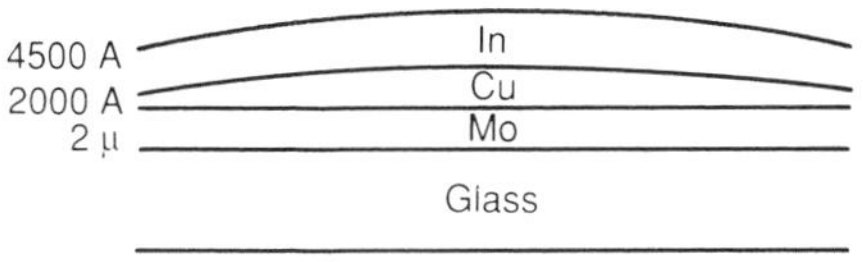

**Figure 5.8:** Schematic of Cu and In layering on Mo/glass by evaporation.

After the deposition of the Cu layer, the film is thoroughly cleaned and rinsed with DI water. This is followed by the electrodeposition of of 4500 Å of In. Deposition is carried out in a solution of 0.28 M indium sulfamate $(In(SO_3NH_2)_3)$ with a constant 20 mA/cm² of current applied for approximately 45 seconds. The equation governing the deposition from this electrolyte is:

$$Cathode: \quad In^{+++} + 3e^- \rightarrow In$$
$$Anode: \quad In \rightarrow In^{+++} + 3e^-$$

with the cathode efficiency of the indium sulfamate electrolyte being around 95%.

### 5.2.1C.3 Evaporation of Cu-In

Evaporation of Cu and In thin metallic films has been used in many industrial applications. The evaporation of Cu and In for selenization to form CuInSe₂ is carried out at around $10^{-6}$ to $10^{-7}$ torr. A thin layer of Cu is first evaporated onto the Mo-coated substrate, followed by the evaporation of a thin In layer (Figure 5.8). The resulting glass/Mo/Cu/In structure is then reacted in an atmosphere of H₂Se at around 400°C to form CuInSe₂. The composition of the CuInSe₂ film is determined by the localized thickness of the Cu and In layers and the selenization time and temperature. The localized thicknesses of the Cu and In films depend on the temperature of the evaporation sources and the substrate, the orientation of the evaporation source(s), the angular dependence of the deposited materials, and the source distance from the substrate. The Cu/In ratio is controlled by adjusting these parameters to achieve a nearly uniform Cu/In ratio across the substrate.

The most commonly used substrate for the evaporation method is Mo-coated glass. As with the other Cu-In deposition techniques, there is concern for adhesion problems at the Mo/glass and Mo/CuInSe₂ interfaces caused by the stresses generated by selenization.

Typically, the evaporation of a 2000 Å Cu layer is carried out with a 99.999% purity Cu evaporation source. The source is held at 1350°C to produce a 0.5 to 1.0 $\mu$m/min deposition rate. The substrate is held at room temperature.

The next layer is 4500 Å of In. The In source is also 99.999% purity. It is operated at a temperature of 1100°C, with the substrate at room temperature, which results in a deposition rate similar to that of the Cu.

The evaporation rates of both the Cu and the In are related to the temperature of the evaporation sources in a complex, non-linear manner. Therefore, the control of the deposition rate and uniformity must be constantly monitored to produce uniform films. The rate of deposition for both the In and Cu atoms is also affected by the cosine dependence of the flux onto the substrate. The relative position of the substrate with respect to the In source must be close to the relative position of the previous Cu source to minimize fluctuations in the Cu/In ratio at each location on the substrate.

### 5.2.1C.4 Selenization

The selenization step for fabricating $CuInSe_2$ layers from a Cu-In film (whether formed by sputtering, electrodeposition, or evaporation) consists of reacting the Cu-In stacked layers with a Se-containing atmosphere. As stated earlier, the most successful processes employ a heat treatment in an atmosphere containing 10%–15% $H_2Se$ in Ar. Although ARCO Solar has several patents in which selenization is discussed (Choudry et al., 1986, Love and Choudry, 1984, Ermer and Love, 1987), the only detailed reports of a selenization process that have resulted in efficient $CuInSe_2$ devices are those of ISET (Kapur et al., 1987, 1988). Their films are ramped from around 25°C to 400°C and held at 400°C for about an hour and a half. The temperature is then ramped down to 25°C. The equation governing the selenization reaction is:

$$Cu + In + 2H_2Se \rightarrow CuInSe_2 + 2H_2$$

This reaction results in the incorporation of Se into the Cu-In layer and the formation of single-phase $CuInSe_2$ film. Volume expansion is on the order of three times as the Se is added. The resulting $CuInSe_2$ films are nominally 2 $\mu$m thick. The underlying Mo film is subjected to great stresses during selenization. It is this stress that creates the significant problems that are associated with the adhesion of the Mo/glass and the Mo/$CuInSe_2$ interfaces.

The Se is incorporated in the film to near-stoichiometric composition. Annealing temperature and time are the determining factor in the film quality. The final stoichiometry can be adjusted by modifying the Cu/In ration in the Cu-In stacked metal film. Control of the film quality can be adversely effected by: In loss from the $CuInSe_2$ through the escape of $In_2Se$ caused by excessive selenization time and/or elevated annealing temperatures; and lack of complete Se incorporation into the Cu-In stacked layer because of insufficient selenization time and/or reduced annealing temperatures.

### 5.2.1D  R&D Issues

### 5.2.1D.1  Materials

The materials required for the electrodeposition of $CuInSe_2$, namely DI water, $H_2SO_4$, Mo, $CuSO_4$, and $In(SO_3NH_2)_3$ and the 99.99% purity Cu and In anode bars, can be purchased. They, as well as the 99.999% pure evaporation source material, are all available in the quantity required for small production facilities. The 99.99% pure Cu and In targets required in magnetron sputtering are readily available in small targets, and no problems are seen in the scale up of these targets to any size required for plant operation. The glass-coating industry routinely uses large-area targets such as these for architectural glass coating.

Although the Cu and In products required for the sputtering, electrodeposition, or evaporation techniques are all available, it is a cost disadvantage for sputtering that the source material must be fabricated into targets. Targets add cost and cannot be 100% utilized. Recycling target material requires additional cost for processing.

No studies have been done to establish whether high purity is absolutely necessary or to what level laboratory quality restrictions can be reduced, reducing materials costs.

Existing facilities can supply the required quantity and purity (99.5%) of $H_2Se$ that a selenization process would require. In fact, the major impurity in commercial $H_2Se$ is sulfur in the form of $H_2S$. This reacts with the Cu and In metal layers during the selenization process, forming $CuInS_2$. The $CuInS_2$ is incorporated into the $CuInSe_2$ bulk as a photoactive species and has not been found to be particularly harmful to device performance.

The availability of the raw Mo, Cu, Se, H, Ar, S, and N starter materials are fully adequate for very large scale production purposes. There is also a sufficient world supply of In for multigigawatt annual production of solar cells. However, 99.99% pure In is around 15 times as expensive as the same quality of Se and 150 times as expensive as the same quality of Cu. There is some concern that a sudden increase in the current usage rate may have an adverse effect on the supply of In and result in fluctuating prices for all In-containing materials, targets, and electrolytes. In the first six months of 1987 the cost of In increased 346% (from a very low price) due to a major increase in the demand for electronic applications. The price stabilized at that point. Price stability is a significant issue for all $CuInSe_2$ processes, as In could affect the cost-of-energy if In prices escalated radically due to an In demand/capacity mismatch.

It is estimated that Cu and In material utilization (evaporated or sputtered material/material incorporated in films) will be from 70%-80% for sputtering and evaporation. The feedstock material in the sputtering targets may only be 50% sputtered before targets require recasting. However, the remaining feedstock material would be recast. The cost of the recasting process beyond the cost of the added feedstock material has not been established, but is not expected to be critical. Electrodeposition uses nearly 100% of all feedstock (Cu and In).

A 2 $\mu$m $CuInSe_2$ film from selenization requires a 2000 A film of Cu and 4500 A film of In. These thicknesses have 1.87 $g/m^2$ of Cu, 3.45 $g/m^2$ of In, and 4.65 $g/m^2$ of Se. One may assume Cu-In utilization rates of 75%, 75%, and 95% for sputtering, evaporation and electrodeposition, respectively. A utilization rate of 45% is assumed for Se. With a price of \$300/kg for 99.99% In, \$2/kg for 99.99% Cu, and \$17/kg for 99.99% Se, the cost of the absorber material for sputtering and evaporation would be \$1.58/$m^2$. The cost of the absorber material for electrodeposition would be \$1.29/$m^2$. Somewhat lower cost from better material utilization is an advantage for the electrodeposition process.

These estimates are based on the currently required $CuInSe_2$ thickness of 2.0 $\mu$m. Cost reductions may be realized in the future with thinner $CuInSe_2$ layers.

### 5.2.1D.2 Deposition Rate

The electrodeposition rate of Cu is around 1 $\mu$m/min, and the deposition of the In layer is around 0.6 $\mu$m/min. The layer thicknesses discussed earlier were 2000 Å and 4500 Å, for Cu and In respectively. This results in a total time to electrodeposit the Cu and In layers of about one minute.

Laboratory-scale single-target sputtering rates for Cu and In have been close to 3 $\mu$m/hr for high quality films. However, the deposition rates for the magnetron

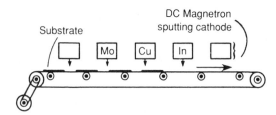

**Figure 5.9:** In-line DC magnetron sputtering targets including Mo, In, and Cu targets as options.

sputtering of metals such as Cu and In can be as high as $1~\mu m/min$ for a $1~m^2/min$ line if multiple in-line magnetrons are used (see Figure 5.9), which would make rates of Cu and In film production approximately the same as for electrodeposition. Although the sputtering rates of some industrial processes can be tens of $\mu m/min$ for single targets, it is unlikely that such rates would produce high quality photovoltaic films since higher rates produce metal layers of a much lower density. In the case of In, high rates could also be difficult to achieve because In melts and spatters easily. However, in comparison to evaporation, sputtering has advantages of control, since sputtering rates are proportional to the power delivered to the source.

Selenization can take about 3.5 hours. Although this is a long time, high throughput is possible because many modules can be annealed at one time. Thus unit production rate is *not* expected to be limited by selenization time.

### 5.2.1D.3 Module Definition Issues

In general the deposition of Cu, In or Se by either electrodeposition, sputtering or evaporation does not create any significant module definition problems. With certain exceptions (see below), the problems associated with the fabrication of CuInSe₂ modules using selenization of Cu and In layers are the same as one would encounter in the fabrication of CuInSe₂ modules regardless of the techniques used to fabricate the absorber layer.

Electrodeposition has particular problems of uniformity of layer thickness with large substrates. The amount of material electrodeposited depends on current flow through each point on the substrate. Variations in conductivity and a cumulative voltage drop across the substrate surface result in inhomogeneous electrodeposited layers. The surfaces that can be electroplated onto are limited in their size and shape by their sheet resistances. However, large conductive substrates can be scribed into long, thin strips, and plated using a fixture that contacts the lengthwise edge of each conductive strip. It is fortunate that the most advantageous cell dimensions for plating are also those which are the most advantageous for module configuration (Figure 5.10). (This is due to the fact that the cells in a module are also limited by their sheet resistances.)

### 5.2.1D.4 Yield/Uniformity

The yield and uniformity of electrodeposited CuInSe₂ cells is uncertain at this time. Careful attention will be required to the Mo contact geometry to minimize problems with current and voltage variations across the cathode. Fabrication in the electroplating industry is not felt to be a strong indicator of the potential yield for

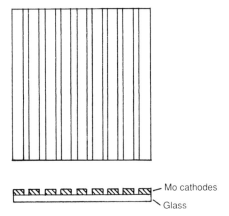

**Figure 5.10:** Electroplated module cathode configuration.

semiconductor-grade films, as the uniformity, quality requirements, and limitations
of the two industries are not directly comparable. However, based on the current
reproducibility and quality of laboratory PV devices, it is felt that the prospects are
good for a high yield and good product uniformity.

Sputtering has been found by the glass-coating industry to give a better yield
and uniformity as the scale of production is increased. In actual PV manufacturing,
several in-line magnetrons can be used to increase uniformity, minimize error, and
allow for maintenance. An overlap of the substrate of from 10%–20% by each
magnetron target field is required in order to compensate for cosine-dominated edge
effects. Advanced designs may yield uniformity of layer thicknesses within 1%, which
is quite adequate for the required stoichiometric control.

The control of the uniformity of large-area thin films of Cu and/or In with
evaporation is much more difficult. Due to the geometry of point or line evaporation
sources and their nonlinear rate with temperature, complex controls are required.
The deposition rate of the In and Cu atoms at any point on the substrate surface is
also effected by the cosine dependence of the flux onto that substrate. The relative
position of the substrate with respect to the In source must be nearly the same as the
relative position of the previous Cu source in order to insure that the correct Cu/In
ratio is approached at each location on the substrate. In large area applications,
multiple point sources of In would be lined up directly behind their corresponding
multiple point sources of Cu to insure that a moving substrate would see almost the
same Cu/In ratio at each location.

### 5.2.1D.5 Capital Equipment

Electrodeposition requires little capital equipment. Most costs are associated
with pumps and heaters. Electrodeposition systems are modular and can be added
to easily with increased production requirements. Their low cost and modular nature
also allow for smaller production facilities to be installed with lower initial capacity.
This allows production capacity to be increased in an incremental fashion in response
to demand.

Sputtering equipment is relatively expensive. But high sputtering rates mean that unit costs can be reasonable. Magnetron sputtering equipment has been proven and refined within several large-scale industries.

DC magnetron sputtering systems are more easily scaled-up than evaporators or RF sputtering systems. The fact that they can operate in the milli-torr range also offers a significant cost advantage over other, higher vacuum deposition techniques such as evaporation. Evaporation equipment is relatively expensive.

### 5.2.1D.6 Labor

Labor is not intensive in the electrodeposition processes for the deposition of Cu and In films. The electroplating industry very easily lends itself to automation, and its process lines are simple and scalable. The labor costs associated with this approach are in the compositional monitoring and control of the electrolytic baths.

Sputter deposition of Cu-In films has labor requirements that are lower than evaporation and are close to electrodeposition. The control of sputtering is very automatable due to its linear deposition rates. Control and monitoring of the working gas pressures and flow rates, and the maintenance and replacement of sputtering targets are the only operator requirements.

Labor requirements for a process line for evaporation are greater than those for a sputtering process. A line feed of the evaporation source would allow a reduction of the labor requirements in the evaporation system. As stated above, evaporation is nonlinear and requires close control, presumably increasing labor costs.

Physical vapor deposition systems lend themselves to compatible automated incorporation with a glass fabrication process line. As-fabricated glass is pristinely clean. If such glass can be directly fed into sputtering or evaporation systems, several cleaning steps could be avoided, thereby reducing labor and other costs.

The only significant labor requirements for any of these processes is associated with the selenization process. Costs are associated with the batch loading and unloading of the Cu-In films from the selenization chamber and the monitoring and control of the $H_2Se$ safety equipment. Even though this step is automatable to some degree, significant concern over $H_2Se$ toxicity is sufficient to require an operator's attention.

### 5.2.1D.7 Power Needs

The power requirements for the electrodeposition of $CuInSe_2$ are low compared to other deposition technologies. The estimated power requirements to electrodeposit the Cu and In metal films may be from half, to an order-of-magnitude, lower than either sputtering or evaporation. This extremely low requirement is primarily due to the high efficiency of the electrodeposition process, the low deposition temperature requirements, and the high material utilization.

Estimates for the power required for both the evaporation and magnetron sputtering of Cu and In films vary significantly. Accurate estimates are not available at this time. The efficiency of the sputtering system (the ratio of the number of sputtered species produced, to the power supplied) is high, resulting in a relatively efficient use of the power required to deposit the films. It is generally assumed that the power requirements will be higher for evaporation than for sputtering processes due to the added power for source heaters, pumps, and chamber and substrate cooling.

The major portion of power requirement for selenization is in the heaters. The pumping and control and/or fabrication of Se-containing gases will also require power, but these are not expected to be significant in comparison to the chamber heating.

In all cases, an in-line glass fabrication capability would also allow the process heat from glass-making to be used in the down-line sputtering, evaporation, and/or selenization steps needed to make $CuInSe_2$ devices.

### 5.2.1D.8 Continuous Processing

Sputtering, evaporation, and electrodeposition are all easily adapted to continuous processing. Sputtering is now more prevalent than electrodeposition in the coating industry, although both are used extensively. These processes have been proven as reliable, scalable and flexible. Many scale-up and continuous processing problems have been solved by the glass coating industry.

### 5.2.1D.9 Safety and Waste Disposal

The safety issues with regard to the large-scale fabrication of $CuInSe_2$ appear to center around those processes which use $H_2Se$. Monitoring, scrubbing, containment, and waste disposal are required for these processes. Safety equipment, and increased technical and administrative labor support, and the required safety programs will effect the cost of energy. However, Fowler (Fowler et al., 1986) has indicated that these costs may be as low as $0.74/m^2$ (under 1 cent/watt). Safety problems can be reduced to some extent by the on-site production of $H_2Se$ or the application of non-$H_2Se$ selenization techniques. However, if $H_2Se$ is used, safety issues associated with $H_2Se$ will be a factor, no matter how small the quantity. Current research has devoted significant attention $H_2Se$ safety. These efforts have resulted in the use of state-of-the-art monitoring equipment and the redundant use of negative pressure processing chambers in conjunction with 1.0M KOH scrubbers.

$H_2Se$ is a flammable, highly toxic poison (Sax and Lewis, 1986). It is very irritating to the skin, eyes and mucous membranes when inhaled. It can cause severe damage to the lungs and liver. The threshold limit value (TLV) for an eight-hour exposure in air is only $0.05$ mg/m$^3$. The storage and transport of $H_2Se$ pose specific safety concerns. Though $H_2Se$ can be handled safely with proper precautions (Fowler et al., 1986), the cost of these safety measures must be added to the cost of energy in the $CuInSe_2$ modules. And although the Fowler estimates of these costs are under $1/m^2$, there is as yet no active demonstration of low cost.

Elemental Se is not as severe a concern as $H_2Se$. However, elemental Se is a poison when inhaled or taken intravenously. When heated, Se compounds decompose and emit toxic fumes. The TLV value for an eight-hour exposure in air is $0.2$ mg/m$^3$. The enclosed nature of deposition systems for sputtering and evaporation will most likely handle safety concerns from this form of Se during deposition. However, due to the possible contamination of the pumping systems, and possible efforts in the recovery of excess Se films on the walls of the deposition chamber, care should be taken in the cleanup procedures. Attention will also be needed in the scrubbing of Se from the exhaust gases and pump oil. It is felt that in its elemental form, Se is not a significant safety concern in fabricating $CuInSe_2$ modules.

$H_2SO_4$, used in electrodeposition, is a poison when inhaled, and is extremely corrosive and toxic to tissue. It is a very serious eye irritant. The TLV for an eight-hour exposure in air is 1 mg/m$^3$. An open vat system such as might be used in electrodeposition may require attention to the $H_2SO_4$ levels in the air. However, $H_2SO_4$ is easily handled safely within industry, and is not a significant safety concern in fabricating $CuInSe_2$ modules.

Elemental Cu is not a significant hazard. However, in the form of Cu dust it is a medical problem and can be a poison if ingested. Caution is urged against inhaling Cu dust or fumes. The TLV for an eight-hour exposure to Cu dust in air is 1 mg/m$^3$. The enclosed nature of deposition systems for electrodeposition, sputtering, and evaporation systems will most likely handle all safety concerns from metallic Cu during deposition. It is felt that in its elemental form, Cu is not a safety concern in fabricating $CuInSe_2$ modules.

Elemental In is also of little concern. Its safety hazard is mainly due to the fact that it can be highly toxic when introduced intraperitoneal or intravenously. It is also a moderate irritant when ingested. The TLV for an eight-hour exposure in air is 0.1 mg/m$^3$. The enclosed nature of deposition systems for sputtering and evaporation obviate most safety concerns from In during deposition. However, during clean-up and recovery activities of the deposition chamber, some care should be taken.

### 5.2.1E  Discussion of R&D Issues

Areas which remain a concern for the selenization processes are: back contact adhesion, In availability and cost, safety associated with large volume $H_2Se$ processes, and $CuInSe_2$ uniformity.

Most organizations with selenization efforts in the fabrication of $CuInSe_2$ have reported some adhesion problems. It is assumed that the root-cause of this is the poor mechanical locking between the unselenized Cu and the Mo layers and the three-fold expansion of film thickness during Se incorporation. Some possible approaches to reduce stress and peeling are the investigation of alternative metal back contacts and ion bombardment of Mo films during deposition. Further research is required to investigate these or alternate Mo deposition techniques to insure the required relief in the film stresses.

Possible issues with the availability and cost of In need to be addressed. As the demand for In rises with increased $CuInSe_2$ production, costs may fluctuate to levels that adversely affect the cost-of-energy. This is considered a problem of matching demand and capacity and should be minimized if In producers can plan for the increased production of $CuInSe_2$. In the long-term, it may be important to reduce the $CuInSe_2$ layer thickness to below 1 $\mu$m to put less pressure on In supplies for multi-gigawatt annual production (Zweibel et al., 1986).

The use of large volumes of $H_2Se$ for selenization has particular safety hazards and some costs associated with it. Monitoring, scrubbing, containment, and waste disposal are all significant safety issues. These can be ameliorated to some degree by the on-site production of $H_2Se$. If $H_2Se$ of sufficient purity could be produced on-site, it would minimize the magnitude of the safety risk and associated costs. However, serious consideration of $H_2Se$ safety will still be required. Alternative methods for selenizing Cu and In films should be investigated to develop second-generation manufacturing approaches that have no $H_2Se$.

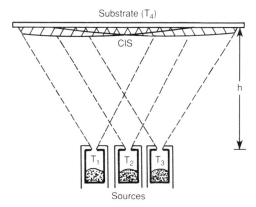

**Figure 5.11:** Schematic of the three source evaporation system. The Cu, In, and Se sources are held at a fixed distance from the substrate (H) and at different temperatures $T_1$, $T_2$, and $T_3$ respectively with the substrate held at temperature $T_4$.

The uniformity and stoichiometry of CuInSe₂ also remain issues for the production of large area thin films of CuInSe₂. The investigation of cathode geometry and cell definition for electrodeposition; source configuration and defined flux rates for PVD; and target geometry and quantity for sputtering are important in addressing these problems. A 90% yield on CuInSe₂ modules of over 10% efficiency will most likely be needed for large-scale fabrication of CuInSe₂ systems. This puts a demand on process design and assumes continuing emphasis in this R&D area.

The kinetics involved in the selenization process are not well understood. To reduce back contact problems, and to refine the selenization process, further research is needed in this area. In addition, the device performance of CuInSe₂ cells is not well understood. More research needs to be done to provide models that can relate materials properties with device performance. When these are available, knowledge of process kinetics will be invaluable in improving device performance.

### 5.2.1F  Physical Vapor Deposition of CuInSe₂ Films

Physical vapor deposition (PVD) processes are those which involve the atom-by-atom transfer of Cu, In, and Se species to the substrate for the formation of single phase CuInSe₂. There are three basic PVD processes: evaporation, ion plating, and sputtering. The formation of CuInSe₂ may involve one or a combination of these three PVD processes. Evaporation and sputtering have had reported successes in the fabrication of near-10% CuInSe₂ devices.

In the evaporation process (Figure 5.11), Cu, In, and Se vapors are produced by either direct resistance heating, radiation, electron beams, laser beams, or arc discharge of three separate sources. The sources are arranged so that the constituent elements arrive at the substrate in the appropriate proportion to form a composite thin film with the correct CuInSe₂ stoichiometry.

In sputtering processes (Figure 5.12), atomic Cu and In species are produced through their kinetic ejection from sputtering targets. This ejection is a result of bombardment of the target surface by energetic, non-reactive ions, usually $Ar^+$. Ejection is caused by momentum transfer between the impinging ions and the atoms

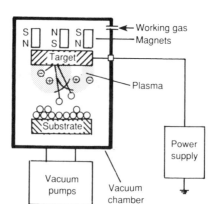

**Figure 5.12:** Schematic of a magnetron sputtering system. Ar$^+$ ions are ejecting the target species which condense on the adjacent substrate.

of the target surface. The material is ejected from targets of either Cu, In, CuInSe$_2$, In$_2$Se$_3$, or Cu$_2$Se by RF-magnetron or DC-magnetron sputtering. A reactive gas such as H$_2$Se can be introduced as the source of the Se.

Reactive sputtering involves the sputtering of Cu and In in an atmosphere of H$_2$Se. The process is similar to that of magnetron sputtering except the Se source is now provided by H$_2$Se. The Cu and In atoms ejected from the targets react with the H$_2$Se gas to form In$_2$Se$_3$ and Cu$_2$Se at, or near, the substrate surface. These react on the surface during deposition to form CuInSe$_2$.

An additional approach which shows promise is a hybrid process combining evaporation and sputtering. Cu and In are sputtered from magnetron targets and Se is thermally evaporated simultaneously to form CuInSe$_2$ on the surface of a Mo-coated glass substrate. This process requires close attention to source positioning and is carried out at around $10^{-5}$ torr. Quality CuInSe$_2$ films have been fabricated (Rockett et al., 1988).

### 5.2.1G History and Present Status

Boeing (with DOE and SERI funding) began work on polycrystalline thin film CuInSe$_2$ solar cells in 1978 (Chen and Mickelsen, 1980, Mickelsen and Chen, 1980). The concept of using a low-resistivity Cu-rich CuInSe$_2$ base layer (0.01 ohm-cm) followed by a high resistivity In-rich top layer ($10^3$ ohm-cm) for the fabrication of efficient CuInSe$_2$ cells was first reported by Boeing following their earlier work (Figure 5.13). These efforts resulted in a patent (Mickelsen and Chen, *Methods for Forming Thin-Film Heterojunction Solar Cells from I − III − VI$_2$ Chalcopyrite Compounds, and Solar Cells Produced Thereby*, U.S. Patent 4,335,266) in June 1982.

The Boeing process used a heterojunction structure based upon a two-layer $n$-type CdS film and a two-layer $p$-type CuInSe$_2$ film. The CuInSe$_2$ layers were prepared by simultaneous elemental evaporation (Figure 5.14). The stoichiometry was modified to form a Cu-rich base layer and an In-rich top layer. The CdS was also deposited in two layers. Both layers were evaporated from CdS powder, and elemental In was used as a dopant. The higher doping level of the top layer of CdS distinguished it from the undoped bottom layer. Early results included

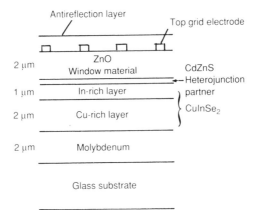

**Figure 5.13:** Schematic of a two-layer CuInSe₂ cell.

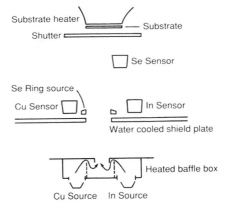

**Figure 5.14:** Schematic of the Boeing evaporation system.

a world record efficiency of 9.4% under simulated AM 1 illumination. This work also identified an important improvement in cell efficiency associated with a post-deposition heat treatment in oxygen (Mickelsen and Chen, 1981). By 1982, Boeing had reported another world record device efficiency of 10.6% with photocurrents in excess of 35 mA/cm$^2$ under simulated AM 1 illumination (direct spectrum). These devices demonstrated $V_{oc} = 400$ mV with high quantum yields ($> 0.8$) in the spectral range from 600 nm to 900 nm. These advances were in part due to Boeing's use of a mixed $Zn_x Cd_{1-x}S$ layer (Mickelsen and Chen, 1982, Mickelsen et al., 1984). The CuInSe₂ cells were also subjected to stability testing in an open laboratory environment at elevated temperatures for 9300 hours with no observed degradation. This was the first significant evidence of the stability of CuInSe₂ for terrestrial solar cell applications.

Boeing has continued to improve the compositional and structural uniformities of their CuInSe₂ films. Alloying of the films with Ga has resulted in a record cell efficiency of 12.5% (total area 0.99 cm$^2$) with $V_{oc} = 555$ mV, $J_{sc} = 34.2$ mA/cm$^2$, and FF $= 65.7\%$ (Figure 5.15). Their research also led (1986) to a 9.6% efficiency for a 91 cm$^2$ monolithically interconnected 4-cell CuInSe₂ submodule with a

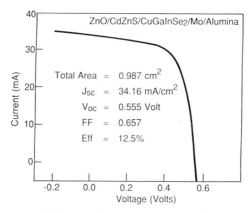

**Figure 5.15:** I-V curve for the Boeing 1 cm$^2$ CuInSe$_2$ cell.

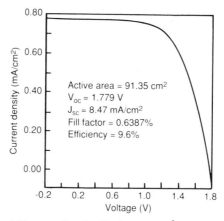

**Figure 5.16:** I-V curve for the Boeing 91 cm$^2$ CuInSe$_2$ submodule.

$V_{oc}$ = 1.78 mV, $J_{sc}$ = 8.47 mA/cm$^2$, and FF = 63.9%, shown in Figures 5.16 and 5.17, respectively.

Based on Boeing results, SERI funded the Institute of Energy Conversion (IEC) at the University of Delaware to begin work on the evaporation of CuInSe$_2$ in 1982. Their process used Kundsen effusion cells for three-source elemental evaporation to form CuInSe$_2$ (Figure 5.18). By 1984, IEC had achieved an 8% efficiency for CuInSe$_2$/CdS cells on both Pt and Mo back contacts (Birkmire et al., 1984 a,b). Small area (0.09 cm$^2$) CuInSe$_2$ device efficiencies subsequently reached a reported 12.3% (active area, under direct ELH 87.5 mW/cm$^2$ illumination) with $V_{oc}$ = 430 mV, $J_{sc}$ = 37.0 mA/cm$^2$, and FF = 67.5% (Birkmire et al., 1985, Hegedes et al., 1986, Jackson et al., 1987, Rocheleau et al., 1987). A SERI-verified 1 cm$^2$ device reached an efficiency of 10.1% (total area) with $V_{oc}$ = 445 mV, $J_{sc}$ = 34.9 mA/cm$^2$, and FF = 64.6% (Figure 5.19).

The deposition of CuInSe$_2$ thin films by co-sputtering from Cu and In planar magnetron sources in an Ar + H$_2$Se working gas is being investigated by the University of Illinois (Figure 5.20). This work was originated by John Thornton with SERI

**Figure 5.17:** Picture of the Boeing 91 cm$^2$ CuInSe₂ submodule.

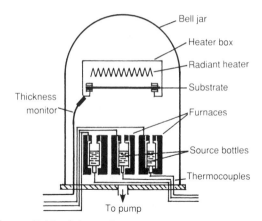

**Figure 5.18:** Schematic of the IEC evaporation system.

support in June 1982 at Telic/Dart & Kraft of Santa Monica CA (Thornton et al., 1983). In October 1983, Thornton moved to the Coordinated Sciences Laboratory of the University of Illinois, where SERI-supported R& D was continued (Thornton et al., 1984 a,b, Thornton and Cornog, 1984). Near-stoichiometric coatings deposited on glass and Mo-coated substrates by this process have been found to have resistivities, Hall mobilities, absorption coefficients, bandgaps, surface topographies, grain sizes and oxygen heat treatment behavior comparable to films produced by three-source evaporation methods (Thornton et al., 1986 a,b, Thornton and Lommasson, 1986, Lommasson et al., 1986). However, it has not been possible by reactive sputtering to deposit In-rich films (In/Cu about 1.5) of the type commonly used as a top

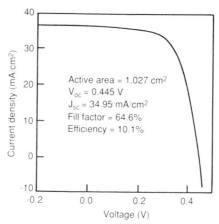

**Figure 5.19:** I-V curve for the IEC 1 cm² CuInSe₂ cell.

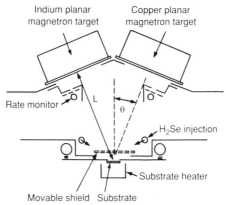

**Figure 5.20:** Schematic of a reactive sputtering system. Targets are carefully arranged with respect to distance ($L$) and angle ($\theta$) to the target.

layer in the preparation of two-layer device quality films by the evaporation method. Apparently an In rejection mechanism that occurs during the deposition process in $H_2Se$ reactive sputtering has made it difficult to achieve In compositions beyond the amount that can be incorporated into the CuInSe₂ chalcopyrite phase. Inability to make "Boeing-like" two-layer CuInSe₂ films has limited the cell efficiencies to about 6% (Thornton, 1987, Thornton et al., 1987, Rockett et al., 1988).

SERI initiated internal efforts in the evaporation of CuInSe₂ in 1982. Our efforts were patterned after the Boeing two-layer structure (Noufi et al., 1984). The basic device consisted of two CuInSe₂ layers, varying in stoichiometry, and two layers of CdS. SERI has investigated the deposition of both $n$- and $p$-type material with mobilities up to 150 cm²/V-s and carrier concentrations ranging from $10^{13}$–$10^{20}$ cm$^{-3}$. Reports on the investigation of these two layer films (Noufi and Dick, 1985; Noufi and Powell et al., 1985; Noufi et al., 1985) indicated a $p$-type, highly resistive, CuInSe₂ bilayer in which the two individual layers were almost completely mixed.

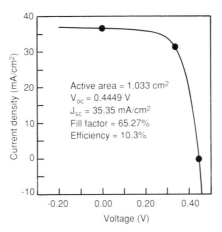

**Figure 5.21:** I-V curve for the SERI 1 cm² CuInSe₂ cell.

But the films had a surface layer of about 0.2–0.4 $\mu$m which was semi-insulating and Cu poor. SERI also investigated the effect of oxidation/reduction heat-treatments on the defect chemistry and efficiency of CdS/CuInSe₂ cells. Recent SERI results (Noufi et al., 1987) have shown that the off-stoichiometric films, both Cu-poor and Se-poor, are rich in native defects such as $V_{Cu}$ (acceptor), $In_{Cu}$ and $In_i$ (donors) and $V_{Se}$ (donor) and were highly compensated. During air annealing, oxygen has been shown to interact with these donor defects, reducing their density and making the material more $p$-type. In certain intrinsic or slightly $n$-type films, the interaction with oxygen was shown to convert the conductivity to $p$-type. Oxygen was felt to be either strongly chemisorbed, or forming coordinated bonds with $In_{Cu}$ on the grain surfaces. SERI's efforts in the fabrication of CuInSe₂ cells (Figure 5.21) have resulted in a 1 cm² cell efficiency of 10.3% (direct AM 1.5) with $V_{oc} = 445$ mV, $J_{sc} = 35.4$ mA/cm², and FF = 65.3%.

The Instituto Militar de Engenharia in Brazil reported compositional and structural studies of CuInSe₂ thin films in 1984. The films were deposited by co-evaporation of the constituent elements in high vacuum in 1984 (Dhere et al.). The morphology, composition and crystallographic structure of films grown at a substrate temperature of 350°C and with thicknesses in the range 0.15–1.0 $\mu$m were studied. Characterization included Auger electron spectroscopy, x-ray and electron diffraction, and optical absorption spectroscopy. Stoichiometric CuInSe₂ thin films were deposited with the chalcopyrite structure and with grain sizes in the range 0.2–0.6 $\mu$m. Band gaps were near 1.02–1.04 eV.

The Royal Signal and Radar Establishment in Great Britain first reported research in the vacuum evaporation of CuInSe₂ thin films in 1979. The electrical and optical properties of these thin films were investigated. Films were chalcopyrite if the substrate temperature was kept below 250°C (Fray and Lloyd, 1979).

The growth of coevaporated CuInSe₂ by the Institute of Microwave Technology in Sweden was first reported by Stolt in 1985 (Stolt et al., 1985a). Their substrate temperatures range from 350° to 450°C in studies of the composition of films made with different deposition conditions. Experiments with double-layered CuInSe₂ and Cu diffusion into CuInSe$_2$ were also reported. Evaporation from three resistively heated elemental sources was controlled by a quadrupole mass spectrometer (Stolt et al., 1985b). Indication of Cu-segregation to the surface by Cu-rich selenides was found (Niemi and Stolt, 1987).

Varela reported research on the Cu, In and Se co-evaporation for the deposition of CuInSe$_2$ thin films by the University of Barcelona (Spain) in 1985. These films were deposited within a wide composition range around stoichiometry. The optical band gaps were reported to be dependent upon the Cu-In ratio and showed a minimum at stoichiometry. These dependencies were correlated with the variations of the crystalline lattice constant with the Cu-In ratio. A bandgap of around 1.0 eV was also found to vary with the Cu-In ratio. A bandgap of around 1.0 eV was also found to vary with the Cu-In ratio (Varela et al., 1985, 1986).

The University of Stuttgart (F.R.G.) first reported the preparation of Cu(Ga$_y$In$_{1-y}$)Se$_2$ films over a range of $y$ from 0 to 1 by evaporation of the single elements in 1987. A nearly linear variation of the bandgap from 1.04 eV (CuInSe$_2$) to 1.68 eV (CuGaSe$_2$) was obtained. Heterojunctions with (Zn, Cd)S as the window material and having different Zn/Cd contents yielded efficiencies up to 5.8% for CuGaSe$_2$, 9.3% for CuInSe$_2$ and 3% for CuGa$_{0.56}$In$_{0.44}$Se$_2$, respectively (Dimmler et al. 1988, Klenk et al., 1988).

Newcastle Polytechnic (Great Britain) first reported on the two-source evaporation of thin films of $p$-type CuInSe$_2$ in 1981. Don et al. (1981) said that the films were generally deficient in Cu and always stratified in composition. The stratification was identified as being responsible for poor reproducibility found in their CdS/CuInSe$_2$ cells. The coevaporation of Cu$_3$Se$_2$-InSe and Cu$_3$Se$_2$-In$_2$Se$_3$ in a hot-wall system included digital rate monitoring. Homogeneous, single phase films of CuInSe$_2$ were formed by this process using Cu$_3$Se$_2$ and In$_2$Se$_3$, with films 0.1–1.0 $\mu$m thick (Don et al., 1981). Films were also made by three-source evaporation of the elements onto a glass substrate in 1985. The crystal structure was reported as chalcopyrite for substrate temperatures above 400°C. The optical properties of the films depended on the Cu/In ratio, but usually showed a direct gap of 1.02 eV (Don et al., 1985). The crystal structure of the films was studied by X-ray and electron diffraction. It was found that the chalcopyrite phase occurred with a substrate temperature of at least 400°C for Cu-rich films and 450°C for In-rich ones. An additional phase of cubic structure with a lattice parameter of 5.73 Å was observed in very Cu-rich layers and was identified as Cu$_{2-x}$Se (Don et al., 1986).

### 5.2.1H Cell Structures

All groups fabricating CuInSe$_2$ solar cells by PVD use a structure that is closely related to the one shown previously (Fig. 5.13). Devices consist of a conductive back-contact deposited on an inexpensive substrate such as glass. The metal of choice for the back contact is Mo, although many research organizations are investigating alternatives. An approximately 2 $\mu$m thick film of Mo is deposited on the glass substrate. The CuInSe$_2$ layer is deposited on the Mo, usually following the two-layer Boeing approach. This step is followed by the deposition of an $n$-type heterojunction partner such as CdZnS, and $n$-type window such as ZnO, and grids and possible antireflection coatings.

### 5.2.1I Process Description

#### 5.2.1I.1 Evaporation

The three source evaporation of Cu, In and Se to form $CuInSe_2$ is carried out at around $10^{-5}$ to $10^{-6}$ torr to minimize collisions in the line-of-sight transfer of the vaporized species to the substrate. Each species is evaporated at specific temperatures, and the flux is directed toward a metal-coated substrate in a specific proportion. The individual atoms condense on the substrate at the required rates to form a polycrystalline $CuInSe_2$ thin film. The composition of each portion of the $CuInSe_2$ film is determined by the substrate temperature and the temperature of each of the constituent sources. The uniformity of each of the constituents across the substrate is controlled by the distance from the substrate, geometry of the sources, and the uniformity of the temperature across the substrate.

The most commonly used substrate for evaporation is Mo-coated glass, although some research has been carried out on Mo-coated alumina, and Ti contact layers have been used. The Mo layer forms a contact to the deposited $CuInSe_2$ with a resistance of around $5 \times 10^{-5}$ ohm-cm$^2$. Adhesion problems at the Mo/glass or Mo/$CuInSe_2$ interfaces have not been significant in PVD processes.

The evaporation of all three species is carried out with 99.999% purity materials. The substrate and In source temperatures are varied to adjust the stoichiometry of the $CuInSe_2$ film. The deposition rate is about 3.6 $\mu$m/hr. The Cu, In, and Se sources may be open boats or Knudsen effusion cells. The material utilization of a Knudsen source is much higher.

The stoichiometry of the $CuInSe_2$ layers is controlled by the variation in the In source and the substrate temperatures. The first (Cu-rich) layer is deposited with the substrate temperature at 300°C and the In source temperature at 1090°C. The second (In-rich) layer is deposited with the substrate temperature at 450°C and the In source temperature raised to around 1, 150°C, which increases the proportion of In impinging on the substrate. Measured separately, the first (Cu-rich) layer is highly conductive. The second (In-rich) layer is highly resistive or $n$-type. The blended $CuInSe_2$ film achieved by thermal diffusion when the two layers are deposited sequentially has a bandgap near 1.0 eV and a resistivity of about $10^3$ ohm-cm.

The evaporation rates of both the Cu, In, and Se are related to the temperature of the evaporation sources in a non-linear manner. Therefore, the control of the deposition rate, stoichiometry, and uniformity must be constantly monitored to produce quality $CuInSe_2$ films.

#### 5.2.1I.2 Reactive Sputtering

In reactive sputtering Cu and In are simultaneously magnetron sputtered in an atmosphere of Ar and 15% $H_2Se$ at $10^{-5}$ torr to maximize the line-of-sight transfer of the atoms to the substrate. The sputtered atoms, which are mostly neutral, react on the substrate to form single-phase $CuInSe_2$. The composition of the $CuInSe_2$ film is determined by the Cu and In sputtering rates, the $H_2Se$ temperature and pressure, and the substrate temperature. The substrate is typically a Mo film deposited on glass. The sputtering targets must be arranged with care. The atoms combine on the surface of the Mo layer to form thin films of $Cu_2Se$, $In_2Se_3$, and/or $CuInSe_2$. Although the sputtering of $CuInSe_2$ has been carried out with both DC-magnetron and RF-magnetron sputtering processes, there has not, as yet, been a near-10% $CuInSe_2$ device from the sputtering of Cu, In, Se, $CuInSe_2$, $In_2Se_3$, or $Cu_2Se$.

The Cu is sputtered from a 99.99%-purity target. If the target is supplied with 1 amp at 1 kV, the deposition rate is 1.0 $\mu m/min$. The substrate is held at the ambient temperature. At this deposition rate, the target temperature is held around 200°C with standard water cooling.

The In target is also sputtered from a target of 99.99%-purity. If the target is supplied with an equivalent 1 amp at 1 kV, the In deposition rate is 1.0 $\mu m/min$. Care must be taken to insure that the target temperature is kept well below 150°C so that the In does not melt. The indium temperature and target geometry are controlling factors in the deposition rate for this target. The substrate temperature is kept at ambient.

The Cu and In sputtered from the magnetron targets react with the $H_2Se$ gas to form $In_2Se_3$ and $Cu_2Se$ at, or near, the substrate surface. The $In_2Se_3$ and $Cu_2Se$ in turn react during the deposition process to form $CuInSe_2$.

During the sputtering process, poisoned surface layers consisting of selenide material sputtered from other surfaces in the system are formed on both targets. The surface layers on the In target are thinner than those on the Cu target. The sputtering fluxes from the In target are, therefore, more responsive to changes in the discharge current and less affected by the target poisoning. Accordingly, composition control is achieved by adjusting the current to the In source. The Cu source is operated with constant current. The $H_2Se$ injection rate is held constant and is selected to provide an excess of Se at the substrate surface.

### 5.2.1I.3 Annealing

Most processes for the PVD of $CuInSe_2$ use a post-deposition heat treatment in air to optimize performance. The heat treatment varies from 15–30 minutes at 200°–220°C, to over eight hours at 200°C. The heat treatment can also be applied before or after subsequent cell fabrication steps such as the deposition of a window layer. For a review, see Zweibel (1985).

### 5.2.1J Problems and R&D Issues

### 5.2.1J.1 Materials

The materials required for the evaporation of $CuInSe_2$ films, namely 99.999% purity Cu, In, and Se source materials, are available in the quantity required for small production facilities. The 99.999% pure Cu and In targets required in magnetron sputtering of $CuInSe_2$ films are readily available in small area targets, and few problems are expected in the scale up of these targets to any size required by plant operation. The glass coating industry routinely uses large area targets such as these for architectural glass plating.

No studies have been done to establish if existing purities are absolutely necessary for either evaporation or sputtering, or to what level this laboratory quality restriction can be reduced. Cost reductions would occur from loosening the material purity requirements. Although the Cu and In products required by either reactive sputtering or evaporation techniques are all available, it is a cost disadvantage for sputtering that the source material must be fabricated into targets.

H₂Se is an extremely toxic gas. For safety, it would be best manufactured on-site to reduce the stored quantity that a facility would have on hand.

It is estimated that Cu and In material utilization will be from 70%–80% for reactive sputtering and evaporation. An aggressive recovery effort for the waste Cu and In from the sputtering or evaporation equipment would be required to raise utilization further. The cost of recovery programs is not known. In the case of sputtering targets, full recovery of In does not avoid the cost of repeated fabrication of targets.

A 2 $\mu$m thick film of CuInSe₂ would require 1.87 g/m² of Cu, 3.45 g/m² of In, and 4.65 g/m² of Se. Let us assume an optimistic utilization rate of 75% for both sputtering and evaporation. (A utilization rate of 45% is assumed for H₂Se in the reactive sputtering process.) If the price is \$300/kg for 99.99% In, \$2/kg for 99.99% Cu, and \$17/kg for 99.99% Se, the cost of the absorber material (excluding source/target preparation) for evaporation would be \$1.51/m² and for reactive sputtering would be \$1.58/m².

These estimates are based on the currently required thickness of 2.0 $\mu$m in CuInSe₂ cells produced. Cost reductions would be realized with thinner CuInSe₂ layers. More importantly, thinner devices would reduce the potential sensitivity of CuInSe₂ to market fluctuations in In prices and availability.

### 5.2.1J.2 Deposition Rate

Deposition rates for three-source evaporation of CuInSe₂ are nominally 0.06 $\mu$m/min. These rates are very slow when compared with other CuInSe₂ deposition processes. This is a significant cost-driver for the evaporation process.

Although laboratory reactive sputtering rates for Cu and In targets are in the range of 1–3 $\mu$m/hr, line-production rates may be as high as 1 $\mu$m/min as discussed earlier. In general, sputtering rates can be another 10 times higher than this, but it is uncertain what effect such rates would have on the film quality since higher rates have been known to produce porous metal films. In the case of In, high rates are difficult to achieve because the In targets melt at high currents/temperatures. But even with the lower rates, multiple in-line sputtering targets allow the metals to be deposited at high rates and with good process control. Compared to evaporation, sputtering has advantages of control, since the sputtering rates of the species are proportional to the power delivered to the source rather than being an exponential of the source temperature.

### 5.2.1J.3 Yield/Uniformity

The large area yield and uniformity of evaporated CuInSe₂ cells is uncertain. Generally it is considered more difficult than for sputtering. In evaporation systems, open boat evaporators with feedback control from flux monitors, and calibrated orifice sources with feed-forward control, are used to provide Cu and In fluxes. Unfortunately, these "point" sources produce non-uniform fluxes. Control problems for coating large areas are exacerbated by the relatively high temperatures needed to evaporate Cu.

Due to the inhomogeneous fluxes emerging from point or line evaporation sources and their nonlinear deposition rates with temperature, complex controls are required. There is not sufficient operating experience to establish production values for yield and uniformity. But based on the uniformity of Boeing test devices (9.6%

for 91 cm$^2$; see above), it is felt that the prospects are good for a moderately-high yield and reasonable uniformity for large substrates.

Sputtering has been found by the glass-coating industry to give a better yield and uniformity as the scale of production is increased. Reactive magnetron sputtering systems for the deposition of CuInSe$_2$ on Mo-coated glass can be designed to be very large. In actual PV manufacturing, several sets of in-line magnetrons (see Figure 5.20) could be used to increase uniformity, minimize error, and allow for maintenance. An overlap of the substrate by the magnetron target field from 10%–20% is required in order to insure uniformity of film deposition. This reduces edge effects. With 2m × 5m substrates, uniformity of layer thicknesses can be within 1%; the question will be cost tradeoffs to achieve this.

### 5.2.1J.4 Capital Equipment and Continuous Processing

Evaporation and sputtering equipment is relatively expensive. But the potentially higher deposition rates for reactive sputtering insure that unit costs are reasonable. In general, sputtering and evaporation equipment have been proven and refined by several industries. The magnetron sputtering systems can be fabricated in virtually any configuration, but industrial participation in developing these process lines has not yet emerged.

D.C. magnetron sputtering systems are more easily scaled up than evaporators or RF sputtering systems. The fact that they can operate effectively in the milli-torr range also offers a significant cost advantage over other, higher vacuum deposition techniques.

### 5.2.1J.5 Power Needs

Evaporation is the PVD process with the highest power requirements. Evaporation source heaters as well as chamber and substrate temperature control require significant power.

The power required for the sputtering of CuInSe$_2$ is less than for evaporation but still significant. The efficiency of the sputtering system (ratio of the number of sputtered species produced, to the power supplied) is high, resulting in an efficient use of the power required to deposit the films.

### 5.2.1J.6 Safety and Waste Disposal

Safety issues for large-scale PVD of CuInSe$_2$ do not appear to be a major concern. However, reactive sputtering using H$_2$Se suffers from the same issues as the selenization processes discussed in earlier sections. Monitoring, scrubbing, containment, and waste disposal for these processes will be essential. The cost of safety equipment, and increased technical and administrative labor support, and the required safety programs will effect the cost of energy for reactive sputtering. However, a study by Brookhaven (Fowler et al., 1986) has indicated that these costs may be as low as $0.74/m$^2$. Problems can be reduced to some extent by the on-site production of H$_2$Se or the application of other (non-H$_2$Se) selenization techniques. However, if H$_2$Se is used, safety issues will be a factor, no matter how small the quantity (see Section 5.2.1D.9, selenization safety).

### 5.2.1K  Discussion of R&D Issues

Areas which remain of concern for PVD processes are: the cost of In, the cost of equipment, deposition rates, process control and $CuInSe_2$ uniformity, and safety associated with large volumes of $H_2Se$ in reactive processes.

The possible adverse affects of market/capacity-induced fluctuations in the cost of In have been previously discussed (Section 5.2.1D.1).

Achievable uniformity and stoichiometry of $CuInSe_2$ remain issues for the PVD production of large area thin films of $CuInSe_2$. The investigation of material purity requirements, source configuration, and defined flux rates for evaporation; and material purity requirements and target geometry for sputtering are important in addressing these problems.

The three-source evaporation of $CuInSe_2$ has very low deposition rates. Continuous, complex control of evaporated fluxes is required to produce quality $CuInSe_2$. In addition, evaporators are not readily scaled up, and significant capital costs may be incurred to achieve production capacity. These are areas of significant concern for evaporation systems for depositing $CuInSe_2$.

### 5.2.2  CuInSe₂: Other Deposition Processes

Numerous deposition methods that have not yet been fully investigated show promise. Others are of interest in terms of how they yield insights into $CuInSe_2$ materials properties.

### 5.2.2A  Electrodeposition

$CuInSe_2$ cells have been fabricated in several ways using electrodeposition for the deposition of the absorber layer. Electrodeposition has been made using: i) a single electrolyte bath to form a Cu-In-Se alloy, which is heat treated in an inert atmosphere, ii) one electrolyte bath to form a Cu-In alloy, which is then heat treated in an atmosphere of $H_2Se$, iii) two separate electrolyte baths to first deposit a Cu layer and then an In layer, which are then heat treated in an atmosphere of $H_2Se$ (discussed in a previous section), iv) three baths to deposit layers (Cu-In-Se), which are then heat treated in an inert atmosphere, v) two separate electrolyte baths to first deposit a $Cu_2Se$ layer and then an $In_2Se_3$ layer which are then heat treated in an inert atmosphere, or vi) two separate electrolyte baths to first deposit a Cu-In layer and then a Se layer, which are then heat treated in an inert atmosphere.

The first reported fabrication of $CuInSe_2$ by electrodeposition, by Weizmann Institute in Israel (Bhattacharya, 1983), describes the deposition of $CuInSe_2$ thin films by electroplating. ARCO Solar began their first efforts to electrodeposit Cu and In layers in the fall of 1980. This effort resulted in a patent (Kapur et al., Patent 4581108, 1986). Weizmann Institute continued efforts in the electrodeposition of both Cu-In and Cu-In-Se films from a single bath until 1986. During 1984, electrodeposition work was also reported by Southern Methodist University. They deposited Cu and In films from two separate electrolytes and selenized both of the Cu-In layers together and individually between deposition steps.

Since 1984, electrodeposition research has been carried out by International Solar Electric Technology (Kapur et al., 1985), Banaras Hindu University, India (Singh et al., 1984), Central Electrochemical Research Institute, India (Lokhande,

1987 a,b), McGill University, Canada (Qiu and Shih 1987, Shih and Qiu 1987), and the Solar Energy Research Institute (Pern 1987 a,b,c).

The researchers who have worked on the various processes listed above are as follows:

i) One electrolyte Cu-In-Se alloy, heat treated in an inert atmosphere, Weizmann Institute (Israel), Banaras Hindu University (India), Central Electrochemical Research Institute (India), McGill University (Canada), SERI

ii) One electrolyte Cu-In alloy, heat treated in $H_2Se$, Weizmann Institute (Israel)

iii) Two electrolytes to deposit a Cu layer and an In layer, heat treated in $H_2Se$, International Solar Electric Technology, Southern Methodist University, ARCO Solar, and Weizmann Institute (Israel)

iv) Three layers (Cu-In-Se) from separate baths, heat treated in inert atmosphere, International Solar Electric Technology

v) Two electrolytes $Cu_2Se$ then $In_2Se_3$ layers, heat treated in inert atmosphere, Weizmann Institute (Israel), SERI, and University of Texas at Arlington

vi) Two separate electrolyte baths to first deposit a Cu-In layer and then a Se layer, heat treated in inert atmosphere, Weizmann Institute (Israel)

### 5.2.2A.1 Present Status

i) The electroplating of $CuInSe_2$ from a single bath has been investigated by at least five organizations with only limited success. This approach usually involves the use of one electrolyte bath to deposit a Cu-In-Se alloy which is heat treated in an inert atmosphere.

Weizmann Institute (Israel) - In 1983 Bhattacharya reported the first progress on the electrodeposition of $CuInSe_2$ thin films. The process included the electrodeposition of $CuInSe_2$ from a single electrolyte bath of Cu ions, In ions, and $SeO_2$ at room temperature. The films were then heat treated at 600°C in an argon atmosphere. Subsequent work on these films was also reported by Bhattacharya (1986) at SERI, and University of Texas at Arlington. In 1986 Hodes reported this research results in the plating of Cu-In-Se alloys from two solutions of Cu and In ions (Hodes 1985, 1986). One bath used $SeO_2$ as its Se source and the other used $Na_2(K_2)SeSO_3$. The Weizmann deposition of $CuInSe_2$ directly from a ternary bath met with only limited success and was eventually discontinued.

Banaras Hindu University (India) - Singh first reported on this research in 1984 (Singh et al.) Their efforts have focused on the deposition and investigation of $n$-type $CuInSe_2$ layers from Cu, In, and Se ions at low current. The photoelectrochemical cells initially had a $V_{oc}$ of only 140 mV and a $J_{sc}$ of only 0.5 mA/cm² at 64 mW/cm² illumination. In 1986 they reported on $n$-type $CuInSe_2/Ag$ Schottky-barriers (Singh et al., 1986 a,b). Deposition of the $CuInSe_2$ was carried out on Ti substrate from an electrolyte of $CuCl_2$, $InCl_3$, and $SeO_2$ in a 6:5:10 ratio at 22°C. Optical absorption indicated a bandgap of 1.08 eV and a $10^3$ ohm-cm resistivity.

Central Electrochemical Research Institute (India) - In 1987, Lokhande reported on pulse electroplating of $CuInSe_2$ films onto Mo and Ti substrates from an acidic electrolyte bath of $CuCl_2$, $InCl_3$, and $SeO_2$ (Lokhande, 1987 a,b). The best films from these baths were deposited at 840 mV at 25°C with 5-msec-on and 10-msec-off pulses on the Ti substrate. The films were observed to have a polycrystalline nature, but their morphology was found to be different for different Cu/In ratios. Crystallinity was found to improve with a 60 min., 300°C heat treatment in an argon atmosphere. However, significant losses of Se increased as the heat treatments rose above 250°C.

McGill University (Canada) - In 1987, Qiu reported on the heat treatments of electrodeposited Cu-In-Se films deposited from a single electrolytic bath. These films were heat treated at near 450°C in an evacuated ($10^{-4}$ torr) tube for 20–30 min. CuInSe₂ cells were fabricated by depositing low resistivity ($10^{-3}$ ohm-cm) CdS on to the films. The morphology of the films was improved by the heat treatment, but Se was lost during the heat treatment. It was also concluded that the Cu/In ratio of the films could be controlled by the electrodeposition process (Qiu and Shih 1987, Shih and Qiu 1987). AM 1 efficiencies of their devices were around 3%, limited by low voltages (near 200 mV).

SERI - Pern first reported in 1987 on research for the deposition of CuInSe₂ on a Mo-coated alumina substrate from a single electrolyte bath (Pern et al., 1987 a,b,c). The process bath consisted of $CuSO_4$, $H_2SeO_3$, and indium sulfamate in an aqueous solution of ethylenediamine dihydrochloride at a pH of 1.70. The deposition of the Cu-In-Se films was followed by a heat treatment at 200°–250°C for 30 minutes in a tube furnace and a subsequent heat treatment of 30-60 minutes in 400°C flowing argon. Cells were fabricated from these films with evaporated CdS window layers. The best cell parameters achieved by this process were a $V_{oc}$ = 320 mV, a $J_{sc}$ = 13.2 mA/cm², and a FF = 0.45. Research efforts at SERI are continuing.

ii) Weizmann Institute (Israel) used one electrolyte to form a Cu-In alloy, which they subsequently heat treated in a $H_2Se$ atmosphere. Hodes initially reported on their efforts to electroplate Cu-In alloys on a Ti substrate in 1985 and in 1986 (Hodes et al., 1985 a,b; Hodes and Cahen, 1986). Their efforts were focused on electrodepositing a Cu-In alloy from aqueous solutions of $Cu^+$ and $In^{3+}$ ions, and thiourea. These films were then heat treated in a $H_2Se$ atmosphere. Efficiencies of 1.5% were measured using a liquid polysulfide electrolyte/semiconductor junction, with 750 nm quantum efficiencies of 0.67 measured at short circuit. Problems with the heterogeniety of the films were assumed responsible for these low efficiencies. A later report (Lokhande and Hodes, 1987, Cahen and Hodes, 1987) described their conclusions that CuInSe₂ films formed from the selenization of Cu-In alloy layers (annealed at from 300°–640°C) were suffering compositional changes due to vapor transport of $In_2Se$ away from the film at the higher temperatures. The resulting films had an excess of $Cu_xSe$ on the surface, and the bulk CuInSe₂ was reported as being 5%-10% Cu-rich.

iii) At least four organizations (ARCO Solar, Weizmann Institute, Southern Methodist University, International Solar Electric Technology) have been involved in the fabrication of CuInSe₂ through the electrodeposition of a Cu layer followed by an In layer, and then heat treating the layered films in a $H_2Se$ atmosphere. The efforts of each of these organizations has been discussed in the earlier section on selenization (Section 5.2.1).

iv) International Solar Electric Technology is investigating the fabrication of CuInSe₂ films by the electrodeposition of three layers (Cu-In-Se) from separate baths followed by a heat treatment in an inert atmosphere.

v) SERI and the University of Texas at Arlington - Bhattacharya reported (Bhattacharya and Rajeshwar, 1986) research for electroplating $CuInSe_2$ from two electrolyte baths, $Cu^{++}$ with $SeO_2$ and $In^{3+}$ with $SeO_2$. The first electrolyte bath deposited a $Cu_2Se$ layers, the second a $In_2Se$ layer. The layered films were then heat treated in an inert atmosphere.

vi) Weizmann Institute (Israel) has reported efforts to electrodeposit $CuInSe_2$ from two electrolytes in which the first bath deposited a Cu-In alloy layer and the second deposited a Se layer. Hodes reported electroplating a Cu-In alloy followed by the electrodeposition of a Se layer from a $SeO_2$ electrolyte (Hodes and Cahen, 1987). These stacked films were then heat treated in an argon atmosphere at $500°C$. It was reported that the resulting $CuInSe_2$ films were limited to less than $1,000$ Å thick.

### 5.2.2B Sputtering

Sputtering is a versatile technique for the deposition of thin films. Besides those methods discussed previously, sputtering can be used to make $CuInSe_2$ by: i) the sputtering of Cu, In, and Se films for subsequent heat treatment, ii) the sputtering of $Cu_2Se$ and $In_2Se_3$ for subsequent heat treatment, or iii) the sputtering of $CuInSe_2$ from a bulk material. None of these approaches have resulted in particularly efficient devices, however. These processes can use either RF-sputtering targets or DC-sputtering targets, and our summary of sputtering is presented in those two categories.

Work in RF-sputtering has been mainly confined to two organizations, Brown University and Parma University (Italy). Brown has investigated $CuIn_yGa_{1-y}Se_zTe_{2(1-z)}$ alloy films for lattice constant matching with CdS window layers. Both Brown and Parma have reported 4%-5% efficient $CuInSe_2$ devices.

Brown University - Loferski and coworkers first reported on sputtering research in 1978 (Loferski et al., 1978 a,b). Their efforts centered on the formation of a $n$-type CdS heterojunction with complex $p$-type alloys of the ternary materials $CuInSe_2$ and $AgInS_2$. Alloys were chosen to allow a close match of the lattice constant with a CdS window layer. The bandgap of the alloy layer tended to be near 1.5 eV. Thin films of $CuInSe_2$ were also prepared separately by RF-sputtering from targets which were fabricated from $CuInSe_2$ powders. To obtain $CuInSe_2$ films of chalcopyrite structure, a careful selection of target properties and sputtering conditions was needed. Brown University was able to obtain grain sizes up to 1 $\mu$m, resistivities in the range of 0.3 to 2.0 ohm-cm, and Hall mobilities up to about 6 $cm^2$/V-s (Piedoszewski et al., 1980 a,b). CdS films were deposited by RF-sputtering and by evaporation. Cells efficiencies up to 5% were obtained with the evaporation of the CdS onto RF-sputtered $CuInSe_2$ films. Further work on alloy films was reported in 1981. Thin films of $CuInSe_2$ and alloys of $CuIn_yGa_{1-y}Se_zTe_{2(1-z)}$ were investigated. The values for $y$ and $z$ were again chosen to produce alloys having the same lattice constant as CdS. Materials with bandgaps of about 1.25, 1.35 and 1.45 eV were deposited by RF-sputtering. These films were reported as having properties in the ranges typical of $CuInSe_2$ in good devices (Loferski et al., 1981). Work was also carried out on a minority carrier mirror concept using a $CuInSe_2$ and a $CuGaSe_{0.9}Te_{1.1}$ multilayer structure. Cells had reported AM 1 efficiencies of about 4% (Kwietniak et al., 1982, Loferski et al., 1984).

University of Madrid (Spain) - Santamaria reported the investigation of RF sputtered CuInSe$_2$ films in 1986 (Santamaria et al.). These nearly stoichiometric chalcopyrite films were sputtered from a CuInSe$_2$ target with an excess of 5% Se in an atmosphere containing a small amount of hydrogen at temperatures up to 430°C. The films consisted of grains in the 1 $\mu$m size range and had resistivities around $10^4$ ohm-cm.

Parma University (Italy) - Romeo reported research by Parma University in 1985 (Romeo et al.). CuInSe$_2$ films were grown by magnetron RF sputtering grain sizes larger than 30 $\mu$m. They were deposited on glass substrates covered by a proprietary thin metallic layer. The CuInSe$_2$ targets were prepared by melting Cu, In and Se in a high pressure furnace using B$_2$O$_3$ as an encapsulant. CuInSe$_2$ film quality was reported to depend on substrate temperature, sputtering power, substrate bias, target stoichiometry and deposition rate. CuInSe$_2$/CdS devices were prepared by flash-evaporating low resistivity CdS on CuInSe$_2$. Solar cells exhibited high photocurrents (close to 40 mA/cm$^2$ at a 100 mW/cm$^2$ illumination) and a maximum efficiency close to 4%. By 1986 the average grain size of the CuInSe$_2$ films was reported as larger than 50 $\mu$m. The films were deposited on glass substrates covered by an ohmic contact layer of Pb, Al, or Mo. In the first case, a 2 $\mu$m thick layer of lead was made by a proprietary process (Romeo et al., 1986 a,b). CuInSe$_2$/CdS thin film solar cells were also prepared by RF sputtering on an Al-covered glass substrate. The Al back contact layer, deposited by an electron gun, had an average grain size of 100 $\mu$m. The Mo back contact and the CuInSe$_2$ and CdS films were then sputtered in sequence in the same sputtering chamber without breaking the vacuum. The CdS film, deposited at 200°C substrate temperature had a resistivity of about 1 ohm-cm. This was achieved by introducing a small amount of H$_2$, which served to increase the S-vacancies in the film. The CuInSe$_2$/CdS sputtered cells exhibited efficiencies around 4%–5% (Romeo et al., 1987 a,b).

DC Sputtering has been studied extensively by ARCO Solar. The first reported work by ARCO Solar in DC sputtering occurred in 1985 (Ermer et al., 1985). CdS/CuInSe$_2$ solar cells were fabricated with efficiencies up to 6.1% on 4 cm$^2$ using CuInSe$_2$ thin films deposited by room-temperature DC magnetron sputtering of Cu$_2$Se and In$_2$Se$_3$ followed by a non-vacuum anneal. The Cu/In atomic ratio in as-sputtered films was graded by adjusting the relative sputtering rates of Cu$_2$Se and In$_2$Se$_3$. The films were annealed in Ar/H$_2$Se mixtures at 400°C for 1 to 2 hours. After annealing, no gradient in the Cu/In ratio was detected. Junctions were fabricated by vacuum evaporation of a high/low resistivity CdS window layer, In/Ag grids, and an SiO antireflection layer. Ermer reported that the CuInSe$_2$ deposited by this process did not require the elevated temperatures in vacuum necessary for reactive processes, and that it was more easily scaled than RF processes. ARCO Solar was awarded a European patent application for this process (Ermer and Love, 1987).

The New Energy Development Organization (Japan) has also reported research on the DC triode sputtering of CuInSe$_2$ films in 1987 (Kobayashi et al., 1987). They found that the films increased in (112) orientation with increased voltage to the sputtering target and increased substrate temperature. Low resistivity films were deposited below 250°C substrate temperature. This was found to be the optimum preparation condition with a target voltage of 300 V. The films had a band gap of

1.0 eV, and were $p$-type with a resistivity of $4 \times 10^4$ to $2 \times 10^{-3}$ ohm-cm. Films deposited at a substrate temperature of more than 300°C and target voltages of 300 V were $n$-type.

### 5.2.2C Chemical Spraying

Spraying offers a significant promise of low cost deposition if high-quality, thin CuInSe₂ layers can be fabricated. The basic processes for the chemical spray deposition of CuInSe₂ is the spraying of CuInSe₂ or the spray pyrolysis of Cu, In and Se compounds with a subsequent heat treatment of the films to form optimized CuInSe₂. In most reported cases to date, groups have chosen to use spray pyrolysis rather than direct spraying of the compound.

There are eight organizations with reported efforts in the chemical spray pyrolysis of CuInSe₂ films: SRI International, University of Science and Technology (France), Brown University, Stanford University, Indian Institute of Technology (India), the University of Simon Bolivar (Venezuela), Yunan Normal University (China), and Radiation Monitoring Devices.

• SRI International - Bates first reported the deposition of 1 $\mu$m films of CuInSe₂ by spray pyrolysis in 1982 (Bates et al., 1982 a,b). These films were deposited on a glass substrate from an aqueous solution consisting of ammonium hydroxide (NH₄OH), CuCl, InCl₃, $n$-dimethylselenourea and HCl with a pH of 3. The spraying step was carried out at substrate temperatures that were varied between 225° and 300°C. The films were then heat-treated in H₂ + Se for 75 minutes at 600°C, and all displayed the chalcopyrite structure. Bates reported that the heat treatments of the films were performed in order to produce conductivities and crystal structures for photovoltaic applications. Bates found that CuInSe₂ gave the sphalerite structure under most spray conditions, but a short (10 min) heat treatment of a film with a starting Cu/In/Se ratio of about 1:1:4 at temperatures of 400° to 600°C produced the chalcopyrite structure (Bates et al., 1983). In 1984 SRI changed its Cu source from CuCl to CuCl₂. CuInSe₂ films with grain sizes of 0.5–2 $\mu$m were prepared from solutions with various pH levels and Cu/In ratios. The substrate temperature was varied between 225° and 300°C. All films prepared with CuCl₂ were chalcopyrite, whereas only Cu-rich CuCl based films had exhibited this phase. The appearance of $Cu_{2-x}Se$ and/or $Cu_2Se$ as a second phase was found to be depend strongly on solution stoichiometry, pH, temperature, and substrate. Two-layer structures consisting of a Cu-deficient layer on top of a Cu-rich layer were produced without the presence of any second phases on both glass and Mo-coated glass. Absorption spectra were measured and were found to improve when CuCl₂ was used instead of CuCl (Abernathy et al., 1984a). It was also found that twice the amount of Se in the spray solution improved film quality (Abernathy et al., 1984b) These programs resulted in a cell with reported 2.5% efficiency. The glass/Mo/CuInSe₂/CdS/In structure had low shunt resistances, believed to be the result of pinhole shorts (Mooney and Lamoreaux, 1984, Mooney et al., 1984, Abernathy and Cammy, 1985, Mooney and Lamoreaux, 1986).

• University of Science and Technology (France) - Touzel reported the fabrication of films by the University of Science and Technology using chemical spray pyrolysis in 1985 (Touzel et al.). The structural, optical and electrical properties of their films were reported to be related to variations in the substrate temperature and the ionic ratio Cu/In/Se in the solution. Their investigations indicated that the crystallographic, electrical and optical properties of this material were strongly related to the fabrication conditions employed, particularly the substrate temperature and the atomic ratio Cu/In/Se. The best cells resulted when the substrate was kept at 200°–290°C, the Cu/In ratio was less than 1, and the Se/Cu ratio was 2–4 (Bougnot et al., 1986). Two-layer CuInSe₂ films were deposited by spray pyrolysis followed by a CdS window layer. The first CuInSe₂ layer (approximately 1.7 $\mu$m thick) was deposited with a resistivity of about $10^4$ ohm-cm. The second layer (approximately 1.4 $\mu$m thick) was deposited on the first with a resistivity of about 0.1 ohm-cm. The small device had a $J_{sc}$ of about 32 mA/cm² and an efficiency of around 2.8% (Duchemin et al., 1986). Further work has been carried out (Duchemin et al., 1987), but improved device performance has not been reported.

• Brown University - Gorska first reported work by Brown University in the spray pyrolysis of thin films of CuInSe₂ in 1980 (Gorska et al.). CuInSe₂ films deposited by spray pyrolysis were reported as being of the sphalerite or chalcopyrite structure, depending on the substrate temperature. Stoichiometric CuInSe₂ thin films were deposited by this process over glass substrates and over conducting substrates such as Al or Cr-Au-coated glass slides. CdS/CuInSe₂ devices were prepared by vacuum evaporation of CdS films over the sprayed CuInSe₂ films. Best cell AM 1 parameters were reported as a $V_{oc}$ of 330 mV, a $J_{sc}$ near 2.2 mA/cm², and a fill factor of around 0.32 (Sarro et al., 1984).

• Stanford University - Pamplin reported a Stanford research effort in the spray pyrolysis of CuInSe₂ in 1979 (Pamplin and Feigelson). CuInSe₂ and related compounds and alloys, both $p$- and $n$-type, were deposited in thin, polycrystalline sphalerite layers with various resistivities. Further efforts were carried out (Pamplin and Brian, 1979), but no device performance was demonstrated, and no subsequent progress has been reported.

• Indian Institute of Technology (India) - Agnihotri first reported on this research in 1983 (Agnihotri et al.). The spray pyrolysis of $p$-CuInSe₂ thin films with the chalcopyrite structure was demonstrated with film resistivities varying from 0.1 to about 100 ohm-cm by varying the Cu/In ratio from 1.2 to unity. The films deposited by this process had a predominantly chalcopyrite structure with a (112) orientation suitable for CdS/CuInSe₂ solar cells. X-ray photoelectron spectroscopy studies showed that the CuInSe₂ films did not contain $Cu_2Se, In_2O_3$ and $In_2Se_3$ phases (Raja Ram et al., 1985). $Cd_{1-x}Zn_xS$/CuInSe₂ devices fabricated with these films had reported efficiencies in the range of 2.3%. The best cell had the following parameters: $V_{oc}$ = 305 mV, $J_{sc}$ = 32 mA/cm², FF = 0.32, area = 0.4 cm², and efficiency = 3.14% (Raja Ram et al., 1986).

• University of Simon Bolivar (Venezuela) - Tomar reported on efforts for the preparation of a thin film ZnO/CuInSe₂ heterojunction solar cell using spray pyrolysis in 1982. The junctions obtained from these films were rectifying. A device shown a $V_{oc}$ of 0.3 V, a $J_{sc}$ of 23 mA/cm², a fill factor of 0.29 and a conversion efficiency of over 2% (Tomar and Garcia, 1982).

• Yunan Teachers University (China) - Wang reported on the fabrication of a single-phase CuInSe₂ film with chalcopyrite structure by spray-deposition in 1986.

The films were spray deposited on a glass substrate from a solution containing $CuCl_2, InCl_3$, and $(NH_2)2CSe$. It was reported that when the In/Cu ratio was greater than 1, the film was $n$-type; otherwise it was found to be $p$-type. A $CuInSe_2$/CdS/superstrate cell was prepared by spray-deposition of a 0.3 $\mu$m layer of $SnO_2$/F, a 3 $\mu$m layer of CdS, a 2 $\mu$m layer of $CuInSe_2$, and a Ag film on glass. It was reported that the device demonstrated a wide spectral response in the wavelength region 0.35–1.15 $\mu$m (Wang and Zhao, 1986).

• Radiation Monitoring Devices - Squillante reported in 1985 on the spray deposition of $CuInSe_2$ films (Squillante et al. 1985). The films were sprayed onto a $n$-CdZnS/ITO/glass substrate and heat treated at 250° to 400°C in an Ar atmosphere with a Se overpressure. The best devices fabricated in this study achieved a 1% efficiency.

### 5.2.2D Evaporation

Cu, In, and Se vapors can be produced by direct resistance heating (thermal), electron beams (E-beam), flash evaporation, and RF-heating of the sources. Thermal coevaporation of $CuInSe_2$ and Cu, In evaporation followed by selenization have been discussed previously.

Several research organizations have chosen flash evaporation for the deposition of $CuInSe_2$. This involves the flash heating of compound materials to the temperature of disassociation. Investigators include the University of Parma (Italy), Karl-Marx University (G.D.R.), MASPEC Institute (Parma, Italy), the Indian Institute of Technology (India), and the Warsaw Technical University (Poland).

• University of Parma (Italy) - Canevari first reported on the flash evaporation of $CuInSe_2$ films by the University of Parma in 1983 (Canevari et al.). Low resistivity $p$- and $n$-type $CuInSe_2$ thin films were obtained in a simple flash-evaporation system in which a separate Se source was used to control the amount of Se in the films. Low resistivity $p$-type films were obtained by using a Se/$CuInSe_2$ flux ratio greater than 3. The films were (112) oriented, with an average grain size larger than 50 $\mu$m. The $CuInSe_2$ was deposited on 7059 glass substrates and covered with a 2 $\mu$m lead film. Cells had a low resistivity CdS layer on top of the $CuInSe_2$ layer and exhibited photocurrents close to 40 mA/cm² under 100 mW/cm² illumination. Maximum efficiencies were about 4% (Romeo et al., 1986).

• Karl-Marx University (G.D.R.) - Horig first reported research efforts at Karl-Marx University on the flash-evaporation of $CuInSe_2$ thin films in 1978 (Horig et al.). One-micron-thick $CuInSe_2$ films, found to have a direct band gap of 1.02 eV, were deposited on GaAs substrates at substrate temperatures of 422°–547°C. All the films were partly polycrystalline (Schumann et al., 1981). These films were grown to investigate the $CuInSe_2$ material using Rheed techniques (Tempel et al., 1981). Films with thicknesses in the range of 1.5 $\mu$m were also deposited onto glass by single-source evaporation. Source temperatures above 1,127°C were found to be necessary to prepare single-phase and nearly stoichiometric $CuInSe_2$ films (Neumann et al., 1983). The layers were found to be single phase and crystallized in the chalcopyrite structure only. The composition of the films varied with substrate temperature. In the range of about 427°C, a nearly stoichiometric composition was found (Schumann and Kuhn, 1984).

• MASPEC Institute (Parma, Italy) - Salviati has also reported on CuInSe$_2$ thin films deposited by flash evaporation onto mica substrates. Both $n$- and $p$-type films were deposited. This research indicated that the film properties were strongly influenced by growth condition such as source temperature, geometry, quality of the elements, and other factors (Salviati et al. 1983).

• Indian Institute of Technology (India) - Pachori reported in 1986 on the flash-evaporation of CuInSe$_2$ thin films at the Indian Institute of Technology. Single phase chalcopyrite films were anamolously high optical gaps of approximately 1.15 eV were obtained (Pachori et al., 1986). Sridevi also reported that the as-deposited CuInSe$_2$ films on glass substrates were amorphous. Subsequent annealing in a Se atmosphere produced polycrystalline films (Sridevi et al., 1986). Finely powdered, prereacted, polycrystalline CuInSe$_2$ was prepared for the flash evaporation of the CuInSe$_2$ onto a glass substrate. The optical absorption measurements of the films in the spectral range 700-1300 nm indicated a band-gap of around 1.02 eV. All films were $p$-type (Sridevi and Reddy 1986).

• Warsaw Technical University (Poland) - Trykozko reported in 1984 on the research at Warsaw Technical University on the flash evaporation of CuInSe$_2$ from bulk material to form thin films (Trykozko et al., 1984). X-ray diffraction studied confirmed the chalcopyrite structure of these polycrystalline films. Films were grown to support work investigating the photoconductivity of CuInSe$_2$ thin films (Trykozko et al., 1986).

E-beam evaporation is being investigated by the Warsaw Technical University (Poland) and New South Wales University (Australia). The use of RF heating for the evaporation of bulk CuInSe$_2$ has been studied by the University of Salford (Great Britain) and Newcastle upon Tyne Polytechnic (Great Britain).

• Warsaw Technical University (Poland) - Trykozko reported on the electron-beam evaporation of thin films of CuInSe$_2$ from a single source of bulk material in 1986. X-ray diffraction studies confirmed the chalcopyrite structure of these polycrystalline films (Trykozko et al., 1984, 1986).

• New South Wales University (Australia) - Szot reported in 1984 on the research efforts by New South Wales University in the co-evaporation of CuInSe$_2$ (Szot and Haneman, 1984). The ternary thin film semiconductors CuInSe$_2$ and In doped CdS were prepared by co-evaporation of the elements using electron bombardment heating techniques. The films were characterized by energy/wavelength dispersive quantitative compositional analysis, x-ray diffraction, optical transmittance, SEM surface morphology and electrical measurements.

• University of Salford (Great Britain) - Tomlinson reported in 1980 on the research by Salford University on the RF-heated evaporation of CuInSe$_2$ films. CuInSe$_2$ powder was used as the source material and was held at temperatures below 1,200°C during deposition. The deposited films contained little or no Cu, and elemental analysis indicated a range of compositions which corresponded to In-rich In$_2$Se or a multiphase mixture (Tomlinson et al., 1980).

Two universities, both in Japan, have reported on the evaporation of bulk CuInSe$_2$ material (Tohoku University and Ehime University).

• Tohoku University (Japan) - Kokubun reported in 1977 on the Tohoku University research in the thermal evaporation of CuInSe$_2$ films. The evaporation of these films was carried out from a compound CuInSe₂ source and an elemental Se source. The photovoltaic properties of heterojunctions prepared by depositing CuInSe₂ on CdS substrate crystals were investigated. The optimum substrate temperature was found to be between 300° and 340°C. The efficiencies abruptly decreased when the Se source temperature was below 180°C. Devices fabricated by this process had reported efficiencies of 5.6% (Kokubun and Wada, 1977).

• Ehime University (Japan) - The first efforts reported by Ehime University were in 1986 (Isomura et al., 1986). Thin films of CuInSe₂ were prepared by vacuum evaporation from the synthesized bulk material. Both $p$- and $n$-type vacuum-evaporated films were obtained by controlling Se vapor pressure and substrate temperature. The vacuum-evaporated films showed a fundamental absorption similar to that of bulk crystals. A shift of the fundamental absorption edge was observed for $p$-type evaporated films depending upon the hole concentration. Photoconductivities were observed for the $n$-type evaporated films. Some samples showed high anomalous photoresponse beyond the fundamental absorption edge.

### 5.2.2E Screen Printing

Screen printing of CuInSe₂ involves the sintering of a thinly deposited paste of either Cu, In, and Se powder, or of a finely ground powder of CuInSe₂ bulk material. Fabricating thin CuInSe₂ layers – essential for low cost – is a general problem with this approach. Two organizations have reported results in screen printed CuInSe₂ films: University of Simon Bolivar (Venezuela) and Matsushita Battery Industrial Co. (Japan).

• University of Simon Bolivar (Venezuela) - Garcia first reported on this research in 1982. CdS/CuInSe₂ layers were successively deposited by screen printing and sintering of semiconductor-containing pastes. Indium chloride was used as a dopant in the CdS paste. Cells had reported efficiencies of around 3% (Garcia and Tomar, 1982).

• Matsushita Battery Industrial Company (Japan) - Arita reported on this effort in 1988. The process employs the use of a fine powder of Cu, In, and Se mixed with distilled water and ethylene glycol monophenylether. The grain size of the powder was 1.5 $\mu$m. Films were screen printed onto a borosilicate glass and sintered in N₂ atmosphere at 700°C. $p$-type films were reported with resistivities of 1.1 ohm-cm, carrier concentrations of $6.4 \times 10^{18}$ cm$^{-3}$, and an optical bandgap of 1.0 eV. Efficiencies of less than 1% were observed (Arita et al., 1988).

### 5.2.2F Sintering

Sintering of Cu-In-Se films to form CuInSe₂ is carried out at high temperatures. The films to be sintered can be deposited by any process which does not result in a single chalcopyrite phase film. Only two organizations (other than those discussed above) have reported work on the sintering of CuInSe₂ films, Banaras Hindu University (India), and La Habana University (Cuba).

• Banaras Hindu University (India) - This work on sintered bulk pellets to form CuInSe₂ films was first reported by Janam in 1983. Their research into the growth and structural characteristics of the photovoltaic behavior of the CdS/CuInSe₂ thin film heterojunction concluded that better conversion efficiency resulted with the sphalerite phase of the CuInSe₂ films due to a smaller lattice mismatch with the

CdS base. The degree of ordering in CuInSe$_2$ phase was also found to affect the performance of heterojunctions. Heterojunctions having the disordered sphalerite type CuInSe$_2$ structure were reported as having exhibited a much lower efficiency than those embodying the ordered sphalerite CuInSe$_2$ phase (Janam and Srivastava, 1983, 1985). These films were fabricated by vacuum evaporation of $n$-CdS thin films on a sintered $p$-CuInSe$_2$ base. The PV activity of these heterojunctions crucially depended on the grain size of CuInSe$_2$. The larger grain polycrystalline material was more easily obtained by the inexpensive sintering process, whereas it was impossible to obtain large grain (greater than 1 $\mu$m) when CuInSe$_2$ was prepared in the form of a thin film (Janam and Srivastava, 1986).

• La Habana University (Cuba) in cooperation with Parma University (Italy) - Leccabue first reported research on the sintering of bulk pellets to form CuInSe$_2$ films in 1985. The rectifying $n$-CuInSe$_2$/Au contacts were prepared by a vacuum evaporation of Au on $n$-type sintered CuInSe$_2$/Au contacts were prepared by a vacuum evaporation of Au on $n$-type CuInSe$_2$ samples in order to evaluate the electro-optical properties of these devices. Low photovoltaic efficiencies were reported and attributed to the presence of the insulating layer, surface states, and high series resistance. Good quantum efficiencies were reported, around 60% in the 0.72–1.24 $\mu$m range (Leccabue et al., 1985). The fabrication of $p$-CuInSe$_2$/$n$-CdS heterojunctions was obtained by the evaporation of In-doped CdS thin film on these sintered CuInSe$_2$ substrates. Certain sintering conditions resulted in large grain CuInSe$_2$ polycrystals with controlled electrical properties. The best reported efficiencies were about 6.9% (Vigil et al., 1987).

### 5.2.2G  Layer Annealing

The annealing of layered thin films of Cu, In, and Se has been investigated by two organizations. The University of l'Etat (Belgium) has investigated the laser annealing of sandwiched layers into films of CuInSe$_2$. Newcastle Polytechnic (Great Britain) is studying the thermal annealing processes for fabricating CuInSe$_2$ films from layers of Cu, In, and Se.

• University of l'Etat (Belgium) - Laude reported on the laser annealing of vacuum evaporated Cu-In-Se layers to form CuInSe$_2$ in 1986. The technique of laser irradiation started with the fabrication of Cu/In/Se sandwiches (around 1,000 Å each) in a 1:1:2 atomic proportion. Laude used transmission electron microscopy, electron diffraction, optical spectrometry, resistivity measurements and Raman spectroscopy to establish the presence of CuInSe$_2$ with the chalcopyrite structure. No other phases were observed in any of the deposited samples. The optical absorptance of these films indicated a direct optical band gap of 0.95 eV. Crystallites obtained on free-standing films reached 20 $\mu$m. Several binary systems such as Cu-Se, In-Se and In$_2$Se$_3$ were also briefly studied, and a model for the ternary phase formation was proposed (Laude et al., 1986).

• Newcastle upon Tyne Polytechnic (Great Britian) - Thermal annealing of thin sandwich layers of thermally evaporated In and Se and Cu was reported by Knowles in 1988. The layers were nominally 44 nm, 92 nm, and 20 nm respectively, and were arranged in three repetitive stacks of this configuration. The resulting 9-layer film was then thermally annealed at 500°C for 30 minutes at $10^{-6}$ torr. $n$- and $p$-type

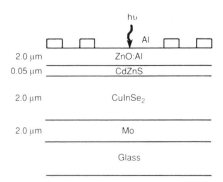

**Figure 5.22:** CuInSe$_2$ cell structure.

films were formed and were reported as having satisfactory optical and electrical properties (Knowles et al., 1988). Work in this area is being pursued.

### 5.2.3 Other Layers Used in CuInSe$_2$ Devices

Most highly efficient CuInSe$_2$ cells are fabricated in a similar configuration (Figure 5.22). The CuInSe$_2$ film is deposited on a metalized substrate, usually Mo-coated glass. The CuInSe$_2$ is followed by the deposition of a thin heterojunction partner and a conductive, transparent window layer. Grids are deposited in some cases, and antireflection coatings are added in fully optimized devices.

Care must be taken to choose a good mechanical and electrical connection between the back-contact and the CuInSe$_2$ layers. Although Mo is a satisfactory ohmic contact, adhesion between the CuInSe$_2$ and Mo has been a problem in some processes (see below).

The transparent heterojunction-partner/window-structure deposited onto the CuInSe$_2$ usually consists of CdS, CdZnS, ZnSe/AnO, or (Cd, Zn)S/ZnO. The techniques used in the deposition of these final layers are limited by properties of the CuInSe$_2$ substrate. For instance, electrodeposition techniques would be difficult due to a non-uniform voltage distribution across a polycrystalline thin film cathode. Electrodeposition of the window material would likely occur down grain boundaries, possibly causing shorts or unpredictable areal uniformity. Sputter-deposition would also be limited by the CuInSe$_2$. High energy particles from sputtering could damage the CuInSe$_2$ surface. Many techniques for the deposition of the window layer also require substrate temperatures which adversely affect the CuInSe$_2$ structure or result in the interdiffusion of the two layers. In most cases, substrate temperature is best kept below 250°C to insure the integrity of the CuInSe$_2$ film.

Some investigation of possible antireflection coatings for CuInSe$_2$/Cd(Zn, S) cells has been carried out at Boeing (Devaney et al., 1987). These antireflection coatings would require re-optimization for the case of CuInSe$_2$/CdS films coated with ZnO, which has a different index of refraction.

### 5.2.3A History, Present Status, and Process Description

### 5.2.3A.1 Ohmic Back-Contacts

The most commonly used substrate for CuInSe$_2$ is Mo-coated glass. Several other contact materials such as platinum (IEC) and Ag, Cu, Au, and Ni have been studied (Matson et al., 1984, Russel et al., 1982). Research in this area is continuing, but most of these contacts are not used in general practice. The deposition of the approximately 2 $\mu$m thick Mo films has been carried out by thermal evaporation, electron-beam evaporation, and sputtering; sputtering predominates. All of these processes have produced films with satisfactory resisitivity, but with different microstructures. In order to address the problem of adhesion for the selenization processes (see discussion in Selenization Section 5.2.1C.4 on adhesion problems in selenization) and to attempt to reduce module costs, future research is anticipated in (1) understanding the microstructure of Mo films by different deposition processes and (2) new materials for ohmic back contacts to CuInSe₂.

### 5.2.3A.2 Window Materials

### 5.2.3A.3 CdS

CdS has been extensively used in the fabrication of solar cells. It is relatively easy to deposit by either evaporation, spray pyrolysis, sputtering, MBE, VPE, CSVT, CVD, screen printing, solution growth (dip-coating), anodization, or electrophoresis. However, the deposition of CdS layers for near 10% efficiency CdS/CuInSe₂ devices has been mainly confined to evaporation, sputtering, and solution growth techniques. With a bandgap of 2.42 eV and a lattice constant of 5.818, it is a good heterojunction partner for CuInSe₂, which has a bandgap of 1.01 eV and a lattice constant of 5.782 (Chopra and Das, 1983).

Until recently, evaporation had been the CdS deposition method used most commonly in the fabrication of near-10% efficiency CdS/CuInSe₂ cells. The substrate temperature required by this process is 200°–250°C, which does minimal harm to the CuInSe₂. The CdS films have been typically 2–5 $\mu$m thick, with a deposition rate of around 1–3 $\mu$m/min. Evaporated films have undoped resistivities of $1$–$10^3$ ohm-cm, with carrier concentrations of near $10^{16}$ cm$^{-3}$. In doped (.01%–.005%) layers of CdS have had resistivities of about $10^{-3}$ ohm-cm, with carrier concentrations of around $10^{18}$ cm$^{-3}$.

Although CdS has had wide use in the fabrication of CdS/CuInSe devices since 1974 (Wagner et al., 1974, Shay et al., 1975, Kazmerski et al., 1976), Boeing's initial work (Mickelsen and Chen, 1980, and Chen and Mickelsen, 1980) resulted in the first CuInSe₂/CdS patent. Their work described the Boeing concept of a 0.5–1.5 $\mu$m layer of undoped CdS followed by a 2–4 $\mu$m layer of In-doped CdS as the window material and heterojunction partner on a CuInSe₂ absorber layer (Mickelsen and Chen, June 1982, U.S. Patent 4,335,266). These layers were deposited by open boat evaporation with a substrate temperature of 150°–200°C (Mickelson and Chen, 1981, 1982). Recently, Boeing has shifted toward thin CdZnS and ZnO as the heterojunction and window materials for CuInSe₂ devices (see below).

The Institute of Energy Conversion deposits 1 $\mu$m CdS layers by effusion cell evaporation of 99.99% purity CdS. The source temperature is around 900°–1,000°C, with the top layer doped $10^{-3}$ ohm-cm with In. The deposition rate is as high as 1.0 $\mu$m/min (Birkmire et al., 1984 a,b).

Romeo has investigated the use of low resistivity ($10^{-1}$ ohm-cm) CdS. Films were deposited on glass substrates by r.f. sputtering. The CdS was deposited in an Ar $-$ H₂ atmosphere at substrate temperatures in the range 100°–200°C (Romeo et al., 1987). The films were highly transparent in the wavelength region between

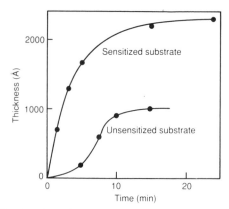

**Figure 5.23:** Growth kinetics for solution growth of films deposited on sensitized and unsensitized substrates (Chopra et al., 1982).

1.6 $\mu$m and the absorption edge of CdS (0.52 $\mu$m). Other work with evaporated CdS has been carried out by groups involved in the fabrication of CuInSe₂ devices (Coutts, 1982, Fraas and Ma, 1977, Shih and Qiu, 1987, and Weng, 1979). Vigil also evaporated In-doped CdS thin film on a sintered CuInSe₂ substrates (Vigil et al., 1987).

Solution growth of CdS involves the use of a Cd source (cadminum chloride, cadmium sulfate, or cadmium acetate) and a sulfur source (sodium thiosulfate or thiourea). The sulfur source is decomposed in an alkaline aqueous solution to yield $S^{2-}$ which reacts with the $Cd^{2+}$ from the Cd source to form CdS. This process takes place at a temperature of around 90°C and does not harm the CuInSe₂ substrate. Typical film resistivities range from $10^7$ ohm-cm to $10^9$ ohm-cm. Resistivities are reduced to around 50 ohm-cm with vacuum annealing (Danaher et al., 1985). The film thicknesses are normally kept very low, in the range of a few hundred angstroms to maximize short-wavelength transmission. Good control of very thin films is possible because of the low deposition rates (0.1 $\mu$m/min; Figure 5.23) (Chopra and Das, 1983.)

### 5.2.3A.4 ZnCdS

ZnCdS is another promising window material closely related to CdS. It is an alloy of CdS and ZnS, which form a solid solution ($Zn_xCd_{1-x}S$) over the entire compositional range. These alloy films exist in single phase structures with varying optical and electrical characteristics as $x$ varies from 0.0 to 1.0. They have been deposited by evaporation, solution growth, spray pyrolysis, and sputtering.

The bandgap of $Zn_xCd_{1-x}S$ can be adjusted from 2.40 eV ($x = 0.0$) to 3.38 eV ($x = 1.0$) (Figure 5.24), but the range of general interest for CuInSe₂ applications has been from $x = 0$ to 0.2, where the optical bandgap is from 2.45 to 2.6 respectively. The limitation to low Zn content is because carrier concentrations and mobilities decrease with increased Zn (Figure 5.25). The higher sensitivity of high concentration Zn films has so far precluded their use. However, high-Zn-content films may be useful when used as very thin transparent layers, since the problem

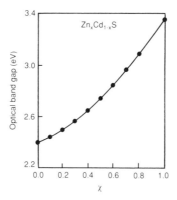

**Figure 5.24:** Bandgap vs. Zn concentration ($x$) for $Zn_xCd_{1-x}S$ (Das et al., 1978).

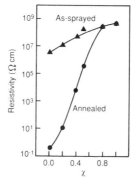

**Figure 5.25:** Resistivity of As-deposited and annealed $Zn_xCd_{1-x}S$ films vs. Zn concentration ($X$) (Banerjee et al. 1978).

of increased sheet resistance may be overcome by the deposition of a top layer of highly transparent and conductive material such as ITO or ZnO (see below).

IEC began work in $CuInSe_2$ using CdS windows, but they started using $Zn_xCd_{1-x}S$ window materials in 1982 to increase performance. Typical cells had 1–1.5 $\mu$m of $Zn_{.16}Cd_{.84}S$ deposited onto the $CuInSe_2$ layer. The ZnCdS was vacuum evaporated from a CdS source with a source temperature of around 950°C and from ZnS with a source temperature of around 1050°C. Films were In doped to around 0.02 ohm-cm and deposited on the $CuInSe_2$ substrate at 200°C. A conductive ITO layer was deposited on the ZnCdS to allow for good current collection.

The University of Parma in Italy has investigated the flash evaporation of In doped ZnCdS films (Canevari et al., 1983). The films prepared by this process had been found to have optical and electrical properties needed for good $CuInSe_2$ devices. They have been deposited at substrate temperatures from 160° to 220°C.

Although Boeing shifted to the use of two-source evaporated ZnCdS films for the $CuInSe_2$ cells in the early 1980's (Mickelsen and Chen, 1984), their most recent use of ZnCdS has been carried out with solution growth methods. Solution growth has been used by Boeing to make very thin ZnCdS to improve cell blue response. The

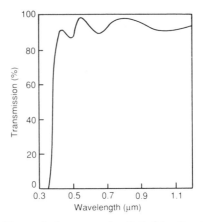

**Figure 5.26:** Transmission curve for ZnO (Brodie et al., 1980).

ZnCdS film is used in conjunction with a conductive ZnO window (see below). The ZnCdS film thicknesses are kept very low, in the range of a few hundred angstroms. Very high efficiency devices (12.5% total area) with improved current densities have been made by this approach.

### 5.2.3A.5 ZnO

ZnO films have been deposited by spray pyrolysis, MOCVD, and RF reactive magnetron sputtering. Undoped films generally have an as-deposited resistivity of about $10^2$ ohm-cm, which decreases to about $10^{-3}$ ohm-cm when annealed in a vacuum (Shanthi, 1981) or $H_2$ atmosphere (Aranovich et al., 1979). Films deposited by spray pyrolysis have reached resistivities of $10^{-4}$ ohm-cm (Chopra and Das, 1983). MOCVD-deposited ZnO films have been made at 280°–350°C with conductivities to 50/ohm-cm (Roth and Williams, 1981). RF magnetron sputtered films of ZnO have been prepared with conductivities to 500/ohm-cm (Webb et al., 1981).

ZnO films have a bandgap of about 3.3 eV, with a demonstrated optical transmission for sprayed films of about 85% (Chopra and Das, 1983). RF magnetron sputtered films have reached a transmission of 90% in the spectral range 4000 Å–8000 Å (shown in Figure 5.26, Webb et al., 1981).

The short-wavelength spectral response of thin film CuInSe$_2$ devices is improved substantially by a thin (< 500 Å) CdS or CdZnS layer and a ZnO conducting window layer. The ZnO acts as an unoptimized antireflection coating and permits photons of wavelength above 360 nm to be absorbed in the CuInSe$_2$. The thin CdZnS does not block a significant fraction of the incident light from reaching the junction region. This approach has been exploited by Boeing and ARCO Solar to make highly efficient CuInSe$_2$ devices (Mitchell and Liu, 1988, Mitchell et al., 1988).

ARCO Solar has used MOCVD to grow ZnO as the window material for their CuInSe$_2$ cells described in their 1988 patent (Vijayakumar et al., U.S. Patent 4751149). Their preferred window configuration was a thin (< 500 Å), undoped CdS layer deposited onto a CuInSe$_2$ substrate followed by a 1 $\mu$m ZnO conducting window layer (Potter, 1986, Potter et al., 1985). They have measured a 25% photocurrent enhancement over comparable devices with thicker CdS. ARCO Solar's

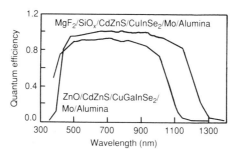

**Figure 5.27:** Improved blue response of a ZnO/CdZnS/CuIn(Ga, Se)₂ cell compared with an earlier CdZnS/CuInSe₂ cell.

high efficiency devices (12.9% active area, SERI measured, and 14.1% reported) are made using thin CdS and ZnO (Mitchell et al., 1988).

IEC has deposited ZnO by RF magnetron sputtering of a cast ZnO target, 0.2% aluminum oxide doped. The deposited films are typically 0.7–1.0 $\mu$m thick and have a resistivity of 0.05 ohm-cm. The transmission of these films in the 700 nm range is 90% (private communication Birkmire, 1988).

Low resistivity thin films of ZnO have also been prepared by Qiu using RF sputtering of an In doped ZnO target (Qiu and Shih, 1986). The resistivity of the ZnO films deposited on CuInSe₂ crystals decreased with increasing In content in the ZnO films.

Recent work by Boeing on a ZnO/CdZnS structure has also demonstrated the increased blue response and improved overall performance resulting from the ZnO/CdZnS window configuration (Figure 5.27). The Boeing devices were made by depositing a 500 Å CdZnS layer on a CuIn.₇₅Ga.₂₅Se₂ substrate by solution growth techniques. A 2 $\mu$m thick ZnO layer was then deposited by reactive RF sputtering of an Al₂O₃ (2%) doped Zn target in an O₂ atmosphere (the cell performance characteristics are shown in Figure 5.15).

The University of Simon Bolivar in Caracus, Venezuela has been successful in depositing ZnO films by spray pyrolysis (Tomar and Garcia, 1982). However, temperatures of 400°C were needed to achieve suitable material characteristics, which may damage a CuInSe₂ substrate. Using this technique, with the spray pyrolysis of the CuInSe₂ layer, their best devices were under 2% efficient ($V_{oc} = 0.3$ V, $J_{sc} = 23$ mA/cm², FF = 0.29).

### 5.2.3A.6 ZnSe

With SERI funding, the Jet Propulsion Laboratory carried out research into the use of ZnSe as a heterojunction partner for CuInSe₂ cells (Stirn and Nouhi, 1985). Using reactive magnetron cosputtering of the Zn and the In dopant in a Ar/H₂Se atmosphere, JPL deposited ZnSe thin films on glass that showed resistivities as low as 20 ohm-cm at deposition temperatures as low as 120°C (Nouhi and Stirn, 1986, 1987). To make devices, the ZnSe was deposited on CuInSe₂/Mo, and a conductive ZnO layer was deposited atop the ZnSe layer. Device results were encouraging, with $V_{oc}$ up to 430 mV. Improved global AM 1.5 $J_{sc}$ (37.4 mA/cm²) was also observed.

IEC (Hegedus et al., 1986) and ARCO Solar have also studied the potential of using ZnSe to replace CdZnS. ARCO Solar mentioned the use of ZnSe in a patent

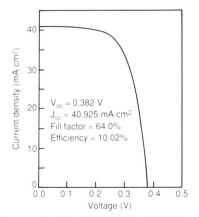

**Figure 5.28:** I-V curve for the ARCO Solar ZnO/ZnSe/CuInSe₂ cell.

(Choudary et al., 1986, U.S. Patent 4611091) as a layer in place of the thin CdS or CdZnS. Building on the JPL research, ARCO Solar has worked with Stanford University to produce 10% efficient $ZnO/ZnSe/CuInSe_2/Mo$ devices (Figure 5.28) with a very thin (50 Å) ZnSe layer.

### 5.2.3B  Problems and R&D Issues

CuInSe₂ devices are evolving toward the use of a thin heterojunction partner (500 Å or less) of CdS, CdZnS, or ZnSe, followed by a 2 $\mu$m ZnO window layer. Similarly, Mo/glass remains the preferred substrate.

### 5.2.3B.1  Materials

The materials required for the deposition of the other layers in the CuInSe₂ cell (Mo, Cd, S, $O_2$ and Zn) are all readily available, and a sufficient world supply exists of all of these materials for gigawatt production of solar cells.

No studies have been done to establish what material purity is necessary for high quality ZnO and CdZnS. It may be expected that the material purities now used in laboratories can be relaxed.

Low material utilization rates and high feedstock costs for MOCVD growth are felt to be significantly disadvantageous in the fabrication of ZnO window layers for CuInSe₂ devices. Solution growth techniques also have a poor material utilization rate, but the feedstock materials are not expensive and their use in the deposition of the very thin CdZnS layer does not represent a significant material loss.

A 2 $\mu$m thick film of ZnO, a 500 Å film of CdZnS, and a 2 $\mu$m thick film of Mo would require 37 mg/m² of Cd, 9 g/m² of Zn, 13.4 mg/m² of S, and 20.4 g/m² of Mo. Let us assume a utilization rate for the Cd, Zn, and S of 20% for the solution growth of CdZnS, and a utilization rate for ZnO and Mo of 95% (with recovery). If the price for these materials is $18/kg for 99.99% Cd, $1.30/kg for 99.99% Zn, $0.15/kg for 99.99% S, and $7.50/kg for 99.99% Mo, the materials cost of the back-contact layer and the window layer would be minimal around $0.17/m². 

These estimates are based on the currently used thicknesses of 2.0 $\mu$m each of ZnO and Mo, and a 500 Å thick CdZnS layer required in present near 10% efficient $ZnO/CdZnS/CuInSe_2$ cells. Some minor cost reductions may be realized in the future with thinner ZnO and Mo layers, or possibly with the use of other window layers or back contacts.

### 5.2.3B.2 Deposition Rate

Deposition rates for the solution growth of a 500 Å CdZnS layer are about 800 Å/min. Solution growth techniques for the growth of ZnO are being investigated by researchers in the field but are not available for solar cell applications as yet, and deposition rates have not been established.

A 2 $\mu$m ZnO layer can be grown by MOCVD at bout 200 Å/min in the temperature range below 300°C. Higher rates have been achieved but require substrate temperatures that could potentially damage the CuInSe₂ film.

The 2 $\mu$m Mo and ZnO layers can be deposited at rates of around 0.3 $\mu$m/min by evaporation and 1 $\mu$m/min by RF sputtering.

Deposition rates of other processes for the Mo, Cd(Zn)S, and ZnO layers have not been demonstrated in the fabrication of CuInSe₂ devices. Based on the most rapid processes for growth of each layer, around 4.6 minutes would be required for the deposition of the Mo contact layer and the ZnO/CdZnS layers.

### 5.2.3B.3 Capital Equipment and Power Needs

Solution coating is very inexpensive costs mainly involve the bath tanks. Sputtering equipment is relatively expensive. But the high deposition rates for sputtering insure that unit costs are reasonable.

Solution growth requires an insignificant amount of power; the highest power requirements are associated with pumps and filters. The power required for the sputtering of ZnO is equivalent to the requirements for the sputter deposition of the CuInSe₂ absorber layer discussed earlier.

### 5.2.3B.4 Continuous Processing

Both solution growth and sputtering techniques are easily adapted to continuous processing. The solution growth is a wet chemical method which can be incorporated into a process line, and the sputter deposition of both the ZnO and Mo layers is very compatible to glass fabrication techniques. The current set of processes may result in large area, quality performance modules; however, the compatibility in an integrated system of solution growth, evaporation, sputtering, and selenization processes will have to be addressed.

### 5.2.3B.5 Safety and Waste Disposal

The safety issues with regard to the large scale deposition of Mo back contacts and ZnO/CdZnS window layers do not appear to be of major concern. Elemental sulfur and oxygen are not safety concerns for the fabrication of CdZnS or CdS films for CuInSe₂/ZnO/CdZnS solar cells, and Mo, and Zn are all relatively safe materials.

Cd is used in minute quantities to make CdS and CdZnS layers. Yet it is a safety concern that needs careful attention. Cd in the form of sulfide, sulfate, or oxide are all potential carcinogens. Many Cd compounds are oral poisons. But Cd is so irritating to the human digestive system that it usually causes sudden violent reactions, and very little Cd is allowed to be absorbed into the body. Fatalities are normally averted due to this reaction. Cd dust is a fire and explosive hazard, and the inhalation of fumes and dust of Cd compounds can affect the human respiratory system and kidneys. The TLV for an eight-hour exposure in air is 0.05 mg(Cd)/m$^3$.

The deposition of CdS or CdZnS by solution growth processes significantly reduces the hazards associated with the inhalation of Cd dust or fumes. However, workers will require monitoring to insure safety with respect to other forms of contamination, especially with aggressive Cd recovery and equipment cleaning procedures. Other than these concerns, Cd is not felt to be a problem for the fabrication of CdZnS or CdS films for CuInSe$_2$/ZnO/CdZnS solar cells.

Mo has been shown to be a poison in animals when introduced intraperitoneal. There have been no human fatalities associated with Mo. However, caution is urged against the inhalation of Mo dust or fumes. The TLV for an eight-hour exposure is 10 mg/m$^3$. The enclosed nature of deposition systems for Mo films will satisfy most safety concerns for exposure related to inhalation.

Although the toxicity of Zn compounds vary, they are generally low. Elemental Zn when heated emits ZnO fumes which can cause respiratory problems if inhaled. Such fumes have caused fatal lung damage in some cases. ZnO has a TLV for an eight-hour exposure in air of 5.0 mg/m$^3$. Although Zn exposure is not considered to be cumulative, Zn salts in large doses, or repeated small doses, can cause health problems. The enclosed nature of deposition systems for ZnO films will satisfy most safety concerns for the Zn exposure related to inhalation, and the deposition of CdS or CdZnS by solution growth will reduce other Zn hazards.

### 5.2.3C  Discussion of R&D Issues

Areas which remain a concern for the back contact, heterojunction partner, and window layer are: the deposition rates and capital equipment costs of the Mo and the ZnO layers; the continuous process compatibility of deposition processes involved in the fabrication of ZnO/CdZnS/CuInSe$_2$ modules; and the effect of the Mo back-contact layer deposition technique on the cell adhesion and performance.

The techniques used in the deposition of Mo, CdS, CdZnS and ZnO have not exhibited very high deposition rates to date. In the case of the solution growth of CdZnS or CdS, the process has minimal capital equipment costs and layers are very thin; so rates are not significant. The present growth rates for the Mo and ZnO deposition by RF sputtering might result in increased per unit costs.

The solution growth process for the deposition of CdZnS or CdS may not be easily incorporated into other process lines. It may be technically difficult to incorporate an aqueous solution method for the growth of one layer with some of the most promising deposition processes for the fabrication of CuInSe$_2$ modules (e.g., sputtering and selenization).

To date, the investigation of back contacts and heterojunction window layer configurations has progressed rather slowly, due to the rapid changes in the CuInSe$_2$ technology and the need for focused research on the various CuInSe$_2$ deposition techniques. With the maturity of this technology, more efforts can now be directed toward the optimization of the other layers and processes for their deposition.

It is possible that the Mo back-contact layer deposition process has some effect on the CuInSe$_2$ cell mechanical stability and its performance. Problems associated with adhesion in selenized cells may be addressed by the investigation of both new back-contact materials and the effect of Mo deposition technique on the structural integrity and cell performance of CuInSe$_2$ devices.

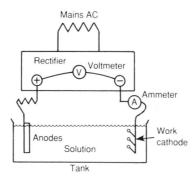

**Figure 5.29:** Schematic of electrodeposition process.

## 5.3 CdTe

### 5.3.1 Electrodeposition

Electrodeposition is an electrochemical process used in manufacturing to make coatings and to refine raw ores. The passage of current through a liquid electrolyte results in the deposition of ions that is essentially proportional to the current flow (Figure 5.29).

In the simplest form of electrodeposition, ions dissolve from the anode, cross the electrolyte, and deposit on the cathode as current flows. The ions to be plated may also be introduced (Continuously or in batches) via the electrolyte solution. The equation governing electrodeposition is:

$$g = Iet/96,500$$

where $g$ is grams of material deposited, $I$ is current in amperes, $e$ is the chemical equivalent weight (number of charges per atom to deposit the material), and $t$ is time in seconds. Thus the amount of material deposited is essentially the amount of electric charge that has been conducted divided by a constant. For the deposit to be uniform across its entire area depends on having nearly equivalent current flow at all points on the cathode surface. This often requires specially designed, multiple contacts. Perfect area uniformity (especially at the edges) can be an issue in PV module fabrication by electrodeposition.

In practice, there are several phenomena which control electroplating. Materials will only plate onto the cathode at certain voltages, and this plating depends on their local concentration and the characteristics of the surface of the cathode. Plating rate depends on current density, but due to the need to replenish ions near the cathode, this current density can be limited. To replenish the ions near the cathode as they are plated, may require stirring of the electrolyte to supplement natural diffusion. Heat and chemical additives may be introduced to assist (or slow) the rates and to alter growth characteristics. In addition to diffusion properties controlling ionic replenishment, the ions may complex with others in the electrolyte. This can strongly affect replenishment and also the voltage dependence of the plating. Purposeful complexing can also be used to improve plating characteristics.

As bath characteristics are altered, the morphology of the grown material can vary from amorphous patches to highly ordered polycrystalline grains. Multiple elements and compounds may be deposited if concentrations, voltages, currents, additives, etc., are tailored appropriately. In fact, compound semiconductor deposition can be easier than elemental deposition because the compound-formation can result in a lower total energy for the system, thus stabilizing the deposition in a way that stimulates stoichiometric formation.

For general reviews of electroplating, see F. A. Lowenheim (1978), Electroplating: Fundamentals of Surface Finishing and F. A. Lowenheim (ed.) (1974), Modern Electroplating.

Metals such as Cu and Cd, and alloys such as bronze, have commonly been electrodeposited in industrial processes. However, the fabrication of compound semiconductors by electrodeposition is almost uniquely the result of efforts involved in developing low-cost solar cells such as CdTe (G. F. Fulop, R. M. Taylor, 1985, see p. 197). This new branch of electrodeposition owes much to the impetus provided by PV R&D.

### 5.3.1A  History

Danaher et al. (1978) and Panicker et al. of the University of Southern California (1978) prepared CdTe by cathodic electrodeposition using acidic electrolytes. Danaher et al. deposited the CdTe from a stirred solution onto a sanded and degreased titanium substrate, producing a PV efficiency of about 2%-4%.

Panicker et al. discussed the general possibility of both anodic and cathodic electrodeposition. They disclosed a range of process conditions for depositing material of varying stoichiometry from a stirred solution of $CdSO_4$ and $TeO_2$. The University of Southern California group of Kröger and Panicker published much of the field's pioneering literature in 1978: R. A. Kröger (1978) Panicker et al. (1978), Panicker (1978).

Following the lead of Panicker et al., Monosolar (under a SERI/DOE contract) formed heterojunction cells as reported in the DOE PV Systems Program Summary, P. 102, December 1978 and other reports from the DOE contract (EX-76-C-01-2457). The contract was with Monosolar and University of Southern California.

At about the same time, a group at Ametek began work which resulted in three patents in the early eighties: G. Fulop et al., *CdTe Schottky Barrier PV Cell* (1981); *Method of Making a PV Cell* (1981); and G. Fulop, (1982).

At the University of Missouri-Rolla, J. L. Boone and P. Van Doren began work in the fall of 1978. They selected cathodic deposition based on the work of Kröger (1978) and Panicker et al. (1978). Their DOE/SERI contract began in 1979, the results of which are reported in their final report (Boone and Van Doren, 1982).

### 5.3.1B  Present Status

Electrodeposition has become one of the most successful methods of making CdTe for solar cells. There are two industrial groups who fabricate high-efficiency thin-film CdTe cells by electrodeposition: Ametek and British Petroleum.

Ametek has had a long, continuous participation in CdTe electrodeposition. They began in the late seventies by making $n$-type CdTe MIS cells which attained efficiencies of about 8.6% (G. Fulop et al., 1982b). Subsequently, however, they recognized the limitations of MIS structures and shifted to a heterojunction approach

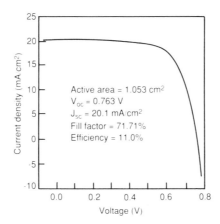

Active area = 1.053 cm²
$V_{oc}$ = 0.763 V
$J_{sc}$ = 20.1 mA/cm²
Fill factor = 71.71%
Efficiency = 11.0%

**Figure 5.30:** 11% electrodeposited CdTe cell by Ametek (SERI AM 1.5 Global).

using $p$-type CdTe deposited on $n$-CdS/tin-oxide-coated glass. After about three years of development, they were able to report efficiencies over 10%, including an 11%-efficient, 1 cm² cell measured at SERI in April 1988 (Figure 5.30).

The Ametek cell includes a new $p$-ZnTe final layer to facilitate contacting and stability (Meyers, 1986). For larger devices (near 100 cm²), active area efficiencies over 9% have been reported by Ametek. Devices with $p$-ZnTe contacts tested under illumination at 70°C were stable for over 3000 hours. The apparent stability of the new Ametek design is crucial to the future of the CdTe technology because all previous devices had unstable contacts.

The British Petroleum effort in CdTe followed their purchase of Monosolar and its electrodeposition technology. (This was the result of BP's purchase of a controlling interest in SOHIO, which itself had bought Monosolar.) At present, BP still reports annually to DOE on its intentions to commercialize the Monosolar patents, which originated during the DOE/SERI support of that small business. Such commercial intentions were required by the DOE contract to keep title to the patents.

Monosolar worked closely in the mid-seventies with the USC team in the initial development of electrodeposited CdTe. Building on the USC work, B. M. Basol and O. M. Stafsudd (both originally at UCLA) deposited and characterized $n$-CdTe Films (Basol and Stafsudd, 1981a and 1981b). Schottky barriers with Au were also investigated (Basol, 1980; Shkedi and Rod, 1980; Basol et al., 1980). The group began work in heterojunction $p$-CdTe structures early in the eighties (Basol et al., 1982). In September of 1983, several heterojunction cells from Monosolar were measured at SERI at over 9% efficiency, the highest measured here for CdTe at that time. Among them were high-efficiency $Hg_x Cd_{1-x}Te$ cells made by electrodeposition, the first of their kind. Monosolar investigated these alloys to promote contacting, since HgCdTe is more $p$-type (and thus more easily contacted) than CdTe. In 1986 (Basol and Tseng, 1986), the Monosolar group reported their highest efficiency devices at 10.6%.

Monosolar patents (all now controlled by BP) included: F.A. Kröger, R.L. Rod, M.P.R. Panicker, U.S. Patent 4,400,244, 1983 (covers electrodeposition procedures); B.M. Basol, E.S. Tseng, R.L. Rod, U.S. Patent 4,388,483, 1983 *Thin Film*

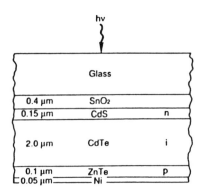

**Figure 5.31:** Electrodeposited CdTe cell: Typical structure.

*Heterojunction Cells and Methods of Making the Same*; B.M. Basol, U.S. Patent 4,456,630, 1984 *Method of Forming Ohmic Contacts*; B.M. Basol, E.S. Tseng, D.S. Lo, U.S. Patent 4,629,820, 1986 *Thin Film Heterojunction PV Devices*.

In Middlesex, England, British Petroleum continues the Monosolar effort develop electrodeposited CdTe. They are reportedly making high-efficiency (over 10%) CdTe devices as well as prototype 8-watt square-foot submodules.

### 5.3.1C Cell Structures

The cell structure of electrodeposited devices is similar in most cases (Figure 5.31). Glass is the 'superstrate' on which conductive, transparent SnO₂ and thin (up to about 1500 Å), resistive CdS is deposited. This forms the cathode for the CdTe electrodeposition. About 2 μm of CdTe is electrodeposited and then heat treated at bout 400°C to form an active junction with the CdS. Back ohmic contact differs between groups, with Au, Ni, ZnTe, HgTe, and Cu-doped carbon paste or ITO in use.

### 5.3.1D Process Description

A process similar to the one originated by Panicker and Kröger (Kröger et al., 1983) was used by Monosolar. Basol (1988) described the most up-to-date version of the Monosolar process. He stated that CdTe was deposited from an acidic solution containing $Cd^{2+}$ and $HTeO_2^+$. The electrolyte contained about 0.5 M CdSO₄ and about 30 parts per million (ppm) of $HTeO_2^+$. The pH of the solution was adjusted to about 1.6 using H₂SO₄. Two anodes were used, one of Te, the other of carbon. They were switched into the circuit alternately. When the Te anode was on, $HTeO_2^+$ was formed at the anode and was added to the solution. When the current was switched to the inert Ca anode, the Te compound was consumed at the cathode, forming CdTe. This was needed to control the $HTeO_2^+$ concentration in the solution. The deposition reaction was thought to be as follows:

$$HTeO_2^+ + 3H^+ + \rightarrow Te + 2H_2O$$

$$Cd^{2+} + Te + 2e^- \rightarrow CdTe$$

Both reactions took place simultaneously at cathode potentials between $-0.2$ V and $-0.65$ V.

The plating procedure had the following basic steps:

1. 0.5 molar solution obtained by disolving $CdSO_4$ in DI water
2. Solution purified for about 2 hours using a platinum cathode at $-620$ mV (with respect to an Ag-AgCl reference electrode) and a solution temperature of $85°$–$90°$C
3. pH adjusted to about 1.6, and 30-40 ppm of $HTeO_2^+$ added by applying a potential of 500 mV to the Te anode.
4. Contact made to the cathode and CdTe plated onto the cathode (glass/$SnO_2$/CdS); the *quasi rest potential* kept just below the value at which the deposition of pure Cd would take place.

Monosolar pioneered the deposition of HgCdTe. The only variations in the process were (a) the addition of $Hg^{2+}$ ions using $HgCl_2$ in the electrolyte after the introduction of $HTeO_2^+$ (Step 3) and (b) a cathode potential of $-0.4$ V to $-0.65$ V. As the $Hg^{2+}$ ions were depleted from the solution, less Hg was incorporated in the films, causing a gradient. A possible way to avoid a gradient would involve the continuous introduction of $Hg^{2+}$ into the electrolyte.

Deposition current depended on the Te concentration in the electrolyte, stirring, and temperature. Typically it was about 0.3–0.5 mA/$cm^2$ in the Monosolar approach, leading to a slow deposition rate of about 0.72–1.1 $\mu$m/hr. Deposition of a 2 $\mu$m film took about two hours.

Ametek has used a variation of the two-electrode method in which they have both Cd and Te anodes. Because these are designed to supply the needed Cd and Te on a continuous basis, their bath can run uninterrupted for long periods.

Every reported solar cell made by electrodeposition has undergone a post-deposition heat treatment. The procedure varies, but is usually done at about 400°C in air (or other ambient) for an hour or more. The purpose is to produce quality polycrystalline grains of CdTe. Device efficiencies may be improved from under 5% to over 10%.

### 5.3.1E Problems and Issues

A major challenge of all thin-film deposition methods is to achieve large-area uniformity. Fortunately, there are reaction kinetics in compound electrodeposition that favor such uniformity. Specifically, the process favors the deposition of the compound CdTe instead of an off-stoichiometric alloy of the metals. Compound formation is a stable state that lowers the system's total energy. Thus the electrodeposition of stoichiometric, high-quality CdTe is usually favored everywhere on the cathode.

However, one problem is to achieve uniform coating of peaks and valleys on imperfectly smooth cathodes such as found on CdS/$SnO_2$-glass. Such imperfections result in local variations in conductivity which would normally mean a buildup or absence of electrodeposit. Since the deposition of stoichiometric CdTe is favored over the metals, even the rough areas are coated with enough CdTe to minimize pinholes and shunts.

Resistance naturally causes current to be less at positions distant from the cathode's contact electrodes. This results in a difficulty of perfectly coating areas that are distant from the electrodes. Again, the tendency to deposit CdTe everywhere prevents this from being a serious problem. Deposits are thinner but still nearly stoichiometric at more isolated positions. Nonstoichiometric or patchy deposits are avoided because (1) the rates (currents) are slow, reducing the voltage drop across the cathode (i.e., minimizing thickness variation) and (2) the thin but resistive CdTe layers are not shunts. In fact, the process is generally rather insensitive to the anode/cathode geometry, which should make scale-up more forgiving.

### 5.3.1E.1 Materials

One may purchase starting material of elemental Cd and Te that is 99.999% pure and successfully form CdTe by electrodeposition. No studies have been done that prove this purity is needed. The electrodeposition process has near-unity material utilization, which (1) keeps down the costs of materials, (2) makes waste disposal easier, and (3) reduces the impact on Te availability. Material requirements for 2 $\mu$m layers are 5.4 g/m$^2$ of Cd and 6 g/m$^2$ of Te. Unlike films made by almost any other process, even thinner CdTe films may be accessible via electrodeposition.

At 99.999% purity for elements bought in small (1 kg) quantities, cost is about $85 and $250/kg respectively for Cd and Te; Cd and Te costs would be $0.5/m$^2$ and $1.5/m$^2$, respectively, or about $2/m$^2$ total for 2 $\mu$m films. This is quite acceptable; costs should be even lower for material bought in larger quantity.

### 5.3.1E.2 Deposition Rate

Currently, the rate of deposition via electrodeposition is relatively slow: about 1 $\mu$m/hr. This means that if high throughput is desired, large areas must be plated simultaneously, requiring larger equipment. Good electroplated devices have been made in the lab at higher rates (over 2 $\mu$m/hr). Increased rates could be achieved if work were done to modify the electrolyte composition, increase stirring rates, and optimize anode/cathode geometry. For thinner layers, rate would be less of a factor. In fact, a slow rate generally allows for better film coverage and uniformity.

### 5.3.1E.3 Module Definition

Ametek has reported small (100 cm$^2$) submodule prototypes with reasonable efficiency (9.3% active area). They have encountered some difficulties with cell delineation and interconnection which are expected to be resolved in the near future, with the use of laser-scribing.

### 5.3.1E.4 Yield and Uniformity

Yield is quite good because the process favors stoichiometric compound formation. Data on large areas is not available yet. However, small-area yields have been good. Slow deposition rates, small currents, and compound formation all favor good uniformity and yield.

### 5.3.1E.5 Capital Equipment

Electrodeposition baths are simple and inexpensive. However, since deposition rate is slow, large or numerous baths are needed, driving this cost up. One bath large enough for 10 1-ft$^2$ modules might cost about $25K. A furnace suitable for annealing several square meters of panels might cost about $25K.

### 5.3.1E.6 Labor (O&M)

Labor could be an expensive item unless the process becomes faster or fully automated. Because of the slow rates, there are more machines and attendants needed to run them. If automated, costs could come down, and various schemes exist for large scale, continuous electrodeposition.

### 5.3.1E.7 Power Needs

Power requirements scale with the amount of material deposited and are very small. In fact, power requirements are mainly for the pumps, filters, and bath heaters rather than the actual electrodeposition, which needs only about 5 W-hr/$m^2$ (0.005 kWh/$m^2$ or about 0.05 cents/$m^2$).

### 5.3.1E.8 Bath Stability

Bath stability appears very good; in some cases, e.g., at Ametek, baths have been run continuously for very long periods (over two years) without any alteration in device performance.

### 5.3.1E.9 Safety and Waste Disposal

Doty and Meyers (1988) have published a paper on the safety advantages of a CdTe-based PV manufacturing plant. They report that all starting materials are easily handled (elemental Cd and Te; laboratory reagents). No toxic gases are used in their processing. Material utilization is near-unity, reducing effluent. Effluent can be purified and recycled. Thus although Cd is a general concern, it appears that it is handled well by their approach. World consumption of Cd is about 4500 metric tons annually (Bureau of Mines, 1988). About 50 metric tons of Cd would be needed for a gigawatt of electric capacity. This is a small fraction of global consumption and indicates that the contribution of a PV manufacturing facility to increased Cd use would have minimal impact.

### 5.3.1F Discussion/R&D Issues

Electrodeposition of CdTe has several inherent advantages:

- Good uniformity in existing devices
- Near-unity material utilization
- Thin layers (2 $\mu$m) and thinner ones possible (under 1 $\mu$m)
- Low-cost equipment and minimal power demands

However, there are some issues:

- Large area cathode design and uniformity have not been demonstrated
- Rates are low (about 2 $\mu$m/hr), which increases equipment and labor costs
- Less than 99.999% purity may work but has not been studied
- Continuous processes are needed

None of these issues appear to be show-stoppers. From a design standpoint, cathode design and continuous processing of large areas are of the most interest for manufacturing.

Deposition rate and layer thickness also appear to be areas where R&D-driven improvements would have valuable payoffs in: reduced equipment, labor, and material costs; lessened Te availability issues; and easier waste disposal.

Labor costs may be a weakness until addressed through rate improvements and continuous processes.

Because of the method's high material utilization rates, material cost, availability, and waste disposal are topics in which electrodeposition may have unique advantages over other methods. The others more naturally result in thicker layers and more waste.

### 5.3.2 Chemical Spray of CdTe

In the chemical spraying process, a solution containing the necessary Cd and Te is sprayed onto a heated substrate where a reaction may take place, excess material is driven off, and CdTe can be grown with the desired properties. Some forms of spraying (e.g., spray pyrolysis) include a reaction between carrier chemicals to form the resulting CdTe; however, other types of spraying exist in which the sprayed material is a slurry containing the final product (e.g., the CdTe). Then no reaction takes place on the substrate. Instead, the carrier chemicals are driven off, and the CdTe grains are regrown in a post-deposition heat treatment (about 500°C). The process of slurry spraying CdTe is the one covered in this section. The CdTe slurry and a carrier gas are introduced to a nozzle where droplets are formed and sprayed evenly on a substrate. Gas pressure is used to control the deposition rate.

In practice, several parameters determine the quality of sprayed films: spray rate, droplet velocity, substrate temperature, spray geometry, and the composition and chemistry of the sprayed solution.

#### 5.3.2A History and Present Status

Most research concerning the spray-deposition of CdTe has involved spray pyrolysis. Several groups (e.g., Radiation Monitoring Devices and University of Missouri at Rolla) engaged in this research under SERI/DOE funding in the late seventies and early eighties without attaining proof-of-concept device efficiencies. At that point, the SERI/DOE program de-emphasized the spraying approach. Subsequently, however, a private group, Photon Energy, entered the field and made substantial progress (Jordan, 1988). SERI/DOE now supports some of Photon Energy's R&D on a cost-shared basis.

Photon Energy makes high-efficiency and large-area, thin-film CdTe devices (Albright et al., 1988). They have reported a small cell of 12.3% efficiency. This device had a state-of-the-art voltage of 783 mV and FF of 63%. A near-square-foot CdTe panel (838 cm²) made by Photon Energy had been measured by SERI at 6.1 W (7.3% aperture-area efficiency) under standard conditions (Figure 5.32), the highest efficiency for CdTe panels of that size. They have also made a prototype 4-ft² panel, the largest monolithic CdTe panel in the world. Photon Energy has made public its plans to go into the production of CdTe panels. They project 12 MW capacity by 1991 and sale prices under $3/W.

#### 5.3.2B Cell Structure

The cell structure of the Photon Energy device is the usual glass/SnO₂/CdS/ CdTe/contact design (Albright et al., 1988) adopted by most CdTe groups. Conductive, transparent SnO₂ (0.5–1.5 μm) and CdS (6 μm) are deposited on glass. They form the substrate on which the CdTe is chemically sprayed. About 6 μm of CdTe is deposited. Back ohmic contact can be Cu-doped graphite or graphite with some other dopant. Other proprietary contacts are also being investigated.

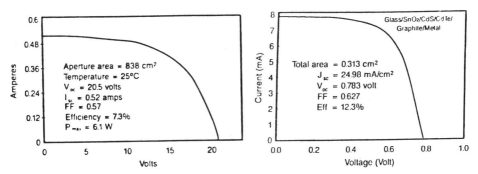

**Figure 5.32:** 7.3% efficient sprayed 1-ft$^2$ CdTe module and 12.3% small area cell by Photon Energy.

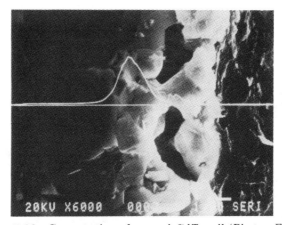

**Figure 5.33:** Cross-section of sprayed CdTe cell (Photon Energy).

### 5.3.2C Problems and Issues

A major challenge of all thin-film deposition methods is to achieve pin-hole free, uniform films. Thickness uniformity of Photon Energy modules is good: variation is less than 7% across a square-foot substrate. Since sprayed CdTe is deposited from a slurry in stoichiometric form, achieving near stoichiometry on the substrate is assured. However, proper doping density is sensitive to subtle changes in local stoichiometry and doping density.

Pin-holes pin-holes are a problem because sprayed material tends to be rather porous. Existing films have pin-holes. The main effect is not on performance (which is degraded slightly), but on limiting the choice of back electrodes. Some conductive electrodes penetrate pin-holes and cause shorts. However, certain granular electrodes, such as doped graphite paste, do not penetrate the pin-holes and can result in effective performance. New electrode materials (e.g., ZnTe and HgTe) that may be significant in achieving long-term stability must be adapted to this requirement. Stability issues also result because of possible migration through pin-holes.

Film doping uniformity is a crucial performance issue which is affected most by temperature control during post-deposition anneals. Thermal gradients across the substrate must be avoided. Demonstration of state-of-the-art module efficiencies (10%) remains a challenge for this process.

### 5.3.2C.1  Materials

One may purchase elemental Cd and Te that are 99.99% pure and successfully make quality CdTe by chemical spray. In fact, Photon Energy sees little effect from impurities in CdTe. The extent to which such impurities may influence efficiencies of fully optimized devices remains an open question.

The chemical spray process has a material utilization rate that is limited only by the geometry of the spraying apparatus. Material utilization is expected to be over 90% in manufacturing.

In large quantities (99.99% pure), the cost of Cd ($>200$ pounds) is about \$33/kg; Te (1,000 lb) about \$66/kg. For a 6 $\mu$m CdTe layer (assuming 90% utilization), 20 g/m$^2$ of Cd and 22 g/m$^2$ of Te are consumed. The cost of materials for a 6 $\mu$m, 1 m$^2$ layer of CdTe made by spraying would be about \$0.66 for Cd and \$1.46 for Te, or a little over two dollars per m$^2$. It is possible that thinner layers of CdTe could be made, reducing this cost further.

### 5.3.2C.2  Module Definition Issues

Module delineation into long, thin (0.93 cm) strip cells presents no unusual problems and is accomplished by laser scribing. Active-area to aperture-area ratio on existing modules is about 91%, indicating good control of intercell spacing.

### 5.3.2C.3  Uniformity and Yield

Yield has been quite stable and high in making 1 ft$^2$ modules of nearly 5 W performance (C. Hobbs, private comm., 1988). A yield over 90% is expected for full-scale operation.

### 5.3.2C.4  Deposition Rate

Throughput is controlled by deposition rate and yield. Spray depositing a 6 $\mu$m CdTe layer takes about 3 minutes (2 $\mu$m/min). This is a very high rate and has an extremely positive impact on both capital equipment and labor costs per unit of production.

### 5.3.2C.5  Capital Equipment

Capital equipment consists of the spraying apparatus and a large, batch furnace for heat-treating. For a small plant (3 MW$_c$), a sprayer and oven costing about \$60-\$120K would be adequate. These costs are unusually low because the spraying and oven equipment are quite simple. Costs for making the CdTe layer are perhaps the lowest among all the technologies for making thin films.

### 5.3.2C.6  Labor, Power Needs, and Continuous Processing

For a non-automated 3 MW plant, about 3 people would be needed to run the CdTe equipment. Costs of about \$3/m$^2$ could be expected. In an automated plant, this figure could be much less.

Power needs for spraying and heat-treating are about $1/m^2$, most of which is needed for the post-deposition regrowth.

There should be no problem scaling the spraying process for continuous, automated processing, thus increasing its usability for large-scale PV. Even in small sizes ($1 ft^2$), the existing process is accomplished with a moving substrate, which lends itself to continuous processing in larger-throughput scenarios.

### 5.3.2C.7 Safety and Waste Disposal

Starting materials are CdCl and Te and common chemicals. There are no toxic feedstock gases used during processing. Reactants and volatiles produced during processing need control, however, care must be taken to flow all effluent into concrete holding tanks. Waste products can be reacted in the holding tank with sodium and other materials to form disposable solids. Coated glass waste is scraped to remove CdTe and then can be disposed without difficulty (e.g., sold to a Cu refinery).

### 5.3.2D Discussion/R&D Issues

Chemical spray of CdTe has several inherent advantages:

- High deposition rate–2 $\mu$m/min; complete film in 3 minutes
- Very low capital equipment
- Low labor
- Ease of continuous processing and automation

These advantages mean that low cost should be readily achievable, making spraying an attractive option for scale-up to manufacture.

However, there are some areas for improvement:

- Six-$\mu$m CdTe layers are presently required and may impact Te availability
- Temperature stresses (500°C) may limit structural options for front contacts (e.g., $SnO_2$)
- Porosity and pin-holes limit the choice of back electrodes, impacting stability
- Thick CdS is used in the existing process line and blocks several $mA/cm^2$ of current
- Very high efficiency and yield in larger areas have not been demonstrated

Thus the potential issues of spraying are two-fold: Te availability and efficiency limits. The spraying process is clearly low-cost. Performance limits and stability are the issues.

Cost projections based on automated, large-scale spraying processes suggest that very low costs ($30–$50/m^2$) could be achievable for sprayed CdTe panels. These kind of cost goals single out spraying as perhaps the most cost-effective film fabrication method being commercialized in photovoltaics.

### 5.3.3 Screen Printing/Sintering

In the screen printing/sintering process (Figure 5.34), a paste containing the necessary Cd and Te is applied ('squeegeed') through a fine screen to a desired substrate. The printed substrate is then sintered (heated) to form large-grained CdTe and to drive off excess chemicals. Thick films (over 5 $\mu$m) tend to result from this process.

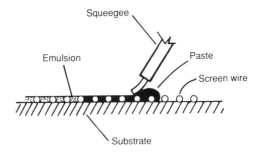

**Figure 5.34:** Schematic of screen-printing process.

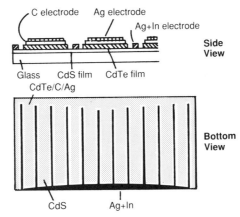

**Figure 5.35:** Screen printed CdTe cell typical structure (Matsushita Battery) (Ikegami 1988).

### 5.3.3A  History and Present Status

Matsushita Battery (Japan) has been successful in developing screen-printed CdTe, and they are the dominant group using the technique. They began work in the late 1970s and now claim very high efficiency, small-area cells (over 12%); square-foot modules at 5.4% efficiency; and a stable encapsulation scheme for outdoor applications (Nakano et al., 1986). Recently, they began production of small cells for calculators marketed by Texas Instruments.

### 5.3.3B  Cell Structure

The cell structure of the Matsushita Battery device is shown in Figure 5.35. Borosilicate glass is the substrate on which CdS (30 $\mu$m) is screen printed using CdS paste. This forms the substrate for the printed CdTe. Unlike devices made by almost all other groups, there is no conductive tin-oxide between the CdS and the glass. About 5–10 $\mu$m of CdTe is deposited on the CdS. Back ohmic contact is Cu-doped (50–100 ppm) carbon paste with a Ag layer to increase conductivity. The unusual electrode geometry shown in Figure 5.35 is unique to Matsushita and places important limits on the efficiency of their modules.

### 5.3.3C Process Description

Nakayama et al. (1980) and Ikegami (1988) describe a procedure in which Cd and Te powders (99.999% pure) are used as starting materials and mixed in equimolar ratio with distilled water and crushed to a grain size of 0.5 $\mu$m in an agate ball mill. After drying, a paste is formed by mixing 100 g of the power with 0.5 g of $CdCl_2$ powder and an appropriate amount of propylene glycol. The paste is then applied to a 400-mesh stainless steel screen, which is placed atop the glass/CdS substrate rising within a continuous belt furnace. The paste is printed on the substrate and dried, after which it is heated within a ventilated alumina case in nitrogen at 620°C for an hour to form the CdTe film.

Previously, Matsushita Battery had used CdTe powder as a starting material. The CdTe crystal was grown in a sealed glass tube by the Bridgman method near the melting point (1,092°C). However, they found that they could successfully synthesize CdTe from Cd and Te powders at temperatures below 300°C, allowing them to avoid the costly melt growth of CdTe starting material.

### 5.3.3D Problems and Issues

### 5.3.3D.1 Materials

One may purchase starting materials of elemental Cd and Te that are 99.999% pure and successfully form quality CdTe by screen printing. At 99.999% purity for elements bought in small (1 kg) quantities, cost is about $85 and $250/kg respectively, for Cd and Te.

The screen printing process has a material utilization rate that can be quite high: we will assume 90%.

Seven microns of deposited CdTe (assuming a 90% utilization rate) require 23 $g/m^2$ of Cd and 26 $g/m^2$ of Te feedstock. The cost of materials for a 7 $\mu$m, 1-$m^2$ layer of CdTe made by screen printing/sintering would be about $8.50, which is high. For Cd and Te bought in larger quantities and of lower purity, costs closer to $33 and $66/kg for Cd, Te would be available. Materials costs for the 7 $\mu$m CdTe layer would then be about $2.50/$m^2$.

### 5.3.3D.2 Module Definition Issues

Module delineation presents some special problems that have hampered the progress of screen printing/sintering. Unlike other approaches for making CdTe modules in which a highly conductive window like $SnO_2$ is the top layer (below the glass), Matsushita Battery uses thick, more resistive CdS instead. Because the CdS is much less conductive than $SnO_2$, modules must have many thinner strip cells to avoid series resistance losses. Their strip cells are about 6 mm wide, about half the size of other CdTe cells. Also (see Figure 5.35), the Matsushita Battery design includes a top contact 'finger' between each strip cell, reducing active area. In addition, the relatively thick CdS (30 $\mu$m) and CdTe layers are not as easily scribed as the thinner layers of other techniques, also adding to active area lost during module delineation. In fact, Matsushita Battery's 30 cm by 30 cm modules have only 559 $cm^2$ of active area, or 62%-a much lower percentage than in competing techniques (which are above 90%). This has a very severe negative effect on module efficiency.

Matsushita Battery proposes several solutions to this problem. They are developing a finer printing technique that will allow them to reduce the space between strip cells. They also plan to investigate the use of conductive $SnO_2$, even to the extent of eliminating the CdS if possible. This would allow them to have fewer, wider strip cells, and they could adopt a more conventional interconnection scheme without the added top contact finger between cells. However, use of $SnO_2$ may require lower processing temperatures than the 625°C now used by Matsushita Battery to sinter their CdTe.

### 5.3.3D.3 Processing Issues

Throughput is controlled by deposition rate and yield. Screen printing is a rapid process similar to spraying. The one-hour sintering can be done in very large quantities.

Capital equipment consists of the screen printing/sintering apparatus and a large, batch furnace for heat-treating. These can be very low in cost.

Power needs for screen printing/sintering and heat-treating should be minimal (about $1/m^2$). Most power would be for the post-deposition regrowth.

There should be no problem scaling screen printing/sintering for continuous, automated processing, thus increasing its usability for large-scale PV. Even in small sizes (1-ft$^2$), the existing process is done with a moving substrate, which lends itself to continuous processing in larger-throughput scenarios.

### 5.3.3D.4 Safety and Waste Disposal

Starting materials are Cd and Te powders and common chemicals. Details concerning process safety remain little-known, but the volatiles may be of concern. Since the approach seems to require more material, increased materials-related safety/health issues may result, especially those with Cd.

### 5.3.3E Discussion/R&D Issues

The advantages of screen printing are its ease of operation, low capital costs, low labor costs, and potential for large-scale automated manufacture. It is clearly a very low cost, scalable method. However, it has some serious drawbacks:

• Active module area is very low and must be increased before the approach can be successful

• Thick CdS reduces current densities by about 30%; but given the existing high-temperature processing regime (625°C), process-resistant alternatives might be difficult to find.

• Materials requirements are greater than for other processes (especially for electrodeposition), since printed layers are generally thicker. Cost could be impacted, as could Te availability.

The problems associated with performance limits at the cell and module levels will have to be addressed successfully before screen-printing can be considered a viable technique for large-scale, low-cost PV.

### 5.3.4 Noncommercial Methods of Fabricating CdTe

The methods now being used to commercialize CdTe are not the only ones by which CdTe has been successfully grown, especially in small areas. There are three additional methods by which cells near 10% or better have been made:

close-spaced sublimation (CSS), chemical vapor deposition (CVD), and metal organic CVD (MOCVD). But there is some question whether any of them, with perhaps the exception of CSS, is capable of low-cost processing.

One of the first major groups to have commercial intentions in CdTe, Kodak, attempted to develop CSS to near commercial-readiness early in the 1980s. Two others–ARCO Solar and Chu at Southern Methodist University (now at University of South Florida)–have been very successful with CSS or its close relative, close-spaced vapor transport (CSVT).

Close-spaced sublimation is a process in which a CdTe source is heated and transported by sublimation to a nearby substrate which is held at a slightly lower temperatures. During transport, the CdTe dissociates into Cd and $Te_2$. These condense and recombine on the cooler substrate, forming a thin film. Because the source and substrate are parallel and very close, the film has properties that are very similar to the source CdTe material. CSS is carried out in a partial vacuum. A closely related method is close-spaced vapor transport (CSVT), in which a carrier gas is introduced and aids in material transport from source to substrate. This process was originally developed by Nicoll (1963) to grow GaAs on Ge. Rates for CSS and CSVT can be quite high (about 1 $\mu$m/min), but temperatures are high as well (near 550°C), exposure to which could degrade layers deposited before the CdTe.

Chemical vapor deposition and metal organic chemical vapor deposition are quite similar except for choice of starting chemicals and temperature regimes. In both cases, a suitable set of feedstock gases are chosen so that when they are passed over a substrate at some elevated temperature, they break down into their constituents and CdTe forms on the substrate. In the case of CVD, typical starting materials can be elemental Cd and $Te_2$ vapors carried by hydrogen or helium and deposited at about 580°C (Chu, 1988a,b). Starting materials for MOCVD can be dimethylcadmium and diethyltelluride deposited at about 400°C (PV Program Branch Annual Report 1988, 125).

### 5.3.4A Cell Structure

As with other work in CdTe, the most successful cell structure has been glass/SnO$_2$(or ITO)/CdS/CdTe/contact, wherein the transparent conductor is deposited first. Layer thicknesses can vary by group or method (see below), as can contact material. In a few cases, variants of the inverse geometry, substrate metal/CdTe/CdS/ITO structure, have been investigated.

### 5.3.4B CSS History and Present Status

Kodak pioneered the CSS process for CdTe and achieved marked success (Tyan, 1980 and 1988). They were able to report over 10% CdTe cells as early as 1982 (Tyan and Perez-Albuerne, 1982) and eventually reached 10.9% (Tyan and Perez-Albuerne, 1984). They scaled up the process to small submodule size, achieving 8.5% on 32 cm$^2$ (Tyan and Perez-Albuerne, 1984). However, Kodak chose not to commercialize their CSS CdTe effort. Subsequently, they have reported infrequently.

Both ARCO Solar and the Chu have since used CSS and CSVT successfully. ARCO Solar began work in the early 1980s and were able to report cells of 9.5% on 4 cm$^2$ (Mitchell et al., 1985). They used the conventional CdTe cell structure with a thin CdS layer only 400 Å thick. They also experimented with a

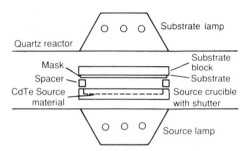

**Figure 5.36:** Schematic of the close-spaced sublimation process.

glass/$SnO_2$/CdTe/contact structure (i.e., one in which the CdS was eliminated) in order to maximize current density. Using this, they were able to demonstrate 10.5% cells with very high currents (28.1 mA/cm$^2$) but at voltages (0.663 V) 50–100 mV lower than with CdS. (The extremely high current measurement may be considered somewhat anomalous, because quantum efficiencies over 1.0 were measured for short wavelengths.) Despite these promising experiments, little has been done to follow up on the valuable idea of eliminating the CdS from CdTe cells. ARCO Solar ceased work in CdTe in the mid-1980s.

Under subcontract to SERI, Professor Ting Chu at Southern Methodist University (now at University of South Florida) has had success using CSS and CSVT to make high-quality CdTe devices. At the same time, Chu has made several advances in both processing and cell structure, addressing some of the problems in the CdTe technology and the CSS method (see below). One of Chu's CdTe cells was measured at SERI in April 1986 at 10.5% efficiency (1.2 cm$^2$) under standard conditions. This was then the highest efficiency 1 cm$^2$ CdTe cell that SERI had measured. (It has since been surpassed by Ametek electrodeposited cells of 11% and Photon Energies 12.3% cell).

### 5.3.4C  CSS Process Description

Figure 5.36 shows a typical geometry for the CSS method. The most critical parameters are the spacing between source and substrate and their temperatures and the temperature difference between them. When the spacing between the source and substrate is less than a few percent of the dimensions of the substrate, the material transport conditions are largely independent of the conditions elsewhere in the system (Chu, 1988b). Kodak used a spacing of 1–2 mm. Chu (1988b) used 1 mm. Mitchell et al. (1985) used less than a 0.5 cm.

Preparation of source material for CSS is an essential and cumbersome first step. Kodak (Tyan and Perez-Albuerne, 1982) used hot-pressed polycrystalline disks of 99.99% CdTe (Ventron) as source material. Early in his work, Chu (1988a,b) directly combined Cd and Te with different degrees of Cd deficiency or combined a stoichiometric mixture of Cd and Te with dopant such as Sb or As. Synthesis required repetition of a high-temperature process (1,000°C) taking several days. When transported, the resulting cast material sublimed unevenly causing problems with reproducibility and bringing into doubt the scale-up potential of the process. An alternative approach was used by Mitchell et al. (1985). They used powdered

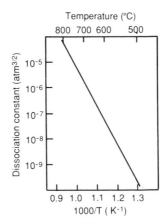

**Figure 5.37:** The dissociation pressure of CdTe in the temperature range $500°-800°$C.

CdTe starting material, forming a block of source material for CSS by transporting CdTe from the powder and sintering it. They make no comment on reproducibility. In later work, Chu (1988a,b) modified this method, transporting from a powder source to an interim CdTe film on glass from which the CdTe was retransported to the final device. Developing a convenient, unchanging source for CSS is a problem that has not been resolved.

Perhaps the most essential CSS parameters are source and substrate temperature and the difference between them, which is what drives the CdTe transport. All groups are in agreement in choosing about $700°$C for the source temperature and $600°$C for the substrate temperature (Tyan, 1988; Chu, 1988a,b; Mitchell et al., 1985). At this temperature, transported material is in the form of a gas consisting of Cd and $Te_2$. Figure 5.37 (Chu 1988b) shows CdTe dissociation pressure in the temperature range of interest.

The ambient in the system is important because it affects CdTe material properties, the rate of CdTe sublimation, and the mean-free-path of the sublimated species. Tyan (1980 and 1988) used the CSS (partial vacuum; 1–2 torr) but claimed the need to include a small amount of oxygen during deposition to make the best devices. Without oxygen, Kodak's devices behaved as buried homojunctions. Tyan attributed a greater $p$-type activity to oxygen. Hsu et al. (1986) studied this question and found that oxygen atoms in CdTe have an ionization energy of about 0.1 eV but do not behave as simple acceptors.

Other groups did not find essential the addition of oxygen during deposition. Chu (1988b) used CSS at 10–20 torr and found no difference in performance with or without oxygen. Mitchell et al. (1985) at ARCO Solar used several ambients, including growth in atmospheric pressure hydrogen or helium, a simplification from the standpoint of fabrication cost. The ARCO Solar group found that small amounts of oxygen could be useful under certain conditions. The role of oxygen in the process was not well understood. Anthony et al. (1983) at Stanford investigated clean $H_2$, He, and Ar at a flow-rate of 30 cc/min in 50 mtorr environment, but they also could not clarify the role of oxygen.

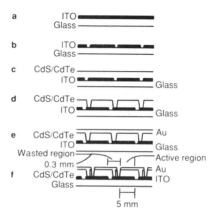

**Figure 5.38:** Fabrication steps of an integrated module made by CSS (Kodak).

### 5.3.4D  CSS Problems and Issues

A major challenge of all thin-film deposition methods is to achieve pin-hole free, uniform films of the proper composition. This manifests itself in CSS in terms of the need to make relatively thick films. One exception was Tyan (1988), who was able to fabricate high-efficiency cells with films 4 μm thick. Chu (1988b), however, required 10–15 μm. Mitchell et al. grew films of about 10 μm thickness.

### 5.3.4D.1  CSS Materials

One may purchase starting CdTe material that is 99.99% pure and successfully make quality CdTe by CSS (Isett, 1983).

The CSS process has a material utilization rate that is near unity due to the close proximity of the source and substrate. Unused source material can be remilled and recycled. Material utilization could be over 90% in manufacturing.

In large quantities (99.99% pure), the cost of Cd ($>$200 pounds) is about \$33/kg; te (1,000 lb) about \$66/kg. Assuming a 5 μm CdTe layer and assuming 90% utilization, 16 g/m$^2$ of Cd and 18 g/m$^2$ of Te are consumed to make a m$^2$ layer. The cost of materials for a 5 μm, 1 m$^2$ layer of CdTe made by CSS would be about \$0.54 for Cd and \$1.20 for Te, or about \$1.75 per m$^2$. If layers are 10 μm, cost would be about \$3.50/m$^2$.

### 5.3.4D.2  CSS Module Definition Issues

Kodak was able to demonstrate some initial success with submodule design using CSS. Their 32 cm$^2$ submodule was 8.5% efficient. Module delineation into strip cells was successfully accomplished (Tyan and Perez-Albuerne, 1984) with laser scribing (Figure 5.38). Active-area to aperture-area ratio was over 91% (including borders), indicating good control of intercell spacing.

### 5.3.4D.3  CSS Uniformity, Yield, and Deposition Rate

Producing near-stoichiometric CdTe films by CSS is relatively easy because of the high temperatures and the tendency of the transport process to reproduce the composition of the source of material. However, achieving precise conductivity control across a substrate is complicated by difficulties in achieving perfectly even temperature.

A real uniformity for module production will require larger area source material and a source-substrate geometry with precise temperature control. No actual yield has been established for larger areas and higher throughput.

Throughput is controlled by deposition rate and yield. One of the advantages of CSS is its high deposition rate of about 1–5 $\mu$m/min (Tyan and Perez-Albuerne, 1982). It is this high rate (and short exposure time) that allows the predeposited layers to withstand the high temperatures of CSS.

### 5.3.4D.4 CSS Capital Equipment and Continuous Processing

Few studies are available to estimate the capital equipment requirements of CSS. However, the main equipment needs may include means of synthesizing or forming source material (e.g., a high-temperature furnace in which to synthesize the CdTe) and a continuous belt furnace on which the substrates can be moved past source material at elevated temperatures.

A problem for continuous processing will be to develop methods of feeding source material while insuring reproducibility. Even though continuous processing appears possible, no demonstration has been achieved.

### 5.3.4D.5 CSS Safety and Waste Disposal

Some safety issues exist concerning the high-temperature CdTe source material synthesis. If synthesis is improperly done, explosion can occur. Otherwise, CSS uses solid starting materials, reducing hazards substantially over manufacturing scenarios that require toxic gases. Issues relating to Cd feedstock and effluents are minimized in relation to processes requiring or producing liquids or gases.

### 5.3.4E CSS Discussion/R&D Issues

CSS of CdTe has several inherent advantages:

- High deposition rate–1–5 $\mu$m/min; complete films in a few minutes
- Good material utilization
- Only moderate vacuum equipment needs compared to more complex processes

However, there are some area for improvement:

- Source material synthesis is a burden
- Temperature stresses at 550°–600°C may limit structural options for front contacts (e.g., $SnO_2$) (but have not, in the past)
- Thick (over 1,000 Å) CdS is used in the existing process line and blocks several mA/cm$^2$ of current
- CdTe thickness is over 5 $\mu$m, perhaps over 10 $\mu$m
- Scalability to high-throughput and good uniformity has not been demonstrated and may be difficult

Thus the potential issues of CSS mainly concern its possible problems with scale-up and manufacturability. Some issues remain uncertain, such as the apparent need for thick (10 $\mu$m) films, as stated by existing CSS groups.

### 5.3.4F CVD and MOCVD: History and Present Status

Small cells of over 9% efficiency have been made using CVD and MOCVD, but no devices of larger area have been demonstrated. Much of the work in this area (SMU, Honeywell, JPL, and Georgia Tech) has been supported by SERI funds.

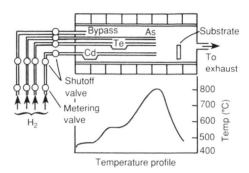

**Figure 5.39:** Schematic of electric furnace and temperature profile of CVE method.

Chu at SMU used CVD since the early 1980s to make CdTe solar cells. In 1985 Chu was able to report 8.2% efficiency on small cells (1 cm²). Instead of the 'inverted' structure (glass superstrate) common to most other approaches, Chu used a geometry in which the CdTe was deposited on graphite or tungsten-coated graphite substrates. However, these substrates were found to affect the conductivity of the CdTe by autodoping, especially with carbon. Chu (1988b) switched to the inverted structure and was able to achieve higher efficiencies (9.85%). But Chu de-emphasized this work because of high temperatures and low rates (see below), which tended to impair the properties of previous layers, i.e., the $CdS/SnO_2/glass$. Besides his on-going work in CSS (above), Chu is now investigating MOCVD of CdTe and its alloys.

Honeywell (Schafer, 1988) investigated MOCVD of CdTe cells from 1985 to 1987. Schafer was able to report a lower temperature (under 300°C) MOCVD process for CdTe as well as the fabrication of HgTe and ZnTe.

In 1987, Richard Stirn at Jet Propulsion Laboratory began to use MOCVD to make CdTe and higher gap CdTe alloys. Stirn did this in close collaboration with Ametek, who supplied JPL with $glass/SnO_2/CdS$ substrates and then finished the cells with ZnTe/contact after JPL deposited CdTe. Progress has been rapid, and JPL has recently reported 9.4% efficient CdTe devices made by MOCVD. Professor Ajeet Rohatgi at Georgia Institute of Technology is also investigating the potential of MOCVD for making CdTe and its alloys (Rohatgi et al., 1988). Rohatgi also reports the fabrication of 9.7% CdTe cells by MOCVD.

### 5.3.4G  CVD and MOCVD: Process Descriptions

Figure 5.39 shows a typical geometry and temperature profile for the Chu CVD process. Chu (1988b) has recently called this process 'combination of vapors of elements' (CVE) because no carrier compounds are involved.

As can be seen from the figure, the hydrogen carrier gas is passed over heated Cd and Te. Cd and Te₂ gases are carried to the substrate, which is kept at an elevated temperature (about 550°C). Helium can also be used as the carrier gas. Conductivity control is done by doping with As ($AsH_3$) or with P($PH_3$), or by controlling the Te/Cd ratio of the film (reduced Cd produces *p*-type films). Figure 5.40 shows the variation of *p*-type resistivity produced by changes of the Te/Cd ratio in the reaction gas mixture over the substrate (Chu, 1988a,b). Chu (1988) also experimented with

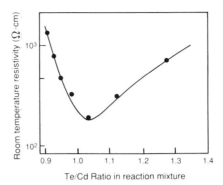

**Figure 5.40:** Electrical resistivity of $p$-CdTe as a function of the Te/Cd molar ratio in the reaction mixture.

the purposeful introduction of oxygen into the gas stream and found a small increase in $p$-type conductivity.

Chu found that nucleation of his CVD films was slow. Growth of continuous films required rather large thicknesses. His CdTe layer thicknesses were about 15–20 $\mu$m, with grain size being about 10 $\mu$m. Deposition rate was about 0.5 $\mu$m/min. Films required about half an hour for fabrication.

Schaffer (1988) at Honeywell used a lower temperature (300°C) Te metal organic source material (dihydrotellurophene) in order to carry out deposition at temperatures that would not harm the glass/$SnO_2$/CdS substrate. Other starting materials were dimethylcadmium and Hg vapor (when needed). Growth rate was under a $\mu$m/hr. Thin layers (1–3 $\mu$m) were grown. The resulting CdTe layers were resistive.

JPL's Stirn (PV Program Branch Annual Report, 1988) used dimethylcadmium and diethyltelluride as starting materials and deposited at about 410°–440°C. JPL's near 9.4% reported efficiencies were encouraging evidence that the previously processed layers (glass/$SnO_2$/CdS) could withstand substantial temperatures for the required times. Rates were low (about 1 $\mu$m/hr) and layers thin (about 2 $\mu$m).

### 5.3.4H  CVD and MOCVD: Problems and Issues

Perhaps the most critical limitations for the CVD process is its temperature (over 550°C) and the thick films required to eliminate pin holes. Although the temperature is lower than both CSS and screen-printing/sintering, exposure is longer, causing unwanted interdiffusion and alloy formation.

With MOCVD, temperatures are characteristically lower (400°C). The problem with MOCVD is the cost of starting organometallics and slow deposition rates.

### 5.3.4H.1  CVD and MOCVD: Materials

One may purchase starting material of elemental Cd and Te that are 99.999% pure and successfully make quality CdTe by CVD. Experiments have not been done to show whether this level of purity is required. With careful design, the CVD process can attain a reasonable material utilization rate. Material utilization is expected to be about 75% in manufacturing. MOCVD is expected to be similar.

For CVD, one may purchase elemental material. Assuming lower purity (99.99%), the cost of Cd (>200 pounds) is about \$33/kg; Te (1,000 lb) about \$66/kg. In a 15 $\mu$m CdTe layer (assuming 75% utilization), 60 g/m² of Cd and 67 g/m² of Te would be required to make 1 m². The cost of materials for a 15 $\mu$m, 1 m² layer of CdTe made by CVD would be about \$2 for Cd and \$4.42 for Te, or about \$6.40 per m². This is higher than for other CdTe fabrication processes because of the greater thickness of the CdTe.

For MOCVD, cost is driven by the organometallics, which currently may cost up to \$10/g. Whether costs would drop for large-volume requirements is at present unknown. If they did not, organometallics alone would put costs for fabricated 1 $\mu$m films (a minimum thickness) at over \$30/m² (even with 100% material utilization). This would be prohibitively high.

### 5.3.4H.2 CVD and MOCVD: Uniformity, Yield, and Deposition Rate

With proper reactor design, uniformity should not be a problem in gas flow processes like CVD and MOCVD.

Throughput is controlled by deposition rate and yield. Yield should be good, but deposition rate can be too slow for the thicknesses required to-date. For CVD, deposition rate is about 0.5 $\mu$m/min. For 15 $\mu$m films, processing would take about 30 minutes. For MOCVD, rates are much slower (about 1 $\mu$m/hr), but thicknesses are also less (about 1.5 $\mu$m), so processing takes about 90 minutes. These long depositions have a negative impact on capital equipment and labor costs per unit of production.

### 5.3.4H.3 CVD and MOCVD: Safety and Waste Disposal

Safety issues may be of concern for MOCVD and less so for CVD. Using elemental sources for CVD eliminates any need for toxic feedstock gases, although it does necessitate careful handling during processing. The MOCVD of CdTe requires toxic organometallics, but these are in liquid form and are thus easier to handle than the toxic gases familiar from III-V MOCVD.

### 5.3.4I CVD and MOCVD: Discussion/R&D Issues

CVD and MOCVD of CdTe have several inherent advantages:

- Uniform films
- Process technology for scale-up should be straightforward

However, there are some problem areas:

For CVD in particular

- High temperatures for long periods (half an hour) could degrade properties of layers (CdS and SnO$_2$) deposited before the CdTe
- Thick layers cause excess materials costs

For MOCVD:

- Cost of organometallics may be prohibitive
- Deposition rate is slow, reducing throughput and raising costs per unit of production
- Toxic feedstocks are used.

### 5.3.5 CdTe Devices: Other Layers

Nearly all CdTe cells are made with a similar, superstrate geometry which begins with glass. On the glass is coated a transparent oxide, usually tin oxide (SnO$_2$) or

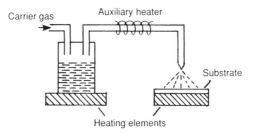

**Figure 5.41:** Spray deposition.

ITO, which (1) acts to allow light to pass through and (2) provides a conductive top electrode to collect electrons. Usually there is no top grid to supplement the tin oxide. Other highly conductive, transparent materials (e.g., ZnO) could be candidate window materials if they were found to be less costly or result in improved device performance. At present, tin oxide has had the best combination of stability (to subsequent processing) and performance.

After the tin oxide, an $n$-type heterojunction partner for the subsequent CdTe layer is deposited. Usually this is CdS, which makes an effective junction with CdTe. The next layer is the CdTe itself. After that, two structures predominate: a more typical one in which (after heat treatment and/or chemical etch) a conductor such as Ni or graphite with Cu is deposited to make contact; or a newer design, developed by Ametek, in which $p$-ZnTe and then a metal contact are deposited.

In all these processes, care must be taken to respect the integrity of the previous layers by avoiding chemical attack or extreme temperatures that would cause chemical or diffusion damages. However, these boundary conditions are not very limiting in practice. For instance, several groups have observed that CdS and CdTe can form solid solutions at high temperatures and improve cell performance.

### 5.3.5A History, Present Status and Process Description

### 5.3.5A.1 Window Materials and Heterojunction Partners

### 5.3.5A.2 $SnO_2$

Tin oxide has been deposited by spraying, CVD and MOCVD, and sputtering. Reviews of processes to make tin oxide include Dawar and Joshi (1984) and Mooney and Radding (1982).

*$SnO_2$ Spraying:* Tin oxide of commercial quality can be made by spraying a solution of a tin chloride ($SnCl_4$) in water, alcohol, and organic solvent on a hot substrate (above 400°C). The solution is carried to an atomizer by a gas such as argon, nitrogen, or air. The atomizer sprays it on a heated substrate, where it breaks down pyrolitically, and a simple reaction with water leaves behind the desired compound (Figure 5.41; and 5.42). Conductivity can be increased by doping with $SbCl_3$ or HF. Fluorine doping has produced the highest conductivities. Spraying has been widely used because it is easy and economical.

*$SnO_2$ CVD and MOCVD:* The chemical vapor deposition of $SnO_2$ is based on the use of volatile compounds which condense/react on a heated substrate (Figure 5.42). Various heaters and flow rates are adjustable, as is the geometry of the reaction chamber and substate. Many workers have reported growing films by this method,

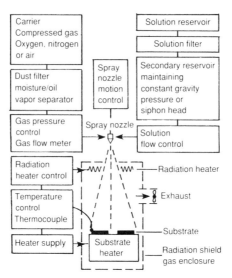

**Figure 5.42:** Deposition of $SnO_2$ by CVD.

including Liversey et al. (1968) and Kane et al. (1975). Kane et al. (1975) used dibutyl tin diacetate as the organometallic feedstock, nitrogen as the carrier gas, and oxygen as the reactant. The substrate was held at 420°C. Subsequently, they added Sb dopant using antimony pentachloride. Later, Baliga and Gandhi (1976) added phosphine to dope the tin oxide and succeeded in reducing resistivity significantly. More recent workers have incorporated F as the *n*-type dopant using $Ch_3CHF_2$ (Kato et al., 1987). Other solutions have been hydrous $SnCl_4$ and $SbCl_3$ dopant (Varma et al., 1984). These appear to have safety and cost advantages over previous alternatives.

In an effort to develop textured tin-oxide for light-trapping, Gordon et al. (1987) investigated several approaches. They used tetramethyl tin with oxygen (20–80 mole%) using trifluorobromomethane (a freon) 5-mole% dopant (590°C substrate temperature) to produce hazy films with pyramidal surface structures that were effective in light-trapping. Growth rate was about 150 Å/sec. Subsequently, they developed a dimethyetin dichloride precursor (Gordon et al., 1987) which they reported can be 15 times less expensive and much less hazardous (lower toxicity, volatility and flammability) than tetramethyl tin. Also, growth rates increased threefold.

*SnO₂ Sputtering:* Dawar and Joshi (1984) review this process for making tin oxide (Figure 5.43). One key objection to sputtering has been the potential for substrate damage by energetic ions during deposition. However, since the $SnO_2$ is deposited on glass (or $SiO_2$ in some cases) this is not a factor. Sputtering of tin oxide has been done with metal-oxide targets and with metal targets in the presence of added oxygen.

Other methods of depositing tin oxide include dip-coating, electroless deposition, vacuum evaporation, and ion plating. Since economics play a strong role in choosing a final production method, new methods that are clearly low cost, such as

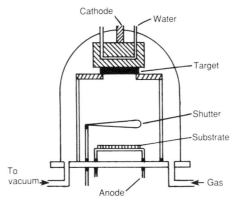

**Figure 5.43:** Sputtering system.

dip-coating, are of potential interest. Dip coating (see below for more) can be carried out at room temperature and depends on the controlled reaction of anions and cations and the precipitation of the desired material. Quality dip-coated tin oxide films have been made by Chopra and Das (1983), Ravindra and Sharma (1985), and others.

### 5.3.5A.3  Cadmium Sulfide

CdS has been deposited by many processes, including spraying, dip-coating, electrodeposition, screen printing, vacuum evaporation, and sputtering.

*CdS Spraying*: Chamberlin and coworkers at Photon Power (now Photon Energy) used spray pyrolysis to make CdS in 1962 (Hill and Chamberlin, 1964, U.S. Patent; Chamberlin and Skarman, 1966). Work in CdS spraying during the seventies included Jordan (1975), Bube and coworkers (Wu et al., 1972; Bube et al., 1977), Savelli and coworkers (Bougnot et al., 1976), and Chopra and coworkers. Chopra (1982) has published a review of spray pyrolysis.

Spray pyrolysis usually involves spraying an aqueous solution of soluble salts of the atoms of the desired compound onto a hot substrate. The droplets undergo endothermic (pyrolytic) decomposition on the substrate. Volatile by-products escape, and the constituent atoms recombine as the desired compound. However, in many cases, intermediate compounds are formed and play a significant role in grain growth and stoichiometry. Sometimes impurities from the solution are unintentionally included, affecting structure and optoelectronic properties. Sintering may be needed to improve the quality of sprayed films. A detailed schematic spray system is shown in Figure 5.44 (Chopra and Das, 1983). Large area uniformity can be attained by moving the spray head, substrate, or both. Banerjee (1978) claims that due to the random nature of atomized deposition, sprayed CdS films can be pin-hole free when they are as thin as 0.1 $\mu$m.

The most common solutions (Chopra and Das, 1983) for spraying CdS are a water-soluble Cd salt ($CdCl_2$) and a sulfo-organic salt (Thiourea, $(NH_2)_2CS$) in a dilute solution. CdZnS alloys have been prepared as well (Banerjee, 1978) with the proper inclusion of a Zn compound.

Ametek has used a narrow-gap pyrolitic reactor to spray CdS at 450°C to a thickness of 0.15 $\mu$m. They use an aerosol containing $CdCl_2$ and thiourea (Meyers,

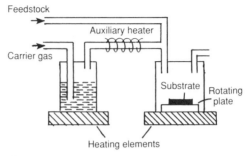

**Figure 5.44:** Sprayed CdS deposition system.

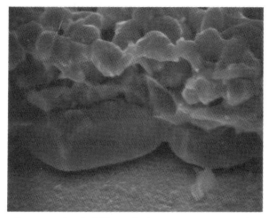

**Figure 5.45:** Large-grained CdS made by slurry spraying (Photon Energy).

1986). Ametek's CdTe cells have been measured as high as 11% at SERI with CdS layers made by this method. They have not been able to produce the thinner CdS (300 Å) needed to improve cell blue response using this method.

A variation of the spray pyrolysis method–involving no pyrolysis–is used by Photon Energy (Albright et al., 1987) to make CdS. A slurry of semiconductor powder and carrier liquid such as propylene glycol and $CdCl_2$ is sprayed on the substrate (e.g., glass/$SnO_2$) at room temperature. After drying, the film is compressed mechanically (5,000–10,000 psi) and regrown in a heated (530°C) ambient of nitrogen and 1%–3% oxygen. The presence of $CdCl_2$ flux assists regrowth. High-quality CdS of substantial grain size (over 5 $\mu$m) has been grown in this manner and incorporated in Photon Energy's CdS/CdTe cells and modules (Figure 5.45).

*CdS Dip Coating:* 'Dip coating' is a term covering several different chemical solution-growth methods of making CdS layers that do not include spraying or the flow of an electric current. The dip-coating method was pioneered by Bode and coworkers (Bode 1963), and subsequent work was done by Kitaev (Kitaev et al.,

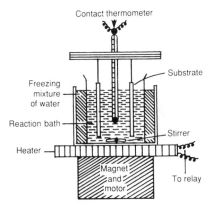

**Figure 5.46:** Solution growth.

1965), Nagao and Watanabe (1968), Chopra (Sharma et al., 1976; Chopra et al., 1982), and Danaher (Danaher et al., 1985) and their coworkers.

A schematic apparatus is shown in Figure 5.46 (Chopra and Das, 1983). After cleaning, the substrate is dipped vertically into the solution, which is stirred. Danaher et al. (1985) rotates the substrate within a solution flowing parallel to the axis of rotation. The solution can consist of thiourea or thioacetamide in alkaline salts of Cd such as $Cd(CH_3COO)_2$ (Kaur et al., 1980) or $CdCl_2$ (Danaher et al., 1985). In addition, $NH_4$ and ammonia can be present (Danaher et al., 1985). When the ion product of the metal and chalcogenide ions is more than the solubility product of the corresponding chalcogenide, the CdS is deposited in an ion-by-ion condensation. Temperatures are about 90°C. Initial deposit on a clean surface can be slow but speeds up as nucleation occurs. Surfaces can be 'sensitized' to improve initial deposition rates. As the process continues, deposition slows, until precipitation from the solution equals dissolution from the film, and deposition ceases. A CdS powder forms in solution, on the walls of the chamber, and can also coat the adherent CdS film. It can be removed by ultrasonic cleaning. Small concentrations of impurities and dopants can be incorporated during growth (Danaher et al., 1985) and detected by SIMS, Danaher et al. (1985) found that annealing at 450°C for four hours in $H_2$ with Cd (or In) vapor improved film properties. It lowered resistivity from $10^9$ ohm-cm (700 ohm-cm under illumination) to 0.04 ohm-cm. Lowering of resistivity was attributed to oxygen desorption and incorporation of dopant.

Very thin (under 500 Å), uniform and adherent films can be obtained from this process, which makes it of interest in the effort to make CdS that does not block the UV spectrum. ARCO Solar has such a patent (Choudary et al., 1986) for thin CdS on $CuInSe_2$ cells and has achieved near optimal blue response in their $CuInSe_2$ devices. Boeing has developed dip-coated CdZnS, also for their $CuInSe_2$ cells, by adding $ZnCl_2$. (These are covered earlier in the $CuInSe_2$ dip-coating section).

*CdS Electrodeposition:* The electrodeposition process has been described previously (CdTe section). Kröger (1978) gave the conditions needed to cathodically deposit CdS. Baranski and Fawcett (1980) deposited CdS on conducting substrates using a Cd-salt ($CdCl_2$) and S dissolved in nonaqueous solvent such as DMSO.

Monosolar originally used electrodeposition to deposit CdS on ITO/glass substrates (Panicker et al., 1978; Kröger and Rod, 1979, UK Patent 1,532,616). These CdS layers were used to make the first high-efficiency CdTe cells using electroplated CdS. The Monosolar CdS layers were as thin as 500 Å, though thinner layers (useful for improved blue response) caused shunting and lower voltages (Basol et al., 1982). Basol used cadmium sulfate (CdSO$_4$) or CdCl$_2$ (emphasized in later work) and sodium thiosulfate as the source materials for the CdS. pH was between 1.5 and 4. The deposition voltage was given as about $-0.65$ V; temperature of the stirred electrolyte was 90°C. Current density was about 0.2 mA/cm$^2$ (Basol et al., 1983, patent) but was later reduced to about 0.03 mA/cm$^2$ (Basol, 1988). After a post-fabrication heat treatment, the thin CdS had a resistivity of about 200 ohm-cm (Basol, 1984a).

*CdS Screen Printing*: Matsushita screen prints CdS (and CdTe) to make high-efficiency cells. Their process has been described previously in the section on CdTe. A good review of their work can be found in Ikegami (1988). CdS paste for screen printing is made using about 100 g 99.999% CdS powder with grain sizes of 1–2 $\mu$m mixed with 10 g of CdCl$_2$ powder and propylene glycol. Matsushita screen prints on bare (no tin oxide) borosilicate glass, dried for 1 hour at 120°C. The film is then sintered in an alumina box in nitrogen at 690°C for 90 minutes. The CdCl$_2$ promotes particle fusion and grain regrowth. CdS thickness is about 30 $\mu$m, and grains are large. This thick CdS layer reduces the current density of the Matsushita cells by about 6 mA/cm$^2$ compared with designs exploiting thin CdS.

*Evaporated CdS*: Evaporation of CdS powder has long been favored as a method of making CdS films in the laboratory. It may also be considered a potentially low-cost method for manufacturing because it can be done at a relatively high rate (about 1 $\mu$m/min) (Russell et al., 1984). The Institute of Energy Conversion, University of Delaware is an acknowledged leader in the evaporation of CdS, especially using an effusion cell design for the sources (Hall et al., 1980). Another advantage of vacuum evaporation is that it is done at relatively low substrate temperatures (about 200°C).

Professor Ting Chu (1988) has deposited thin (350 Å) CdS by evaporation followed by an in situ etch that reduces the CdS layer thickness. To produce thinner layers in a controlled manner, the CdS is evaporated at a lower source temperature (800°C instead of 900°–950°C), thus at a lower rate (0.12 $\mu$m/hr). His cells have shown quantum efficiencies of over 50% at 400 nm, proving that the CdS is not fully opaque to ultraviolet light. The in situ etch in hydrogen that thins the CdS also improves junction performance. Many workers have observed that annealing CdS in hydrogen removes oxygen from the films and increase mobility and carrier concentration by an order of magnitude.

*CdS by Other Methods*: Several other methods, including sputtering, CVD, and anodization, have been used to make CdS. Anodization is a process in which CdS is fabricated on a surface by the anodization of Cd with a solution containing Na$_2$S in NaOH. The Cd can be electrodeposited cathodically. However, imperfect anodization allows the possibility that a Cd layer will not be fully reacted and remain as an imperfection between the CdS and the tin oxide substrate.

Nicolau (1985) describes a procedure in which a bath deposition process similar to 'dip coating' can be used to sequentially deposit Cd$^{+2}$ and S$^{2-}$, which are then

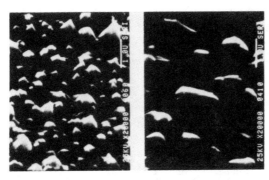

**Figure 5.47:** A critical step in the fabrication of high-efficiency CdTe devices is a post-deposition heat treatment at about 400°C that regrows and fuses grains (Ametek) (before: left; after: right).

reacted to form CdS. The process is exceedingly slow and laborious, but for very thin layers it might potentially be of interest.

Several authors (Williams et al., 1979; Pande et al., 1984) describe the electrophoresis of CdS. It is the deposition of charged particles from a solution to an electrode. CdS can be made by passing $H_2S$ through $Cd(CH_3COO)_2$ and water to form a colloid. Deposition rate can be high, but powdered films may result. Post-deposition treatments can make the films more adherent and electrically active.

### 5.3.5A.4 Back Contacts

No metal has a work function higher than that of $p$-CdTe (about 5.7 eV when doped $p$-type). Thus it is difficult to form a true ohmic contact to $p$-CdTe (Bube, 1988). The usual procedure is to either (1) form a low-resistance tunnel junction by doping the CdTe heavily or (2) find other, nonmetals that can form true ohmic contacts. One of the drawbacks of the tunnel junction approach is the well-known difficulty of stably making $p$-CdTe highly conductive. The combination of a chemically altered $p^+$-CdTe back surface (for high conductivity) and a diffusive metal contact has been the cause of much instability in CdTe devices.

In practice, CdTe contacting is a rather empirical procedure that is done in several steps (Zweibel and Hermann, 1985). Usually there is a heat treatment of the CdTe material that takes place before contacting. This heat treatment makes the CdTe $p$-type or in some way activates the junction. Ametek observes grain-regrowth of the CdTe during this anneal (Figure 5.47). Then there may be an etch procedure (usually with a potassium dichromate) to prepare the exposed CdTe surface. The purpose is to make it more highly conductive $p$-type for the formation of a tunnel junction contact. The CdTe is Cd-depleted; a second etch then removes residual $TeO_2$ oxide. Contact metal (Au, Ni, stainless steel, Cu) or conductive pastes may then be applied.

A different procedure has been adopted by Ametek. They use an '$n$-$i$-$p$' structure in which they avoid CdTe contacting by sandwiching $i$-CdTe between $n$-CdS and $p$-ZnTe. The metal contact is applied to the $p$-ZnTe. Curing heat treatments are carried out to complete the device.

Examination of contacting will be broken down into the above-described procedures: (1) etch and metal and (2) ZnTe (or HgTe) and metal. Heat treatments are part of both procedures.

*Etch and Metal:* A classic approach was carried out by Basol (1988) and coworkers at Monosolar. Their as-deposited CdTe films were highly resistive $n$-type ($10^5$ ohm-cm). To prepare efficient devices, they developed a 'type-conversion' procedure (Basol et al., 1983; Basol, 1984a; Basol et al., 1985). The CdTe films (on glass/ITO/CdS) were heat-treated at 400°C for 8–12 minutes. Before heat treatment, quantum efficiency was low–but maximum in the long wavelengths. This is the kind of quantum efficiency one might find from a deeply buried homojunction. After annealing, response became strong and maximum in the short wavelengths, as would be expected of a heterojunction between $p$-CdTe and $n$-CdS (Basol, 1988). This indicated that the heat-treatment resulted in type-conversion of the CdTe. They then created a Te-rich ($p^+$) back surface by two etches (Basol, 1984b). The first consisted of an oxidizer such as chromic acid to produce a Te-rich surface 300 Å–500 Å thick. Then a hydrazine etch removed tellurium oxide, which is highly resistive. A contact metal such as Ni was then applied.

Tyan and coworkers at Kodak used an etch consisting of $HNO_3$ and $H_3PO_4$ (Tyan, 1982) to produce a Te-rich surface. Matsushita has used a variation of these methods wherein a carbon paste doped with Cu (50–100 ppm) is applied as the back electrode. They claim (Ikegami, 1988) that the Cu diffuses into the CdTe during anneal at 400°C (30 minutes) and produces the CdS/CdTe junction by making the CdTe $p$-type. The ambient is $N_2$ with 1–1.5 mole % $O_2$. The anneal also makes the surface of the CdTe $p^+$, allowing for a good tunnel-junction contact. Ikegami (1988) states that their encapsulated CdTe modules have been tested for up to 800 days with no change in performance.

Photon Energy's contact procedure (Albright et al., 1988) is a variation of the Matsushita method using carbon paste. They had previously used a Te metal electrode, which they abandoned because of diffusion-induced instability. They are presently experimenting with pastes doped with Te compounds including but not confined to those developed by Ametek and Chu (below).

*ZnTe and HgTe:* ZnTe doped highly $p$-type has a work function of about 5.7 eV, which is almost the same as that of $p$-CdTe. Thus it can be a true ohmic contact with $p$-CdTe. Highly $p$-type ZnTe can also induce a drift field across an intrinsic CdTe layer in a $p$-$i$-$n$ design. Both of these options are useful for avoiding contact degradation in CdTe cells. Ametek has used the '$n$-$i$-$p$' design ($n$-CdS/$i$-CdTe/$p$-ZnTe) to make high efficiency (11% measured at SERI) cells.

As mentioned above (see Figure 5.47), Ametek heat treats their glass/SnO₂/CdS/CdTe devices in air for about 20 minutes at 400°C (Meyers, 1986) to improve CdTe morphology and reduce defects. They then etch the CdTe surface with a mild solution of Br:methanol before $1,000$ Å of $p$-ZnTe are deposited by evaporation.

The materials used by Ametek to contact $p$-ZnTe are graphite, Ni, Cu, or Al. Before application, an etch and heat treatment are done to maximize the ZnTe properties and to prepare the exposed surface. The metals are deposited by evaporation.

International Solar Electric Technology (ISET) has developed a two-stage electrodeposition and annealing method of making conductive ZnTe (Basol and Kapur, 1988, private communication). They electrodeposit the metal layers separately then anneal, causing a reaction that forms ZnTe. Cu doping has also been achieved, with highly transparent films having 0.1 ohm-cm resistivity.

Professor Ting Chu has developed another material, HgTe, as a true ohmic contact to p-CdTe (Chu, 1988). HgTe has an electron affinity of 5.9 eV, greater than that of p-CdTe (5.7 eV). Chu was able to achieve very low contact resistance, equivalent to those found with gold contacts. (These latter are very conductive but unstable due to the mobility of the Au atoms.) His 10.5% efficient cells (SERI-measured) had a HgTe contact.

### 5.3.5B  Problems and Issues

Perhaps the major challenge for most of these non-CdTe layers is to gain–via experience–the reproducibility needed for production yield to be high. In some cases, there must be further optimization of processes and performance. Cost is an issue with only a few.

### 5.3.5B.1  Materials

Tin oxide can be made from relatively common, low-cost chemicals. The same is true of CdS. In addition, the CdS used in very high-efficiency devices will eventually be less than 0.05 $\mu$m thick, minimizing cost. However, at present, there are groups hampered by high costs associated with thick CdS.

Screen-printed CdS (Ikegami, 1988) made by Matsushita is about 30 $\mu$m thick. The materials cost of this layer is high, but more importantly it causes problems for device delineation and UV current loss. More than likely, screen printing will never be a preferred method for depositing thin CdS.

The other materials (graphite paste, Cu, Al, Ni) are either common or are used in minimal amounts.

Module definition issues are minimal except for specific cases such as the unusual glass/thick CdS structure used by Matsushita. Because of the thickness (30 $\mu$m) and temperature (690°C) of screen-printed CdS, serious cell definition problems exist for Matsushita's CdTe modules. They have a very poor active/total area ratio of about 0.6 (Ikegami, 1988). Until the high-temperature CdS deposition step is replaced (so that they can use tin oxide as the first layer on the glass), their efficiency achievements will be limited to small-area cells.

### 5.3.5B.2  Deposition Rate

Most of the processes being considered have high rates, making capital equipment expenditures low per unit of production.

For tin oxide, Chopra (1983) suggests that spray pyrolysis rates are about 0.1 $\mu$m/min.

For CdS, Chopra and Das (1983) suggests that spray pyrolysis rates are about 0.1 $\mu$m/min.

For CdS, Chopra and Das (1983) say that deposition rates for spray pyrolysis are about 500 Å/min. The deposition rate of dip-coating can vary (100 Å–1,000 Å/min), allowing for controllable, very thin layers. Basol (1988) gave the deposition rate of their electrodeposition process as a slow 15 Å/min. Russell et al. (1984) estimates the production rate of CdS evaporation at over 1 $\mu$m/min.

### 5.3.5C  Discussion/R&D Issues

Perhaps the most significant issue remaining for windows and contacts of CdTe devices is that of maximizing blue response by the use of thin CdS or some alternative. The pressure to provide very thin but pin-hole free CdS is a stringent requirement that will reduce the number of options for making CdS in CdTe device manufacturing.

Some of the most costly steps in making CdTe modules occur during contacting and metalization. For instance, Ametek's use of evaporated ZnTe is one of their most capital-intensive steps. A new process to replace evaporation of ZnTe would be of use and is being investigated. The ISET electrodeposition and annealing approach (above) is an example. The subsequent metalization layers are also areas in which lower cost processes would be valuable.

In general, all processes will need to be upgraded and optimized as manufacture of CdTe begins. However, they have already been designed for high-throughput and low capital equipment cost. Progress toward very low cost CdTe modules should continue to be strong.

# References

Abdul-Hussein, N.A.K., A.N.Y. Samaan, R.D. Tomlinson, A.E. Hill, H. Neumann (1985), *Electrical properties of RF sputtered thin films of CuInSe₂. I. Experimental results*, Cryst. Res. & Technol. (Germany) **20**, No. 4, pp. 509-14.

Abernathy, C.R., C.W. Bates, Jr., A.A. Anani, B. Haba, G. Smestad (1984), *Production of single phase chalcopyrite CuInSe₂ by spray pyrolysis*, Appl. Phys. Lett. **45**, No. 8, pp. 890-2.

Abernathy, C.R., C.W. Bates, Jr., A.A. Anani, B. Haba (1984), *Kinetic effects in film formation of CuInSe₂ prepared by chemical spray pyrolysis*, Thin Solid Film **115**, No. 1, L41-3.

Abernathy, C.R., R. Cammy (1985), *Chemical spray pyrolysis of copper indium diselenide thin films and devices*, Univ. Microfilms, Int., 153 pp.

Agnihotri, O.P., P. Raja Ram, R. Thangaraj, A.K. Sharma, Atul Raturi (1983), *Structural and optical properties of sprayed CuInSe₂ films*, Thin Solid Films **102**, No. 4, pp. 291-7.

Albright, S.P., D.K. Brown, J.F. Jordan (1987), *Method and apparatus for forming a polycrystalline monolayer*, European Patent Application, 87114766.6 (assignee Photon Energy).

Albright, S.P., Singh, V.P., Ackerman, B. (1988), *High efficiency large-area CdTe panels*, Annual Technical Progress Report, June 1987-June 1988, SERI/DOE (in press).

Anthony, T.C., A.L. Fahrenbruch, R.H. Bube (1984), *Growth of CdTe by close-spaced vapor transport*, J. Vac. Sci. Technol. A 2 (3), 1296-1302.

Aranovich, J., A. Ortiz, R.H. Bube (1979), *Optical and electrical properties of ZnO films prepared by spray pyrolysis for solar cell applications*, J. Vac. Sci. Technology **16**, pp. 994.

Arita T., N. Suyama, Y. Kita, S. Kitamura, T. Hibino, H. Takada, K. Omura, N. Ueno, M. Murozono (1988), *CuInSe₂ films prepared by screen-printing and sintering*, Matsushita Battery Industrial Co., 20th IEEE PV Specialists Conference - 1988.

Banerjee, A. (1978), Ph.D. Thesis, Indian Institute of Technology, New Delhi.

Banerjee, A., P. Nath, V.D. Vankar, K.L. Chopra (1978) *Properties of $Zn_x Cd_{1-x}S$ films prepared by solution spray technique*, Phys. Stat. Sol. (a) **46**, pp. 723.

Baranski, A.S., Fawcett, W.R. (1980), J. Electrochem Soc. **127**, 766.

Basol, B.M. (1988), *Electrodeposited CdTe and HgCdTe solar cells*.

Basol, B.M. (1988), Solar Cells **23** # 1=2, 69-89.

Basol, B.M. (1984), U.S. Patent 4,456,630, *Method of forming ohmic contacts*.

Basol, B.M. (1984a), *High efficiency electroplated heterojunction solar cell*, J. Appl. Phys. **55** (2), 601-603.

Basol, B.M. (1984b), *Method of forming ohmic contacts*, U.S. patent 4,456,630.

Basol, B.M. (1980), Ph.D. UCLA

Basol, B.M., S.S. Ou, O.M. Stafsudd (1985), J. Appl. Phys. **58**, 3809.

Basol, B.M., O.M. Stafsudd (1981a), Solid State Electronics, **24**, 121.

Basol, B.M., O.M. Stafsudd (1981b), Thin Solid Films **78**, 217.

Basol, B.M., O.M. Stafsudd, R.L. Rod, E.S. Tseng (1980), *Proceedings 3rd EC PV Solar Energy Conference*, pp. 878.

Basol, B.M., E.S. Tseng (1986), *Mercury cadmium telluride solar cell with 10.6% efficiency*, Appl. Phys. Lett 48 (14), 946-948.

Basol, B.M., E.S. Tseng, D.S. Lo (1986), U.S. Patent 4,629, 820, *Thin film heterojunction PV device*.

Basol, B.M., E.S. Tseng, R.L. Rod (1983), U.S. Patent 4,388, 483, 1983, *Thin film heterojunction cells and methods of making the same*.

Basol, B.M., E.S. Tseng, R.L. Rod, S. Ou, O.M. Stafsudd (1982), *Ultrathin electrodeposited CdS/CdTe heterojunction with 8% efficiency*, *Proceedings of the 16th IEEE PV Specialists Conference*, 805-808.

Bates, C.W. Jr., K.F. Nelson, S. Atiq Raza, J.B. Mooney, J.M. Recktenwald, L. Macintosh, R. Lamoreaux (1982), *Spray pyrolysis and heat treatment of CuInSe₂ for photovoltaic applications*, Thin Solid Films **88**, No. 3 pp. 279-83.

Bates, C.W. Jr., M. Uekita, K.F. Nelson, C. Abernathy, J.B. Mooney (1982), *The effect of pH on the production of chalcopyrite CuInSe₂ prepared by spray pyrolysis*, Conference Record of the Sixteenth IEEE Photovoltaic Specialists Conference - 1982, 27-30 Sept. 1982, San Diego, CA, pp. 1427-8.

Bates, C.W. Jr., M. Uekita, K.F. Nelson, C.R. Abernathy, J. B. Mooney (1983), *Effect of pH on the production of chalcopyrite CuInSe₂ prepared by spray pyrolysis*, Appl. Phys. Lett. **43**, No. 9, pp. 851-2.

Bhattacharya, R.N. (1983), *Solution growth and electrodeposited CuInSe₂ thin films*, Journal Electrochemical Society, (130:10), pp. 2040-2.

Bhattacharya, R.N., Rajeshwar, K.R. (1986), *Electrodeposition of CuInX (X=Se,Te) thin films*, Solar Cells (Switzerland), (16), pp. 237-43.

Bilaga, B.J., Gandhi, S. (1976), J. Electrochem. Soc. **123**, 941.

Binsma, J.J.M., Van Der Linden (1982), *Prepartion of thin CuInSe₂ films via a two-stage process*, Thin Solid Films, 97, pp. 237-243.

Birkmire, R.W., L.C. Dinetta, J.D. Meakin, J.E. Phillips (1984), *CuInSe$_2$/CdS-CdTe/CdS polycrystalline tandem solar cells*, Inst. of Energy Conversion, Delaware Univ., Newark, DE, USA, Seventeenth IEEE Photovoltaic Specialist Conference, 1406-7.

Birkmire, R.W., R.B. Hall, J.E. Phillips (1984), *Material requirements for high efficiency CuInSe$_2$/CdS solar cells*, Inst. of Energy Conversion, Delaware Univ., Newark, DE, USA, Seventeenth IEEE Photovoltaic Specialists Conference, 882-6.

Birkmire, R.W., J.E., Phillips, L.C. DiNetta, J.D. Meakin (1985), *Thin-film tandem solar cells based on CuInSe$_2$*, Inst. of Energy Conversion, Delaware Univ., Newark, DE, USA, Sixth E. C. Photovoltaic Solar Energy Conference 270-5 London, England.

Bloss, W.H., F. Pfisterer, H.W. Schock (1988), *Polycrystalline II-VI-related thin film solar cells*, in K. W. Böer (ed), Advances in Solar Energy, Plenum Press and American Solar Energy Society, Boulder, Colorado.

Bode, D.E. (1963), *Proc. Natl. Elec. Conf.* **19**, 630.

Boone, J.L., P. Van Doren (1982), *Investigation of chemical spray and electrodeposition techniques...Final Report*, University of Missouri-Rolla.

Bougnot, J., S. Duchemin, M. Saveilli (1986), *Chemical spray pryolysis of CuInSe$_2$ thin films*, Centre d'Electron, de Montpellier, Univ. des Sci. et Tech. du Languedoc, France. Sol. Cells (Switzerland) **16**, pp. 221-36.

Bougnot, J., M. Perotin, J. Marucchi, M. Sirkis, M. Savelli (1976), *Proc. 12th IEEE PV Specialists Conf.* Baton Rouge, 519.

Brodie, D.E., R. Singh, J.H. Morgan, J.D. Leslie, C.J. Moore, A. E. Dixon (1980), *Proc. of 14th IEEE Photovoltaic Specialists Conference* p. 468.

Bube, R.H. (1988), *CdTe junction behavior*, Solar Cells **23** # 1-2, 1-18.

Bube, R.H. (1988), *Thin film polycrystalline solar cells*, Solar Cells **23** # 1-2, 1-18.

Bube, R.H., F. Buch, A.L. Fahrenbruch, Y.Y. Ma, K.W. Mitchell (1977), IEEE Trans. Electron Devices, ED-24, 487.

Bunshah, R.F., Schwartz, M. et. al. (1982), *Deposition technologies for films and coatings*, Noyes Publications, pp. 385-453.

Bureau of Mines (1988), *Cadmium*, Minerals Commodities Summaries.

Cahen, D. (1984), *Electrodeposited layers of copper indium sulfides (CuInSe$_2$, CuIn$_5$S$_8$) and coper indium diselenide (CuInSe$_2$)*, Prog. Cryst. Growth Charact. (10) *Proc. Int. Conf. Ternary Multinary Compd., 6th*, pp. 345-51.

Cahen, D., G. Hodes (1987), *Ternary adamantine materials for low-cost solar cells*, SERI Subcontractor Final Report, IL-5-04132-1.

Call, R.L., N.K. Jaber, K. Seshan, J.R. Whyte (1980), *Structural and electronic properties of three aqueous-deposited films: CdS, CdO, ZnO*, Solar Energy Materials 2, 373-380.

Canevari, V., U. Emiliani, N. Romeo, G. Sberveglieri, L. Zanotti (1983), *Low resistivity and high transparency Zn$_x$Cd$_{1-x}$S thin films grown by flash evaporation*, University of Parma (Italy), Thin Solid Films **106**, No. 4, pp. L91-4.

Canevari, V., N. Romeo, G. Sberveglieri, L. Zanotti, M. Curti (1983), *Low resistivity N- and P-type CuInSe$_2$ thin films grown by flash evaporation*, presented at the *Proceedings of the Joint Meeting of the CNR Fine-Chemicals Finalized Pro-

*gramme and the Italian Association for Crystal Grown*, Milan, Italy, 1-3 Dec., Mater. Chem. and Phys. **9**, No. 1-3, pp. 205-11.

Chamberlin and Skarman (1966), J. Electrochem. Soc. **113**, 86.

Chen, W.S., R.A. Mickelsen (1980), *Thin-film CdS/CuInSe₂ heterojunction solar cell*, Proc. Soc. Photo-Opt. Instrum. Eng. **248**, pp. 62-9.

Chen, Y.W., J.A. Turner, R. Noufi (1985), *Preliminary studies on the low-cost preparation of chalcogenide semiconductors from solution depositions*, Appl. Phys. Communications 4, No. 4, p. 241-252.

Chopra, K.L., Das, S.R. (1983), Thin Film Solar Cells, Plenum Press, NY, pp. 259-265.

Chopra, K.L., R.C. Kainthla, D.K. Pandya, A.P. Thakoor (1982), Physics of Thin Films **12**, Academic Press, N.Y.

Choudary, U.V., Y. Shing, R.R. Potter, J.H. Ermer, V.K. Kapur (1986), U.S. patent 4,611,091, *CuInSe₂ thin film solar cell with thin CdS and transparent window layer*, assignee Atlantic Richfield Co.

Chu, T.L. (1988a), *Thin film CdTe solar cells by two chemical vapor deposition techniques*, Solar Cells **23** # 1-2, 31-48.

Chu, T.L. (1988b), *Thin film CdTe solar cells*, Final Technical Report May.

Chu, T.L. (1988), *Thin film CdTe solar cells*, Final Technical Report May 1985-May 1988, DOE/SERI (in press).

Chu, T.L. (1985), *Thin film CdTe solar cells*, Final Technical Progress Report Oct. 1983-May 1985, Southern Methodist University for the Solar Energy Research Institute, Golden, CO.

Chu, T.L., S.S. Chu, S.C. Lin, J. Filipowicz (1984), *Photoelectrical properties of CuInSe₂ thin films*, presented at the proceedings of the Sixth International Conference on Ternary and Multinary Compounds, Caracas, Venezuela, 15-17 Aug., Prog. Cryst. Growth and Charact. **10**, pp. 361-4.

Chu, T.L., S.S. Chu, S.C. Lin, J. Yue (Sept. 1984), *Large grain copper indium diselenide films*, Journal Electrochemical Society, (131:9), pp. 2182-5.

Coutts, T.J. (1982), *High efficiency solar cells with CdS window layers*, Newcastle Polytechnic, Thin Solid Films **90**, No. 4, pp. 451-60.

Danaher, W.J. et al. (1978), *Photoelectrochemical cell with CdTe film*, Nature **271**, No. 5641, pp. 139.

Danaher, W.J., L.E. Lyons, G.C. Morris (1985), *Some properties of thin films of chemically deposited cadmium sulfide*, Solar Energy Materials, 12, pp. 137-148.

Das, S.R. (1978), Ph.D. Thesis, Indian Institute of Technology, Delhi, India.

Dawar, A.L., Joshi, J.C. (1984), *Review: semiconducting transparent thin films: their properties and applications*, J. Mat. Sci. **19**, 1-23.

Devaney, W.E., R.A. Mickelsen, W.S. Chen, V.E. Chen, B.J. Stanbery, J.M. Stewart, F.W. Lytle, A.F. Burnett (1987), *Cadmium sulfide/copper ternary heterojunction cell research*, SERI Subcontract Final Report, SERI/STR-211-3230.

Dhere, N.G., M.C. Lourenco, R.G. Dhere (1984), *Composition and structure of CuInSe₂ thin films prepared by vacuum evaporation of the constituent elements*, Sol. Cells **13**, No. 1, pp. 59-65.

Dimmler, B., H. Dittrich, H.W. Shock, *Structure and morphology of evaporated bilayer and selenized CuInSe₂ films*, University of Stuttgart, Twentieth IEEE Photovoltaic Specialists Conference - 1988.

Dimmler, B., H. Dittrich, H.W. Shock (1988), *Structure and morphology of evaporated bilayer and selenized CuInSe₂ Films*, University of Stuttgart, Twentieth IEEE Photovoltaic Specialists Conference.

DOE PV Systems Program Summary (1978), DOE/ER-0075, pp. 102.

Don, E.R., S.R. Baber, R. Hill (1981), *Copper indium diselenide for CdS:CuInSe₂ solar cells*, Newcastle Upon Tyne Polytech. Newcastle Upon Tyne, England, Third E. C. Photovoltaic Solar Energy Conference 897-901.

Don, E.R., R.R. Cooper, R. Hill (1985), *The structural, optical and electrical properties of copper indium diselenide and their dependence on stoichiometry*, presented at the Sixth E. C. Photovoltaic Solar Energy Conference, London, England, 15-19 April, Reidel, xxxv+1104 pp.

Don, E.R., R. Hill, G.J. Russell (1986), *The structure of CuInSe₂ films formed by co-evaporation of the elements*, Solar Cells **16**, pp. 131-42.

Doty, M., P. Meyers (1988), *Safety advantages of a CdTe based PV module plant*, in AIP Conference *Proceedings 166, PV Safety, Jan 19-20, 1988*, Denver, CO (SERI), W. Luft, ed., New York, pp. 10-18.

Duchemin, S., V. Chen, J.C. Yoyotte, C. Llinares, J. Bougnot, M. Savelli (1987), *Sprayed CdS/CuInSe₂ solar cells: first results*, Seventh E. C. Photovoltaic Solar Energy Conference, 615-19.

Duchemin, S., V. Chen, J.C. Yoyotte, C. Llinares, J. Bougnot, M. Savelli (1986), *Sprayed cadmium sulfide/copper indium selenide (CuInSe₂) solar cells: first results*, E. C. Photovoltaic Sol. Energy Conf., 7th, pp. 615-19.

Ermer, J.H., R.B. Love, A.K. Khanna, S.C. Lewis, F. Cohen, *CdS/CuInSe₂ junctions fabricated by DC magnetron sputtering of Cu₂Se and In₂Se₃*, presented at the Conference Record of the Eighteenth IEEE Photovoltaic Specialists Conference, 21-25 Oct. 1985, Las Vegas, NV, pp. 1655-8.

Ermer, J.H., R.B. Love, A.K. Khanna, S.C. Lewis, F. Cohen, *CdS/CuInSe₂ junctions fabricated by DC magnetron sputtering of Cu₂Se and In₂Se₃*, Arco Solar Inc., Chatsworth, CA, USA, 18th IEEE Photovoltaic Specialists Conference, pp. 1655-8.

Fowler, P.K., D.G. Dobryn, and C.M. Lee (1986), *Control of toxic gas release during the production of CuInSe photovoltaic cells*, Brookhaven National Laboratory, MIT/BNL-86-4.

Frass, L.M., Y. Ma (1977), *CdS thin films for terrestrial solar cells*, Hughes Research Labs, J. of Crystal Growth (Netherlands) **39**, No. 1, pp. 92-107.

Fray, A.F., P. Lloyd (1979), *Electrical and optical properties of thin P-TA films prepared by vacuum evaporation from the chalcopyrite CuInSe₂*, presented at the Fourth International Congress on Thin Films, Loughborogh, England, 11-15 Sept., Thin Solid Films **58**, No. 1, pp. 29-34.

Fulop, G., J. Betz, P. Meyers, M. Doty (1981), U.S. Patent 4,260,427, *CdTe Schottky Barrier PV Cell*.

Fulop, G., J. Betz, P. Meyers, M. Doty (1981), U.S. Patent 4,261,802, *Method of making a PV cell*.

Fulop, G., J. Betz, P. Meyers, M. Doty (1982a), U.S. Patent 4,345,107.

Fulop, G., M. Doty, P. Meyers, C.H. Liu (1982b), *High efficiency electrodeposited CdTe Solar Cells*, Appl. Phys. Lett. 40(4), 327-328.

Fulop, G.F., R.M. Taylor (1985), *Electrodeposition of semiconductors*, Ann. Rev. Mater. Sci., 15, pp. 197-210.

Garcia, F.J., M.S. Tomar (1982) *A screen printed CdS/CuInSe$_2$ solar cells*, J. Appl. Phys. Suppl. (Japan), pp. 535-8. *Proceedings of the 14th Conference (1982 International) on Solid State Devices, 24-26 Aug. 1982*, Tokyo, Japan.

Gordon, R.G., J. Proscia, K. Gustin, J. Chapple-Sokol, D. Strickler, R. McCurdy, *Optimization of transparent and reflecting electrodes for amorphous silicon solar cells*, Final Technical Report Oct. 1986-Nov. 1987, Prepared by Harvard University for the Solar Energy Research Institute under contract XB-7-06110-1.

Gorska, M., R. Beaulieu, J.J. Loferski, B. Roessler, J. Beall (1980), *Spray pyrolysis of CuInSe$_2$ thin films*, Sol. Energy Mater. **2**, No. 3, pp. 343-7.

Grindle, S.P. (July 1979), *Preparation and properties of CuInS$_2$ thin films produced by exposing RF-sputtered Cu-In films to an H$_2$S atmosphere*, Appl. Phys. Lett., 35(1), pp. 24-26.

Hall, R.B., R.W. Birkmire, J.E. Phillips, J.D. Meakin (1980), *Proceedings of the 3rd EC PV Solar Energy Conference*, Cannes, France, Ed W. Palz. 1094.

Hegedus, S.S., J.D. Meakin, B.N. Baron, J.A. Miller (1986), *Laboratory safety procedures for processing II-VI and related compounds for thin film photovoltaics*, Inst. of Energy Conversion, Delaware Univ., DE, USA, SERI Photovoltaics Safety Conference.

Hill, J.E., R.R. Chamberlin, U.S. Patent 3 148084 (1964).

Hodes, G., D. Cahen (Jan.-Feb. 1986), *Electrodeposition of CuInSe$_2$ and CuInS$_2$ films*, Solar Cells (Switzerland), (16), pp. 245-54.

Hodes, G., T. Engelhard, D. Cahen (1985), *Electroplated CuInSe$_2$ and CuInSe$_2$ layers: preparation and physical and photovoltaic characterization*, Thin Solid Films (Switzerland), (128:1-2), pp. 93-106.

Hodes, G., T. Engelhard, D. Cahen, L.L. Kazmerski, C.R. Herrington (June 1985), *Electroplated CuInS$_2$ and CuInSe$_2$ layers: preparation and physical and photovoltaic characterizatin*, Thin Solid Films (Switzerland), (128:1-2), pp. 93-106.

Hodes, G., T. Engelhard, D. Cahen, L.L. Kazmerski, C.R. Herrington (April 1985), *CuInS$_2$, CuInSe$_2$ and CuIn$_5$S$_8$ layers prepared by electrodeposition*, Sixth E. C. Photovoltaic Solar Energy Conference 846-9 1985 London, England, pp. 846-9.

Hodes, G., T. Engelhard, C.R. Herrington, L.L. Kazmerski, D. Cahen (1984), *Electrodeposited layers of copper indium sulfides (CuInSe$_2$, CuIn$_5$S$_8$) and copper indium diselenide (CuInSe$_2$)*, Prog. Cryst. Growth Charact. (10) *Proc. Int. Conf. Ternary Multinary Compd., 6th*, pp. 345-51.

Horig, W., H. Neumann, H. Sobotta, B. Schumann, G. Kuhn (1978), Thin Solid Films **48**, No. 1, pp. 67-72.

Hsu, T.M., R.J. Jih, P.C. Lin, H.Y. Ueng, Y.J. Hsu, H.L. Hwang (1986), *Oxygen doping in close-spaced-sublimed CdTe thin films for PV cells*, J. Appl. Phys. 59 (10), 3607-3609.

Ikegami, S. (1988), *CdS/CdTe solar cells by screen printing/sintering*, Solar Cells **23** # 1-2, 89-107.

Isett, L.C. (1983), *Deep level impurities and current collection in CdS/CdTe thin film solar cells*, Appl. Phys. Lett. 43 (6), 577-579.

Isomura, S., A. Nagamatsu, K. Shinohara, T. Aono (1986), *Preparation and some semiconducting properties of CuInSe₂ thin films*, Solar Cells (Switzerland) **16**, pp. 143-53.

Jackson, S.C., B.N. Baron, R.E. Rocheleau, and T.W.F. Russell (1987), *A chemical reaction model for physical vapor deposition of compound semiconductor films*, Inst. of Energy Conversion **33**, No. 5, pp. 711-21.

Janam, R., O.N. Srivastava (1983), *Growth and structural characteristics of the chalcopyrite semiconductor CuInSe₂*, Dept. of Phys., Banaras Hindu Univ., Varanasi, India, Cryst. Res. and Technol. (Germany) **18**, No. 12 1475-81.

Janam, R., O.N. Srivastava (1986), *Sintered p-CuInSe₂/n-CdS heterojunction solar cells*, Dept. of Phys., Banaras Hindu Univ., Varanasi, India, J. Phys. D (GB) **19**, No. 5 L85-7 14.

Janam, R., O.N. Srivastava (1985), *Structural characteristics of CuInSe₂ thin films and its influence on PV activity of CdS/CuInSe₂ thin film solar cells*, Dept. of Phys., Banaras Hindu Univ., Varanasi, India, Sol. Energy Mater. (Netherlands) **11**, No. 5-6 409-17 Jan.-Feb. 1985.

Jordan, John (1988), *Large-area CdS-CdTe photovoltaic cells*, Solar Cells **23** # 1-2, 107-114.

Jordan, J.F. (1975), IEEE PV Specialists Conf. Scottsdale, Arizona, May 6-8, pp. 508-514.

Kane, J., H.P. Schweizer, W. Kern (1975), J. Electrochem Soc. **122**, 1144.

Kapur, V.K., B.M. Basol, E.S. Tseng (June-August 1987), *Low cost methods for the production of semiconductor films for CuInSe₂/CdS solar cells*, Solar Cells (Switzerland), 7th Photovoltaic Advanced Research and Development Project Review Meeting, 13 May 1986, Denver, Colorado, (21), pp. 65-73.

Kapur, V.K., B.M. Basol, E.S. Tseng (October 1985), *Low cost thin film chalcopyrite solar cells*, Proceedings Eighteenth IEEE Photovoltaic Specialists Conference, pp. 1429-32.

Kapur, V.K., B.M. Basol, E.S. Tseng (Sept. 1986), *Preparation of thin films of chalcopyrites for photovoltaics*, Ternary & Multinary Compounds, *Proceedings of the 7th International Conference Snowmass*, Colorado, pp. 219-24.

Kapur, V.K., B.M. Basol, E.S. Tseng (February 1987), *High-efficiency, copper ternary, thin-film solar cells*, SERI Subcontractor Annual Report, SERI/STR-211-3030.

Kapur, V.K., B.M. Basol, E.S. Tseng, R.C. Kullberg, N.L. Nguyen (September 1988), *High-efficiency, copper ternary, thin-film solar cells*, SERI Subcontractor Annual Report, SERI/STR-211-3226.

Karmerski, L.L., M.S. Ayyagiri, F.R. White, and G.A. Sanborn (1986), J. Vac. Sci. Tech., 13, 139.

Kato, Y., M. Hyodo, M. Misonou, H. Kawahara (1987), *A new tin oxide film with large grains and its application to α-Si solar cells*, Technical Digest of the International PVSEC-3, Tokyo, Japan, 1987, P-II-26, 611. R. G. Liversey, E.

Kaur, I., D.K. Pandya, K.L. Chopra (1980), *Growth kinetics and polymorphism of chemically deposited CdS Films*, J. Electrochem Soc: Solid State Science and Technology **127**, 4, 943-948.

Kitaev, G.A., A.A. Uritskaya, S.G. Mokrushin (1965), Soviet J. Phys. Chem. **39**, 1101.

Klenk, R., R.H. Mauch, R. Menner, H.W. Shock (1988), *Wide bandgap Cu(Ga,In)Se$_2$/(Zn,Cd)S heterojunctions*, University of Stuttgart, Twentieth IEEE Photovoltaic Specialists Conference.

Knowles, A., H. Oumous, M.J. Carter, R. Hill (1988), *Properties of copper indium diselenide thin films produced by thermal annealing of elemental sandwich structures*, Newcastle upon Tyne Polytechnic, Twentieth IEEE Photovoltaic Specialists Conference.

Kobayashi, S., S. Sasaki, F. Kaneko, Bin Wang, T. Maruyama (1987), *Preparation of CuInSe$_2$ thin films by DC triode sputtering methods*, Trans. Inst. Electron. Inf. & Commun. Eng. C (Japan) **J70C**, No. 2, pp. 143-51.

Kokubun, Y., M. Wada (1977), *Photovoltaic effect in CuInSe$_2$/CdS heterojunctions*, Dept. of Electronic Engng., Tohoku Univ., Sendai, Japan, Jpn. J. Appl. Phys. (Japan) **16**, No. 5 879-80.

Kröger, F.A. (1978), *Cathodic deposition and characterization of metallic or semiconducting binary alloys or compounds*, J. Electrochem. Soc. **125** # 12, 2028-2034.

Kröger, F.A., R.L. Rod (1979), UK Patent 1,532,616 (Feb. 28, 1979).

Kröger, F.A., R.L. Rod, M.P.R. Panicker (1983), U.S. Patent 4,400,244, (covers electrodeposition procedures)

Kwietniak, M., J.J. Loferski, R. Beaulieu, R.R. Arya, E. Vera, L. Kazmerski (1982), *Thin film heterojunction CdS/Cu ternary alloy solar cells with minority carrier mirrors*, Brown Univ., Providence, RI, USA, 4th E. C. Photovoltaic Solar Energy Conference, 727-31.

Landis, G.A. (1986), *Cost assessment for electrodeposition process for production of CuInSe$_2$, CuGaSe$_2$, and CdTe solar cells*, Brown University.

Laude, L.D., M.C. Jollet, C. Antoniadis (1986), *Laser-induced synthesis of thin CuInSe$_2$ Films*, Sol. Cells **16**, pp. 199-209.

Leccabue, F., D. Seuret, D. Vigil (1985), *Sintered n-CuInSe$_2$/Au Schottky Diode*, Appl. Phys. Lett. **46**, No. 9 pp. 853-5.

Loferski, J.J., C. Case, M. Kwietniak, P.M. Sarro, L. Castaner, R. Beaulieu (1985), *Comparison of properties of thin films of CuInSe$_2$ and its alloys produced by evaporation, RF-sputtering and chemical spray pyrolysis*, Appl. Surf. Sci. (Netherlands) **22-23**, pt. 2, pp. 645-55. *Proceedings of the International Conference on Solid Films and Surfaces, 27-31 Aug. 1984,* Sydney Australia.

Loferski, J.J., M. Kwietniak, J. Piekoszewski, M. Spitzer, R. Arya, B. Roessler, R. Beaulieu, E. Vera, J. Shewohun, L.L. Kazmerski (1981), *Thin film heterojunction solar cells based on n-CdS and p-Cu ternary alloys of the type CuIn$_2$Ga$_{1-y}$Se$_{2z}$Te$_{2(1-z)}$ (RF sputtering)*, *Proceedings of the Fifteenth IEEE Photovoltaic Specialists Conference-1981, 12-15 May 1981,* Kissimmee, FL., pp. 1056-61.

Loferski, J.J., J. Shewchun, B. Roessler, R. Beaulieu, J. Piekoszewski, M. Gorska, G. Chapman (1978), *Investigation of thin film cadmium sulfide/mixed copper ternary heterojunction Photovoltaic Cells*, Brown Univ., Providence, RI, USA, Thirteenth IEEE Photovoltaic Specialists Conference.

Loferski, J.J., J. Shewchun, B. Roessler, G. Vitale, M. Gorska, J. Piekoszeskwi, J. Beall, R. Beaulieu, Y. Ercil (1978), *Thin-film cadmium sulfide/mixed copper ternary heterojunction solar cells. Technical progress report, September 30, 1977-September 30, 1978*, DOE/ET/20415-4, 35 pp.

Lokhande, C.D. (1987), *Pulse plated electrodeposition of copper indium diselenide films*, Journal Electrochemical Society, (134:7), pp. 1727-9.

Lokhande, C.D., G. Hodes (1987), *Preparation of CuInSe₂ and CuInSe₂ films by reactive annealing in H₂Se or H₂S*, Solar Cells, (21), pp. 215-24.

Lokhande, C.D. (1987), *Pulse plated electrodeposition of copper indium diselenide films*, Journal Electrochemical Society, (134:7), pp. 1727-9.

Lokhande, C.D., G. Hodes (1987), *Preparation of CuInSe₂ and CuInSe₂ films by reactive annealing in H₂Se or H₂S*, Solar Cells, (21), pp. 215-24.

Lommasson, T.C., A.F. Burnett, M. Kim, L.H. Chou, J.A. Thornton (1986), *Chalcopyrite CuInSe₂ films prepared by reactive sputtering*, Proceedings of the 7th International Conference, Snowmass CO. Solar Energy Res. Inst., ARCO Solar, Boeing Electron Co., Ternary and Multinary Compounds, edited by S. K. Deb and A. Zunger, Pittsburgh, PA: Master. Res. Soc., 572 pp.

Lowenheim, F.A. (1978), Electroplating: Fundamentals of Surface Finishing, McGraw-Hill Book Co., NY, 594 pp.

Lowenheim, F.A. (ed.) (1974), Modern Electroplating, John Wiley and Sons, NY.

Lyford, H. Moore (1968), J. Phys. E. **1**, 947.

Matson, R.J., O. Jamjoum, A.D. Buonaquisti, P.E. Russel, L.L. Kazmerski, P. Sheldon, R.K. Ahrenkiel (1984), *Metal contacts to CuInSe₂ (Solar Cell Applications)*, SERI, Solar Cells (Switzerland) **11**, No. 3, pp. 301-5.

Melsheimer, J., D. Ziegler (1983), *Thin tin oxide films of low conductivity prepared by CVD*, Thin Solid Films **109**, 71-83.

Meyers, P.V. (1986), *Polycrystalline CdS/CdTe/ZnTe n-i-p solar cell*, 7th European Community PV Solar Energy Conference, 1211-1213.

Mickelsen, R.A., W.S. Chen (1987), *Development of thin-film CuInSe₂ solar cells*, High Technol. Center, Boeing Electron Co., Seattle, WA, USA, Ternary and Multinary Compounds 7th International Conference 39-47.

Mickelsen, R.A., W.S. Chen (1984), *Polycrystalline thin-film CuInSe₂ solar cells*, Boeing Aerospace Co., Seattle, WA, USA, Sixteenth IEEE Photovoltaic Specialists Conference, 781-5 1982.

Mickelsen, R.A., W.S. Chen (1982), *Polycrystalline thin-film CuInSe₂ solar cells*, Boeing Aerospace Co., Seattle, WA, USA, Sixteenth IEEE Photovoltaic specialists Conference - 1982 781-5.

Mickelsen, R.A., W.S. Chen (1981), *Development of a 9.4% efficient thin-film CuInSe₂/CdS solar cell*, Proceedings of the Fifteenth IEEE Photovoltaic Specialists Conference, Kissimmee, FL., pp. 800-4.

Mickelsen, R.A., W.S. Chen (1980), *High photocurrent polycrystalline thin-film Cd/CuInSe₂ solar cell*, Appl. Phys. Lett. **36**, No. 5, pp. 371-3.

Mickelsen, R.A., W.S. Chen, Y.R. Hsiao, V.E. Lowe (1984), *Polycrystalline thin-film CuInSe₂/CdZnS solar cells*, Boeing Aerospace Co., Seattle, WA, USA, IEEE Trans. Electron Devices (USA) **ED-31**, No. 5 542-6.

Mickelsen, R.A., B.J. Stanbery, J.E. Avery, W.S. Chen, W.E. Devaney (1987), *Large area CuInSe₂ thin-film solar cells*, Boeing High Technol. Center, Seattle, WA, USA, Nineteenth IEEE Photovoltaic Specialists Conference 744-8.

Mitchell, K.W.C. Eberspacher, Cohen, F., Avery, J., Duran, G., Bottenberg, W. (1985), *Progress towards high efficiency, thin film CdTe solar cells*, Proceedings of the 18th IEEE PV Specialists Conference, Las Vegas, Nevada, Oct. 21-25, 1985.

Mitchell, K., C. Eberspacher, J. Ermer, K. Pauls, D. Pier, D. Tanner (1989), *Single and tandem junction CuInSe₂ technology*, PVSEC-4: Fourth International PV Science and Engineering Conference, Sydney, Australia (in press).

Mitchell, K., C. Eberspacher, J. Ermer, D. Pier (1988), *Single and tandem junction CuInSe₂ cell and module technology*, ARCO Solar, Twentieth IEEE Photovoltaic Specialists Conference.

Mitchell, K., H.I. Liu (1988), *Device analysis of CuInSe₂ solar cells*, ARCO Solar, Twentieth IEEE Photovoltaic Specialists Conference.

Mooney, B., S.B. Radding (1982), Ann. Rev. Mater. Sci. **12**, 81.

Mooney, J.B., R.H. Lamoreaux (1984), *Spray pryolysis of electronic materials*, Proc.-Electrochem. Soc., *83-7 (Proc. Symp. High Temp. Mater. Chem., 2nd)*, pp. 89-93.

Mooney, J.B., R.H. Lamoreaux, C.W. Bates Jr. (1984), *Spray pyrolysis of cadmium sulfide/copper indium selenide cells. Final report, 1 June 1982-30 November 1983*, SERI/STR-211-2279, Energy Res. Abstr., 9(13).

Mooney, J.B., R.H. Lamoreaux (1986), *Spray pyrolysis of CuInSe₂*, Solar Cells (Switzerland) **16**, pp. 211-20.

Nagao, M., S. Watanabe (1968), Japan J. Appl. Phys. **7**, 684.

Nakano, A., S. Ikegami, H. Matsumoto, H. Uda, Y. Komatsu (1986), *Long-term reliability os screen printed CdS/CdTe solar cell modules*, Solar Cells **17**, 233.

Nakayama, N., H. Matsumoto, A. Nakano, S. Ikegami, H. Uda, T. Yamashita (1980), *Screen printed thin film CdS/CdTe solar cell*, Japanese Journal of Applied Physics **19**, No. 4, pp. 703-712.

Neumann, H., B. Schumann, E. Nowak, A. Tempel, G. Kuhn (1983), *Influence of source composition on the properties of flash-evaporated thin films in the Cu-In-Se System*, Cryst. Res. and Technol. **18**, No. 7, pp. 895-900.

Nicolau, Y.F. (1985), *Solution deposition of thin solid compound films by a successive ionic-layer absorption and reaction process*, Applications of Surface Science 22/23, 1061-1074.

Nicoll, F.H. (1963), J. Electrochem. Soc. **110**, 1165.

Niemi, E., L. Stolt (1987), *ESCA analysis of CuInSe₂ and CuGaSe₂ thin films*, Inst. of Microwave Technol., Stockholm, Sweden, Ternary and Multinary Compounds 7th International Conference 87-92.

Noufi, R., R. Axton, D. Cahen, S.K. Deb (1984), *CdS/CuInSe₂ heterojunction cell research*, Solar Energy Res. Inst., Golden, CO, USA, Seventeenth IEEE Photovoltaic Specialists Conference 927-32.

Noufi, R., J. Dick (1985), *Compositional and·electrical analysis of the multilayers of a CdS/CuInSe₂ solar cell*, Solar Energy Res. Inst., Golden, CO, USA, J. Appl. Phys. (USA) **58**, No. 10 3884-7.

Noufi, R., R. Powell (1985), *Thin-film copper indium diselenide/cadmium sulfide solar cells*, Solar Energy Res. Inst., Golden, CO, USA, Sixth E. C. Photovoltaic Solar Energy Conference 761-7 London, England.

Noufi, R., R.C. Powell, R.J. Matson (1987), *On the effect of stoichiometry and oxygen on the properties of CuInSe₂ thin films and devices*, Solar Energy Res. Inst., Golden, CO. USA, Sol. Cells (Switzerland) **21**, 55-63.

Noufi, R., P. Souza, C. Osterwald (1985), *Effect of air anneal temperature on the spectral response of CdS/CuInSe₂ thin-film solar cells*, Solar Energy Res. Inst., Golden, CO, USA, Sol. Cells (Switzerland) **15**, No. 1 87-91 Sept. 1985.

Nouhi, A., R.J. Stirn, A. Hermann (1986), *Preliminary results on CuInSe₂/ZnSe solar cells using reactive sputter-deposited ZnSe*, Jet Propulsion Laboratory, 7th PV Ar&D Review Meeting.

Nouhi, A., R.J. Stirn, A. Hermann (1987), *CuInSe₂/ZnSe solar cells using reactive sputter-deposited ZnSe*, Jet Propulsion Laboratory, Nineteenth IEEE Photovoltaic Specialists Conference - 1987, pp. 1461-5.

Pachori, R.D., A. Banerjee, K.L. Chopra (1986), *Flash-evaporated thin films of CuInSe₂*, presented at the *Proceedings of the Symposium on Thin Film Science and Technology*, India, 9-11 Jan., Bull. Mater. Sci. **8**, No. 3, pp. 291-6.

Pamplin, B., R. Brian (1979), *Spray pyrolysis of ternary and quaternary solar cell materials*, Prog. Cryst. Growth Charact., 1(4), pp. 395-403.

Pamplin, B., R.S. Fiegelson (1979), *Spray pyrolysis of CuInSe₂ and related ternary semiconducting compounds*, Thin Solid Films **60**, No. 2, pp. 141-6.

Pande, P.C., G.J. Russell, J. Woods (1984), *The properties of electrophoretically deposited layers of CdS*, Thin Solid Films 121, 85-94.

Panicker, M.P.R. (1978), PhD. Thesis, USC

Panicker, M.P.R., M. Knaster, F.A. Kröger (1978), *Cathodic deposition of CdTe from aqueous electrolytes*, J. Electrochem. Soc. **125**, pp. 566-572.

Pern, F.J., J. Goral, R. Matson, T.A. Gessert, R. Noufi (1987), *Device quality thin films of CuInSe₂ by a one-step electrodeposition process*, *Proc. 8th PV AR & D Review Meeting*, Denver, Colorado.

Pern, F.J., R. Noufi, A. Mason, J. Goral, A. Swartzlander, R. Matson, A. Nelson (1987), *One-step electrodeposition of polycrystalline thin film CuInSe₂*, Polycrystalline Thin Film Program Review Meeting, Lakewood, Colorado.

Pern, F.J., R. Noufi, A. Mason, A. Swartzlander (1987), *Fabrication of polycrystalline thin films of CuInSe₂ by a one-step electrodeposition process*, *Proc. 19th IEEE PV Specialists Conference*, New Orleans, Louisiana.

Photovoltaic Program Branch (1988), Annual Report FY 1987, SERI/PR-211-3299 (DE88001155).

Piekoszewski, J., J.J. Loferski, R. Beaulieu, J. Beall, B. Roessler, J. Shewchun (April-1980), *RF-sputtered CuInSe₂ thin films*, Sol. Energy Mater, Netherlands **2**, No. 3, pp. 363-72.

Piekoszewski, J., J.J. Loferski, R. Beaulieu, J. Beall, B. Roessler, J. Shewchun (1980), *RF-sputtered CuInSe₂ thin films*, Brown Univ., Providence, RI, USA, Fourteenth IEEE Photovoltaic Specialists Conference 980-5.

Potter, R.R. (1986), *Enhanced photocurrent ZnO/CuInSe₂ solar cells*, ARCO Solar, Solar Cells **16**, pp. 521-7.

Potter, R.R., C. Eberspacher, L.B. Fabick (1985), *Device analysis of CuInSe₂/ (Cd,Zn)S/ZnO solar cells*, ARCO Solar, Eighteenth IEEE Photovoltaic Specialists Conference, 1985, pp. 1659-64.

Qiu, C.X., I. Shih (March-April 1987), *Effect of heat treatment on electrodeposited copper indium selenide (CuInSe₂) Films*, Solar Energy Materials (Netherland), (15:3), pp. 219-23.

Qiu, C.X., I. Shih (1986), *Properties of zinc oxide/copper indium diselenide heterojunctions*, Proceedings of the Northeast Regional Meeting of Mettall, Society, edited by Green, Martin, L. Metall. Soc., pp. 379-83.

Raja Ram, P., R. Thanagaraj, O.P. Agnihotri (1986), *Thin films CdZnS/CuInSe₂ solar cells by spray pyrolysis*, Proceedings of the Symposium on Thin Film Science and Technology, 9-11 Jan. 1985, India. Bull. Mater. Sci. (India) **8**, No. 3, pp. 279-84.

Raja Ram, P., R. Thangaraj, A.K. Sharma, O.P. Agnihotri (1985), *Totally sprayed CuInSe₂/Cd(Zn)S and CuInS₂/Cd(Zn)S solar cells*, Solar Cells (Switzerland) **14**, No. 2, pp. 123-31.

Ravindra D., J.K. Sharma (1985), *Electroless deposition of SnO₂ and Sb-doped SnO₂ films*, J. Phys. Chem. Solids **46** # 8, 945-950.

Rocheleau, R.E., J.D. Meakin, R.W. Birkmire (1987), *Tolerance of CuInSe₃ cell performance to variations in film composition and the implications for large cell manufacture*, Inst. of Energy Conversion, Delaware Univ., Newark, DE, USA, Nineteenth IEEE Photovoltaic Specialists Conference, 972-6.

Rockett, A., T.C. Lommasson, J.A. Thornton (1988), *Depositon of CuInSe₂ by the hybrid sputtering-and-evaporation method*, University of Illinois, Twentieth IEEE Photovoltaic Specialists Conference.

Rohatgi, A., S.A. Ringel, J. Welch, E. Meeks, K. Pollard, A. Erbil, C.J. Summers, P.V. Meyers, C.H. Liu (1988), *Growth and characterization of CdMnTe and CdZnTe polycrystalline thin films for solar cells*, Solar Cells **24**, 185-194.

Romeo, N., A. Bosio, V. Canevari, D. Seuret (1987), *Low resistivity CdS thin films grown by r.f. sputtering in an Ar-H₂ atmosphere*, Dipartimento di Fisica, GNSM-CISM, Parma, Italy, Sol. Cells **22**, No. 1, pp. 23-7.

Romeo, N., A. Bosio, V. Canevari, L. Zanotti (1986), *Copper indium selenide (CuInSe₂)/cadmium sulfide thin film solar cells by r.f. sputtering*, E. C. Photovoltaic Sol. Energy Conf., 7th, pp. 656-61.

Romeo, N., A. Bosio, V. Canevari, L. Zanotti (1987), *CuInSe₂/CdS thin film solar cells by RF sputtering*, Consiglio Nazionale delle Ricerche-Centro Interuniv., Parma, Italy, Seventh E. C. Photovoltaic Solar Energy Conference, 656-61.

Romeo, N., V. Canevari, G. Sberveglieri, A. Bosio, L. P. Zanotti, *Large grain (112) oriented CuInSe₂ thin films grown by RF sputtering*, Parma Univ., Italy, Eighteenth IEEE Photovoltaic Specialists Conference, 1388-92.

Romeo, N., V. Canevari, G. Sberveglieri, A. Bosio, L. Zanotti (1986), *Growth of large-grain CuInSe₂ thin films by flash-evaporation and sputtering*, Solar Cells (Switzerland) **16**, pp. 155-64.

Roth, A.P. and D.F. Williams (1981), *Properties of zinc oxide films prepared by the oxidation of diethylzinc*, J. Appl. Phys. **52**, pp 6685.

Russel, P.E., O. Jamjoum, R.K. Ahrenkiel, L.L. Kazmerski, R.A. Michelsen, W.S. Chen (1982), *Properties of Mo-CuInSe₂ interface*, SERI, Applied Physics Letters (USA) **40**, No. 11, pp. 995-7.

Russell, T.W.F., B.N. Baron, R.E. Rocheleau (1984), *Economics of processing thin film solar cells*, J. Vac. Sci. Technol. B 2 (4), 840-844.

Salviati, G., D. Seuret (1983), *N- and P-type CuInSe₂ thin films deposited by flash evaporation*, Thin Solid Films **104**, No. 3-4, pp. L75-8.

Samaan, A.N.Y., N.A.K. Abdul-Hussein, R.D. Tomlinson, A.E. Hill, D.G. Armour (1981), *Evaluation of some physical characteristics of CuInSe₂ thin films produced by RF sputtering*, Fourth International Conference on Ternary and Multinary Compounds, Tokyo, Japan, 27-29 Aug 1980. Jpn J. Appl. Phys. **19:3**, 15-22.

Santamaria, J., E. Iborra, I. Martil, G. Gonzalez-Diaz, J.M. Gomez de Salazar, F. Sanchez-Auesada (1986), *RF sputtered copper indium selenide (CuInSe₂) thin films in argon, hydrogen atmospheres*, Comm. Eur. Communities, E. C. Photovoltaic Sol. Energy Conf., 7th, pp. 1223-5.

Sarro, P.M., R.R. Arya, R. Beaulieu, T. Warminski, J.J. Loferski, W. Palz, and F. Fittipadi (1984), *Preparation and properties of copper indium diselenide thin films deposited on conducting substrates by chemical spray pyrolysis for photovoltaic cells*, 5th E. C. Photovoltaic Sol. Energy Conf., *Proc. Int. Conf.*, pp. 901-7.

Sax, I.N., R.J. Lewis (1986), *Rapid guide to hazardous chemicals in the workplace*, Van Norstand Reinhold Company, N.Y.

Schafer, D.E. (1988), *MOCVD techniques for CdTe solar cells*, Final Subcontract Report 15 February 1985-1 April 1987, SERI/STR-211-3218 (DE88001163).

Schumann, B., G. Kuhn (1984), *Growth structure, stoichiometry of epitaxial Cu-III-VI₂ Films on CaF₂ substrates*, Cryst. Res. and Technol. **19**, No. 8, pp. 1079-84.

Schumann, B., H. Neumann, E. Nowak, G. Kuhn (1981), *Influence of substrate surface polarity on epitoxial layer growth of CuInSe₂ on GaAs*, Cryst. Res. and Technol. **16**, No. 6, pp. 675-80.

Shanthi, E. (1981), Ph.D. Thesis, Indian Institute of Technology.

Sharma, N.C., D.K. Pandya, H.K. Sehgal, K.L. Chopra (1976), Mater. Res. Bull. **11**, 1109.

Shay, J.L., et al. (1975), *Efficient CuInSe₂/CdS solar cells*, Appl. Phys. Lett. **27**, pp. 89-90.

Shih, I., C.X. Qiu (1987), *Characteristics of preliminary cells fabricated using electroplated CuInSe₂ films*, McGill University, Nineteenth IEEE PV Specialists Conference, pp. 1291-4.

Shkedi, Z., R.L. Rod (1980), 14th IEEE PV Specialists Conference, pp. 472.

Singh, R.P., N. Khare, S.L. Singh (1984), *Thin film photoelectrochemical solar cell*, National Academy Science Letters (India), (7:11), pp. 347-50.

Singh, R.P., S.L. Singh (1986), *Electrodeposited semiconducting CuInSe₂ films.-II. Photo-electrochemical solar cells*, Journal Physics (GB), (19:9), pp. 1759-69.

Singh, R.P., S.L. Singh, S. Chandra (July 1986), *Electrodeposited semiconducting CuInSe₂ Films. - I. prepartion, structural and electrical characterization*, Journal of Physics (GB), (19:7), pp. 1299-309.

Squillante, M.R., L. Moy, G. Entine (1985), *Low cost, sprayed copper indium diselenide solar cell research, September 15, 1982-June 30, 1985*, DOE/ER/130026-T2, Energy Res. Abstr. 1985. 10(23).

Sridevi, D., J.J.B. Prasad, K.V. Reddy (1986), *Preparation and characterization of flash-evaporated CuInSe₂ thin films*, presented at *Proceedings of the Symposium on Film Science and Technology*, India, 9-11 Jan. 1985, Bull Mater. Sci. (India) **8**, No. 3, pp. 319-24.

Sridevi, D., K.V. Reddy (1986), *Electrical and optical properties of flash evaporated CuInSe₂ thin films*, Indian J. Pure & Appl. Phys. **24**, No. 8, pp. 392-6.

Stirn, R.J., A. Nouhi (1985), *Investigation of reactive magnetron sputter deposited ZnSe for solar cell application, a Final Subcontract Report*, SERI/STR-211-2750, Jet Propulsion Laboratory, Pasadena CA.

Stolt, L., J. Hedstrom, M. Jargelius, D. Sigurd (1985), *CuInSe₂ for thin film solar cells*, presented at the Sixth E. C. Photovoltaic Solar Energy Conference, London, England, 15-19 April 1985, Reidel, xxxv+1104 pp.

Stolt, L., J. Hedstrom, D. Sigurd (1985), *Co-evaporation with a rate control system based on a quadrupole mass spectrometer*, J. Vac. Sci. & Technol. **3**, No. 2, pp. 403-7.

Szot, J., D. Haneman (1984), *Preparation and characterization of CuInSe₂ and CdS films*, Sch. of Phys., New South Wales Univ., Kensington, NSW, Australia, Sol. Energy Mater. (Netherlands) **11**, No. 4 289-98.

Tempel, A., B. Schumann, K. Kolb, G. Kuhn (1981), *Structural investigations of CuInSe₂ epitaxial layers on (100)- and (100)-oriented GaAs substrates*, J. Cryst. Growth **54**, No. 3, pp. 534-40.

Thornton, J.A. (1986), *Recent advances in sputter deposition*, Dept. of Mater. Sci., Illinois Univ., Urbana, IL. Surf. Eng. **2**, No. 4, pp. 283-92.

Thornton, J.A., D.G. Cornog (1984), *CuInSe₂ solar cell research by sputter deposition. Semiannual report, February 1, 1983-March 31, 1984*, SERI/STR-211-2203, TELIC Corp., Santa Monica, CA, 56 pp.

Thornton, J.A., D.G. Cornog, R.B. Hall, J.D. Meakin, S.P. Shea (1983), *Preparation of CuInSe₂ films by reactive co-sputtering*, Conference on Materials and New Processing Technologies for Photovoltaics, San Francisco, CA, *Proceedings - Electrochem. Soc*, Vol. 83-11.

Thornton, J.A., D.G. Cornog, R.B. Hall, S.P. Shea, J.D. Meakin (1984), *An investigation of reactive sputtering for depositing copper indium diselenide films for photovoltaic applications*, Conference Record of the Seventeenth IEEE Photovoltaic Specialists Conference, 1-4 May 1984, Kissimmee, FL, New York: IEEE, pp. 781-5.

Thornton, J.A., D.G. Cornog, R.B. Hall, S.P. Shea, J.D. Meakin (1984), *Reactive sputtered copper indium diselenide films for photovoltaic applications*, J. Vac.

Sci. and Technol. **2**, No. 2, Pt. 1, pp. 307-11. *Proceedings of the 30th National Symposium of The American Vacuum Society, 31 Oct.-4 Nov. 1983*, Boston, MA.

Thornton, J.A. (1987), *Fundamental processes in sputtering of relevance to the fabrication of thin film solar cells.* Dept. of Mater. Sci., Illinois Univ., Urbana, IL. Sol. Cells **21**, pp. 41-54. 7th Photovoltaic Advanced Research and Development Project Review Meeting, 13 May 1986, Denver, CO.

Thornton, J.A., T.C. Lommasson, A.F. Burnett (1986), *CuInSe₂ solar cell research by sputter deposition. Annual Subcontract Report, 1 December 1984-31 December 1985*, SERI/STR-211-2923, 75 p.

Thornton, J.A., T.C. Lommasson, D.G. Cornog (1986), *In-line sputtering systems for depositing CuInSe₂/CdS heterojunctions*, Trans, ASME, J. Sol. Energy Eng. **108**, No. 4, pp. 259-66.

Thornton, J.A., T.C. Lommasson, H. Talieh (1987), *CuInSe₂ solar cell research by sputter deposition: Annual subcontract report, 1 January 1986-31 December 1986*, SERI/STR-211-3174, 87 pp.

Thornton, J.A., T.C. Lommasson (1986), *Magnetron reactive sputtering of copper-indium-selenide*, Solar Cells (Switzerland) **16**, pp. 165-80.

Tomar, M.S., F.J. Garcia (1982), *A ZnO/P-CuInSe₂ thin film solar cell prepared entirely by spray pyrolysis*, Dept. de Fisica, Univ. Simon Bolivar, Caracas, Venezuela, Thin Solid Films (Switzerland) **90**, No. 4 419-23 30.

Tomlinson, R.D., D. Omezi, J. Parkes, M.J. Hampshire (1980), *Some observations on the effect of evaporation source temperature on the composition of CuInSe₂ thin films*, Thin Solid Film **65**, No. 2, pp. L3-6.

Touzel, J., A. Foucaran, V. Chen, S. Duchemin, J. Bougnot, M. Savelli (1985), *Fabrication and characterization of sprayed CuInSe4₂ thin layers*, Sixth E. C. Photovoltaic Solar Energy Conference, 15-19 April 1985, London, England, edited by W. Palz and F. C. Treble, Dordrecht, Netherlands: Reidel, pp. 836-40.

Trykozko, R., R. Bacewicz, J. Filipowicz (1984), *Photoelectrical properties of CuInSe₂ thin films*, presented at the *Proceedings of the Sixth International Conference on Ternary and Multinary Compounds*, Caracas, Venezuela, 15-17 Aug. 1984, Prog. Cryst. Growth and Charact. **10**, pp. 361-4.

Trykozko, R., R. Bacewicz, J. Filipowicz (1986), *Photoelectrical properties of CuInSe₂ thin films*, Sol. Cells **16**, pp. 351-6.

Tyan, Y.S. (1988), *Topics on thin film CdS/CdTe solar cells*, Solar Cells **23**, # 1-2, 19-31.

Tyan, Y.S. (1980), *Polycrystalline thin film CdS/CdTe PV cell*, # 4,207,119; assignee Eastman Kodak Company.

Tyan, Y.S., E.A. Perez-Albuerne (1982), *Efficient thin film CdS/CdTe solar cells*, *Proceedings 16th IEEE PV Specialists Conference*, San Diego, pp. 794.

Varela, M., E. Bertran, J. Esteve, J.L. Morenza (1985), *Crystalline properties of co-evaporated CuInSe₂ thin films*, Thin Solid Films **130**, No. 1-2, pp. 155-64.

Varela, M., E. Bertran, M. Machon, J. Esteve, J.L. Morenza (1986), *Optical properties of co-evaporated CuInSe₂ thin films*, J. Phys. D (GB) **19**, No. 1, pp. 127-36.

Varma, S., K.V. Rao, S. Kar (1984), *Electrical characteristics of Si-tin oxide heterojunctions prepared by chemical vapor deposition*, J. Appl. Phys. **56** (10), 2812.

Vigil, O., D. Seuret, F. Leccabue, L. Hernandex (1987), *Sintered p-CuInSe$_2$/n-CdS photovoltaic heterojunction*, La Habana Univ., Cuba., Mater. Lett. (Netherlands) **6**, No. 3 85-8.

Wagner, S. et al. (1974), *CuInSe$_2$/CdS Heterojunction photovoltaic detectors*, Appl. Phys. Letts. **25**, pp. 434-435.

Wang, D., G. Zhao (1986), *Properties of copper indium selenide (CuInSe$_2$) films deposited by spray pyrolysis and copper indium selenide (CuInSe$_2$)/cadmium sulfide solar cells*, Chinese Journal of Solar Energy, 7(1), pp. 114-18.

Webb, J.B., D.F. Williams, M. Buchanan (1981), *Transparent and highly conductive films of ZnO prepared by R. F. reactive magnetron sputtering*, Appl. Phys. Lett. **39**, pp. 640.

Weng, T.H. (1979), J. Electrochem. Soc. **126**, No. 10.

Williams, E.W., K. Jones, A.J. Griffiths, D.J. Roughley, J.M. Bell, J. H. Stevens, M. J. Huson, M. Rhodes, T. Costic (1979), *Proc. 2nd EC PV Solar Energy Conference*, Berlin, 874.

Wu, C.S., R.S. Fiegelson, R.H. Bube (1972), J. Appl. Phys., 43, 756.

Zweibel, K. (1985), *Post-deposition heat treatments of CdS/CuInSe$_2$ solar cells*, SERI/PR-211-2468.

Zweibel, K., A. Hermann (1985), *CdTe solar cells*, Photovoltaics–SPIE **543**, 119-126.

Zweibel, K., B. Jackson, Hermann (1986), *Comments on 'Critical Materials Assessment Program'*, Solar Cells **16**, pp. 631-634.

# Word Index

absorption air conditioner, 18, 51.

acceptance angle, 466.

accessible by road, wind turbines, 117.

adherent CdS film, 500, 559.

adjustable-pitch blade tips, 88.

adjustable-tilt south facing flat plate, 159.

advanced concepts, 288, 398.

advanced heat transfer fluids, 253.

advanced heterostructures, 400.

advanced PV module technologies, 146.

advanced receiver coolants, 229, 230.

advanced solar cell structures, 422.

advanced system experiments, 289.

advantage of topography, 107.

aerodynamics controls, 93.

air annealing, 507.

aircraft warning beacon, 185.

airfoils designed, 102, 126.

Alabama Power Company, 129, 153, 154.

AlGaAs grown on GaAs, 417.

allowable module cost, 395.

Altamont Pass wind turbines, 72, 84, 89, 110.

alternator problems, 109, 249.

aluminum blades, 96.

aluminum membranes, 229.

ambipolar Auger coefficient, 454.

Ametek cell, 535, 548.

amorphous-silicon thin-film PV, ,129, 154, 167.

annealing effect, 168, 523.

annual capacity factor, 88, 134.

annual conversion efficiency, 223, 295, 344.

annual lease fee, 110.

annual performance, 292, 343, 376.

annual wind speed, 91.

antireflection coating, 491, 524.

APS Saguaro station, 28.

APS/PG&E conceptual designs, 44.

Arbutus, 82.

architectural glass plating, 510.

ARCO Solar CuInSe$_2$, 488.

ARCO Solar process, 488.

ARCO Solar, 131, 513, 528.

Arizona Public Service Company, 8, 24, 27, 41, 42, 46.

Arizona Solar Energy Office, 149.

Arizona State University, 149.

array dc efficiency, 156, 157.

array-related issues, 186.

Ascension Technology, 164.

assessments of turbulence, 97.

assessments PV technology, 176, 199.

Auger coefficient, 456, 457, 459, 461.

Auger electron spectroscopy, 507.